£53.00

3 8

D1460820

Library of Congress Cataloging-in-Publication Data

Ecological risk assessment / editor and principal author, Glenn W.
  Suter II ; contributing authors, Lawrence W. Barnthouse . . .
  [et al.].
      p.   cm.
    Includes bibliographical references and index.
    1. Ecological risk assessment.   I. Suter, Glenn W.
  II. Barnthouse, L. W. (Lawrence W.)
  QH541.15.R57E25 1993
  574.5′222--dc20   92-25474
  ISBN 0-87371-875-5 (alk. paper)

10 53 56

LEWIS PUBLISHERS
121 South Main Street, Chelsea, Michigan 48118

PRINTED IN THE UNITED STATES OF AMERICA
1 2 3 4 5 6 7 8 9 0
Printed on acid-free paper

Dedication

*To Prometheus (forethought), surely the patron deity
    of risk assessors;
He collected all hazards and closed them in a box;
He warned his brother Epimetheus (afterthought) to
    keep the hazards contained;
Epimetheus allowed Pandora (all-giving) access to the
    box without proper instruction;
She released the hazards;
And only Prometheus was punished by Zeus.*

*Also to those forethoughtful environmental scientists
    whose efforts have provided the bases for eco-
    logical risk assessment;
May they all escape the bureaucratic Zeuses of this
    world.*

Preface

I write a manifesto and I want nothing . . . in principle I am against manifestos, as I am against principles.

Tristan Tzara, *Dada Manifesto*

The reader of this volume will discover that, although it is intended as a text, there is much of the manifesto to it. Therefore, in the interest of openness and honesty with the reader, I will state in advance the program of the manifesto. I believe that assessments of risks to nonhuman biota have not been taken as seriously as they might in part because we ecologists, environmental toxicologists, and environmental chemists have not taken them seriously enough ourselves. There are a variety of historical and institutional reasons for this that need not be discussed. However, risk methodologies for routinely predicting the weather, failures of engineered systems, and insurance claims have a rigor that is found in only the most exemplary ecological risk assessments. The fact that, as of this writing, the U.S. Environmental Protection Agency has no agency guidance for ecological risk assessment is in large part due to the absence of formal ecological assessment methods for which to provide guidance.

This situation will be remedied not by a change of terminology from hazard assessment or impact assessment to risk assessment, but by carefully thinking about what ecological assessors do and what their assessments represent. In addition, those who use the results of assessments must be clearer about what they want out of ecological risk assessments, and need to provide assessors with more time and support to develop the needed assessment tools. Among the questions that need to be asked more explicitly and answered more clearly are the following:

1. What exactly are we trying to protect and to what extent should it be protected? In the jargon, what is the endpoint? Too many assessments have an undefined or vaguely defined goal. They are like experiments without hypotheses.
2. How do the data relate to the endpoint? When we say that the fathead minnow is a surrogate species, what do we mean? When we say that a toxicity test in a mesocosm is a test of ecosystem effects, do we mean that all ecosys-

tems will respond in the same way? What models might relate death of a fathead minnow in a jar or net production of a 3×3 meter pond to the production of a population of largemouth bass or of a reservoir ecosystem?

3. How are the biota exposed to the pollutant? The answer should include consideration of temporal and spatial dynamics and of routes of exposure. Too often the route of exposure and the duration of exposure in the field are assumed to be identical to those of the available test data and the spatial distribution is undefined.

4. What aspects of the assessment are uncertain, how uncertain are we, and how does that uncertainty influence the risk?

5. What alternate assessment approaches or models might supplement my preferred approach or model, thereby potentially increasing confidence in the validity of my results?

This text is intended for advanced students in the environmental sciences and for practicing environmental scientists. It assumes a basic knowledge of ecology, toxicology, chemistry, mathematics, and statistics. However, it assumes no knowledge or experience in risk assessment. Because much of the terminology in this field will be unfamiliar to many readers, and because many of the arguments among risk assessors turn out to be semantic, a glossary is provided.

The treatment of topics in this text is not balanced. Risks to aquatic systems are emphasized because more ecological risk assessments are concerned with aquatic than terrestrial systems, and because the body of data and models is far richer for aquatic systems. I believe that the balance of terrestrial and aquatic examples reflects the current state of environmental toxicology, chemistry, and assessment.

Risk assessments that predict the effects of new chemicals and effluents are emphasized (Chapters 3–9) relative to those that describe the extent, magnitude, and causation of effects that began in the past (Chapter 10). This is because the standard risk assessment paradigm is designed for predictive assessments, and because most of the currently available tests and models were designed to support predictive assessments.

The text is concerned with risks from chemicals and mixtures of chemicals, although other human actions such as logging and extreme natural events such as wildfires can and should be the topics of ecological risk assessments. The exception is a brief chapter on risks of exotic organisms (Chapter 13), which is included because introductions of organisms are regulated in much the same way as introductions of new chemicals.

Finally, assessment of effects of a chemical exposure is emphasized (Chapters 7–9) relative to assessment of transport and fate of chemicals resulting in exposure (Chapters 4–6). This is because the methods for fate modeling, given physical-chemical data, are quite well developed relative to those for predicting effects, given toxicity test data. For example, standard fate models are available from the U.S. Environmental Protection Agency but not standard ecological effects models. Therefore, the chapters on exposure are relatively short and conceptual and refer the reader to available texts and manuals.

This text presents and provides references to a number of models and approaches that are immediately applicable. However, its chief expected benefit is introduction of a different way of assessing the ecological effects of human actions to an audience that will develop and expand the field. A hoped-for benefit is recognition that ecological risk assessment, like other types of risk assessment, is an important and complex applied science that is worthy of support by private industry, governmental agencies, and environmental advocacy groups, and is worthy of the best efforts of the best students and environmental scientists.

Glenn W. Suter II
Oak Ridge, Tennessee
April, 1992

Contents

PART I. INTRODUCTION TO ECOLOGICAL RISK ASSESSMENT

PART II. EXPOSURE ASSESSMENT FOR PREDICTIVE RISK ASSESSMENTS

PART III. EFFECTS ASSESSMENT

PART IV. UNCONVENTIONAL
ECOLOGICAL RISK ASSESSMENT

Glenn W. Suter II is a Research Staff Member in the Environmental Sciences Division of Oak Ridge National Laboratory, where he is involved in the development and application of methods for ecological risk assessment. He received a BS in Biology from Virginia Polytechnic Institute and a PhD in Ecology from the University of California, Davis. Dr. Suter has served on the Board of Directors of the Society for Environmental Toxicology and Chemistry, and has been a member of an International Institute of Applied Systems Analysis Task Force on Risk and Policy Analysis, the EPA/Conservation Foundation's Ecosystem Valuation Forum, an Expert Panel for the Council on Environmental Quality, a Working Group of the Scientific Group on Methodologies for the Safety Evaluation of Chemicals, of SCOPE (Scientific Committee on Problems in the Environment), and the Editorial Board of *Environmental Toxicology and Chemistry*. He served as Rapporteur for the OECD Workshop on Extrapolation of Aquatic Toxicity Data to the Field, and as chairman of the Eleventh ASTM Symposium on Aquatic Toxicology and Hazard Assessment. Dr. Suter has authored or coauthored 45 open literature publications and numerous technical manuscripts and reports. He has received the Environmental Sciences Division's Annual Scientific Achievement Award and has twice been awarded the Martin Marietta Energy System's Technical Achievement Award. His research experience has included development and application of aquatic toxicity tests, soil microcosm tests, and monitoring of the environment at energy facilities.

Acknowledgments

William Adams, Larry Barnthouse, Steve Bartell, Thomas Burns, Willo-dean Burton, Virginia Dale, Donald DeAngelis, James Gillett, Richard Halbrook, Dexter Hinckley, Owen Hoffman, Carolyn Hunsaker, Thomas La Point, William Murdoch, Robert O'Neill, Andy Redfearn, Robert Reed, Frances Sharples, Eric Smith, Arthur Stewart, Frieda Taub, and Monica Turner each reviewed portions of this book. The book has been greatly improved by their contributions. Allen Moghissi, Environmental Protection Agency Project Officer for the first ecological risk assessment project at Oak Ridge National Laboratory, got us off to a good start. He was an example of that rare species of manager who recognizes that development of assessment methods must be supported before assessments can be performed. Chapters 1 and 7 were begun during a six-week sabbatical, for which I thank the Environmental Sciences Division Awards Committee and David Reichle. Larry Barnthouse's intellectual contributions to this book are much greater than is indicated by his authorship of Chapters 2 and 8. Bob O'Neill and Bob Gardner contributed to the early development of ecological risk assessment concepts and models at ORNL. The original methods presented in Chapters 7 and 8 were made possible by Aaron Rosen's unique ability to convert poorly defined ideas and odd data sets into working assessment tools. Finally, Linda made the editor's life a pleasure during the years of development of this book.

Development of this book was supported in part by the U.S. Department of Energy's Oak Ridge National Laboratory which is managed by Martin Marietta Energy Systems, Inc., under Contract DE-AC05–84OR21400. It is ORNL Environmental Sciences Division Publication No. 3929.

ECOLOGICAL RISK ASSESSMENT

PART I

INTRODUCTION TO ECOLOGICAL RISK ASSESSMENT

Because ecological risk assessment is a new field, introductions are particularly important. The first chapter of this introductory section defines ecological risk assessment in the broader context of environmental assessment. The second chapter presents concepts that are important to ecological risk assessment and must be gotten out of the way before getting down to specific methods. The third chapter presents the standard risk assessment paradigm, as adapted to ecological risk assessments. All subsequent chapters discuss components of this paradigm (Chapters 4–9), variations on this paradigm (Chapter 10), or applications of the paradigm to unconventional assessment problems (Chapters 11–13).

1

Defining the Field

Glenn Suter

Risk assessment can be defined as the process of assigning magnitudes and probabilities to the adverse effects of human activities or natural catastrophes. This process involves identifying hazards such as the release of a toxic chemical to surface waters that support fisheries, and using measurement, testing, and mathematical or statistical models to quantify the relationship between the initiating event and the effects. This book describes an approach to using the concepts and methods of risk assessment to improve decisionmaking for protection of the nonhuman environment.

The stimulus for adopting risk assessment as a fundamental component of environmental decisionmaking is the recognition that (a) the cost of eliminating all environmental effects of human activities is impossibly high, and (b) regulatory decisions must be made on the basis of incomplete scientific information (Ruckelshaus 1983, Moghissi 1984). The objective of risk-based environmental regulation is to balance the degree of risk to be permitted against the cost of risk reduction and against competing risks. The need for this balancing is implicit in the language of environmental legislation such as the Federal Insecticide, Fungicide, and Rodenticide Act (FIFRA) and the Toxic Substances Control Act (TOSCA) that speak of protection from "unreasonable risks." Risk assessment has several advantageous properties in environmental decisionmaking:

- It provides the quantitative bases for comparing and prioritizing risks. If, as is usually the case, all alternatives have hazardous properties, it is not possible to make a choice without characterizing the risks.
- It provides a systematic means of improving the understanding of risks. Research can be prioritized by identifying and comparing the uncertainties associated with the characterization of the different steps in the causal chain from the initiating event to the ultimate effect.
- By expressing results as probabilities, risk assessment acknowledges the inherent

uncertainty in predicting future environmental states, thereby making the assessment more credible.

- Risk assessment estimates clear consistent endpoints such as cancer death rates or probability of bankruptcy. This is in contrast to other approaches to ecological assessment which have unstated or ambiguous endpoints such as "ecosystem integrity."

- By basing the estimates of probability and magnitudes of effects on formal quantitative methods, risk assessment provides a means for the parties to environmental decisions to compare the implications of their assumptions and data rather than negotiating on the basis of political clout. The openness and consistency of such methods permits an assurance of fairness and allows thorough scientific review of the bases for decisions. Differences in results expressed as magnitudes and probabilities of effects can be compared to determine whether they really have different implications, given costs and benefits of the alternate actions.

- Risk assessment clearly separates the scientific process of estimating the magnitude and probability of effects (risk analysis) from the process of choosing among alternatives and determining acceptability of risks (risk management). The result of this separation is reduced likelihood of analyses that are biased to fit desired decisions and greater credibility for both the technical and the policy people.

This book focuses on the major problem in ecological risk assessment, estimation of risks posed by chemicals. Although risk assessment can be applied to other hazards such as thermal effluents, flow regulation, resource harvesting, or climatic modification, environmental risk assessment has been employed primarily to deal with chemicals. Several hundred new chemicals are introduced each year, thousands of chemical waste sites are identified, and thousands of new stacks and pipes release chemical effluents into the air and water. The complexity of managing these chemicals virtually necessitates some form of risk assessment to prioritize concerns and rationalize the regulatory process.

This book also focuses on prediction of ecological risks because it is a more common and difficult problem than determining the magnitude of existing effects. However, even assessments of existing effects are often conducted using predictive techniques because monitoring may be too expensive, time-consuming, or insensitive, and often produces ambiguous results (Chapter 10). For example, permitting of existing aqueous effluents in the United States is usually based on water quality standards or on effluent toxicity tests and plume dispersal models rather than on monitoring the receiving ecosystem (EPA 1985b). We emphasize, however, that the concept of prediction in ecological risk assessment is more akin to the meteorologist's 30% chance of rain than to prophecy.

THE SCOPE OF ENVIRONMENTAL ASSESSMENT

We begin by considering the place of ecological risk assessment in the array of activities, termed assessment, that attempt to predict and evaluate the consequences of hazardous actions. We use the term analysis narrowly to refer to formal, usually quantitative techniques of estimating effects. The term assessment is used more broadly to refer to analysis plus risk management, the policy-related activities such as defining issues, determining the significance of risks, and choosing the appropriate action. This distinction is consistent with some authors (e.g., Westman 1985) but others have given the two terms the opposite definitions (e.g., NRC 1983). The terms can be used interchangeably in most phrases with only a broadening or narrowing of perspective. Risk assessment has no particular legal mandate. It is an analytical tradition that takes different forms depending on the particular application, and hence, risk assessments performed for different assessment purposes will use different methods. For this reason, we begin by discussing the major types of legislation-related assessment needs and the assessment approaches that have been applied to them. Later we will explain how the risk perspective can improve the credibility and management utility of these assessments.

Disclosure of Effects

The National Environmental Policy Act of 1969 (NEPA) has been described by the U.S. federal judiciary as "at the very least .... an environmental full disclosure law" (Anderson and Daniels 1973). This requirement is simultaneously the most demanding and most trivial aspect of NEPA environmental impact assessments. It is demanding in the sense that it requires the prediction of all consequences of an action. Thus it requires that the techniques of environmental science be expanded to encompass secondary effects and effects occurring at long physical and temporal ranges. It has often been trivialized in practice by assessments dominated by long catalogs of the communities that will be paved over and species whose habitat will be altered.

Comparison of Actions

Any assessment problem can be formulated as a comparison of alternatives (e.g., should a permit be issued or withheld?), but most assessments focus on selection and justification of a preferred action, rather than on comparison. NEPA contains a specific requirement that alternative actions be considered and compared to the preferred action. This requirement is reinforced by the U.S. Council on Environmental Quality's regulations (CEQ 1986) that made comparison of alternatives the primary focus of environmental impact assessment under NEPA. The purpose of the comparison is to allow decisionmakers to balance environmental effects against other considerations when choosing an action. In many cases, alternatives to the preferred action would not be considered without this provision of NEPA. Comparison of alternatives has been commonly interpreted as disclosure of the effects of each alternate action followed by some summary comparison, possibly including a cost/benefit analysis. However, comparison does not inherently require that effects be predicted and disclosed; it is sufficient that effects be assigned to an ordinal scale. Thus, when comparison is emphasized, there is a tendency to slight prediction in favor of devising systems that place incommensurables such as endangered species' habitat, historic sites, and crop production on a common scale. Although comparison is central to environmental impact assessment, it has been treated as just one way of evaluating acceptability in risk assessment (Whyte and Burton 1980).

Damage Assessment

Some laws require either compensation for damage done by release of pollutants or restoration of the damaged communities in lieu of compensation. For example, U.S. federal or state officials, acting as trustees for natural resources, can seek compensation under the Clean Water Act (CWA) or Comprehensive Environmental Response, Compensation and Liability Act (CERCLA) for damage to natural resources caused by releases of oil or other toxic materials. Cases arising from these laws and civil suits involving pollution require that a damage assessment be performed to determine what magnitude of compensation is necessary and what injuries need to be remediated. These assessments are retrospective rather than predictive in that they must reconstruct the state of the injured environment prior to the pollution release so that it can be compared to the injured state. In addition, the assessment is often performed well after the fact, so that the injured state must be reconstructed as well. Some damages, such as loss of pelagic marine life after an oil spill, are difficult to measure even if a crew of biologists is present. Thus, damage assessments range in complexity from simple counts of dead organisms to rather elaborate combinations of exposure modeling, laboratory toxicity testing, and field monitoring of polluted and clean sites. Damage assessments have generally been ad hoc performances, but a modeling system has been developed by the U.S. Department of the Interior for natural resource damage assessments (DOI 1987a, 1987b).

Prioritization of Hazards

When regulatory activities are initiated with respect to a large, long-standing problem, it is not possible to address all instances immediately. Similarly, not all components of a complex pollutant are equally worthy of study or treatment. Instead, cases and components are addressed in order of decreasing seriousness. Examples include the prioritization of waste sites for cleanup under the Superfund program (EPA 1982a, Hivey et al. 1985), the prioritization of existing industrial chemicals for regulatory consideration under the

Toxic Substances Control Act (NRC 1981a, Walker and Brink 1988), and the prioritization of components of a complex effluent for treatment (Chapter 7). This prioritization does not require that the environmental effects of the waste sites, chemicals, or effluents be predicted, or even that the relative magnitudes of effects be determined. Rather, it is sufficient that they be correctly ordered in terms of the magnitude of effects. A distinct class of assessment models and scoring systems has arisen to deal with the prioritization of hazards.

Dichotomous Regulation

Certain regulatory actions require a Yes or No decision with respect to a hazard. Examples include decisions to allow a particular use of a pesticide under FIFRA, or an industrial chemical under TOSCA or the European Economic Community's equivalent statute, Directive 831/79 on Dangerous Substances. These actions do not necessarily require prediction of the nature or magnitude of effects. It is sufficient to show either that a chemical is innocuous, or that it would be seriously damaging in its proposed use. However, when the expected effects are near the threshold of acceptability, it becomes necessary to clearly define the expected effect, the threshold of acceptability, and the likelihood that the threshold would be exceeded.

Scaler Regulation

Certain regulatory actions require that a decision be made about where, on a scale of exposure concentrations or doses, the boundary between acceptability and unacceptability occurs. Examples include the setting of ambient air quality standards under the Clean Air Act, and the setting of federal water quality criteria and state water quality standards under the Federal Water Pollution Control Act (FWPCA). This process is generally more difficult than dichotomous regulation because it makes fine distinctions between acceptable and unacceptable concentrations. Thus, it requires that science and policy be pushed to their limits in every case, while dichotomous decisions are often routine because chemical uses and effluents are often clearly acceptable or clearly unacceptable.

Explanation of Observed Degradation

While most assessment activities begin with specific actions and attempt to anticipate effects, it is often necessary to perform assessments that begin with observed environmental degradation and attempt to explain it in terms of a causal action. Examples include the decline of some raptorial and piscivorous birds in the 1950s and 1960s, the absence of fish in many Adirondack and Scandinavian lakes, forest decline in Europe and North America, and the decline of the Chesapeake Bay biota. While there is no specific legal mandate for explanatory assessments, they can arise in regulatory contexts. For example, the observation of large numbers of dead birds in areas where some granular pesticides had been used has led the U.S Environmental Protection Agency to reconsider the registration of granular formulations because of the previously unconsidered issue of direct ingestion (Balcomb et al. 1984). These assessments resemble epidemiological assessments (Chapter 10).

Issue Definition and Research Planning

Issue definition is the assessment of a perceived environmental problem that has not yet been demonstrated to have damaged the environment. Examples of issues that have recently required this sort of assessment are acid rain, stratospheric ozone depletion, and climatic warming due to CO_2 and other greenhouse gases. The purpose of issue definition assessments is to summarize the state of knowledge so as to determine whether an issue requires research and risk assessment, or can be safely dismissed. If research is required, these assessments propose research priorities. Because the evidence is typically scant, these assessments depend heavily on expert judgment and are often assigned to prestigious groups such as U.S. National Research Council panels.

A common failing of issue definition assessments is that they focus on scientific issues, rather than the need to assess risks so that policy decisions can be made. The National Atmospheric Precipitation Assessment Program (NAPAP) is a conspicuous example of the dangers of letting science drive assessment. Because NAPAP focused on studying the major scientific issues associated with acid deposition rather than on assessing the risks and benefits associated with alternate control actions, it failed to contribute to the passage of control legislation (ORB 1991, Roberts 1991). The best way to define a new environmental issue is to define the risk assessment that would allow decision-makers to determine the best response. Research could then be chosen for its ability to increase confidence in the results of the risk assessment rather than to resolve scientific debates.

Habitat Assessment

Assessments that determine the suitability of ecosystems as habitat for a species are termed habitat assessments. Habitat assessments provide a broader context for estimation of pollution effects, and provide a means for resource managers to incorporate pollution effects into their management models (Lee and Jones 1982). The most conspicuous example is U.S. Fish and Wildlife Service's habitat evaluation procedure, which provides a framework for determining habitat quality for specific fish and wildlife species (DES 1980) and has been extended in an attempt to consider pollution effects (DOI 1987b). The use attainability analyses performed under the U.S. Clean Water Act (EPA 1983a) also function as habitat assessments. They determine what uses of a water body are attainable, the extent to which pollution prevents those uses, and the level of pollution control that is necessary to allow those uses. They must consider habitat limitations such as frequency of extreme low flows, the natural water quality, and the physical structure of the habitat. The Environmental Requirements and Pollution Tolerance System (ERAPT) is a large relational database on habitat requirements of aquatic species in the upper midwestern U.S. that allows inference about water quality from the species present (Dawson and Hellenthall 1986).

Estimation of Benefits of Protection or Remediation

If the costs of remediating a waste site, building a treatment facility, or denying the use of a product are to be balanced against the benefits to the environment, then those benefits must be specified. Ecological benefits of regulation have seldom been explicitly specified because: (1) the hazard assessment procedures used by most regulators do not identify what is being protected, (2) the state-of-the-art in ecological assessment has not permitted reliable prediction of benefits, and (3) the practice of risk-benefit analysis has been controversial. Formal risk-benefit analysis requires that economists specify the monetary value of environmental resources (Box 3.3). Market values (e.g., the amount that oysters from a particular bed would sell for if the bed were not destroyed) are easily determined, but nonmarket values (e.g., the value of songbirds or landscape aesthetics) can only be estimated by survey techniques (e.g., how much would you feel that you should be compensated for finding the beach littered with dead birds?). Despite these problems, regulatory decisionmakers must make some judgment about the relative risks and benefits, however subjective. Even human cancer deaths get balanced against the costs of regulation (Travis et al. 1987). Environmental decisions can be aided by better specification of the benefits of protective and remedial actions.

Protection of Human Health

One of the most common arguments for studying the effects of contaminants on natural biotic communities has been that they acted as an early warning system for effects on humans. This argument commonly took the form of an analogy to the use of canaries by coal miners to warn of asphyxiating atmospheres. This concept is still potentially valid. For the reasons discussed in Box 1.1, nonhuman organisms are likely to be more affected by chemicals in the environment than are humans. For example, piscivorous wildlife in the Great Lakes Basin have experienced severe toxic effects associated with bioaccumulation of organochloride chemicals which has led some observers to suggest that humans are likely to be experiencing less severe effects (Colborn 1991). It has also been suggested that tumors in fish are indicative of risk to humans (Dawe 1990).

SCHOOLS OF ENVIRONMENTAL ASSESSMENT

The differences in assessment problems noted above, along with differences in the academic training, traditions, and paradigms of the professional groups that conduct particular types of assessments have resulted in rather distinct schools of environmental assessment. Although these schools are not completely isolated, they display considerable divergence in the ways that environmental problems are formulated and analyzed. An appreciation of the very real bases for the diversity of environmental assessment and of the tools developed by various assessment traditions is required to appreciate how risk assessment can contribute to environmental assessment.

Environmental Impact Assessment

Environmental impact assessment (EIA) is the most diverse of the assessment traditions, both because of the diversity of activities that are considered "major federal actions" under NEPA and because NEPA requires that the assessments be performed by multidisciplinary teams. The dominant professions involved in EIA have been ecology and environmental engineering. Because of its legal mandate, EIA is predictive, comparative, and concerned with all effects on the environment. Because of the demand for full disclosure of effects, EIAs have devoted considerably more attention to identifying the full range of affected environmental components, on defining the geographic and temporal extent of effects, and on identifying secondary and tertiary effects than have other assessment traditions. This concern led to the development of a variety of checklists, matrices of activities and environmental components, and other devices for identifying, organizing, and displaying the numerous effects of a complex project (Skutsch and Flowerdew 1976). While EIAs have largely relied on deterministic methods, the requirement that uncertainties be presented (CEQ 1986) suggests that quantification of uncertainty through the use of risk models would be appropriate. Good discussions of environmental impact assessment as practiced in the U.S. and elsewhere are found in Munn (1975), Beanlands and Duinker (1983), and Westman (1985).

Resource Management

Distinct intellectual traditions have developed among managers of fisheries, game, forest, and land resources. These traditions are explicitly predictive because they must assess the consequences of harvesting and other uses. They are concerned with species-specific and site- or region-specific issues, and they typically have large bodies of data available concerning the history and nature of the resource, although the reliability of the data is often questionable. As a result, they tend to rely more heavily on statistical and mathematical models than other assessment traditions. Probabilistic models are becoming increasingly popular among resource managers who are particularly aware of how environmental stochasticity can influence the reliability of their predictions (Walters 1986). Their models have been used in environmental impact assessment, and the adaptive environmental assessment paradigm of Holling (1978) is a generalization of management modeling for environmental assessment. Adaptive assessment consists of (1) developing one or more models of the resource and its interaction with harvesters or other hazards, (2) development of parameter distributions for the models, (3) testing the models against data concerning the history of the resource, (4) performing experiments on the resources (e.g., reducing harvest of a fishery for a few years) to resolve remaining uncertainties, and (5) revising the models and parameter estimates in light of new results (Walters 1986). Management models have seen little use in assessment of chemicals, although their applicability has been pointed out by a number of authors (Waller et al. 1971, Cushing 1979, West et al. 1980, Tipton et al. 1980, Barnthouse et al. 1986, 1987, 1988, 1990). Discussions of resource management models can be found in Buongiorno and Gilles (1987), Starfield and Beloch (1986), and Walters (1986).

Hazard Assessment

Hazard assessment is the most commonly used methodology for analyzing the effects of chemicals on the natural environment. In essence, it consists of comparing the expected environmental concentration (EEC) and the estimated toxic threshold (ETT) and making a judgment as to whether the proposed release is safe, hazardous, or not sufficiently characterized for a conclusion to be reached. It does not use probabilistic methods, and does not attempt to predict the nature or magnitude of effects or compare different chemicals or releases.

Hazard assessment was developed primarily by aquatic toxicologists as a means of applying toxicological information to regulatory or quasi-regulatory situations where a dichotomous decision about release, use, or marketing is required. It was formalized at a 1977 workshop (Cairns et al. 1978) and has since been adopted by wildlife toxicologists but not, in general, by phytotoxicologists. It is based on an iterative process of testing and assessment called tiered testing (Figure 1.1). At each step, the EEC and ETT are estimated, based on data from prior toxicity tests, and measurements of properties of the chemical and are compared. If, based on formal decision criteria or expert judgment, the two concentrations are clearly different, then a decision can be made about the hazard. If the two concentrations are similar, then the decision is deferred and more data are sought. Clearly, this paradigm is not appropriate if the assessor does not have the power to defer a decision and demand more data.

The paradigm for hazard assessment was represented by Cairns et al. (1978) as the diagram shown in Figure 1.2. It shows that as more tiers of testing and measurement are completed, the assessor's confidence about the relative magnitudes of the EEC and ETT increases, and eventually it will become clear that

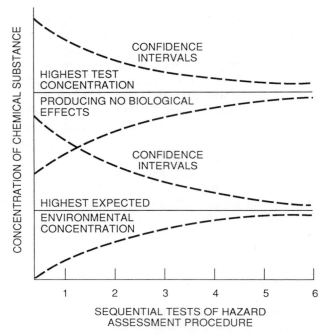

Figure 1.1. Tiered testing and assessment in the hazard assessment paradigm.

the ETT is safely below the EEC or hazardously above it. The confidence intervals shown in the figure are expressions of personal or institutional confidence rather than statistical confidence, and are expressed in practice as safety factors. Safety factors are typically even orders of magnitude such as 10X and 1000X, and are based on expert judgment which may be supported by simple data analyses. This informality in quantitative analysis is acceptable because it is assumed that testing and measurement can continue until the outcome is obvious. Therefore, this paradigm is appropriate when time and resources for testing and measurement can increase indefinitely. It has worked reasonably well in most cases even though resources and time are limited, because most nonpesticide chemicals do not exist in the environment at concentrations that are similar to their threshold for toxic effects. In addition, because hazard assessment does not require predictions or even specification of the assessment endpoint (the type and level of effect that is considered unacceptable) and relies heavily on expert judgment, the assessor has considerable latitude in deciding when to terminate the process.

The agreement about how hazard assessments for aquatic effects should be conducted is sufficient for publication of a consensus standard practice (ASTM 1991). Reviews of hazard assessment practice are found in Maki (1979a), Maki and Bishop (1985), Cairns et al. (1979), Bergman et al. (1986), and Dickson and Rodgers (1986). Examples of specific hazard assessment schemes can be found in Dickson et al. (1979), Schmidt-Bleek and Haberland

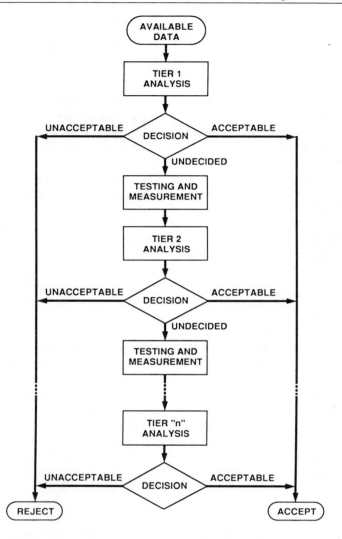

Figure 1.2. A diagrammatic paradigm of hazard assessment (from Cairns et al. 1978).

(1980), Beck et al. (1981), OWRS (1985), Bergman et al. (1986), and Urban and Cook (1986). The relationship of hazard assessment to risk assessment is discussed at length in Suter (1990c).

Scoring Systems

Scoring systems have been developed as one means of prioritizing environmental hazards for further study and assessment. Scoring systems force all cases onto a common scale by assigning scales of subscores to each characteristic of the hazard and the receiving environment that are thought to be relevant to assessment, and then combining these subscores into a final score on a standard scale.

As an example, consider the U.S. Department of Defense's waste-site prioritization model (Smith and Barnthouse 1987). This system, which is portrayed in Figure 1.3, is unusual in that ecological effects are calculated in the same detail as health effects. Scores for ecological and human health effects are calculated for surface and groundwater pathways. Each of the subscores shown is itself composed of subscores, all on a 0–3 scale. Some are quantitative and straightforward, such as the "net precipitation" score, a component of the groundwater pathway score (Table 1.1). Others are qualitative and rather subjective such as the "character of biota/habitats" score, a component of the ecological receptor score (Table 1.2). The final score is a weighted root-mean-square of the four receptor/pathway combinations, with human health effects given a weight of 5 relative to ecological effects.

Scoring systems involve technical issues that do not appear in other types of assessments. Examples include treatment of missing values (do they get the highest, median, or lowest possible score?), weighting to indicate relative importance (e.g., is a human carcinogen twice as important as a fish toxin?), scaling (e.g., logarithmic or arithmetic scales), and procedures for combining subscores (e.g., additive, multiplicative, or root mean square).

Examples of scoring systems for chemicals are the MEGS system (Cleland and Kingsbury 1977) and the Office of Toxic Substance's chemical scoring system (O'Bryan and Ross 1988). A general discussion of chemical prioritization can be found in NRC (1981a). Scoring systems for toxic waste sites have been reviewed by Hivey et al. (1985). The best known of these is the EPA's Hazard Ranking System (EPA 1982a, 1990a). Reviews of the system have revealed characteristics that unintentionally biased its results. For example, it scores effects based on the most hazardous component of the waste, but waste quantity is scored on the basis of all components. Therefore, sites that had a large amount of relatively innocuous material and a small amount of a highly toxic material can receive a high final score.* Because the development of scales and rules for combining scaled variables tend to be somewhat arbitrary, the potential for this sort of bias is high in all scoring systems. Therefore, scoring systems must be carefully tested for their sensitivity to assumptions and ability to correctly classify sites before they are generally applied.

An alternative to conventional scoring systems that has received relatively little use is statistical models. Klee and Flanders (1980) ranked hazardous waste sites using linear discriminant analysis. A "training set" of well characterized sites is assigned to categories on a scale of hazard (i.e., from highly hazardous to nonhazardous), and site variables are used to establish a discrim-

---

*Although the recent revision of the HRS (EPA 1990a) corrected many of the faults of the original model, it left others, including the one discussed here, in place. The EPA (1990a) argued that it is not feasible to consider the relative dose or concentration of the waste constituents.

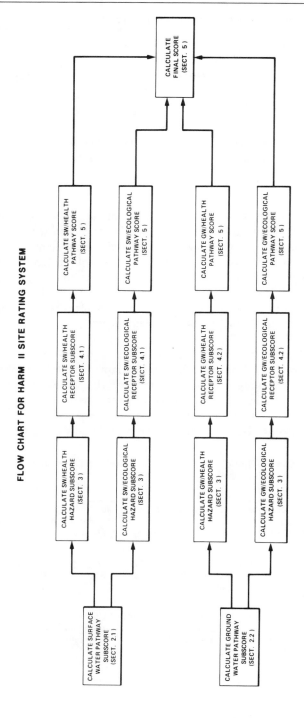

Figure 1.3. Flow chart of a hazardous waste site scoring system, the Defense Priority Model (Smith and Barnthouse 1987).

Table 1.1. Rules for Scoring Net Precipitation in the Defense Prioritization Model[a]

Net Precipitation	Score
< −10 in.	0
−10 to +5 in.	1
+5 to +20 in.	2
> +20 in.	3

[a]*Source*: Smith and Barnthouse 1987.

inant function that separates the categories in the variable space. New sites can then be classified on the hazard scale by applying the discriminant function.

Risk Assessment

Risk assessment is an array of methodologies that has grown out of the actuarial techniques of the insurance industry and is concerned with the estimation of probabilities and magnitudes of undesired events such as human morbidity, mortality, or property loss. Although it has proliferated to include a variety of quantitative and qualitative techniques from engineering, toxicology, epidemiology, sociology, and economics, it is still distinguished by the core concept of estimating "the objectified uncertainty regarding the occurrence of an undesired event" (Willett, 1901, *The Economic Theory of Risk and Insurance*, quoted in Rowe 1977).

Environmental risk assessment deals with risks that "arise in or are transmitted through the air, water, soil or biological food chains to man" (Whyte and Burton 1980). It can be thought of as that part of risk assessment that is concerned with events in the environment, or as that part of environmental impact assessment that formulates problems in terms of risks. While activities identified as environmental risk assessment have been largely concerned with

Table 1.2. Rules for Scoring the Biotic Communities Exposed to Hazardous Waste in the Defense Prioritization Model[a]

Character of Biota/Habitats	Score
Surface water ephemeral or absent; no wetlands; no perennially or seasonally flooded vegetation or vegetation irrigated with groundwater; no critical environments; OR no groundwater discharges within 3 miles (4.8 km) from the site in a downgradient direction.	0
Permanent or intermittent stream, wetlands <3 acres (<1.2 ha.), spring, or coastal marine environment that is not managed for fishing or hunting and that does not constitute a critical environment; small area [<100 acres (<40 ha)] of vegetation irrigated with groundwater.	1
Lake or reservoir; wetland >3 acres (>1.2 ha.); regionally important spawning, nursery, nesting, or feeding grounds, permanent or intermittent stream, tidal estuary, or other aquatic environment that is managed for fish or wildlife; extensive areas [>100 acres (>40 ha)] of vegetation irrigated with groundwater.	2
Critical environment[b]	3

[a]*Source:* Smith and Barnthouse 1987.
[b]Defined in the text of Smith and Barnthouse (1987).

effects of human actions on humans, it also includes effects of natural hazards such as floods and earthquakes and effects on the natural environment (Whyte and Burton 1980; Conway 1982). The study of risks to humans of natural hazards has been termed environmental (or natural) hazard risk assessment (Kates 1978; Petak 1982). We have termed the study of risks to the natural environment, ecological risk assessment (Barnthouse and Suter 1986).

In practice, ecological risk assessment has been the application of the science of ecotoxicology to public policy. Ecotoxicology was defined by Truhaut (1977) as the extension of toxicology to the ecological effects of chemicals. It bears the same relation to environmental toxicology as ecological risk analysis does to environmental risk analysis.

There is little literature on ecological risk assessment; Barnthouse et al. (1982) and Barnthouse and Suter (1986) present attempts to identify and develop methods that have the characteristics of risk assessment but were applicable to ecological effects. Risk models for ecological assessment include Goldstein and Ricci (1981), O'Neill et al. (1982), Suter et al. (1983), Barnthouse et al. (1987, 1990), and Hallam et al. (1990), but probabilistic ecological assessment models that are not associated with the concept of risk date back somewhat further (e.g., O'Neill 1973). The clear endpoints (yield reductions of individual crops) and probabilistic models used by the EPA's National Crop Loss Assessment Network (NCLAN) constitute risk assessment, although risk terminology has not been used. Environmental risk assessment is presented rather broadly in Whyte and Burton (1980), and the Society for Environmental Toxicology and Chemistry has described ecological risk assessment (which they termed environmental risk assessment) and its research needs (Fava et al. 1987). At this writing, a conceptual framework for ecological risk assessment is being prepared by the U.S. EPA and a report is being prepared on ecological risk assessment by the U.S. National Research Council.

Numerous and diverse risk assessment methods have been developed for the protection of human life, health, and property by engineers, economists, health researchers, systems analysts, and others. Familiarity with these methods would reveal many useful applications to ecological assessment. For example, transportation accident risk analysis techniques were used by the U.S. Department of Interior to estimate that there is a 5% chance that more than 20% of southern sea otters would be killed by an oil spill in the next 30 years (Ladd 1986). Because this book discusses the application of only a fraction of risk analysis techniques to ecological problems, the reader is referred to the literature in human health and engineering risk assessment including Rowe (1977), Lave (1982), Clayson et al. (1985), EPA (1984c), Kandel and Avni (1988), Cohrssen and Covello (1989), Paustenbach (1990), and Greenberg and Cramer (1991) as well as the journal *Risk Analysis*, published by the Society for Risk Analysis.

Box 1.1. Why Human Health Risk Assessment Is Insufficient

Risk assessments have emphasized risks to human health and have largely ignored ecological effects. This bias results in part from anthropocentrism and in part from the common but mistaken belief that protection of human health automatically protects nonhuman organisms. The assumption that health risk assessments will be universally protective is justified by the protection of humans from very small risks (one in a million risks of cancer) and the use of conservative assumptions in most health risk assessments. However, there are obvious counter examples, such as the fact that some chemicals that commonly cause severe effects to aquatic organisms such as chlorine, ammonia, and aluminum, pose no risk or negligible risks to humans in drinking water. Nonhuman organisms, populations, or ecosystems may be more sensitive than human for any of the following reasons:

a. Some routes of nonhuman exposure are not credible for humans, including respiring water, drinking from waste sumps, oral cleaning of pelt or plumage, and root uptake.

b. Chemicals are likely to be more toxic to some nonhuman species than to humans simply because there are far more nonhuman species, and some of them are likely to have properties that make them more susceptible than humans. In some cases this sensitivity is due to mechanisms of toxicity that do not occur in humans, such as eggshell thinning by DDT, stomatal closure in plants by sulfur dioxide, and imposition of male sex in snails by tributyl tin. In other cases the cause of the greater sensitivity is unknown, as in the greater sensitivity of birds and nonhuman mammals to chlorinated dibenzo-dioxins.

c. There are mechanisms of action at the ecosystem level such as eutrophication by nutrient chemicals, aquatic anaerobiosis due to degradable organic chemicals, and blockage of light by suspended solids, that have no human analogs.

d. Nonhuman organisms may be exposed more intensely to chemicals, even when the routes of exposure are the same. Any environmental pollution is likely to result in much higher exposure to some nonhuman organism than to humans. Humans (at least in affluent cultures) inhabit closed dwellings, obtain a variety of food from a variety of locations, tend to move among a variety of locations, and in general are not immersed in a particular ambient environment. For example, most humans eat at most a few meals a week that contain fish as one component, and the fish come from a variety of sources, while a heron or river otter eats only fish for nearly every meal and eats the whole fish and not just the relatively uncontaminated muscle.

e. Most birds and mammals have higher metabolic rates than humans, so they receive a larger dose per unit body mass because proportionately they consume more contaminated food, drink more contaminated water, and breath more contaminated air.

f. Some chemicals are designed to kill "pests" and are released to the environment at levels that are lethal to nonhuman organisms by design. In such cases, "nontarget" organisms that are physiologically and ecologically similar to the pest are inevitably affected.

g. Nonhuman organisms are highly coupled to their environments, so that even when they are resistant to a chemical they may experience secondary effects, such as loss of food or physical habitat. In contrast, humans in industrialized countries have alternate sources of food and materials for shelter if a portion of their environment is damaged by chemicals.

These arguments are mitigated somewhat by the fact that we are concerned about extremely low levels of human mortality that cannot be detected, and have no significant consequences for nonhuman populations. They also do not apply to mutagenic effects. Mutations are unacceptable in humans, but natural selection inconspicuously weeds them out of all but the smallest nonhuman populations. However, even for the polychlorinated dibenzo-p-dioxins (PCDDs), which are much-feared and regulated mutagens and carcinogens, environmental exposures have been associated with significant effects on nonhuman organisms without significant human effects, despite careful human monitoring. While the toxic effects on humans of the PCDD release at Seveso, Italy have been limited to some cases of chloracne, rabbits and other herbivorous animals were killed (Wipf and Schmidt 1981; Mastroiacovo et al. 1988). No human effects have been established at the Love Canal dump site which included PCDDs, but field mouse populations have been devastated by sterility and early mortality (Christian 1983; Rowley et al. 1983). The PCDD-contaminated oil at Times Beach, Missouri, killed horses, cats, dogs, chickens, and hundreds of sparrows, but apparently no humans (Sun 1983).

SUMMARY

Risk assessment is a rigorous form of assessment that uses formal quantitative techniques to estimate probabilities of effects on well-defined endpoints, estimates uncertainties, and partitions analysis of risks from decisionmaking concerning significance of risks and choice of actions. Some areas of environmental assessment, particularly the assessment of marine fisheries, already incorporate the properties of risk assessment without consciously adopting the risk assessment paradigm and jargon. Risk assessment approaches can improve the rigor and credibility of environmental impact assessment and environmental regulation. Many of the expert panels that have been convened to define new environmental issues would have produced more useful results if they had used the existing data to perform or at least outline a risk assessment rather than summarize the data in narrative form. Scaling systems would prove more reliable if they ensured that their scales were at least correlated with scales of risk. For example, the failings of the EPA's original Hazard Ranking System, which were notorious (Hivey et al. 1985; Doty and Travis 1990), have been largely corrected by the revised system, which consciously attempts to estimate relative risk (EPA 1990a). Conversely, risk assessment could benefit from incorporation of aspects of other assessment traditions such as the well-validated population models of resource managers, the public scoping of NEPA impact assessments, and the tiered testing and assessment approach of hazard assessment. Ecological assessors have much to learn from risk assessment, but risk assessors have even more to learn from ecologists, ecological toxicologists, resource managers, and environmental chemists before applying their tools to ecological problems.

2

Assessment Concepts

Glenn Suter and Lawrence Barnthouse

This chapter introduces some central conceptual issues in ecological risk assessment. The first section deals with the problem of defining what is to be protected and how measurements relate to valued components and processes. The second section discusses different tools for ecological risk assessment, including various types of models. A number of alternative or complementary approaches are presented, consistent with our philosophy that it is counterproductive to propose a single specific procedure for ecological risk assessment analogous to the standard procedures that exist for nuclear power plant or carcinogen risk assessment. The chapter concludes with a discussion of the relationship of risk to probability and uncertainty.

THE ECOLOGICAL ASSESSMENT PROBLEM

The purpose of ecological risk assessment is to contribute to the protection and management of the environment through scientifically credible evaluation of the ecological effects of human activities. Successfully performing this task requires that we (a) define, in operational terms, just what "the environment" is, and (b) devise methods for characterizing the state of the environment and quantifying expected changes. We must also determine when a "significant" change has occurred and evaluate the importance of uncertainties in our data and models. In human health risk assessment, the basic objective is usually well defined: to estimate the incidence of death and disease resulting from exposure to hazardous agents. For common types of health risk assessments, e.g., for food additives, nuclear power plants, and environmental releases of carcinogenic chemicals, formalized assessment procedures have been developed. In contrast, the objectives and specific procedures for ecological risk assessment are much more difficult to define. Should the emphasis be on ecosystem-level properties such as production and diversity? Should studies be limited to species of aesthetic or commercial value? Should we focus on "representative" species? Furthermore, even when there is agreement on *what* to assess, there may be substantial disagreements among ecologists about *how* to assess.

Endpoint Definition

Any assessment must have defined endpoints. An assessment endpoint is a formal expression of the environmental values to be protected (Suter 1989). A clear statement of its endpoint is as important to an assessment as a clear statement of the hypothesis is to an experimental research project. Defining an assessment endpoint involves two steps: (1) identifying the valued attributes of the environment that are considered to be at risk, and (2) defining these attributes in operational terms. In regulatory situations, the environmental attributes to be protected are broadly defined by statute. For example, the Clean Water Act, section 316b stipulates that a "balanced indigenous population" must be maintained in water bodies receiving thermal plumes from power plants. In NEPA-mandated environmental impact statements, attributes at risk are identified by a process called scoping in which the concerns of the public and governmental bodies are elicited (CEQ 1986).

Because statutes generally are not helpful sources of operational definitions for assessment endpoints (Lave 1982), the second step is left to the judgment of the responsible agencies. In the past, little systematic thought has been given to this critical step. Beanlands and Duinker (1983) identified ill-defined objectives (another way of saying poorly defined endpoints) as the single most common cause of failure of environmental impact assessments in Canada. In U.S. agencies dealing with toxic substances, the practice has been to let the available toxicity tests and models define the assessment endpoints. The EPA has recently recognized the importance of endpoint definition and has sponsored endpoint workshops and reviews (AMS 1987, Mittelman et al. 1987, Suter 1989 and 1990b). Given the diversity of the biological world and the multiple values placed on it by society, there is no universal list of assessment endpoints. We believe, however that there are five criteria that any endpoint should satisfy:

1. societal relevance
2. biological relevance
3. unambiguous operational definition
4. accessibility to prediction and measurement
5. susceptibility to the hazardous agent.

Societal relevance implies that the endpoint should be understood and valued by the public and by decisionmakers. This criterion is necessary because assessments of risks to organisms of little intrinsic interest to the public (nematodes, zooplankton) are unlikely to influence decisions unless they can be clearly shown to indicate risks to biota of direct human interest (e.g., fish, wildlife, crops, and forests). This criterion is controversial because many environmental scientists argue that societal values should not be considered. They use two arguments. The first argument confuses intermediate factors for endpoints. It takes the form: zooplankton and spawning gravels are not valued by the public, but they are biologically important to fish so they should be assessment endpoints. In this case, the assessment endpoint is clearly fish, and the assessment must incorporate the relationship between effects on zooplankton and spawning gravel and effects on fish. The second argument is that biological values should be protected even if there is no demonstrable connection to societal values. However, time and resources to develop risk models and conduct tests is limited. Just as human health risk assessments for chemicals always consider carcinogenicity and seldom consider skin rashes because cancer is more dreaded by the public, assessments of effects of old-growth forest clear-cutting in Oregon focus on the probability of extinction of spotted owls rather than oribatid mites. In theory, we would like to assess risks of rashes and mite population declines, but in the real world we must consider cancer and birds. The argument for protection of all biological entities and processes is rebutted by Orians (1990) in a similar way:

> In practice, however, an insistence on valuing everything equally results in little value being given to anything. Environmental assessment and management are best served when people explicitly choose a limited number of 'valued ecosystem components'. . .

Societal relevance is not a constant. For example, 20 years ago wetlands were not valued by society. To be effective, assessments of risks to a wetland had to make the link between loss of the wetland and valued endpoints such as fisheries, waterfowl, and flood mitigation. Now the values of wetlands are generally recognized, so wetlands preservation is an important assessment endpoint in its own right.

The biological significance of a property is determined by its importance to a higher level of the biological hierarchy. For example, a physiological change is biologically significant if it affects a property of the whole organism such as survival or fecundity; a change in fecundity of individuals is biologically significant if it affects the size, productivity, or other property of the population; and a decrease in the size of a population is biologically significant if it affects the number of species, the productivity, or some other property of the ecosystem. Because of functional redundancy, negative feedbacks, and other compensatory mechanisms, much variability in biological properties can occur at one level without perturbing higher levels. A societally relevant endpoint may or may not also have biological significance. For example, the abundances of African elephants and peregrine falcons are both societally significant and worthy endpoints on that basis. However, while declines of African elephant populations drastically affect savannah ecosystems, the extinction of populations of peregrine falcons had no apparent effect on ecosystem properties. Endpoints that have biological as well as societal significance should have high priority.

Without unambiguous operational definitions, endpoints provide no direction for testing and modeling, and the results of assessments tend to be as ambiguous as the endpoints. Ambiguity is the principal reason why legislatively defined endpoints (e.g., "balanced indigenous population") and common expressions of environmental goals (e.g., "ecosystem health") are often inadequate for assessment purposes. The definition of an assessment endpoint includes an entity such as the Hudson River striped bass population and a property of the entity such as abundance or yield. The entity can be referred to as the endpoint species, endpoint process, etc. The extent to which the endpoint must be defined quantitatively depends on the assessment. A screening assessment to determine which chemicals in a mixed waste must be assessed in detail may not require quantitatively defined endpoints. However, to design a monitoring program that minimizes risks of missing real effects, one must quantify the magnitude of change that must be detected, the area that can be affected before effects are detectable, a time allowed before effects are detected, and statistical confidence (Chapter 12).

Susceptibility results from a potential for exposure and responsiveness to the exposure. When choosing endpoints for a predictive assessment susceptibility may be unknown, but for a retrospective assessment it may be all too apparent. Techniques for identifying potentially susceptible endpoints are discussed in Chapter 3. More detailed discussions of the use of these criteria to select assessment endpoints were presented by Suter (1989, 1990b).

If the response of an endpoint cannot be measured or estimated from measurements of related responses or component responses, it cannot be assessed. The best assessment endpoints are those such as fish and crop production, for which there are well-developed test methods, field measurement techniques, and predictive models.

The things that are measured in a monitoring study or toxicity test are generally referred to as indicators. The formal, usually quantitative, expression of the results of toxicity testing or monitoring of an indicator is a measurement endpoint. Measurement endpoints are the toxicologist's or field biologist's input to risk assessments. They are usually single numbers such as 96-h LC_{50}s derived from toxicity tests and estimates of fish population sizes from mark-recapture studies. However, for risk assessments the most useful measurement endpoints are multidimensional descriptive models such as concentration-response functions, rather than a single number. Examples of assessment endpoints, indicators, and measurement endpoints are presented in Table 2.1.

In most cases, assessment endpoints and measurement endpoints are not the same. Assessment endpoints generally refer to characteristics of populations and ecosystems defined over rather large scales, e.g., forest production over a large geographic area or a fish population in a reservoir. It is usually impractical to directly measure changes in these characteristics as part of an assessment. In many cases, e.g., pesticide registration and toxic chemicals review, assessments must be made *before* any large-scale release can occur. The mea-

Table 2.1. Examples of Assessment Endpoints, Possible Indicators of Effects on Those Assessment Endpoints, and Possible Endpoints for Measurements of Those Indicators

Hazard/ Policy Goal	Assessment Endpoints	Indicators of Effects	Measurement Endpoints
Herbicide used for weed control in southern lakes/ No unacceptable loss of fisheries	Probability of $> 10\%$ reduction in game fish production	Laboratory toxicity to fish	Fathead minnow LC_{50} Larval bass concentration/ mortality function
		Laboratory toxicity to food-chain organisms	*Daphnia magna* LC_{50} *Selenastrum capricornutum* EC_{10}
		Field toxicity to fish	Percent mortality of caged bass
		Populations in treated lakes	Catch per unit effort Size/age ratios by age class
Agricultural insecticide associated with bird kills / No unacceptable reductions in avian populations	Proportion of raptors killed within the region of use	Laboratory toxicity to prey	Rat LD_{50} Japanese quail dietary LC_{50}
		Laboratory toxicity to raptors	Sparrow hawk dietary concentration/response function Japanese quail dietary LC_{50}
		Avian field toxicity	Number of prey carcasses per hectare Number of dead or moribund raptors per hectare
	Increase in the rates of decline of declining bird populations within the region of use	Avian laboratory toxicity	Japanese quail dietary LC_{50} Starling dietary LC_{50}
		Avian field toxicity	Number of bird carcasses per hectare by species
		Trends in populations of declining birds	Rates of decline in areas of use as proportions of reference areas

surements used in ecological risk assessments most often are obtained by local environmental sampling and laboratory testing. Every ecological risk assessment should explicitly specify (1) the assessment endpoint(s), (2) the measurement endpoint(s), and (3) the methods used to extrapolate from the measurement to the assessment endpoints.

Spatiotemporal Scales and Levels of Ecological Organization

Biologists view the world in terms of hierarchical "levels of organization." Populations are composed of individual organisms and organisms are composed of organs and fluids. Communities are groups of populations that interact with each other; ecosystems are communities, together with their physical and chemical environment; and regions are spatial groupings of ecosystems.

We note the existence of levels of organization for three reasons. First, although both measurement and assessment endpoints can be defined at any level, they are not equally important at each level. The organism is the smallest unit that interacts directly with the environment. It is organisms that are exposed to toxic chemicals. The boundary between the organism and the environment divides the domains of environmental chemistry and toxicology, and organism-level responses are the best-described component of toxicology. However, organisms themselves are transitory, and, because they are consumed by humans, they are treated as expendable. The reproducing population is the smallest ecological unit that is persistent on a human time scale, and hence the lowest level that we can meaningfully protect. However, populations do not exist in a vacuum, because they must eat, be pollenized, be sheltered, etc. This requires a community of other organisms of which the population is a part. The response of any population to a stress depends in part on its interactions with organisms, many of whom may themselves be affected by the same stress. The community, in turn, occupies a physical environment with which it forms an ecosystem. The physicochemical environment can itself be susceptible to modification by stressors. Two prominent examples are the modification of soil and surface water pH by acid deposition, and the modification of regional and global climates by atmospheric carbon dioxide releases.

The second reason for considering levels of organization is that the various levels differ dramatically in terms of their practical utility as assessment or measurement endpoints. We have noted above that most assessment endpoints are defined at the population level and higher. However, the feasibility and cost of toxicity testing and monitoring increases with level, while the precision and generality of the results decreases! As shown in Table 2.2, spatial and temporal scales for observations at the different levels are quite different. For the most "relevant" observations, i.e., field observations of populations and ecosystems, the time scale for an experiment may be measured in years, the space scale in km^2, and the cost in millions of dollars! Obviously, the time and resources required for such studies are inconsistent with the requirements of practical schemes for routine risk assessment. The existence of this gap

Table 2.2. Scales of Observation in Toxicity Tests

| Spatial | Temporal | Organizational | |
		Structural	Functional
Organismal-level laboratory tests			
cc-m³	hours-days.	organism	growth reproduction
Microcosm tests			
cc-m³	days-months	organism population, community, food web	production cycling
Field tests			
m²-km²	days-years	organism population, community, food web	production cycling
Environmental monitoring			
m²-km²	years	organism population community food web	production cycling

between measurement feasibility and assessment interest is the principal reason why models play such a prominent role in the assessment techniques described in this book. Some form of statistical or mathematical model is necessary for meaningful extrapolation across scales of time, space, and biological organization.

The third reason for considering levels of organization is that assessors must be aware of the relative spatial and temporal scales of the hazardous actions being assessed (Figure 2.1) and the assessment endpoints (Figure 2.2). For example, spills occur at spatial scales that are similar to those of populations, but their occurrence is much briefer than population processes. Therefore, to assess the risks to populations of a transient event like a spill, it is necessary to emphasize the time-course of processes such as reproduction and recolonization that occur on longer time scales and that determine the magnitude of the population response and its significance relative to natural population variance. For further discussions of the implications of ecological levels of organization see O'Neill et al. (1986), Sigal and Suter (1987), Levin (1987).

Uncertainty

Although all scientific activities are concerned with identifying and reducing uncertainty, uncertainty plays a particularly important role in risk assessment (Suter 1990a). Risk is the probability that a specified harmful effect will occur, or, in the case of a graded effect, the relationship between the magnitude of the effect and its probability of occurrence. If it is certain that the undesired event

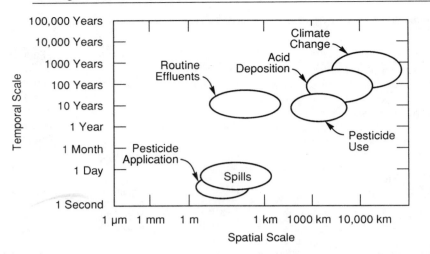

Figure 2.1.   Arrangement of a variety of anthropogenic hazards on spatial and temporal scales.

will occur or that it will not occur (i.e., the probability of occurrence of an exactly specified event is known to be 1 or is known to be 0), there is no risk.

The importance of uncertainty in risk assessment derives from its central role in risk-based decisionmaking (Ruckelshaus 1983); the decisionmaker needs to understand the uncertainties associated with the scientific information on which the decision will be based. Commonly, assessors use conserva-

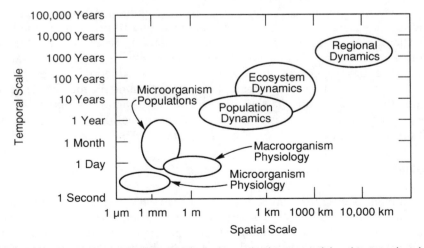

Figure 2.2.   Arrangement of levels of biological organization on spatial and temporal scales.

tive assumptions rather than estimating uncertainty. This practice has several disadvantages from the standpoint of decisionmaking (Paustenbach 1990).

1. It is inconsistent. Even "worst case" assumptions are not consistent; it is always possible to conceive of a worse and more improbable case.
2. Conservative assumptions tend to hide uncertainty and error from the decisionmaker by burying it in the estimates of exposure and effects.
3. Conservatism assumes that there are no societal or environmental costs of regulating false positives. In fact, remediation or regulation often result in intermedia transfers of treated pollutants or replacement of one product with another whose properties are not as well studied.

For these reasons, conservative assumptions generally are not adequate responses to uncertainty. The major exception is the use of conservative assumptions in the screening of hazards, to quickly eliminate chemicals and routes of exposure that are clearly trivial from further assessment.

Uncertainties in risk assessment have three basic sources: (1) the inherent randomness of the world (stochasticity), (2) imperfect or incomplete knowledge of things that could be known (ignorance), and (3) mistakes in execution of assessment activities (error).

Stochasticity refers to uncertainty that can be described and estimated but can not be reduced because it is characteristic of the system being assessed. Even if the universe were ultimately deterministic at the subatomic level, physicochemical "drivers" of ecosystems such as rainfall, temperature, and wind are effectively stochastic at levels of interest in risk assessment. For example, the minimum dilution volume of a stream during the period in which an effluent will be released can not be deterministically predicted. Biological processes such as colonization, reproduction, and death are similarly stochastic. Limits on the precision with which variable properties of the environment can be quantified define the upper limit of the precision of every assessment.

Ignorance refers to lack of knowledge of some aspect of a system that is potentially knowable. In some cases this is a fundamental ignorance of some scientific issue. For example, ignorance of the phenomenon of acid rain made assessments of the environmental effects of sulfur oxide emissions incomplete prior to the late 1970s. This fundamental ignorance results in undefined uncertainty, the "unknown unknowns" that can not be described or quantified (Suter et al. 1987a). More commonly, ignorance is simply a result of practical constraints on our ability to accurately describe, count, or measure everything that pertains to a risk estimate. Examples include the inability to test all toxicological responses of all species exposed to a pollutant and misspecification of the forms of mathematical models used in risk assessments due to ignorance of the system dynamics. Much of the literature on uncertainty in risk assessment pertains to identifying, quantifying, and reducing this kind of uncertainty.

The third source, human error, is an inevitable attribute of all human activities, including risk assessment. Examples include incorrect measurements, misidentifications, data recording errors, data entry errors, and computational errors. Such errors are primarily a quality assurance problem that is outside the scope of this book.

The above discussion is quite general and applies to all types of risk assessments, but the diversity and complexity of ecological systems ensures that the ignorance component in ecological assessments will be very high. Levins (1966) suggested that there is an unavoidable tradeoff between precision, generality, and realism of ecological models. A precise and realistic model will necessarily be limited to a few specific applications, whereas a very general approach applicable to many situations will necessarily be limited in precision. The more realistic the test or model (in terms of inclusion of components), the lower the generality, and vice versa. This has the following implications for ecological assessment:

1. There is no unique "best" model or test system for ecological risk assessment. In general, an assessment approach that minimizes one source of uncertainty will increase other sources.
2. For most assessments, multiple, independent lines of evidence are better than any single approach. Those results that are supported by multiple methods are called "robust."
3. As previously noted, uncertainty increases with the hierarchical level of an assessment endpoint.

Box 2.1. Species Sensitivity Distributions as a Case Study in Uncertainty

Some issues in describing and estimating uncertainty can be illustrated by considering the use of a simple type of model, species sensitivity distributions, to establish environmental quality criteria. The basic assumption of these models is that sensitivity of species is a stochastic variable that can be characterized by fitting a probability density function to test endpoints (e.g., LC_{50}s) for several species (Figure 2.3a). This distribution can be used as a model of individual species to estimate the probability that a particular species will be affected at a prescribed concentration (Chapter 7), or as a model of communities to estimate the proportion of the community that would be protected at a prescribed concentration (Chapter 9). This type of model has been used since 1980 to calculate acute National Ambient Water Quality Criteria for Protection of Aquatic Life (EPA 1980, EPA 1985, Stephan et al. 1985).

The stochasticity of species sensitivities is not the only source of uncertainty involved in this model. There is uncertainty concerning the true distribution of sensitivity. Dutch (Kooijman 1987, van Straalen and Denneman 1989, Aldenberg et al. 1990, BKH 1990, Van Leeuwen 1990) and Danish (Wagner and Lokke 1991) investigators have taken into consideration the fact that the set of n points (tested species) to which the distribution was fit is only one of many

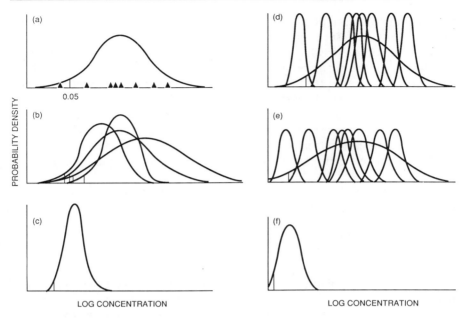

Figure 2.3. Uncertainties in the use of species sensitivity distributions to estimate water quality criteria for a chemical. In all cases, the vertical bar marks the lower fifth percentile of the distribution. (a) A probability density function fit to LC_{50} values for eight species. (b) Probability density functions fit to four samples of eight LC_{50}s. (c) Probability density of the fifth percentiles of species sensitivity distributions for exactly specified LC_{50}s. (d) A probability density function fit to LC_{50}s for eight species, each of which has a specified measurement uncertainty. (e) A probability density function fit to LC_{10}s for eight species, each of which has a specified uncertainty due to measurement and extrapolation from the LC_{50} to LC_{10}. (f) Probability density of the fifth percentiles of species sensitivity distributions for uncertain LC_{10}s.

possible samples of species that might be tested, none of which is the true distribution of all N species in the community (Figure 2.3b). This uncertainty due to sampling from the true distribution is used to derive an estimate of the lower fifth percentile of the estimate of the concentration that is protective of a proportion x of the community (Figure 2.3c). Incorporation of this uncertainty has the advantage that as more testing is done (as n increases) the variance of the distribution of the concentration protecting $x\%$ decreases, thereby raising the criterion. This creates an incentive for regulated industries to do more toxicity testing. This approach has been adopted by the government of the Netherlands (HCN 1989). Various distribution functions (log-normal, log-logistic, and log triangular) and formulae for calculating the uncertainties have been proposed by the various authors (Van Leeuwen 1990). Hence, these models acknowledge uncertainty due to ignorance concerning the true stochastic distribution.

At least two other potentially significant sources of uncertainty concerning the true distribution of species sensitivity are not acknowledged in these models. First, the test endpoint values are treated as constants. Actually, there is uncertainty in these values due to imperfect fit of the statistical model to the test data (e.g., the probit models used to derive LC_{50}s discussed in Chapter 7) and due to variance in the results of replicate tests of the same species. Therefore, the species sensitivity distributions could be fit to distributions of the test endpoints for each species rather than to point estimates of those endpoints (Figure 2.3d). These distributions of the test endpoints represent the stochasticity of responses of different populations of the tested species, variance in the conduct of the test permitted by the protocols, and errors in the conduct of the tests.

Second, the test endpoints are not estimates of concentrations that pose minimal risk to the species. For example, half of the test organisms are expected to die at the LC_{50}. If the goal is to prevent excessive acute lethality rather than to permit it, the LC_{50} values should be extrapolated to LC_{10}, LC_{01}, or other appropriate values. This would shift the distributions on the species values to the left and broaden the distributions due to the uncertainty introduced by the extrapolation (Figure 2.3e). The shift is due to recognition of model uncertainty (i.e., we recognize that the test endpoint is an imperfect model of the assessment endpoint), and the increased variance is due to ignorance concerning the relationship between the test endpoint and the assessment endpoint. The result is a distribution of the lower 5th percentile of species that has a lower mean and greater variance than if LC_{50}s are assumed to have no variance (Figure 2.3 f versus c).

All of these uncertainties are based on the assumption that we know nothing about the expected sensitivity of species. That is, species sensitivities are completely randomly distributed. Potentially, use of prior knowledge concerning the distribution of sensitivity among species might result in a more complex model with reduced uncertainty. For example, one might reduce uncertainty by fitting separate species sensitivity distributions to plants and animals, or to fish and invertebrates, rather than fitting a single unimodal function to what is likely to be a polymodal distribution for many chemicals. Alternatively, rather than assuming that species sensitivity distributions are completely random, one could incorporate knowledge of patterns of sensitivity among species (Kooijman 1987). These might be purely empirical patterns or patterns of taxonomic similarity, of morphological similarity, etc. For example, knowledge that the test species more precisely represent the responses of closely related species than distantly related ones (Chapter 7) potentially allows test endpoints for individual species to represent (with some uncertainty) taxa within the community, rather than simply representing a random species in the entire community. Further, the relative abundance of different taxa in the communities to be protected could be used to weight the contributions of these taxon-specific distributions to the overall distribution of species sensitivities. Alternatively, if response to a chemical is believed to be largely controlled by kinetics, and if test endpoint values are normalized for differences among aquatic species in uptake rates, the residual species sensitivity distribution for acute lethality might have considerably lower variance for many chemicals. These suggestions are hypothetical but plausible.

All of the uncertainties discussed so far follow from the choice of species sensitivity distributions as regulatory models. In addition, there is model error due to uncertainty concerning the relationship between the distributions and the community properties to be protected, as well as the appropriate degree of protection. This uncertainty is expressed as variance in the proportion of species to be protected and the confidence in that degree of protection. If you use the best estimate (50th percentile) of the fifth percentile of the species sensitivity distribution (e.g., Stephan et al. 1985), you are assuming that on average, if you protect at least 5% of species, communities will not be unacceptably harmed. This is probably the case, but sometimes the species lost will be a highly valued species (e.g., a sport fish), or a species that plays a critical role in the community. On the other hand, if you protect all of the species in a community with 95% confidence (e.g., Kooijman 1987), your criterion will nearly always be much more conservative than it needs to be to preserve the value of the community. The choice of a concentration that protects 95% of species 95% of the time (e.g., HCN 1989) intuitively seems like a reasonable compromise, but its basis is more political than scientific (BKH 1990). It is not clear what levels of protection and of confidence are most appropriate, or even if the value of a community is a simple function of the proportion of species remaining. This model uncertainty must be resolved by validation research and by clearer specification of the assessment endpoint. That is, exactly what properties of communities are these regulatory models trying to estimate so that they can be protected? The Dutch have been unusually forthright in specifying that their regulatory endpoint is preservation of species composition (HCN 1989). Validation of various versions of the species sensitivity distribution model as well as other types of community or ecosystem models against this endpoint would be relatively straightforward.

ASSESSMENT METHODS

The methods to be discussed here include physical methods (test systems) and quantitative methods, both statistical and mathematical. As is discussed above, there is no universal method for quantifying ecological risks that will produce precise, general, and realistic results. There are always limitations in the amount of information that can be obtained, either because of time, resources, or absence of fundamental understanding. The alternatives discussed below are, therefore, often complementary means of obtaining the same end: quantification of exposures, effects, and risks.

Physical Models

Physical models are material representations of some object or system that is not itself subject to manipulation, or that cannot be manipulated as easily or with as much control as the model. The most common type of physical model used in environmental assessment is the laboratory test. A *Daphnia* in a beaker and a bean seedling in a flask of hydroponic solution are models of zooplankters in lakes and crops in fields, respectively. Similarly, a shakeflask culture is a model of biodegradation of chemicals by natural communities. The object or system represented is often unclear and a single model may be used to represent different systems in different assessments or even in different phases of a single assessment. For example, a fathead minnow acute lethality test may represent lethal effects on cyprinids, on all fish, or on all aquatic organisms, and may represent effects of toxic exposures of various durations. Large physical models of airsheds and hydrologic systems are sometimes constructed to represent transport processes as an alternative to complex mathematical modeling.

Existing disturbances or pollution sources and the associated receiving environments can serve as physical models of similar sites exposed to similar contaminants or stresses. For example, in the assessment of whether a proposed reservoir would become eutrophic, the most credible evidence was the fact that a similar, nearby reservoir was eutrophic (Goodman 1976). Comparisons to existing affected sites have been termed "analog studies" by the National Research Council (1986).

If one type of environmental disturbance is considered to be analogous to others, then relatively well-developed models and assessment approaches for one disturbance can be used to assess others. For example, the effects of fishing on fish populations have been used as a model for effects of mortality from power plant cooling systems (McFadden 1977) and toxic chemicals (Goodyear 1983, Barnthouse et al. 1987, 1990). To the extent that these actions resemble fishing in that they cause mortality at different rates on different life stages, the mathematical models and computational techniques developed by fisheries scientists can also be employed in risk assessments.

Statistical Models

Statistical models attempt to derive generalizations by using regression, principal components analysis, and other statistical techniques to summarize experimental or observational data. For example, in toxicology, dose-response models are obtained by statistically fitting a continuous function such as the probit to the discontinuous results of toxicity tests of discrete doses. The model assumes that the sensitivities of exposed organisms to toxic chemicals can be characterized by a statistical distribution with a mean and a variance, independent of the mechanism involved in uptake, translocation, and toxicity.

There are three distinct purposes for using statistical models in risk assessment: hypothesis testing, description, and extrapolation. Hypothesis testing was originally developed to determine whether data from controlled studies provided sufficient support for hypothesized relationships between controlled independent variables and the observed responses of dependent variables. However, hypothesis testing has been used in risk assessment to calculate "no-effects" concentrations in toxicity tests (Chapter 7) and comparison of contaminated and reference sites in monitoring studies (Chapter 10). In most cases the test is designed to reject a "null hypothesis," that there is no difference between treatments or sites. The use of hypothesis tests in environmental assessment involves a problem of interpretation that does not occur in classical experimental design: the tradeoff between Type I and Type II error (Parkhurst 1990). Type I error refers to the rejection of the null hypothesis when it is, in fact, true; α is the probability of making a Type I error. Type II error refers to the acceptance of the null hypothesis when it is false; β is the probability of making a Type II error. Type I and Type II errors are referred to as false positives and false negatives, respectively. In classical experimental design it is usually assumed to be preferable to falsely accept the null hypothesis rather than to falsely reject it; hence α is set at a very low value of 5% or 1%. The value of β is usually not constrained and is rarely reported. In risk assessment, however, accepting the null hypothesis when it is false implies concluding that there is no effect when one has, in fact, occurred. Hence, β is at least as important as α. The probability of *correctly* rejecting the null hypothesis, defined as $1-\beta$, is known as the power of a test. Studies of the power of test designs have received significant attention in research related to risk assessment (Vaughan et al. 1982, Peterman and Bradford 1987, Carpenter 1989) and should be specifically addressed in any risk assessment that employs a hypothesis test.

More importantly, many statisticians and practitioners have come to realize that hypothesis testing is not appropriate for most problems in applied science (Yoccoz 1991). It is a convenient crutch that allows the investigator to calculate statistical significance, thereby avoiding the more difficult issues of real-world ecological significance. Because there is no necessary association between the two types of significance, use of hypothesis tests in applied environmental science often leads to absurd results which may not be apparent without careful reanalysis of the data (Suter et al. 1987b, Smith et al. 1990). For example, contaminant concentrations in soil may average more than 10 times the average background concentration and may be above phytotoxic levels but still not be "significantly elevated," in purely statistical terms.

The second use of statistical models is description. For example, a multivariate regression model might be used to describe the results of a pollution monitoring study by regressing concentration against distance downstream and flow rate. Similarly, a multivariate classification method such as principal component analysis might be used to distinguish the sets of natural and pollution-adapted biotic communities within an ecosystem. This description of patterns and relationships in observational data is the major activity of statistical ecologists (Green 1980, Ludwig and Reynolds 1988). The results can be used to frame hypotheses which are then tested in experimental studies, to guide further sampling, or simply to increase understanding of the system.

The third use of statistical models is extrapolation, the use of a model to statistically estimate something other than the data from which it was derived. The most familiar type of extrapolation is extrapolation of a model to conditions outside the range of the data from which the model was derived. This range extrapolation is considered bad practice, but is sometimes necessary in risk assessment. Another more useful type of extrapolation is extrapolation from a type of data that is available to a type that is desired but unavailable. For example, a concentration-response model of a fathead minnow toxicity test describes the response of a laboratory stock under laboratory conditions, but it may be extrapolated to fathead minnows in the field or to fish in general. Note that this data extrapolation, whose opposite is description, is distinct from range extrapolation whose opposite is interpolation. Range extrapolation may be used with data extrapolation. For example, in human health risk assessments, rodent dose-response data are extrapolated to humans (a data extrapolation) and the model is extrapolated to untested dose levels (a range extrapolation). Data extrapolations require that the assessor either assume that the systems between which the extrapolation is being performed respond identically, or use some extrapolation model. In the example, a dose-scaling model is used for the rodent to human extrapolation (Chapter 7). Range extrapolation requires that the assessor assume that the relationships between the dependent and independent variables are the same outside the range of observations as they are within that range. Models that will be used for range extrapolation must be chosen for their behavior beyond the data as well as for their fit to the data.

In assessments of impacts of acidification on surface waters, statistical models have been used extensively to extrapolate from observations on subsamples of lakes and streams to impacts on regional surface water resources (Baker et al. 1990). Exercises of this kind are sometimes referred to as "empirical modeling" (Peters 1983); they are an important part of basic field ecology.

Strictly speaking, a statistical model makes no presumption to explain observations in terms of causal relationships between the independent and dependent variables. The model simply summarizes the relationship between the variables. However, more interpretive weight can be obtained by assigning biological or physical meaning to the fitted coefficients. For example, Breck (1988) showed that the probit and logit dose-response models can be derived as special cases of the Mancini damage-repair model of contaminant toxicity (Mancini 1983), so that the coefficients of the Mancini model can be estimated from standard fits of statistical concentration-response models. Applications such as these may be viewed either as statistical models or as mechanistic models for which statistical methods provide a rigorous means of parameter estimation. The ability of such models to fit real data is supportive of the hypothesized mechanisms, but it would also be supportive of other hypotheses that generate functions with the same form. For example, Koch (1966) pointed out that the lognormal distribution of species abundance can be generated by a variety of biological models but the fact that biological data are commonly fit by that distribution is not a strong support for any of those models because it fits equally well the size distributions of sand and gravel particles, cookie cutters, and planets.

Mechanistic Models

Mechanistic models are what most people associate with the term "model," either positively or negatively. The purpose of a mechanistic model is to describe in quantitative terms the relationship between some phenomenon and its underlying causes. Whereas in a statistical model the fitted coefficients such as the slope and intercept of a regression line have no intrinsic meaning, the parameters in a mechanistic model have real operational definitions and are (at least in principle) amenable to independent measurement. The classical laws of physics such as Newton's laws and Maxwell's equations are mechanistic models. Contaminant transport models used to predict ambient concentrations of contaminants in terms of the physical and chemical processes are also fundamentally mechanistic, although most contain empirical submodels. Biological models can also be mechanistic. For example, toxicokinetic and toxicodynamic models relate the uptake, translocation, and effects of contaminants to fundamental physical and chemical properties of the molecules, rather than to statistically derived dose-response functions and bioaccumulation factors (Menzel 1987, Barber et al. 1988, Parkhurst 1986). A great variety of water quality models quantify the relationship of nutrient dynamics to phytoplankton blooms and other undesirable consequences of eutrophication (Scavia and Robertson 1979).

It has often been argued that mechanistic models are unsuited to environmental studies and environmental impact assessment in general (Lauer et al. 1981, Peters 1985). Obviously, natural systems are so complex that complete mechanistic descriptions on the level of Newton and Maxwell are impossible. Approximations and simplifying assumptions always have to be made, and these necessarily introduce errors. Intuitively, it might seem that a good statistical relationship based on empirical measurements should always be superior to an inevitably imperfect mechanistic model. In spite of these limitations, mechanistic models are useful (even necessary) components of many assessment studies.

As noted above, risk assessments often require prediction of events beyond the range of available empirical data (different temperature, different exposure concentration, different population size), or prediction of future events based on knowledge of today's conditions. Using purely statistical relationships to extrapolate beyond the range of observations requires an assumption that the relationships between dependent and independent variables are the same outside the observed range as within the range. For small-scale perturbations or short-term predictions, this may be perfectly adequate. However, many important assessment problems involve large changes and long-term predictions. Examples include predicting ecological effects of climate change, responses of landscapes to regional air pollution (Hunsaker and Graham 1991), long-term management of fisheries (Walters 1986), and remediation of PCB contamination (Limburg 1986). Models that describe the relationships between variables in terms of causal mechanisms are, at least in principle, more useful for such problems than are statistical relationships among measured variables. If the mechanisms are correctly represented, then predictions should be correct outside as well as within the range of observations. Where there is (as there almost always is) substantial uncertainty about the "true" nature of the mechanisms and the values of critical parameters, astute use of alternative models can provide valuable guidance in making informed management decisions (Walters 1986).

Another important use of mechanistic models is to integrate complex sets of observations made in different times and places. Ecosystem simulation models are often used for this purpose. For example, the Narragansett Bay model (Kremer and Nixon 1978) includes among its variables the abundance and production of phytoplankton, zooplankton, benthic invertebrates, and fish. The model simulates the responses of all of these variables to temperature solar radiation, tidal circulation, freshwater inflow, and nontidal exchanges between the bay and Long Island Sound. The model integrates field and laboratory data collected over many years by many different research groups. Similarly, numerous models of forest stand composition and succession have been developed (Shugart and West 1977, West et al. 1980). These models express forest composition as a function of the physiological properties of individual trees, measured in greenhouses or experimental plots.

Yet another use of mechanistic models is to predict variables or events that are difficult or impossible to measure from more readily measurable variables. For example, Christensen et al. (1976) discussed the problem of predicting the effects of the operation of power plants on the abundance of fish populations. Ideally, one would like to base management decisions on the change (absolute or percent) in the size of the population under different operational regimes. Actual measurement of this change is impossible even in principle, because it would require comparisons between alternative future states, only one of which can actually be realized. With a model that relates the operation of the plant to the reproduction and mortality of the fish, alternative scenarios (with and without plant) can readily be analyzed and compared (Christensen et al. 1976).

The details of mathematical representation and computer programming are well beyond the scope of this book. Readers interested in learning mathematical modeling techniques should consult one of the several available textbooks on this subject, including Hall and Day (1977), Jorgensen (1988), and Swartzman and Kaluzny (1987). For water quality modeling, Thomann (1972) is an excellent text. The journal *Ecological Modelling* publishes a variety of papers on both theory and applications of environmental models.

The mechanistic models of interest for risk assessment can be categorized into two classes: fate models and effects models. *Fate models* simulate the movement and transformations of toxic contaminants in the environment. Processes generally simulated in fate models include the physical movement of particles and dissolved materials, chemical transformations, and exchanges between major compartments of the environment. A detailed account of the use of fate models in risk assessment can be found in Chapter 5 of this book. Food web models are a special subclass of fate models often used to simulate (1) exchanges between biotic and abiotic components of the environment, and (2) transfer of contaminants from prey to predators. The emphasis in food web models is on the biological components of the environment, especially the movement of materials through grazing, predation, and human harvest. Models of DDT (Harrison et al. 1970) and PCB (Thomann and Conolly 1984) bioaccumulation, and of contaminant movement through food chains (Lipton and Gillett 1991) are excellent examples of this type of model.

Effects models simulate the effects of stress on biota. This is by far the most diverse category of models of interest in ecological risk assessment. The "stresses" that have been modeled include human exploitation, environmental contamination on both local and regional scales, and, more recently, climate change resulting from increasing atmospheric carbon dioxide. Effects models corresponding to all of the levels of organization described in this chapter have been developed for various purposes. Organism-level effects models include toxicodynamic models (Mancini 1983, Kooijman and Metz 1984, Lassiter 1985, Lassiter and Hallam 1990) that relate the risk of death of organisms to the uptake and internal concentration of contaminants. Models of energetics and growth (Kitchell et al. 1977, Rice et al. 1983) are probably also relevant to ecological risk assessment, but have not been extensively applied to date.

The population-level models of current or potential use in risk assessment include the many models developed for management of fish and wildlife populations. Logan (1986) and Barnthouse et al. (1987, 1989, 1990) have provided examples of the use of fisheries-derived models in contaminant risk assessment. Although such models have not yet been used to assess risks of contaminants to wildlife populations, Emlen (1989) has reviewed the available wildlife population models and discussed their potential uses. Population models are more extensively discussed in Chapter 8.

Community and ecosystem models are the most diverse of ecological effects models. These models may be site-specific (Kremer and Nixon 1978, Andersen and Ursin 1977) or generic (O'Neill et al. 1976, 1982). Spatial scales considered range from microcosms (Rose et al. 1988) to regions or landscapes (Dale and Gardner 1987). The use of ecosystem models in risk assessment is discussed in Chapter 9.

A few integrated fate/effects models that permit changes in biotic compartments to affect the fate of chemical contaminants have been developed (Bartell et al. 1988). Such models are applicable principally to cases in which (1) biota are a major sink for contaminants (true only rarely), and (2) contamination is sufficient to cause major disruption of the biota.

Model Validation

A final topic that must be addressed to understand the concept of risk assessment is the subject of model validation. The literature contains many misconceptions concerning the definition of the term "validation" and the role validation plays in determining whether or which model to use in a risk assessment. Misconception often arises in the form of the question "is this model valid?" or statements such as "no model should be used unless it has been validated." "Validated," in this context, means either (1) proved to correspond exactly to reality, or (2) demonstrated by thorough experimental tests to make consistently accurate predictions. The principle that only valid models should be used in assessments is intellectually satisfying but irrelevant to actual assessment practice. Mankin et al. (1975) analyzed the concept of model validation and showed that no model can ever satisfy the definitions of validity given above. Because every model contains simplifications, none can ever correspond exactly to reality. Because of these same simplifications, predictions derived from a model can never be completely accurate. This argument applies to all types of models, both statistical and mechanistic.

Some models are clearly more valid than others, in the sense of corresponding more closely to reality and making more accurate predictions. Hence, validation may be more usefully viewed as a process of selecting among alternative approximations than as a proof of truth or falsity. Some authors (e.g., Walters 1986) have proposed using statistical decision theory to develop formal schemes for comparing and selecting among alternative models. Qualitative criteria for model selection can also be formulated. These relate to credibility, rather than validity in the strict sense. Credibility can be established in three ways:

1. experimental testing
2. publication in peer-reviewed journals
3. use in regulatory practice.

Experimental testing is the scientific ideal, but is possible only under relatively limited circumstances. Toxicodynamic models can be readily tested provided the test organisms (e.g., *Daphnia*, rats, fathead minnows) can be reared under controlled conditions. Models of contaminant dynamics in replicable microcosms have also been tested experimentally (Rose et al. 1988). Most ecological models of interest for risk assessment involve space and time scales too large for rigorous experimental testing to be feasible. Even when experimental testing is not possible, submission for peer review is a good test of the value of a model. This at least guarantees consistent internal logic and conformance to generally accepted modeling practices. Successful use in regulatory practice is probably superior even to peer review in enhancing the credibility of a model. Models used to support regulatory decisions often receive much more careful scrutiny than does any journal manuscript! Regardless of how it is done, the job of the risk assessor involves (1) selecting the best model or set of models, and (2) paying careful attention to the multiple uncertainties involved in their use.

The Role of Expert Judgment

Expert opinion also plays an important role in even the most technically complex quantitative risk assessments. Sophisticated methods for eliciting expert judgments about uncertainty have been developed and applied in both engineering and health risk assessments (Morgan et al. 1979, 1985; Bonano 1990). One reason for deemphasizing expert opinion in this book is that opinion is inherently subjective. Different experts inevitably provide different judgments on exactly the same topic. Even on narrowly defined topics such as estimating failure rates for pipes and valves in nuclear power plants, the reliability of expert opinion has been questioned (Martz 1984, Apostolakis 1985, Martz 1985). Because of the vaguely defined nature of many ecological problems and the lack of unity of ecological theory, the diversity of opinion among practitioners of ecological assessment on any given question is enormous. Our principal interest, however, is in demonstrating how data and models can be used to improve the credibility, consistency, accessibility, and reproducibility of ecological risk assessments. Assessors should first examine the available quantitative methods, and fall back on expert judgment when quantification proves infeasible.

Characterization and Quantification of Uncertainty

The most important feature distinguishing risk assessment, as discussed in this book, from impact assessment is the emphasis in risk assessment on characterizing and quantifying uncertainty. As noted above, several distinct sources of uncertainty are of interest in risk assessment. These categories are quite general, and many different approaches to dealing with them have been developed. Much of the technical literature on uncertainty deals with conceptual definitions of probability, frequency, and uncertainty or with the use of expert opinion to quantify uncertainty. Discussions of the theoretical implications of different assumptions about the roles of probability, frequency, and uncertainty in human decisionmaking are provided by Buckley (1983) and Unwin (1986). Although it is clear that different assumptions about the nature of the world and of how humans obtain knowledge of it can lead to different mathematical models of uncertainty, we have seen no evidence that these considerations are of practical relevance in risk assessment. Regardless of whether the Empirical Bayes approach or fuzzy set theory (to name two alternate world views) provides the more accurate theoretical foundation, the basic sources of information are still the same: data and models.

Of particular interest in ecological risk assessment are the three types of uncertainty that contribute to "analytical uncertainty," the uncertainty that contributes to estimating the credibility of a predicted value (Suter et al. 1987a).* They are natural stochasticity, parameter error, and model error. Methods exist for assessing the influence of all three types of uncertainty on quantitative risk estimates. The first two types can be quantified in relatively straightforward ways using either statistical or mechanistic models (Cox and Baybutt 1981, Hoffman and Gardner 1983, Suter 1990). With statistical models, the variances on parameter estimates obtained by fitting models to data provide direct estimates of parameter uncertainty. Suter et al. (1983, 1987b) and Suter and Rosen (1988) used this approach to quantify uncertainty inherent in extrapolating toxicity test results between different test types and species (Chapter 7). Although straightforward in concept, the use of statistics to quantify uncertainty is complicated in practice by the need to consider measurement errors in both the dependent and the independent variables and to combine errors when multiple extrapolations must be made (Linder 1987, Chapter 7). Statistical methods can also be used to quantify spatial and temporal variability for use in risk assessments.

Analogous approaches can be used to quantify parameter uncertainty in mechanistic models. The mathematical approaches can be categorized as sensitivity analysis (Gardner et al. 1981), response surface analysis (Downing et al. 1985), and error or uncertainty analysis (Gardner et al. 1981, Iman and Helton 1988). Because of its generality and ease of application to mechanistic simulation models, Monte Carlo uncertainty analysis is emphasized in this book. This approach has the advantage that parameter estimates and distributions can be obtained by many different methods, including time series analysis of measurements, sampling, and laboratory experimentation. Hence estimates of stochastic parameters such as temperature can be readily combined with error-prone parameters such as respiration coefficients or contaminant sensitivities. The effects on model output of different degrees of uncertainty associated with different parameters, and correlations among parameters can be directly quantified.

The steps involved in a Monte Carlo analysis include (1) defining the statistical distributions of input parameters, (2) randomly sampling from these distributions, (3) performing repeated model simulations using the randomly selected sets of parameters, and (4) analyzing the output (Figure 2.4). Relatively sophisticated methods and computer programs are now available for performing Monte Carlo analysis (e.g., the PRISM code of Gardner et al.

---

*Note that earlier in this chapter we discussed three fundamental sources of uncertainty. In contrast, these are three types of analytical uncertainty that can be distinguished and estimated in practice to contribute to the estimation of the credibility of a prediction.

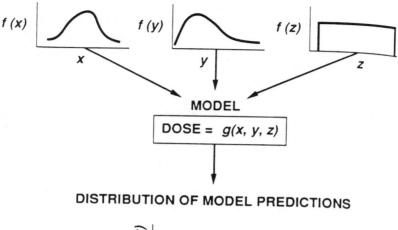

DISTRIBUTIONS OF INPUT VALUES
FOR PARAMETERS x, y, AND z

$f(x)$ $f(y)$ $f(z)$

x y z

MODEL

DOSE $= g(x, y, z)$

DISTRIBUTION OF MODEL PREDICTIONS

$f(x, y, z)$

VALUES OF DOSE

Figure 2.4. A diagrammatic representation of uncertainty analysis of a mathematical model (from Hoffman and Gardner 1983).

1983, and @RISK and Crystal Ball for PC and Macintosh-based spread sheets). Contributors to this book have used uncertainty analysis to quantify risks of toxic contaminants to aquatic ecosystems (O'Neill et al. 1982, O'Neill et al. 1983, Bartell et al. 1983) and to evaluate the relative uncertainties inherent in using different kinds of toxicity test data to quantify risks of contaminants to fish populations (Barnthouse et al. 1988, 1990).

The above discussion assumes that a reasonable risk model exists and thus ignores the most problematic, and potentially the most important, type of uncertainty in ecological risk assessments: model error. Model errors including inappropriate selection or aggregation of variables, incorrect functional forms, and incorrect boundaries. They are not subject to analysis using any straightforward statistical or mathematical technique. There are two possible approaches to dealing with model error. First, one could perform an experiment or collect field measurements to test the model. However, as has been frequently noted (Mankin et al. 1975, Suter et al. 1987a, Reckhow and Chapra 1983), experiments to validate ecological models are difficult to perform, interpret, and extrapolate to other sites or conditions. Another approach involves the use and comparison of alternative models. Different methods and assumptions can be used to derive alternative models for quantifying risks. For example, a statistical extrapolation model can be compared to a mechanistic simulation model (Suter et al. 1984, Barnthouse et al. 1985). If different approaches lead to the same conclusions, confidence that the conclusions are sound is enhanced. Walters (1986) and Reckhow and Chapra (1983) have advocated a more formal "Bayesian" approach to the comparison of alternative models. Using this approach, alternative models are constructed, predictions are derived, and the predictions are statistically compared to observed data. The statistical comparisons are then used to estimate the relative likelihood that each of the alternatives is the "correct" model. The Bayesian approach is limited by the requirement that the alternatives be similar enough in structure that the same data sets can be used to parameterize and test them.

The impossibility of accurately quantifying the magnitude of model error is, along with the impossibility of identifying a "correct" model for any given assessment problem, a reason for the inclusion of several distinct approaches to ecological risk assessment in this book. We view the application of alternative testing and modeling approaches at the individual, population, and ecosystem levels to be complementary rather than competing. It may well be that, due to paucity of data or theory, one or more approaches will be impossible to implement for any given problem, but wherever possible, several approaches should be employed.

PROBABILITY, UNCERTAINTY, AND RISK

Although risk is defined as a probability of an undesired effect, the meaning of probability differs among assessments (Suter 1990a). In conventional human health risk assessments, the probabilities are frequencies expressed as either individual risk or population risk. Individual risk is the estimated frequency of occurrence of the effect among members of an exposed population, which is treated as equivalent to the probability of occurrence of effects in each individual. For example, traffic accident records for the U.S. have been used to calculate that a person traveling 150 miles by car experiences a 10^{-6} risk of dying in a traffic accident (Fischhoff et al. 1981). The product of the probability of occurrence (mean individual risk) and the number of individuals in the exposed population is the population risk (i.e., the expected number of cases). Risk managers must balance individual and population risk to avoid imposing large risks on a few individuals or small risks on very large populations (Travis et al. 1987).

Frequentist concepts of risk are seldom applicable to ecological assessments because the endpoints are levels of effects on population or ecosystem properties, not the fate of their individual components. Also, there are seldom similarly exposed replicate populations or ecosystems from which an assessor could determine a frequency of effects. However, there are exceptions. For example, acid rain assessments have attempted to estimate the risk of acidification of lakes in a region by determining the frequency of acidification in a sample of lakes (Linthurst et al. 1986, Baker and Harvey 1984).

The type of probability that is most applicable to ecological risk assessments is what Bertrand Russell (1948) termed "credibility." The most familiar example of credibility is the weather forecaster's probability of rain. The ecological risk assessor predicting the effects of an effluent must, like a weather forecaster, estimate the occurrence of an unreplicated event. The result is that, given the uncertainty in the data, the stochasticity of the environment, and the realism and precision of the model assumptions, there is a certain estimated probability (credibility) of an effect. If the credibilities are accurate, then they will equal frequencies determined ex post facto. For example, it rains on approximately 30% of the days when weather forecasters predict a 30% chance of rain.

Credibilities can be estimated by the procedures described in the previous section. Using statistical models or error analysis of mathematical models, the assessor generates probability density functions for the chemical concentration, effects endpoint, or other model output. Particular probabilities can then be associated with particular concentrations, effects, etc. Three examples of the use of credibility in risk assessment follow.

Figure 2.5a shows a cumulative probability function for the predicted concentration of a chemical in some environmental medium. Such curves can be generated by probabilistic environmental fate models (Parkhurst et al. 1981, DiToro 1985). The spread of the curve results from both the stochasticity of the environment and ignorance concerning measurable characteristics of the chemical and the environment. The probability of exceeding some benchmark

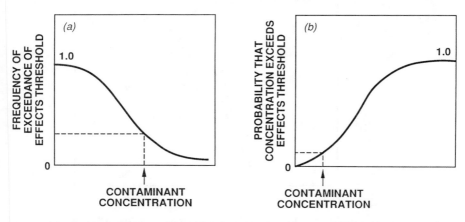

Figure 2.5.   Two applications of the results of probabilistic risk estimates. (a) A probability (cumulative frequency) function is used to estimate the frequency with which a standard or other action level will be exceeded. (b) A cumulative probability function for the threshold of significant effects on a species or biotic community is used to select a concentration with an *x*% chance of exceeding that threshold. (from Suter et al. 1987a)

value such as a water quality standard can be read off the curve, as indicated by the arrow and dashed line.

Similarly, Figure 2.5b shows a cumulative probability function for the effects of a chemical resulting from chemical exposures. Such curves can be generated by probabilistic effects models (Chapters 7–9). The spread in the curve is due to stochastic variance in the sensitivity of organisms, populations, and ecosystems to chemical exposures and to ignorance concerning measurable components of the responses. A standard could be set at a concentration that has a particular probability of exceeding a chosen effects threshold, as indicated by the arrow and dashed line.

Finally, Figure 2.6 depicts probability density functions for exposure and effects. The joint probability of occurrence of a particular concentration in the environment given a particular source term and of occurrence of effects at that concentration is the credibility of occurrence of effects given the source term

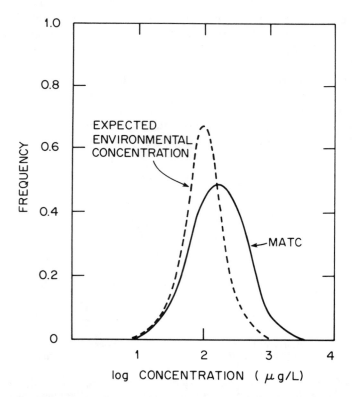

Figure 2.6. Probability density functions for the expected environmental concentration and the estimated effects threshold. The risk that the threshold will be exceeded is estimated by the probability that a value of the concentration function exceeds a value of the effects function.

(Chapter 7). Each distribution can be estimated from statistical or mathematical models.

These uses of probabilistic analysis can help to clarify the relationship between decisionmaking and uncertainty. They can be used to justify a particular degree of conservatism in the face of uncertainty or can be used to justify making additional measurements or conduct additional tests to reduce uncertainty. Such additional data would steepen the cumulative functions (Figure 2.5) or narrow the density functions (Figure 2.6), thereby reducing their overlap. This decreased uncertainty can change the outcome of a decision, and will certainly increase the confidence in a decision. Thus, this approach provides a means of determining the need for more data, and for prioritizing data needs. One would do the research that would do the most to decrease the total uncertainty within the restraints of time and money. In addition, these curves make clear the advantage of estimating the expected effects and associated uncertainties, rather than using worst-case assumptions or arbitrary safety factors. Because there is no objective scale of badness or safety, there is no objective way to compare the defensibility of safety factors or to justify how bad a worst-case must be. Probabilistic analysis provides a means of comparing assumptions, models, and data put forth by the parties in an environmental dispute.

3

Predictive Risk Assessments
of Chemicals

Glenn Suter

Buddy can you paradigm?

Anonymous

The standard paradigm for risk assessment (NRC 1983) and most of the tools developed for risk assessment are applicable to predictive assessments such as those conducted for regulation of new chemicals or new effluents. In this section we describe the steps involved in performing a predictive risk assessment for an individual toxic chemical, chemical mixture, or chemical effluent. Predictive risk assessments for other hazardous agents could have analogous structures (Chapter 13). Assessments of existing contamination may have a similar structure (Chapter 10).

Predictive ecological risk assessments begin with a set of preliminary descriptive activities that define the hazard to be assessed: choosing endpoints, describing the environment, and obtaining the source terms (Figure 3.1). Although these three activities are often performed independently, they should be coordinated so that the assessment endpoints are appropriate to the environment, the scope of the environment is appropriate to the size of the source term, etc. The hazard definition then constrains the exposure assessment (estimation of dispersal through the environment of the chemicals in the source terms and contact with the receptor biota) and the effects assessment (estimation of the interactions of the chemicals with the endpoint organisms or systems). The exposure assessment and effects assessment are then integrated to characterize the level and likelihood of effects. The risk characterization is an input to risk management. There are feedback loops at all stages of this process. For example, the exposure and effects assessments should be performed in a manner that facilitates risk estimation, and the hazard definition should take into consideration the political and economic concerns of the risk manager.

The stages in the paradigm are discussed below. Exposure assessment and effects assessment are only briefly summarized because they are discussed at length in Parts II and III.

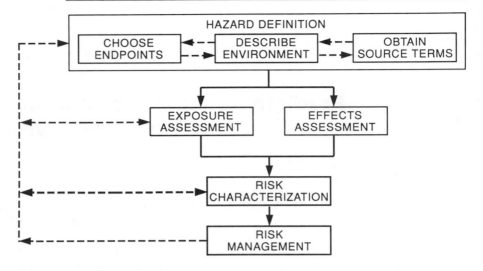

Figure 3.1. A diagrammatic representation of the paradigm for predictive ecological risk assessment. The solid arrows represent the sequential flow of the procedure. The dashed arrows represent feedback, policies, and other constraints on one assessment process by other processes.

CHOOSING ASSESSMENT ENDPOINTS

In choosing assessment endpoints, two general questions must be answered: (1) what valued components of the environment are considered to be at risk [i.e., an entity and a property of the entity (Chapter 2)]; and (2) how should effects be defined. The process of answering the first question has been termed hazard identification (NRC 1983). As was discussed in Chapter 2, the selection of assessment endpoints is constrained by legal mandates, regulatory policy, and public concerns. In addition, the process of hazard identification is often constrained by habits and preconceptions. For example, despite evidence that birds are particularly susceptible, assessments of air pollution effects tend to assume that terrestrial animals are protected by assessments of human health effects (Newman 1979, 1980).

Although legally or politically mandated endpoints must be included in assessments (to the extent that they are sufficiently defined), the assessor should also determine what endpoints are justified on technical grounds. To facilitate this process it can be useful to perform one of the following formal analyses of the relationship of components of the action being assessed and components of the receiving environment.

1. When the action being assessed is complex (involving effluents to multiple media, physical disturbance, etc.), it is helpful to create a matrix of component actions (releasing an aqueous effluent from a process or cooling system, spilling a product during shipment, etc.) and environmental components

(fish, terrestrial plants, the aquatic heterotrophic microflora, etc.) that are potentially affected. Components that are exposed to the effects of each action are then checked off and possibly scored for the intensity of the exposure and relative sensitivity to the toxic agent.

2. To identify which organisms in which media will be most exposed to a chemical it is valuable to perform a receptor identification exercise. It must consist of two steps: (1) performing a qualitative or rapid quantitative exposure assessment to determine what media are most contaminated, and (2) determining what communities, trophic groups, populations, and life stages are most exposed to those media.

3. Indirect effects of an agent can be identified by developing trees of causal linkages between the emissions and various environmental components (Figure 3.2). These are analogous to the fault trees and event trees of engineering risk assessment (Kandel and Avni 1988), but cannot be quantitatively analyzed as probabilistic causal chains because the probabilities of failures of ecological components are not quantifiable like pipes and pumps. Andrewartha and Birch (1984) discuss the creation of trees of causation for ecological systems.

4. Existing data can be reviewed to determine the sensitivity of species or processes to the contaminant or to similar contaminants. These may include data from toxicity testing or from biological monitoring of prior releases.

These four techniques are the most useful aids to hazard identification for ecological effects of chemicals, but other techniques including network models, mapping, and brainstorming may also be useful (Westman 1985).

The second question, how should the assessment endpoint be operationally

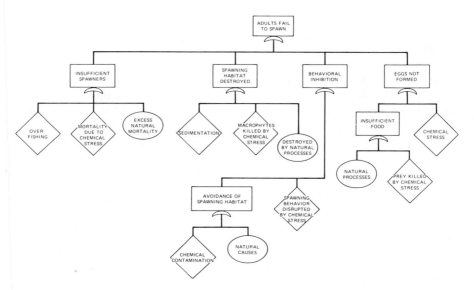

Figure 3.2. Event tree for failure of fish to spawn. This tree would be a branch of a larger tree used to determine how toxicants might cause reductions in fish populations.

defined once hazards have been identified, is often neglected. An operational definition of an assessment endpoint includes a subject (bald eagles or endangered species in general) and a characteristic of the subject (local extinction or percent reduction in range). Finally, the operationally-defined assessment endpoint must be translated into a numeric form. There are a number of ways to express effects on the endpoint, depending largely on the role of the assessor in decisionmaking. If the assessor makes the decision about the significance of effects, then the assessor may choose a threshold for significant effects and present the risk as the probability that the threshold will be exceeded. If a 10% reduction in harvestable yield is judged to be a threshold for significant effects, then the expression of the effect on the endpoint is an $x\%$ probability of a > 10% reduction or, more simply, a conclusion that a > 10% reduction is or is not expected to occur. If the assessor does not make the decision about significance of effects, then the expression of the risk is typically a simple prediction of the expected effect. These take the form: we expect (i.e., there is a > 50% probability) that there will be a $y\%$ reduction (or simply, a significant reduction) in harvestable yield. The decisionmaker then decides whether that effect should be allowed to occur. If the decisionmaker is sophisticated or if the results of the ecological assessment are input to a broader assessment including issues such as human health effects or the costs and benefits of treatment, it is appropriate to express the effects on the assessment endpoint as a function relating probability to magnitude of effects (Figure 3.3).

Unfortunately, in most assessments of toxic effects, assessors have not explicitly defined the assessment endpoint, but rather have used toxicological test endpoints or other measurement endpoints as de facto assessment endpoints (Chapter 2). It is preferable to choose measurement endpoints for their relationship to a defined assessment endpoint. In a few cases a standard measurement endpoint will be equivalent to the assessment endpoint. For example, if the assessment endpoint is prevention of fish kills in farm ponds resulting from pesticide drift, then the 96-h LC_{50} for bluegill might be chosen as its expression. In many cases the selection of test endpoints to approximate the assessment endpoint is largely constrained by the availability of published data, by differences in the quality of available data, or by the need to match the mode and duration of exposure in the test to the exposure scenario. However, when data are abundant or when testing can be prescribed by the assessor, measurement endpoints should be selected on the basis of their statistical form and their expression of the important responses of the population or system of interest.

There are two statistical types of measurement endpoints: (1) those that prescribe a level of effect, and (2) those that are based on hypothesis testing. The first type is obtained by fitting a function to sets of points relating the measured effects (proportion dying, mean weight, etc.) to measurements of exposure (dose, concentration in water, duration, etc.). That multidimensional expression may itself be the measurement endpoint, but more commonly, a single number is derived by inverse regression. That is, the exposure causing a

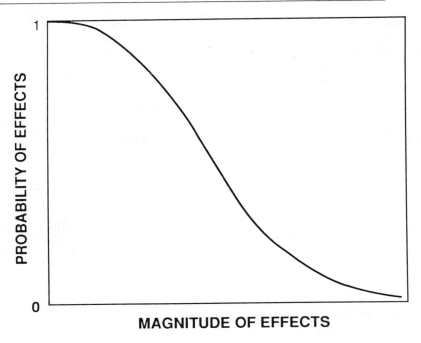

Figure 3.3. A function-relating probability of effects to their magnitude.

particular level of effect is calculated from the fitted model. Examples of this type of endpoint include the LC_{50}, median lethal dose (LD_{50}), median effective concentration (EC_{50}), and lethal threshold concentration (LC_{01}).

The other statistical category of measurement endpoint consists of those that are derived by hypothesis testing techniques. Responses at the exposure concentrations are compared with control (unexposed) responses to test the null hypothesis that they are the same as the control responses. Endpoints of this type include the no-observed-effect-concentration (NOEC) and the lowest-observed-effect-concentration (LOEC).

The disadvantages of measurement endpoints based on hypothesis testing relative to those based on curve fitting have been discussed by Stephan and Rogers (1985).

1. The use of conventional hypothesis testing procedures (with $\alpha = 0.05$ and β unconstrained) implies that it is very important to avoid declaring that a concentration is toxic when it is not, but it is not so important to avoid declaring that a concentration is not toxic when it is.
2. The threshold for statistical significance does not correspond to a toxicological threshold or to any particular level of effect.
3. Poor testing procedures increase the variance in response, and therefore reduce the apparent toxicity of the chemical in a hypothesis test.

4. The results are relatively sensitive to the aspects of test design, including the number of replicates and the number and spacing of concentrations tested.
5. Because no exposure-response model is derived, there is no means of describing the exposure-response dynamics or even of determining whether the data reflect a reasonable toxicological response.

The advantages of hypothesis testing endpoints are that they can be calculated even when the test data are too poor or meager to fit a model, and they allow the assessor to avoid specific decisions about what constitutes a significant level of effect. We feel that hypothesis testing is generally an inappropriate way to calculate endpoints; however, in many cases, the use of such endpoints by the assessor is unavoidable because no other is available.

Although it is common practice to express the assessment endpoint in terms of a measurement endpoint, it is unrealistic, in general, to assume that any test result represents the response of a population or ecosystem in the field. To make this extrapolation, some model must be applied to either the measurement endpoint or the original test data from which the test endpoint was calculated. These models are discussed in Part III.

ENVIRONMENTAL DESCRIPTION

An environmental assessment must begin with a conceptualization of the environment which will be considered to be subject to the effects. For assessments of individual developments, effluents, or chemical releases, the environment will be an actual place. For generic assessments of technologies or chemicals, it will be a reference environment which is representative of sites where release would occur. Reference environments can be arrayed on a scale of abstraction. At one extreme, values are simply assigned to the parameters of the assessment models that are thought to be representative or worst-case. For example, some aquatic assessments represent the environment as simply a dilution factor which is conservatively set to 10 (Beck et al. 1981). This approach is appropriate for quickly screening chemicals. At the opposite extreme, an actual location or set of locations may be chosen to serve as surrogates for the expected release site. This approach requires additional effort in site description which may seem unnecessary, but it assures that the assessment bears some relationship to reality, and helps the assessment team to appreciate the scope of the problem. A reference environment corresponding to an actual location could serve as a site for field tests or for collection of realistic media and organisms for laboratory testing. If an effluent exists similar to the one being assessed, then the realism of the assessment could be increased by using the location of the existing effluent as a hypothesized site of the new release. For example, existing coke oven effluents have been used to represent future coal liquefaction effluents (Herbes et al. 1978).

Reference environments for many generic assessments fall between these extremes of abstraction; they represent a region or have certain siting requirements such as a river large enough for barges and proximity to a source of raw materials, but they are not tied to an actual location. For example, assessments of commercial oil shale and coal liquefaction plants used reference environments in the Green River Formation of western Colorado and the central portion of the Appalachian Coal Basin, respectively (Travis et al. 1983). Another example is the creation of reference environments for modeling the fate of agricultural chemicals based on characteristics of the areas where the crops are grown (Oliver and Laskowski 1986). The appropriate degree of abstraction of the reference environment depends on the type of assessment to be conducted and the degree to which the characteristics of likely release sites can be specified.

The principal considerations in environmental description are the boundaries placed on the environment and characterization of the entities and processes occurring within the boundaries. Boundaries may be defined by either (1) a regulatory or equivalent a priori definition or (2) properties of the assessment problem. Examples of the first type of definition would be requirements that effluents meet some criterion at the edge of the zone of initial dilution, or in the first fully mixed reach of a stream (OWRS 1985). An example of the second type of definition would be defining the boundaries in terms of the area within which the concentration of a chemical is higher than the concentration at which a threshold for toxic effects occurs. This approach requires that the assessor begin the assessment either by making a rough conservative estimate of the outcome of the assessment, or by performing the assessment iteratively until appropriate boundaries are reached. For example, in assessments of effects in streams, stream segments can be added until, in the last segment, the exposure concentration drops below a specified threshold at which risks are negligible. Another example of the second approach to defining boundaries is using the range of a population of concern such as the Hudson River striped bass (Barnthouse et al. 1984). When secondary effects are considered, the setting of boundaries becomes a complicated task which is tied to the process of identification of those effects (Beanlands and Duinker 1983). Less acceptable but commonly used approaches include use of the bounds of applicability of a favored transport model, political boundaries, or a "reasonable" distance such as 1 km.

SOURCE TERMS

Source terms are estimates of the rate and the spatial and temporal pattern of release of a chemical from a source or set of sources. In most cases, source terms are contributed to risk assessments by plant engineers or by the manufacturer's recommendations for use, so they will not be treated in this text. However, environmental analysts need to be aware that the accuracy and precision of source terms vary considerably. Direct emissions from pipes or stacks are generally better characterized than indirect emissions, such as leachates from buried wastes or environmental release of consumer products. Routine emissions are better characterized than irregular emissions, such as spills or emissions from plants operating under upset conditions. The best source terms are those that are based on measurements of emissions of similar sources or products. In many cases the uncertainty concerning the source term is the major source of uncertainty in an ecological risk assessment, because either the content of the waste streams and the efficiency of the waste treatment system, or the volume of use or the pattern of disposal cannot be accurately predicted. Even in the best circumstances (a measured, direct, routine emission) the source term can contain significant uncertainty due to changes in product mix, operating conditions, feed stocks, and other operational variables. In most cases it is difficult to obtain estimates of uncertainties in source terms, but they should be obtained if possible to determine their contribution to the uncertainty component of the risk estimate.

EXPOSURE ASSESSMENT

The process of converting a source term into estimates of contact with or doses to the endpoint organisms or systems is termed exposure assessment (Figure 3.4). It requires that a model of the receiving environment be constructed that includes all media into which the pollutant is released or is likely to partition and to which the endpoint organisms are significantly exposed. These models simulate transport, dilution, transformation, degradation, and partitioning between media, and generate predictions of the temporal dynamics of concentration in the various media of interest. Input parameters to these models are the release rates from the source term analysis, characteristics of the pollutant such as degradation rate and solubility, and characteristics of the receiving environment such as flow rates and wind directions. These parameters should be expressed as probability distributions so that probabilistic fate models can be used to estimate distributions of exposure levels (Chapter 2).

The transformation of the concentrations of the pollutant in specific media into exposure requires knowledge of the natural history, behavior, and physiology of the organisms involved as well as characteristics of the pollutant. In most ecological assessments it is assumed that exposure is equivalent to concentration in a single medium without demonstrating that uptake from that medium is the only significant route of exposure. In some cases, exposure can be complicated, involving multiple routes of exposure or complex media with multiple phases. In addition, exposed organisms may avoid contaminated media, may be attracted to contaminated media (e.g., pesticide-debilitated prey), may lose their ability to detect contamination due to toxic effects, or otherwise modify the exposure process in response to an initial exposure.

A conceptual problem associated with estimation of exposure is the treatment of background exposure from natural sources (e.g., streams with natu-

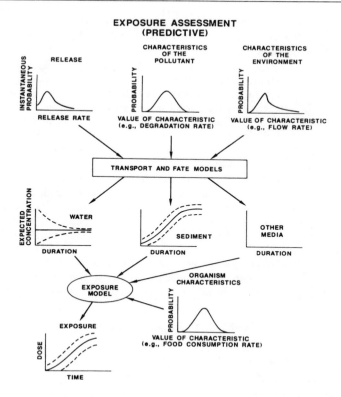

Figure 3.4. A diagrammatic representation of the process of exposure assessment.

rally high metal concentrations) or anthropogenic sources (e.g., upstream emissions). Basing assessments on the incremental exposure alone can result in unpredicted environmental effects of jointly acting agents (Chapter 11), but basing assessments on the total exposure can unfairly penalize the source being assessed relative to other sources contributing to the background exposure. The treatment of incremental exposures relative to background is a policy decision to be made by the risk manager. However, the assessor can aid that decision by clearly identifying background, incremental, and total exposure.

EFFECTS ASSESSMENT

Effects assessment is the process of determining the relationship between exposure to the toxic material and its effects which are potentially hazardous to the assessment endpoint. Effects assessment can be based on ecological epidemiology (i.e., on effects models derived from retrospective assessments of field sites contaminated by other sources of the same chemical), as discussed in Chapter 10. However, most effects assessments are based on toxicity testing. Effects assessments that are based on toxicity testing consist of the following components, represented in Figure 3.5.

1. Toxicity tests are conducted to determine the effects of various combinations of exposure concentration and duration on the frequency or severity of the responses of concern, such as increased mortality and decreased fecundity.
2. Statistical models are fit to the test data (response surfaces are shown in Figure 3.5). An exposure-response model or other numeric summary of the test results (i.e., test endpoint) is selected to represent the toxicological responses in the effects models.
3. Effects models are generated that represent the assessor's assumptions concerning the nature of the relationship between the test endpoints and the assessment endpoint. These may range from an assumption that the test endpoint captures all of the relevant features of the response, to mathematical models simulating the mediation of toxicological responses by population and ecosystem processes.
4. The test endpoints and data concerning relevant population and ecosystem processes are used to parameterize the effects model which is used to derive a function relating the level of effects on the assessment endpoint to exposure.

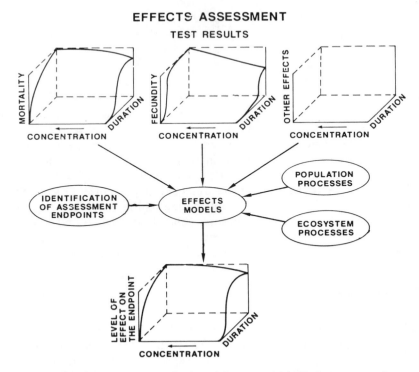

Figure 3.5. A diagrammatic representation of the process of effects assessment.

RISK CHARACTERIZATION

The risk characterization is the summary output of the risk analysis which serves as the risk analysts' input to risk management. The principal component of risk characterization is the integration of the results of the exposure assessment and effects assessment to obtain an estimate of the level of effects that will result from the exposure.

The integration of exposure and effects assessments requires that the dynamics of exposure and effects be expressed in terms of common dimensions. The effects of exposure to a toxicant can be defined by a surface in a four-dimensional space. The dimensions are (1) the concentration of the substance to which organisms are exposed, (2) the duration of the exposure, (3) the proportion of the organisms, populations, or communities responding, and (4) the severity of the effect. (The first two variables are aspects of exposure; the second two are aspects of effects.) In other words, a particular duration of exposure to a particular concentration of a substance will cause a particular proportion of organisms, populations, or communities to experience an effect of a particular severity. As the concentration or duration increases, the proportion of biotic entities experiencing a particular effect increases, and the severity of effects experienced by a particular proportion of the population of entities increases. If all four dimensions are quantified on quantal or continuous scales (see Box 3.1), the realized magnitudes of the variables define a three-dimensional volume in the hypervolume.

Box 3.1. Scaling of the Dimensions of Toxic Effects

1. Concentration is a continuous variable.

2. Duration of exposure is a continuous variable with potential complications:

   a. If exposures are brief and recovery occurs between exposures, duration can be neglected; it is effectively zero and exposure is a function of concentration.

   b. If concentration is effectively constant (i.e., variance in concentration is so small or rapid that temporal variance can be ignored) or exposures occur as independent events with significant duration, negligible temporal variance within events, and recovery of the system between exposures; a simple continuous temporal scale is adequate to estimate the level of effects for either the entire exposure or the individual exposure events.

   c. If effects are induced by discrete and brief exposure events and recovery does not occur between events, the duration of the individual events is negligible and can be neglected so the relevant temporal scale is recurrence frequency.

d. If effects are induced by events of significant duration and if recovery does not occur between events, both the duration of the events and their recurrence frequency must be considered.

e. If concentration is highly variable in time (i.e., exposure cannot be treated either as continuous or as a pattern of events and interevent recovery periods), the concentration and time scales must be integrated either by calculating some integral variable such as the time weighted average concentration or a weighted temporal integral of concentration (e.g., Lefohn et al. 1989), or by using both time and concentration in a toxicokinetic model to estimate internal dose (Chapter 6).

f. Although time is inherently continuous, assessors commonly treat it as a graded variable with two poorly defined categories, acute and chronic (Chapter 3).

3. Proportion responding is a quantal variable.

4. Severity scales may assume various forms. Many are continuous, such as primary production and population biomass or growth rate. Others are quantal, such as population abundance or species richness. All of these parameters are potentially good estimators of assessment endpoints. One might estimate the proportional reduction in productivity (severity) that would not be exceeded by at least 90% of lakes in a region (proportion responding), or the proportion of individuals in the population (proportion responding) with growth less than y (severity). In general, it is desirable to use continuous or quantal severity scales.

Difficulty arises when one wishes to array diverse responses on a severity scale. In such cases, it is necessary to create a graded scale. Very general graded scales may be created to summarize diverse data such as: (a) no observed effect, (b) no observed adverse effect, (c) adverse effects, and (d) frank effects (Dourson 1986). However, because such generic scales have relatively little descriptive or analytical power, it is generally desirable to create categories that are more specific. For example, air pollution effects on plants may be arrayed on the scale: (a) no observed effects, (b) metabolic and growth effects, (c) foliar lesions, and (d) death (EPA 1982d). However, more specific scales may not be consistently ordinal. For example, exposures of plants to some pollutants result in foliar lesions at lower levels than growth decrements. In any case, such graded scales generally cannot be used in quantitative analysis without assigning scores that are likely to be little better than arbitrary [Newcombe and MacDonald's (1991) scored scale of severity of suspended sediment effects is an apparent exception (Figure 3.6)]. In any case, graded scales can serve to illustrate the rate at which effects become qualitatively more severe as exposure increases, or how near a particular exposure level is to inducing a more severe type of effect.

Although all of the four dimensions are measurable in toxicity tests, the response surface is difficult to define or conceptualize, and cannot be graphically portrayed. In practice, because of the large amount of data required to define multiple dimensions, or because some dimensions are not important to the assessment, the number of dimensions used to portray toxic effects is reduced — commonly to one, occasionally to two, and rarely to three (Figure 3.7 through 3.9). This can be done in three ways. First, one of the variables can be ignored or defined away. The most conspicuous example is defining away both effects axes by using statistical thresholds in place of biological effects. Second, the space can be collapsed along one of the axes by holding that variable constant. For example, if exposures of only one duration are consid-

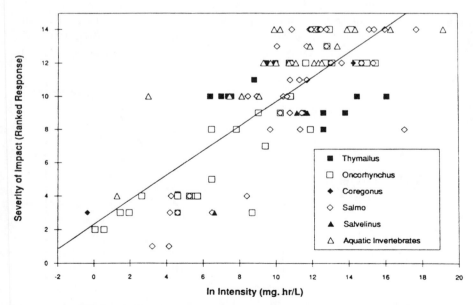

Figure 3.6. Severity of responses of aquatic organisms to suspended sediments (expressed as a numerically-scored qualitative scale of severity) as a function of exposure (expressed as the product of concentration and duration) from Newcombe and MacDonald (1991).

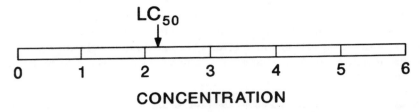

Figure 3.7.   Toxic effects as a function of concentration.

ered relevant, then a concentration-duration-response surface [Figure 3.9 collapses to a concentration-response curve (Figure 3.8)]. Third, variables can be combined to create a new variable. For example, the product of concentration and time can be used as the expression of exposure so that effects can be represented as a function of a single exposure variable (e.g., Figures 3.6 and 3.10).

Assessors often find that the available toxicological data have more or fewer variable dimensions than they desire, or that variables are undefined within the desired ranges. Thus they must collapse, extrapolate, or interpolate variables. The problem of surplus variables occurs most commonly in fields such as phytotoxicity in which, because testing procedures are not standardized, nearly every test has a different duration, and a different response is measured (i.e., variable severity). In such cases, the process of combining different studies to summarize the effects of a substance requires dealing with points scattered across all four variable dimensions when the assessment endpoint may be a one-dimensional air quality criterion. The problem of too few variables frequently arises in highly standardized fields such as fish toxicology. An

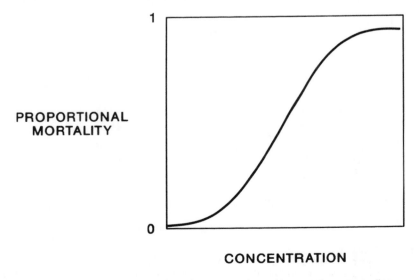

Figure 3.8.   Toxic effects as a function of concentration and proportion responding.

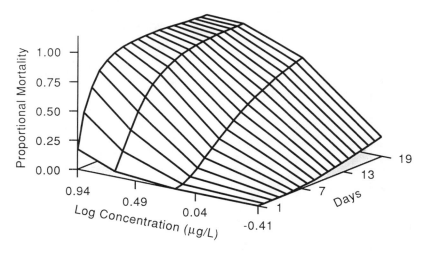

Figure 3.9. Toxic effects as a function of concentration, duration, and proportion responding (data from Nimmo et al. 1977).

assessor often finds that the only ichthyotoxicological information for a substance is a point on a concentration scale (e.g., a 96-h LC_{50}) when the assessment endpoint involves population production (severity), exposures shorter than 96 hours (duration), or the threshold of lethality (proportion responding).

In addition to dealing with the dimensionality of toxic effects, assessors must decide which way of expressing the axes is appropriate. This is in part a matter of scaling the variables appropriately (Box 3.1). However, it is also a matter of choosing appropriate units. For example, concentration can be expressed as total concentration in a medium to which the organism is exposed, concentration in a bioavailable form in the environmental media, body burden, or any of several other possible units.

In the following sections, the treatment of these dimensions in current risk assessment practice is discussed in terms of the number of dimensions used. Although the point of the discussion is the integration of fate and effects dimensions for risk estimation, many of the examples are based on the treatment of exposure and effect dimensions in models fit to toxicity test data, because the form of the test endpoints is commonly allowed to constrain the form of the model used to estimate the assessment endpoint. This fact that the dimensions of risk models are the same as those of toxicity tests has two troublesome consequences that need to be corrected as ecological risk assessment develops. First, exposure duration is used in place of effects duration in the output of risk assessments (Box 3.2). Second, space is not a dimension of conventional risk assessments (Box 3.3).

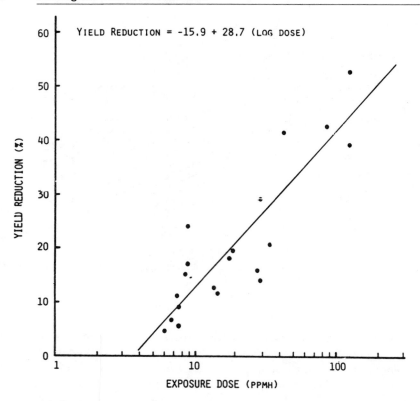

Figure 3.10.   Percent reduction in yield of beans as a function of SO_2 dose in ppm-h (from McLaughlin and Taylor 1985).

Box 3.2. Duration of Exposure Versus Duration of Effects: Induction, Recovery, and Adaptation

Currently, the expression of duration used in ecological risk assessments is duration of exposure and the reported effects are the instantaneous effects levels at the end of exposure (e.g., 96-h LC_{50}). The emphasis on duration of exposure is acceptable in traditional human health risk assessment, which is concerned with estimating the number of deaths or illnesses induced following a given dose (dose rate × time). However, in ecological assessments it is desirable to consider how long an ecosystem is degraded or what the total loss of resources is over time. Therefore, the time scale that should be used in ecological risk estimates and should be of interest to risk managers is duration of effects, not duration of exposure.

The two durations correspond if effects cease immediately after exposure ceases and if recovery is essentially immediate (Figure 3.11a). Often, effects induction ends soon after exposure ceases, but a significant period of recovery is required (Figure 3.11b). In addition, effects may continue to occur after exposure ceases because of time lags in the induction of effects, because residues are remobilized at some later time (either internally as in metabolism of fat reserves during migration, hibernation, etc., or externally as in the resuspension of sediments by floods), or because effects are expressed only at certain points in the life cycle such as reproduction (Figure 3.11c). Delayed

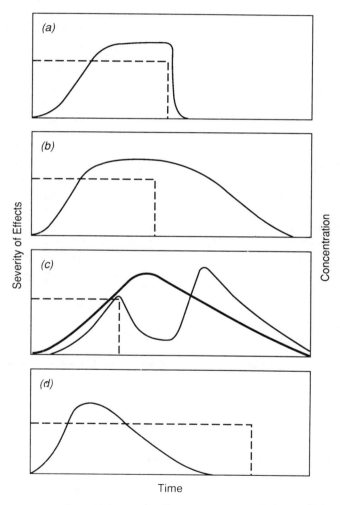

Figure 3.11. Toxic effects as a function of duration of effects (solid lines) contrasted with duration of exposure (dashed line). (a) Effects rapidly decline following cessation of exposure. (b) Induction of effects ceases after cessation of exposure but recovery requires significant time. (c) Effects are induced after cessation of exposure due to lagging responses (thick line) or delayed responses. (d) The system adapts to effects before cessation of exposure.

effects are routinely measured only in single dose wildlife toxicity tests, where the need to wait for effects following an instantaneous exposure is obvious. Finally, ecosystems exposed for long periods may adapt to the exposure so that biological effects end before the exposure does (Figure 3.11d). Use of the duration of exposure to estimate the temporal sum (or time integral) of effects would be reasonable in case (a), but would underestimate effects in cases (b) and (c), and overestimate effects in case (d).

Currently, it is not possible to accurately express ecological risks in terms of duration of effects. The results of toxicity tests are expressed in terms of durations of exposure, and there are no models to extrapolate the test results in time beyond the end of exposure. Delayed effects are known to occur but have not been studied to any extent. Adaptation of individual populations is well documented (Chapter 7), and resilience of ecosystems has been the subject of much modeling and theorizing (DeAngelis et al. 1989). However, neither one is predictable except by citing cases in which similar populations and ecosystems have been exposed to a similar contaminant. There is no successful theory of ecosystem recovery that could serve as a basis for general models (Cairns 1990).

Use of rough estimates of time to recovery based on experience is the best available approach to estimating the duration of effects. There is a large body of literature on recovery of terrestrial ecosystems from logging, farming, and similar physical disturbances, but there is relatively little literature on recovery of aquatic ecosystems, on recovery of any class of ecosystems from toxic effects, or recovery of ecosystems from less than catastrophic effects. However, there are ongoing efforts to summarize existing experience and develop useful models of recovery (Cairns et al. 1977, Cairns 1980, Jordan et al. 1987, Yount and Niemi 1990, Detenbeck et al. 1992). One can conclude that even following severe or catastrophic toxicological damage to stream communities, recovery from a pulse disturbance usually occurs in less than two years (Niemi et al. 1990, Detenbeck et al. 1992).

Box 3.3 The Spatial Dimension

The absence of space as a dimension in ecotoxicological risk assessment seems puzzling upon reflection. Obviously, a spill that could kill 80% of fish in a 10 km river reach constitutes a greater risk than one that would kill that proportion of fish in a 100 m reach. This absence of a spacial dimension results from two factors. First, ecotoxicology and ecological risk assessment are derived from their human health-oriented predecessors, and, because only the number of humans matters (not their spatial distribution), space is not included in the expression of health risks. Second, there is no spatial dimension in toxicity tests, and descriptions of ecological risk have hardly advanced beyond comparison of test endpoints to expected environmental concentrations.

Spatial dimensions of ecological risk are considered in the emerging field of regional risk assessment (Chapter 11). Spatial extent should also be routinely considered in conventional risk assessments. Spatial dynamics are included in many fate models, and a spatially explicit analysis is needed for consideration of avoidance, attraction, and other dynamic aspects of exposure. Spatially explicit risk assessment is limited largely by the available effects models. The organism, population, and ecosystem effects models currently in use nearly always assume homogeneous exposures and responses. Spatially explicit effects models are being developed and the geographic analysis tools used in regional analysis could also be used to portray the distribution of effects at local scales. In the absence of such tools, ecological risk assessments should at least include statements concerning the extent of different levels or types of effects.

One-Dimensional Models

Most ecological assessments depend on one-dimensional models of toxicant-organism interaction, and in most cases that dimension is concentration (Figure 3.7). Because toxicology is historically the science of poisons, the fundamental paradigm of toxic effects is the single lethal dose. To the poisoner, duration of exposure, severities other than mortality, and even the exact proportion responding are unimportant, so the only dimension of interest is concentration. This paradigm is appropriate in weed and pest control and, to a somewhat lesser extent, to unintentional acute poisoning, as when wildlife are sprayed with pesticides.

Use of only the concentration dimension simplifies assessment because the outcome is determined by the relative magnitudes of the exposure concentration and the effective concentration. However, most assessments in environmental toxicology involve some element of time and some concern for the severity and extent of effects, so use of only the concentration dimension requires that these other dimensions be collapsed. This collapsing of the other dimensions is most commonly done by using a standard test endpoint as the effective concentration. Standard test endpoints are used because (1) the assessor does not have the skill, time, latitude, or inclination to develop alternate models; (2) the assessor has a vaguely defined assessment endpoint and is willing to accept the toxicologists' judgments as to what constitutes an appropriate duration of exposure, severity of effect, and frequency of effect; or (3) by luck the assessment endpoint corresponds to a standard test endpoint. The most abundant test endpoints are median lethal concentrations and doses. With these endpoints, proportion responding, severity, and time are collapsed by considering only 50% response, only mortality, and only the end of the test.

Occasionally, time is the basis of a one-dimensional assessment. If the concentration of a release is relatively constant, then the relevant question is how long it can be released without unacceptable risk. For example, a waste treatment plant might be taken off-line for repairs, and the operator would like to know how long he can release untreated waste. The most common temporally defined test endpoint is the median lethal time (LT_{50}). In fact, many releases have relatively constant concentrations and time would be a more useful single dimension than concentration for assessment. However, because time is generally felt to be less important than concentration, it is more often considered in higher dimensional models.

The most common model for risk estimation, the quotient method (Barnthouse et al. 1982, Urban and Cook 1986), relies on a one-dimensional scale. The method consists of dividing the expected environmental concentration by a test endpoint concentration (i.e., determining their relative position on the concentration scale); the risk is assumed to increase with the magnitude of the quotient (Chapter 7).

Two-Dimensional Models

Concentration-Response

The most common two-dimensional model in ecological toxicology is the concentration-response function (Figure 3.8). This preeminence is explained by the desire to show that response increases with concentration of a chemical, thereby establishing a causal relationship between the chemical and the response. In this model, time is collapsed by only considering the end of the test, and either severity or proportion responding is eliminated so as to have a single response variable. Most commonly, proportion responding is preserved and severity is collapsed by considering only one type of response, usually mortality. Severity is most often used when a functional response of a population or community of organisms such as primary production is of interest, rather than the distribution of response among individuals. A probit, logit, or other function is fit to the concentration-response data obtained in the test. Typically, that function is used to generate the standard LC_{50}, EC_{50}, or LD_{50} by inverse regression, thereby reverting to a one-dimensional model and throwing out much information. Because assessments are more often concerned with preventing mortality or at least preventing mortality to some significant proportion of the population, it would be more useful to calculate an LC_1, LC_{10}, or other threshold level, but information would still be thrown away. If both dimensions are preserved, the concentration-response function can be used in each new assessment to estimate the proportion responding at a predicted exposure concentration or to pick a site-specific threshold for significant effects. When only an LC_{50} is available and the assessment requires estimation of the proportion responding, it is possible to approximate the concentration-response curve by assuming a standard slope for the function and using the LC_{50} to position the line on the concentration scale (see Chapter 7). Concentration-response functions are an improvement over being stuck with an LC_{50}, but in many cases the proportion responding is less important than the time required for response to begin.

One improvement in concentration-response functions would be to match the test durations to the exposure durations, rather than using standard test durations and assuming that the exposure durations match reasonably well. Ideally, the temporal pattern of exposure would be anticipated and reproduced in the toxicity tests. For example, if a power plant will periodically "blow-down" chlorinated cooling water, on a regular schedule for a constant time period, that intermittent exposure can be reproduced in a laboratory toxicity test (Brooks and Seegert 1977, Heath 1977).

Time-Response Functions

Time-response functions, like concentration-response functions, are primarily generated in order to calculate a one-dimensional endpoint, in this case the LT_{50}. Time-response functions are useful when concentration is relatively constant but the duration of exposure is variable. With the time-response function defined, it is possible for the assessor to consider the influence of changes in exposure time on the severity of effects or the proportion responding.

Time-Concentration

The most useful two-dimensional model of toxicity that includes time is the time-concentration function or toxicity curve (Sprague 1969). This is created from experimental data by recording data at multiple times during the test and, for each time, calculating the LC_{50}, EC_{50}, or the concentration causing some other response proportion or severity, as described above. These concentrations are then plotted against time and a function is fit (Figure 3.12). Such functions have been advocated by eminent scientists in prominent publications (Sprague 1969, Lloyd 1979), but are seldom used even though they require no more data than is already required by standard methods for flow-through LC_{50}s for fish (ASTM 1991, EPA 1982b). If the response used corresponds to a threshold for significant effects on the assessment endpoint, then this function can be used to determine whether any combination of exposure concentration and duration will exceed that threshold (i.e., the area above a line, as in Figure 3.12).

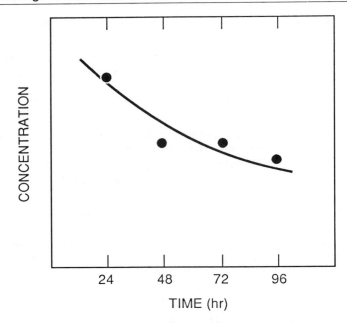

Figure 3.12. Toxic effects as a function of concentration and time. The function is derived by plotting LC_{50}s or other test endpoints against the times at which they were determined (i.e., 24, 48, 72, and 96 hours).

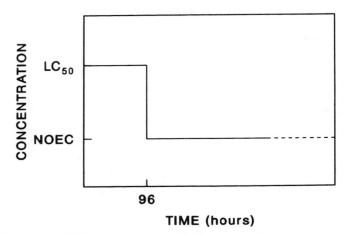

Figure 3.13. Toxic effects as a function of concentration and time, with time expressed as a dichotomous variable, acute and chronic. Acute time ends at 96 hours, the end of a standard acute lethality test for fish.

When time is a concern but temporal test data are not available, it is necessary to approximate temporal dynamics. The simplest approach is to treat the level of effects and the exposure concentration as constants within set categories of time. This approach is used in most environmental assessments. The assessor, faced with toxic effects expressed as standard test endpoints, tries to match the test durations to the temporal dynamics of the pollution in some reasonable manner. This process can be represented by the time-concentration relationship shown in Figure 3.13. Time is divided into acute and chronic categories, and standard acute and chronic test endpoints are used as benchmarks separating acceptable and unacceptable concentrations. However, there are serious conceptual problems with this categorization (Box 3.4).

Box 3.4 Acute and Chronic as Temporal Categories

Ambiguities concerning use of the terms acute and chronic complicate the matching of test durations to exposure durations. Because the terms acute and chronic refer to short and long periods of time, respectively, it is tempting to relate endpoints from acute tests to short exposures, and those from chronic tests to long exposures. However, the terms acute and chronic have acquired additional connotations. Complications arise from the use of these terms to describe severity as well as duration. Acute exposures and responses are assumed to be both of shorter duration and more severe than chronic exposures and toxicities. The implicit model behind this assumption is that chronic effects are sublethal responses that occur because of the accumulation of the toxicant or of toxicant-induced injuries over long exposures. Conversely, because of the cost of chronic toxicity tests, toxicologists have attempted to reduce testing costs by identifying responses that occur quickly and that are severe enough to be easily observed, but that occur at concentrations that are as low as those that cause effects in chronic tests (McKim 1985, Woltering 1984, Birge et al. 1981). As a result, the relationship between the acute-chronic dichotomy and gradients of time and severity has become confused.

This confusion is illustrated by the standard test endpoints for fish. The standard acute endpoint is the 96-hour median lethal concentration (LC_{50}) for adult or juvenile fish (EPA 1982b, ASTM 1991, OECD 1981). The standard chronic test endpoint has been the maximum acceptable toxicant concentration (MATC, also termed the "chronic value"), which is the threshold for statistically significant effects on survival, growth, or reproduction (EPA 1982b, ASTM 1991). Because this chronic endpoint is based on only the most sensitive response, life stages that appeared to be generally less sensitive have been dropped from chronic tests so that those tests have been reduced from life cycle (12 to 30 months) to early life stages (28 to 60 days) (McKim 1985). Tests that expose larval fish for only 11 (Birge et al.1981) or 5–7 days (Norberg and Mount 1985) have now been proposed as equivalent to the longer chronic tests. As a result, the chronic test endpoint for fish is now tied to events of short duration (the presence and response of larvae), whereas the acute endpoint is applicable to exposures of similar duration and to life stages that are continuously present.

Even the severity distinction between acute and chronic tests is not clear. Although the LC_{50} clearly indicates a severe effect on a high proportion of the population, the fact that the MATC is tied to a statistical threshold rather than a specified magnitude of effect means that it too can correspond to severe effects on much of the population (Suter et al. 1987b). For example, more than half of female brook trout exposed to chlordane failed to spawn at the MATC (Cardwell et al. 1977).

It would be advantageous to clarify the distinction between acute and chronic toxicity by restoring the original temporal distinction and expressing effects in common terms. Without that clear distinction, concentration-time functions like Figure 3.13 will be uninformative.

In the absence of good time-concentration information from the test data, a concentration-duration function may be assumed. For example, the first version of the model of marine spill effects for type A damage assessments simply assumed a linear function (DOI 1986, 1987a). Lee and Jones (1982) combined this linearity assumption for temporal effects in acute exposures with an assumption of time independence in chronic exposures to generate the concentration-duration function shown in Figure 3.14. This linearity assumption is chosen for its simplicity rather than any theoretical or empirical evidence. Parkhurst et al. (1981), assuming that LC_{50} values are available for 24, 48, and 96 hours, linearly interpolated between these values and assumed time independence for exposures beyond 96 hours to generate the function shown in Figure 3.15.

These approaches depend on the assumption that temporal dynamics can be treated in terms of various durations of exposure to prescribed concentrations. In reality, organisms are subjected to a continuous spectrum of fluctuations in exposure concentrations due to variation in aqueous dilution volume or atmospheric dispersion, variation in effluent quality and quantity, intermittent release of effluents, and accidental spills or upset effluents. These can be treated conventionally if (1) the time between episodes is sufficient for recovery so they can be treated as independent, (2) the fluctuations are of sufficient frequency and of sufficiently low amplitude that the organisms effectively average them, or (3) certain frequencies predominate, so that temporal categories of exposure can be identified as discussed above. An example of the third

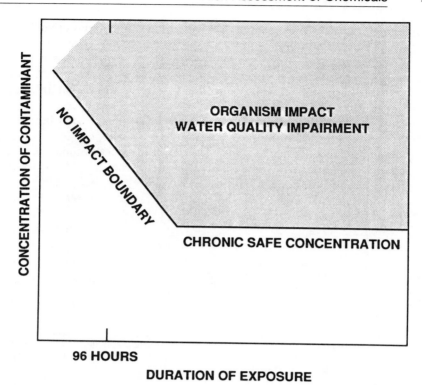

Figure 3.14. Toxic effects as a function of concentration and time, with acute time expressed as an assumed continuous linear function and chronic time expressed as a constant discrete variable (from Lee and Jones 1982).

possibility is Tebo's (1986) categorization of fluctuations in aqueous, point-source effluents as: (1) ponded and well mixed-wastes that are fairly uniform in character; (2) wastes subject to short-term, daily fluctuations; and (3) batch process wastes subject to severe fluctuations.

If it is not possible to characterize fluctuations in one of these ways, one can define a worst-case concentration and duration and assume that if that event does not cause unacceptable effects when it occurs in isolation, then it will also be acceptable when it occurs as part of a history of fluctuating exposures (Figure 3.16). This assumption is commonly adopted in effluent regulation. For example, the worst-case dilution condition for aqueous effluents has traditionally been the minimum flow which occurs for seven days with an average recurrence frequency of 10 years, referred to as the 7Q10. The EPA now recommends use of the lowest one-hour average and four-day average dilution flows that recur with an average frequency of three years (EPA 1985b). These correspond to highest one-hour and four-day exposures. The 3-year recurrence

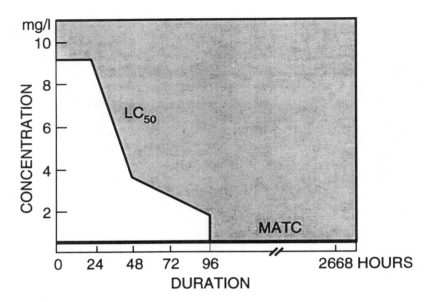

Figure 3.15. Toxic effects as a function of concentration and time, with acute time interpolated between measured values and extrapolated to zero and to the value for chronic time which is expressed as a constant discrete variable (from Parkhurst et al. 1981).

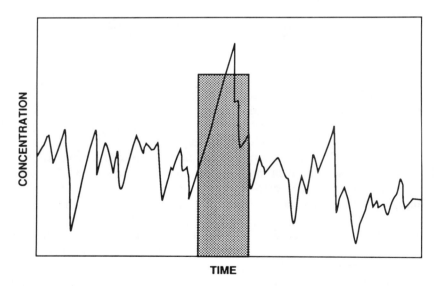

Figure 3.16. A fluctuating ambient exposure concentration (solid line) can be represented in an assessment by a continuous exposure (hatched bar) corresponding to the worst-case peak exposure.

frequency is assumed to allow for recovery of the system so that the peak exposures can be treated as independent events.

The best solution would be to avoid the acute/chronic dichotomy and worst-case assumptions by identifying characteristic temporal patterns of exposures or biological responses, and either conducting toxicity tests to simulate those patterns or classifying test results into the environmentally-based temporal category in which their durations fall. That is, one can scale time to the characteristic temporal scales of the processes determining the risk. For example, exposures to gaseous pollutants from point sources might be classified as (1) plume strikes (an hour or less), (2) stagnation events (hours to a week), and (3) the growing season average exposure. Existing data on concentrations of an air pollutant causing phytotoxic effects might be plotted against time in terms of these categories (Figure 3.17), and then compared to estimated ground-level concentrations for each of the three categories of events. When regulating aqueous effluents, rather than reducing temporal dynamics to a worst-case with a set duration based on hydrology as discussed above, hydrology might be related to the temporal dynamics of the biota. For example, if the most sensitive response to a chemical is mortality of larval fish, which begins within a day of the beginning of exposure to concentrations above the lethal threshold, then effluent limits might be based on dilution of the effluent in the 24-h low flow that occurs during the months in which larval fish are present at the site, with a mean recurrence interval greater than the time it takes for the

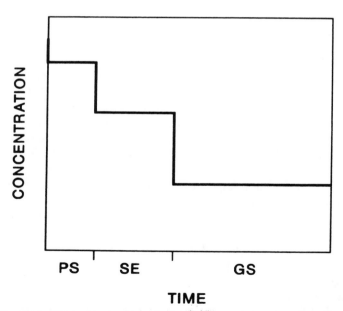

Figure 3.17. Toxic effects as a function of concentration and time, with time expressed as a quantal variable defined in terms of categories of exposure (PS = plume strike, SE = stagnation event, and GS = growing season).

population to recover from a poor recruitment year. Use of this approach in test design would require abandonment of the current standard test durations in favor of a new set of durations that are felt to represent important exposure-response patterns or even guidelines for designing tests to fit particular technologies and sites. In any case, the matching of exposure durations with toxicological endpoints should be based on an analysis of the situation being assessed rather than on preconceptions about acute and chronic toxicity.

Dose-Response Functions

Although dose is defined in a variety of ways, all definitions concern the amount of material taken up by an organism. Traditionally, dose is simply the amount of material that is ingested, injected, or otherwise administered to the organism at one time. Exposure duration is not an issue for this definition (although time-to-response may be), and results are represented as dose-response functions, analogous to the concentration-response functions discussed above. These functions, like the LD_{50} values that are calculated from them, are useful to environmental assessors in cases like the application of pesticides in which very short duration exposures occur due to ingestion, inhalation, or surface exposure.

An alternate definition of dose is the product of concentration and time, or, if concentration is not constant, the integral over time of concentration (Butler 1978). This concept is applied to exposure concentrations as well as body burdens, allowing the calculation of dose-response functions for exposures to polluted media. It has been commonly used in the assessment of air pollution effects on plants, and is referred to as "exposure dose." For example, McLaughlin and Taylor (1985) compiled data on field fumigations with SO_2 of soy beans and snap beans, and plotted percent reduction in yield against dose in ppm-hours (Figure 3.10). Newcombe and MacDonald (1991) found that the severity of effects of suspended sediment on aquatic biota was a log linear function of the product of concentration and time (Figure 3.6).

The most strictly defined form of dose is delivered dose, the concentration or time integral of concentration of a chemical at its site of toxic action. This concept is applied empirically by measuring the concentration of the administered chemical in the target organ or tissue, in some easily sampled surrogate tissue such as blood, or in the whole body in the case of small organisms. By relating effects to internal concentrations, rather than (or in addition to) ambient exposure concentrations, this approach can provide a better understanding of the action of the chemical observed during tests, and it is useful as an adjunct to environmental monitoring. It allows body burdens of pollutants in dead, moribund, or apparently healthy organisms collected in the field to be compared to controlled test results in an attempt to explain effects in the field (Chapter 10). Toxicokinetic models provide a means of predicting dose to target organs from external exposure data (Chapter 6).

Rather than simply being a function of peak body burden (i.e., mg/kg), effects may be a function of the product of body burden and time or, more generally, the time integral of body burden. This variable, termed dose commitment, is used in estimating effects of exposures to radionuclides, and may be appropriate to some heavy metals and other environmental pollutants (Butler 1978).

Three-Dimensional Models

Three-dimensional representations of toxic effects are rare. Only concentration-time-response models can be readily derived from conventional test data (Figure 3.9). Functions of this sort can be derived from the data that are required from flow-through 96-hr LC_{50} tests of fish by ASTM and EPA protocols (ASTM 1991, EPA 1982b). Such models have the obvious advantage of allowing the assessor to estimate the level of effects from various combinations of exposure concentration and duration. The response surfaces developed by Richardson and Burton (1981) for two estuarine species exposed to ozonated water, and by Jensen and Dochinger (1979) for trees exposed to SO_2 are good examples.

Concentration-time-response relationships can also be a useful tool for dealing with diverse data. If test data are diverse with respect to both duration and severity of effects, then both of these dimensions must be related to concentration if regulatory criterion concentrations are to be calculated. Dourson (1986) explained how this is done to estimate allowable daily intake for humans. The various exposures are categorized as having no observed effect, no observed adverse effect, adverse effects, or frank effects. This severity scale is then translated into a series of symbols that are plotted on a dose and time surface. Lines are then fitted by eye separating the categories of severity and establishing the no adverse effect level, adverse effect level, and frank effect level. The same approach has been used by the EPA to set air quality criteria for phytotoxic effects of air pollutants (Figure 3.18).

Four-Dimensional Models

Four-dimensional models are essentially unknown in toxicological assessment. Severity and proportion responding are the two dimensions that are least often considered together. While it is clear that the two aspects of exposure, concentration and duration, are distinct and independent, the distinction between the two aspects of effects, severity and proportion responding, has not been appreciated. It should, however, be enlightening to understand the relationship of these dimensions. For example, the implications for a population of a relatively uniform reduction in the growth rate of its members are quite different from the implications of the same average reduction in growth if it results from severe effects on a few members of the population and no effects on most others. Yoshioka et al. (1986) displayed all four dimensions of toxic effects as a matrix of two-dimensional plots (Figure 3.19).

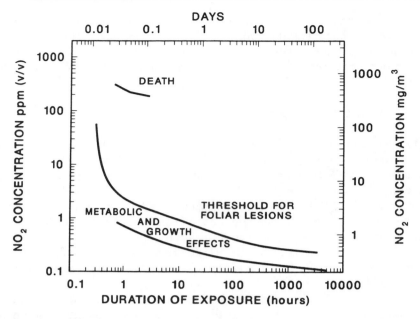

Figure 3.18. Categorization of effects of NO$_2$ on plants arrayed with respect to concentration and duration of exposure (from EPA 1982d).

Derivation of the Risk Estimate

It should be clear from the discussion above, how the integration of exposure and effects leads to a deterministic estimate of the expected level of effect, or a determination that exposure exceeds some threshold for significant effects. For a screening level assessment or in a traditional ecological hazard assessment, these estimates would be sufficient either alone or with safety factors. However, risk assessments should consider the probability densities associated with these estimates in order to determine probabilities of effects or probabilities of exceeding thresholds.

Risk can be described in general as the joint probability that each of the relevant dimensions of the effect on the assessment endpoint will assume a particular set of values corresponding to the assessment endpoint. The probabilities result from the types of uncertainty that are acknowledged and estimated, the models that are used to estimate the values, and the manner in which the uncertainty is propagated through the model, as discussed above. A simple example with a one-dimensional model is presented in Figure 2.4. The expression of effects is a distribution on the concentration scale of the estimate of some prescribed level of effect, and exposure is expressed as the distribution of an expected environmental concentration (EEC). The probability density function for effects results from the uncertainty in the toxicity test, plus any extrapolations that were made to convert the test endpoint to an estimate of

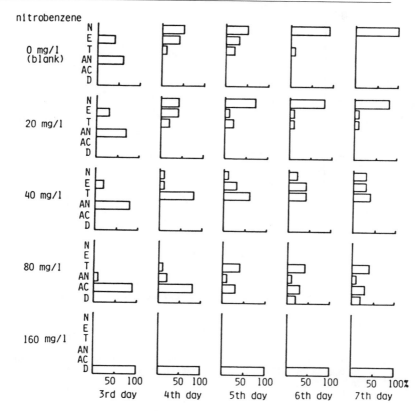

Figure 3.19. The relationship of concentration, duration, severity of response, and proportion displaying the response, displayed as a set of severity versus proportion responding relationships arrayed on concentration and time axes. The responses are: N, normal; E, eyespot; T, tetratopthalmic; AN, anopthalmic; AC, aciphalic; and D, death (from Yoshioka et al. 1986).

the assessment endpoint. The probability density function for the EEC results from propagation of the uncertainties in the parameters through the exposure model. The probability that the prescribed effect will occur is the joint probability of the exposure and the effect, which is roughly the overlap between these distributions. Both the probabilities and the graphic representation of the distributions can be input to the risk management process.

This probabilistic integration of exposure and effects could be simplified, at some loss of information, by simply considering three realizations of each component of the analysis: the most likely case, the best plausible case, and the worst plausible case (e.g., the 50th, 5th, and 95th percentiles). As illustrated in Figure 3.20, the likely, best, and worst exposure estimates would be applied to the likely, best, and worst exposure-response relationships, respectively, to generate likely, best, and worst estimates of effects.

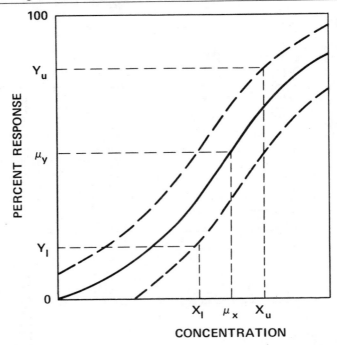

Figure 3.20. A simple model of the uncertainty concerning the level of response resulting from uncertainty in the exposure concentration and the concentration-response function.

Narrative Characterization

In addition to these numeric and graphical results, the risk analyst must prepare a narrative description of the results for the risk manager. This should include not only a description of the results, but also an explanation of how they were derived. This should include (1) descriptions of the models used, their major assumptions, the evidence for their validity, and their degree of acceptance by the scientific community; (2) sources of the test data and other input parameters and the quality of the data; and (3) sources of uncertainty in the analysis and the extent to which they have been quantified. In addition, a context for the assessment results should be provided, including (1) any conflicting evidence and explanations for its existence, (2) alternate credible assumptions and interpretations, (3) alternate results that are obtained if alternate credible assumptions are used, (4) a description of any research that would resolve major uncertainties, and (5) precedent assessments and analogous situations (e.g., similar chemicals with well-characterized risks) that would help to clarify the decision. The content and complexity of this narrative must be based on the technical capabilities and concerns of the risk manager.

RISK MANAGEMENT

Risk management, the process of decisionmaking that attempts to minimize risks without undue harm to other societal values, should be performed independently of risk analysis (NRC 1983); therefore, it is beyond the scope of this book. However, the effectiveness of a risk analysis depends on the successful interaction of the risk analyst and the risk manager. The risk analysts do not perform the decisionmaking process for the risk manager, and the risk manager does not dictate what data, models, and assumptions the analysts use, but they must communicate for the risk analysis to be relevant and the decision to be well-founded.

As discussed above, this relationship should have begun with the definition of the source, environment, and endpoints of concern. Because these activities of the scientist/risk analyst require that value judgments be made that are not based on science, the only way to avoid improper injection of the analyst's values is by interjecting the values of the risk manager through either policy statements or ad hoc judgments. However, risk managers often are not environmental scientists, so the risk analysts are obliged to ensure that the risk manager's judgments are informed by an understanding of the scientific background. In addition, the risk analyst must understand the needs and interests of the risk manager well enough to perform the risk characterization.

Figure 3.21 presents a simple diagram of the environmental risk decisions involved in setting criteria for the allowable concentration of chemical in air, water, or other ambient medium. The ideal input from the risk analysis is a function relating the probability of unacceptable effects on the endpoint of concern to the criterion level, or more likely, functions for each of a number of endpoints. W.D. Ruckelshaus, former director of the EPA, has said that the probability distributions are clearer to decisionmakers than numeric values (Ruckelshaus 1984). From this function, the risk manager may choose a criterion that is unlikely to be overprotective from the upper end of the curve, one that is unlikely to be underprotective from the lower end of the curve, or an intermediate criterion. Although not all risk managers are equally comfortable with probabilities, pressure on regulatory agencies to acknowledge and quantify uncertainty is increasing (Finkel 1990).

In addition to the results of the human health and ecological risk assessments, the decision will be based on a variety of social, political, legal, engineering, and economic considerations. Legal considerations include the phrasing of the law being implemented and any judicial decisions that indicate the appropriate degree of protection. The principal engineering consideration is the availability of practical technologies to meet the criterion. The cost of meeting the criterion and the value of the resource at risk (Box 3.5) may be the subject of a formal risk-benefit or cost-benefit analysis. Other, more nebulous considerations include the degree of public concern and the political philosophy of the currently governing administration. The manager may set a criterion or may defer a decision and require that more testing, measurement, or

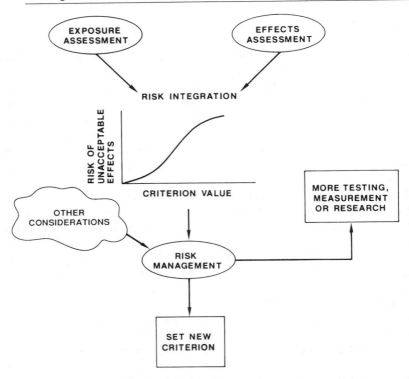

Figure 3.21. Diagram of risk management for setting an environmental quality criterion.

research be performed to reduce the contribution of uncertainty to the slope of the function.

Box 3.5 Valuation of Ecological Resources

In order to perform cost-benefit analyses or to set fines or compensation levels for environmental damage, it is necessary to predict, measure, or estimate the potential or actual loss of value of the system, and then convert that loss into monetary or equivalent units. Many ecologists and environmentalists have objected to the practice of placing monetary values on components of the natural environment, because they feel that economic analyses inevitably undervalue the environment relative to the value of a new chemical or development, or relative to the costs of treatment or cleanup. A counter argument is that decisionmakers inevitably consider economics in their judgments, and, if environmental scientists do not develop methods for monetarizing the values of the environment, decisionmakers are likely to underestimate the economic value of the environment. For example, the U.S. Environmental Protection Agency is required to justify new environmental regulations by performing cost-benefit analyses.

A fundamental argument against valuation of environmental components is that they have intrinsic values that are independent of human values (Rolston 1988). However, even if this premise is accepted, it is impossible for a human to determine what a forest's values are (i.e., would it rather be old-growth or young and vigorously growing?), so a poetic description of the intrinsic worth of a forest is not utilitarian, but it is no less anthropocentric than a dollar value. All value systems devised by humans are inevitably based on human values.

In valuing ecological resources, it is helpful to consider that there are various classes of utilitarian and nonutilitarian values that can be assigned. Each class of environmental values must have its own methods of estimation and valuation.

Utilitarian

Commodity values: Trees, rangeland, commercial fish, and other species and communities are sold or leased, so they have market values that are easily determined by conventional market surveys.

Potential commodity values: Plants and animals are sources of chemicals that may be marketed directly or be synthesized and marketed as drugs, pesticides, etc. In addition, all species constitute genetic resources that may be useful for genetic engineering of crops, livestock, industrial microbe, etc. Hence, all species have potential commodity values that are very difficult to estimate. However, this value is lost only in cases of extinction.

Recreation: Fish and wildlife have recreational value for fishermen, hunters, and bird watchers. Natural areas have recreational value for campers, hikers, etc.

Services of nature: Ecosystems moderate floods, control the local and global climate, degrade and sequester pollutants, reduce soil and nutrient loss, and perform other services that have palpable economic benefits. These can be estimated as costs of replacing the service (e.g., additional waste treatment) or costs resulting from the lack of the service (e.g., flood damage).

Nonutilitarian

Existence: Species and ecosystems have value to many people simply because they exist. Those people may not expect to see a golden lion marmoset or the Arctic National Wildlife Refuge, but they would experience a loss if the marmoset became extinct or the refuge was converted to an oil field.

Aesthetic: The aesthetic experience provided by organisms and ecosystems has a value equivalent to that of paintings or dance performances, but is not so easily quantified.

Cultural: Certain ecosystems and species have cultural significance. This value is illustrated by the greater emotional impact in the United States of the decline of the bald eagle than the concurrent and more serious declines of the peregrine falcon and brown pelican.

Scientific: Both species and ecosystems may have scientific value because they may serve as a particularly apt illustration of an unrecognized property of nature or because they are the subjects of long-term studies which may be disrupted.

Of these, only the commodity values and recreational values are reasonably well characterized by existing methods. Research is needed to determine whether defensible methods can be developed for estimating losses of the other values, and monetarizing those losses. It can be argued that attempts to monetarize nonmarket values are inherently artificial and the balancing of apples and oranges should be performed by decisionmakers rather than economists (NRC 1975). In addition, methods need to be developed for applying monetary values to decisions. Are the values independent and additive; is it appropriate to discount any of these values; how should obligations to future generations be incorporated; etc.? The reader is referred to Desvousges and Skahen (1987), Costanza (1991), and the journal *Ecological Economics* for discussions of these issues.

Finally, it should be emphasized that cost-benefit analysis should not be considered to be the sole reliable basis for decisionmaking. It is one input that describes one particular criterion. In particular, it ignores moral and ethical considerations. Many, if not most, members of modern western cultures would argue that we have moral duties toward the environment that supersede considerations of utility. However, these duties must be balanced against duties toward the poor, the hungry, practitioners of traditional cultures, etc. (Randall 1991). Therefore, environmental values are not absolute, even in moral terms, but must be balanced against competing values. Complexities like these put risk management beyond the scope of formal analysis.

Figure 3.22 represents a dichotomous decision, such as whether an effluent is to be permitted or whether a particular use of a pesticide is to be registered. In this case the level of pollution is set and the input of interest to the risk manager is the distribution of risks of effects of various levels of severity. The risk manager must decide whether the risk is unacceptable and must be reduced or prevented, or whether the risk is acceptable. The manager may

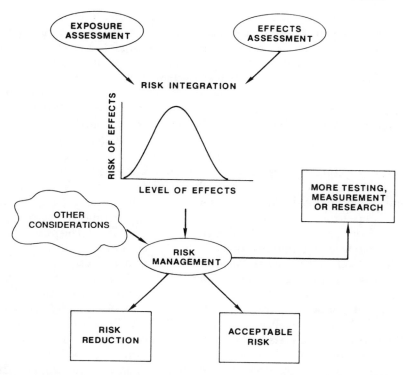

Figure 3.22. Diagram of risk management for making a decision concerning the acceptability of a situation.

emphasize the most likely effects, the peak of the distribution, or the most severe effects that have some specified probability of occurring. Considerations other than risk enter the decision as described above for criterion setting. An additional consideration is the environmental effects of the alternatives. Alternate waste treatment processes not only produce effluents of differing compositions but also produce their own wastes that pose risks and must be disposed of. Similarly, the alternatives to a particular pesticide pose their own risks. For example, the alternative to a pesticide that kills x birds per year may be clean farming which destroys refugia for pests but thereby destroys habitat for y birds. Even if y is much larger than x, clean farming may be more desirable because it is less repugnant to environmental and animal rights groups. The integration of such considerations is nearly always an informal process.

Each of these examples is presented as if there is only one exposure model, one effects model, and one set of assumptions about how to parameterize them and combine them into a risk estimate. This is more nearly the case in human health risk assessments where endpoints, models, and treatment of parameters are standardized. However, it is not the case in ecological risk assessments. One might estimate effects on fisheries with an ecosystem model, a population model, and a statistical model of an organism-level toxicity test. Ecosystem-level effects might be represented by results from a microcosm test, a lake ecosystem model, and a stream ecosystem model. Each risk estimate will have its own assumptions and associated uncertainties and those uncertainties may not be expressed equivalently. The separate lines of evidence must be evaluated, organized in some coherent fashion, and explained to the risk manager so that a weight-of-evidence evaluation can be made.

One approach to presenting alternate lines of evidence is to present them graphically on common axes. The axes should be chosen from the four dimensions of risk estimation so as to clarify the differences in the risk estimates that are most important to the decision. That is, does the decision depend primarily on the concentration at which an effect occurs, the time to recovery, the number of organisms dying, or some combination of axes? The output of the alternate models, assumptions, or data should be plotted on those axes.

An important concept in risk management for human health is that there are levels of risk that are so great that they must not be allowed to occur, no matter what the cost, and levels that are so low that they are not worth bothering with, no matter how little the cost. These are known as de manifestis and de minimis levels, respectively (Travis et al. 1987, Whipple 1987, Kocher and Hoffman 1991). Risk levels between these bounds are balanced against costs, technical feasibility of remediation, and other considerations to determine their acceptability. Because there have been no standard assessment endpoints for ecological effects, there have been no de minimis or de manifestis ecological risk levels (Box 3.6). However, the protection provided to endangered species in the United States provides one basis for de manifestis risk levels because it ostensibly precludes economic and sociopolitical considerations.* Establishment of de manifestis and de minimus risks would allow assessors to quickly dispense with clearly unacceptable or trivial risks after minimal assessments that do not establish the exact expected risk level, or balance costs and benefits. Consideration of what levels and types of ecological effects are clearly and invariably acceptable, and which are clearly and invariably unacceptable would provide risk assessors and risk managers with a good basis for working out the relation between societal values and ecological principles.

A final consideration in risk characterization and management is the need for audiences other than the risk manager to understand the risks and the bases for the decision. These include the public, environmental advocacy groups,

---

*Exemptions from the endangered species regulations are allowed, but that does not preclude treating risk of extinction as de manifestis. The exemptions are for emergencies which are analogous to the more severe emergencies (e.g., wars) that allow governments to kill large numbers of people, even though allowing a small number to be killed, is normally a de manifestis risk. Neither depletion of an endangered species nor death of a large number of people can be justified by the usual balancing of economic costs and benefits.

and regulated parties. Risk communication to the nonprofessional has been a major topic of research and discussion with respect to health risks, but not ecological risks (NRC 1989a). This is understandable because the average citizen is more concerned with his health than with ecological effects. However, when outside parties become concerned about potential ecological effects, the diversity of ecological endpoints and the uncertainty of their valuation make risk communication highly complex. For example, a new pesticide may pose a higher risk to avian populations but lower risk to zooplankton than the pesticide it would replace. An explanation of why the public should accept more dead birds in exchange for fewer dead zooplankton would be complex and not easily conveyed to a diverse audience.

Box 3.6. Conceptual Bases for De Minimis Risks

The concept of de manifestis risk is not controversial because some effects are clearly unacceptable. However, the idea that some exposures to and effects of pollutants are acceptable (de minimis) is controversial. The use of the de minimis concept is based on the following considerations:

1. Some exposures and effects are manifestly trivial. For example, it is absurd to assess effects of exposure to one molecule per organism or to consider restricting a pollutant exposure because it causes a transient physiological response or causes one copepod species to replace another.

2. There are many genuinely significant risks so it is inappropriate to set them aside while assessing the trivial ones. As Alvin Weinberg (1987) remarked, one does not swat gnats while being charged by elephants.

3. Regulatory scientists and decisionmakers do not have infinite time and resources. The benefits from their time and resources should be maximized by quickly setting aside trivial risks.

4. Very low-level exposures and effects are very difficult to accurately quantify. Therefore, any attempt to balance their costs against the costs of treatment or prevention, or the benefits of a product are likely to be confusing or misleading rather than enlightening.

5. Small increases in exposure to naturally occurring chemicals or other hazardous agents are likely to have trivial effects, because of adaptation to the background levels. This is particularly true if the increase is also small relative to variation in the naturally-occurring background levels.

6. Effects that are small relative to natural temporal or spatial variability of the biological parameter are likely to have trivial implications either at other levels of biological organization or for other components of the same level of organization. For example, an apparently large (e.g., 50%) temporary depres-

sion of the decomposition rate is unlikely to significantly affect primary production or abundance of birds because they are adapted to much larger variance in the rate due to wetting and drying cycles and other natural variables.

7. Treating a trivial risk as if it were potentially significant unnecessarily raises public concerns that are not entirely extinguished by a subsequent finding that the risks are small.

In summary, we cannot prevent all human effects on the environment, and we cannot even carefully assess the benefits and costs of all human actions. Therefore, it is desirable to develop criteria for eliminating clearly trivial risks from further consideration.

PART II

EXPOSURE ASSESSMENT FOR PREDICTIVE RISK ASSESSMENTS

In this section we discuss methods for estimating the exposure of organisms to chemical pollutants resulting from a proposed source. The section is composed of three chapters. The first (Chapter 4) deals with estimating the physical/chemical properties of pollutants to determine their behavior in the environment. The second (Chapter 5) deals with estimating the transport and fate of pollutants. The third (Chapter 6) deals with estimating the degree of exposure experienced by organisms in the polluted environment.

4

Environmental Chemistry

Theodore Mill

INTRODUCTION

Several billion tons of natural and synthetic organic chemicals move through the air, water, and soil environments each year. Much of this chemical burden is transformed to other compounds and eventually to carbon dioxide by a complex web of physical, chemical, and biological processes which act on each chemical species in ways that are uniquely determined by its molecular structure. Environmental chemistry attempts to describe and quantify environmental processes and the rates and products of movement and transformations of both organic and nonorganic chemicals released to surface or groundwaters, soils, or the atmosphere. These descriptions form the mechanistic basis for chemical fate and transport models discussed in the next chapter and the quantitative measurements of rates provide the numeric values for model parameters.

This chapter focuses on environmental reactions of organic compounds, including methods for measuring or estimating rate constants. Almost all synthetic nonpolymeric organic compounds can be oxidized to CO_2 or acetate by chemical or biological processes and, from an exposure perspective, are "destroyed." Inorganic compounds also can be oxidized or reduced, with conservation of the core metal or metalloid atoms, but with different biological/toxilogical effects and different exposure problems. Both organic and inorganic compounds can be sequestered by sorption or precipitation, reducing their bioavailability and exposure potential.

Three simplifying assumptions generally underlie predictive and evaluative fate methods : (1) the concentration of a chemical and its loss rate at any location and time is determined by a unique combination of independent input and loss processes. For example, atmospheric acetone is oxidized by HO radi-

ISBN 0–87371–875–5
© 1993 by Lewis Publishers

Table 4.1. Environmental Processes and Parameters

Process (k)	Parameter (E)
Atmosphere	
Photolysis	Light intensity
Oxidation	Oxidant concentrations
Rain out	Precip. rate, sticking coeff.
Transport	Wind velocity
Surface Waters	
Volatilization	Henry's constant, surface roughness
Sorption/bio-uptake	Organic/lipid content of sediments, organisms
Hydrolysis	pH, temperature
Photolysis	Light intensity
Biotransformations	Organism population, nutrient level, temperature, pH

cal, photolyzed by sunlight, washed out in rain, and returned to the atmosphere by volatilizing from surface waters, all with comparable rates; (2) each loss process can be expressed by a simple relation such as Rate = k[C][E], where k is the rate constant for the removal process. Values of k depend on the chemical structure, reaction temperature and phase, and have units of 1/concentration × time; [C] and [E] are the concentrations of the chemical C and environmental property responsible for loss; and (3) values of [E], such as sunlight or wind velocity, often vary over time, but can be well represented by location- and/or season-averaged values based on field measurements.

Thus, the net rate of change in concentration of chemical C is the sum of all equilibrium and kinetic processes that affect the chemical in a specific environmental location, including input from movement or a source.

$$R_T = \Sigma k_n \, [E]_n \, [C] \qquad (4.1)$$

Where R_T is the net rate of change, k_n is the rate constant for the n-th loss or input process, and $[E]_n$ is the environmental property for the n-th process. If $[E]_n$ is a constant, average value, Equation 4.1 simplifies to a simple first-order relation

$$R_T = \Sigma \, k_n' \, [C] \qquad (4.2)$$

Environmental processes and environmental properties believed to play a significant role in transport or transformation under some conditions are summarized in Table 4.1.

Measurements of the loss of a chemical at one location and time, even within this kinetic framework, have limited value for predicting how rapidly the chemical will change at a later time or in another environmental location because it is often impossible to disentangle the composite effects of several independent processes to learn the role each played in the observed loss. Thus the measurements are not extendable to a new time or location where a different set of processes may control loss of the chemical.

Over the past 20 years, a more successful approach has evolved which uses estimates and laboratory measurements of kinetic and equilibrium constants (k or K) for each chemical in each important process, combined with corresponding values for environmental parameters (E) to calculate an effective overall loss or partitioning constant at different times and locations. Often the value of [E] used in the estimate is averaged over time or over a region as large as a hemisphere for some atmospheric properties (Singh, 1977). This procedure gives both overall rates of transformation, movement, or partitioning, and the contribution of each processes to the total loss.

Usually, only one or two processes are important in each phase (water or air), but their relative contribution may change seasonally or diurnally. For example, both volatilization and photolysis of the a-diketone biacetyl are important in surface waters. However, photolysis is the faster process in summer at mid latitudes, whereas volatilization is the dominant loss process in winter when photolysis slows much faster than volatilization (Mill et al., 1984).

Kinetic and Equilibrium Rate Processes

Chemical movement and fate in the environment are controlled by dynamic kinetic and equilibrium processes. The kinetic constants used to evaluate the contribution of different processes in Equations 4.1 and 4.2 can be obtained by appropriate laboratory measurements in which each process can be conducted free (or nearly so) from contributions from other processes. Under these conditions the process

$$C + E = \text{Reaction or Movement} \qquad (4.3)$$

is observed, usually by a change in concentration of C with constant E. Since both chemicals and environmental agents are usually present in the environment at very low concentrations, the rate of their interaction (R_n) is almost always first order in each.

$$R_n = d[C]/dt = k_n [E_n] [C] \qquad (4.4)$$

and this kinetic relation is simplified if E_n is constant during the time interval of the measurement or is averaged over time. Sunlight photon fluxes, for example, are conveniently averaged over a 24-hr day. If E_n is constant,

$$R_n = d[C]/dt = k_n'[C] \ (k_n' = k_n[E]) \qquad (4.5)$$

integration of Equation 4.5 gives a log dependence of [C] on t

$$\ln (C_o/C_t) = k_n't \qquad (4.6)$$

where k_n' is the pseudo first-order constant for the n-th process, and C_o and C_t are chemical concentrations at times 0 and t. A plot of $\ln (C_o/C_t)$ versus time gives the value at k_n' as the slope of the plot. This relationship also provides a useful estimate of persistence, or the time required for loss of half of the original amount of C (half-life)

$$t_{1/2} = \ln 2/k_n' = 0.693 \qquad (4.7)$$

A more complete kinetic picture of environmental fate processes also includes one or more terms for reversible sorption of chemicals to sorbent (S) such as sediments, soils, biomass, or airborne particulate.

$$C_W + S \leftrightarrows C_s \; K_s = \frac{k_s}{k_{-s}} \qquad (4.8)$$

Where C_w is concentration of chemical in the water phase and C_S is the concentration in the sorbed phase

$$C_w + C_s = C_T \text{ (mass balance)} \qquad (4.9)$$

if C reacts in the water phase but not in the sorbed phase (Karickhoff, et al., 1979)

$$R = k_n' \; [C_T]/(K_s[S] + 1) \qquad (4.10)$$

The net effect of sorption is to slow the rate of loss of C_T by the factor $1/(K_s[S] + 1)$; in effect, sorption buffers the loss of C and leads to longer half-lives. Substitution in equation 4.7 gives

$$t_{1/2} = (K_s[S] + 1) \; \ln 2/k_n' \qquad (4.11)$$

Measurement and Estimation of Kinetic and Equilibrium Constants

A rate constant (k) is a numerical value which characterizes one reaction of one set of reaction partners (C and E) at one temperature and is independent of how much C and E are present. Increased temperatures always increase k for thermal reactions such as hydrolysis or oxidation, but not always for photochemical or biological processes. Simple equilibrium constants (K) are ratios of kinetic constants (k_1/k_{-1}), but usually are measured under equilibrium conditions where $R_1 = R_{-1}$, rather than from independent values of k_1 and k_{-1}. Both rate and equilibrium constants for environmental processes can be directly measured under controlled laboratory conditions or they can be estimated from empirically-derived structure activity relations (SARs), based on previous measurements of several similar compounds (Chapman and Shorter, 1972; Lyman et al.,1982; Mill, 1989). In many cases, the needed rate constants are already published in the scientific literature, thus avoiding the need to remeasure or estimate the values, but in cases when no published data are available, the decision of whether to measure or estimate a constant is best answered in the context of the question to be answered.

If approximate values within a factor of 2 to 5 are needed for initial screening to determine which processes are dominant in a specific environment,

SARs are the method of choice. SARs also find application in estimating constants for atmospheric processes for many compounds, where experiments are unusually difficult, as for example in measuring the atmospheric HO radical oxidation constant for a compound with very low vapor pressure. In this case, a very good SAR exists (Atkinson, 1987a) and the direct measurement is often very difficult. However, measurements usually are the only means to provide the most reliable constants which may be needed for a critical exposure assessment, or for processes for which SARs are not available.

SAR estimates often are the first step in developing a plan for laboratory measurements for a new chemical for which no reliable data exist. In this case, the best strategy is to perform a comprehensive set of estimates for all rate and partition constants using SARs where possible and other estimation tools where appropriate. This first tier of estimated rate constants now provides a basis for decision about which expensive laboratory measurements should be conducted to obtain more reliable values. In each section of this chapter, sources of measured rate constants are cited as the primary source of reliable constants.

For existing compounds, many of the important process constants have been measured recently enough to provide data relevant for use in environmental assessments, but nonexperts may have difficulty selecting useful constants from among many kinds of published data. Several criteria are available to help select relevant and valid rate constants for environmental assessments. These include: use of pure or buffered and sterile water at 25°C for hydrolysis and photolysis measurements, use of the appropriate wavelength of UV light for photolysis experiments (if sunlight is not used), use of an appropriate organism or consortium of organisms for biotransformation experiments and, most important, use of the appropriate controls for each kind of experiment to ensure that some other, unwanted, process has not intervened to confound the results.

Data published in reputable scientific and engineering journals and/or conform to recommendations for experimental protocols described under TSCA regulations by the Environmental Protection Agency (EPA, 1985c, 1988) or by the ASTM are the most reliable primary published data. Data contained in databases are less reliable secondary data, but if derived and traceable to primary sources, can be used with the same degree of confidence.

The EPA has developed a set of standardized, scientifically-defensible procedures (protocols) for measuring environmentally-relevant rate and equilibrium constants for important environmental transformation and transport processes to ensure that different laboratories and investigators would have a common set of standard procedures available with which to make reliable measurements of environmental process constants for exposure modeling (EPA, 1985c, 1988). These are the experimental measurement methods of choice for determining environmental constants, and can be reliably applied to the great majority of chemicals.

Acetone is a good example of a high-tonnage commodity chemical for which a wide range of kinetic and equilibrium constants needed for assessment and modeling have been published. Although the processes affecting acetone in the environment are limited, the methods and journal sources for many of these data are exemplary of good environmental chemical practices: Weast (1985) describes most of the physical properties; Snider and Dawson (1985) give a Henry's constant for volatilization, while Leo et al. (1971) have determined the value of K_{ow}. Chemical and biological transformation constants are discussed by Rathbun et al. (1988), Meyrahn et al. (1986), and Atkinson (1986). The net result of comparing all of the published data for acetone is to eliminate sorption and photolysis in water as important processes. Volatilization to the atmosphere is rapid enough that direct photolysis and indirect photooxidation with HO radical compete as the dominant loss processes. Biodegradation in the water column could compete with volatilization under some circumstances.

TRANSPORT PROCESSES: PARTITIONING AND VOLATILIZATION

Flowing water and air move and dilute chemicals as they simultaneously sorb and desorb from sediments or particulate and undergo irreversible transformations to new chemical species. Movement of some chemicals from air to water or soil, and water or soil to air further complicates the overall understanding of environmental fate. As part of any fate modeling effort, a screening panel of transport constants must be included along with transformation constants to help focus on the compartments where major loss processes will occur. Methylene chloride, for example, released into wastewater volatilizes to the atmosphere so rapidly that only processes in the atmosphere are likely to control its loss.

Many physical property and transport constant measurement and estimation methods for organic compounds have been reviewed by Lyman et al. (1982). This handbook and the EPA protocols (EPA, 1985c) are convenient one-stop sources for many of these methods. Partitioning of organic compounds to soils, sediments, biota and particulate, and volatilization from water are the only transport processes discussed here. Sorption and complexation of heavy metal ions are discussed elsewhere by Bodek et al. (1988).

Partitioning and Sorption Processes

Organic chemicals partition to natural solids in several ways, but for most organics, partitioning between water and an organic phase in sediment, soil, or biological tissue by a nonspecific sorption process accounts for most of the sorption observed. Sediments and soils contain 0.1% to 5% by weight of organic carbon as humic material in which organics dissolve, much in the same ways as they dissolve in biological lipids or organic solvents such as n-octanol

(Chiou et al., 1977). The extent of sorption or partitioning is expressed by the partition constant k_n'.

$$K_p = [C]_s/[C]_w \qquad (4.12)$$

The value of K_p is largely controlled by the water solubility of the chemical — the least soluble chemicals have the highest K_p values. The net effect of sorption is to slow the overall rate of loss of a chemical through reactions in water (see Equation 4.10). However, for some compounds, partitioning to soils or sediments is followed by irreversible transformation reaction of the sorbed chemical. For example, polyhalogenated alkanes can be reduced in sediments (Vogel and McCarty, 1987), while many aromatic amines irreversibly react with humic materials.

K_p values for strongly sorbed chemicals such as benzo[a]pyrene or PCBs are often larger than 10^5, whereas K_p values for weakly sorbed chemicals such as acids or amines are < 10. (Karickhoff et al., 1979). Since K_p for a specific compound will vary with the organic or lipid content of the soil, sediment, or biota, a more nearly equal basis for comparing distribution constants can be formulated by normalizing K_p values to the fraction of organic carbon in the sediment (α)

$$K_{oc} = K_p \alpha \qquad (4.13)$$

where α is expressed as mg of C per mg of soil or sediment, and K_{oc} is the normalized distribution constant. A more reliable method to evaluate K_p or K_{oc} uses n-octanol as a standard surrogate lipid phase. K_{ow}, the distribution constant for a compound in a water-octanol mixture, is used widely as a point of departure for estimating sorption to natural sorbants. K_{ow} is related to K_{oc} and the bioconcentration factor (BCF).

$$\log K_{oc} = \log K_{ow} - 0.21 \qquad (4.14)$$

$$BCF = 0.893 \log K_{ow} + 0.607 \qquad (4.15)$$

At this time several thousand values of K_{ow} have been measured, and a reliable computerized system is available for estimating new values from molecular structure (Leo et al., 1971). Equation 4.14 provides a simple way to estimate a difficult-to-measure process constant (K_{oc}) from an easy-to-measure or easy-to-estimate constant, K_{ow}.

Another partition process in sediments involves sorption to clay mineral surfaces via cation exchange of organic cations and metal ions with surface clay cations. Typical sediments have 10–40 milliequivalents (meq) of exchange capacity per 100 g, more than enough to completely retard the flow of these ions at ppm concentrations in groundwaters (Thurman, 1985). Unlike partitioning to organic fractions in sediments or soils, cation exchange is site-specific and limited by available sites and by competition among all cations including protons. Sorption may also occur on silica and metal oxides, but the mechanisms are poorly understood and usually pH sensitive.

K_{ow} and Related Partitioning Measurements and Estimates

Many methods exist for measuring and estimating values of K_{ow} for almost any kind of organic structure. Experimental methods as simple as shaking an octanol solution of the chemical with (usually) a larger volume of octanol-saturated water, separating the phases and analyzing for the chemical in one or both phases works well when $K_{ow} < 10^4$. Less direct methods must be used for large values of K_{ow} ($> 10^4$) such as high performance liquid chromatography (Mackay, 1980). Both methods are included in the EPA protocols (EPA, 1985c, 1988).

K_{ow} is readily estimated for most kinds of organic structures using a fragment-additivity method such as the one developed by Leo et al. (1971) as the CLOGP computer program. To use the computer program, the molecular structure of the compound is entered at the keyboard by so-called line notation or entered on the screen as a two-dimensional line drawing. Molecular connectivity and surface area also are used. The relation between aqueous solubility (S) and K_{ow} provides an additional route for estimating K_{ow} when S is readily measured or vice versa. Many of the S-K_{ow} correlation equations listed and evaluated by Lyman et al. (1982) are reliable only for liquids. One S-K_{ow} correlation equation for aromatics, corrected for solid melting (the last term) is

$$\log S(\mu mol/L) = 7.26 - 1.37 \log K_{ow} - 0.01(T_m - T) \qquad (4.16)$$

where T_m is the melting temperature of solids melting above 25°C.

Equation 4.16 gives an estimated solubility of isopropylbenzene (log K_{ow} = 3.43) of 43 mg/L, in excellent agreement with the measured solubility of 50 mg/L. The estimated solubility of benzofluorene, a highly insoluble solid melting at 208°C, based on Equation 4.16, is 0.77 μg/L, in good agreement with the measured value of 2 μg/L; if the version of Equation 4.16 which is uncorrected for melting point is used, the value is 50 μg/L.

In an environmental context, sorption of chemicals in rivers or lakes having a significant sediment load will slow volatilization of the bulk of the compound, extending the effective half-life for volatilization in the water column by the value of $K_{ow} \times$ (mass sediment/mass water), or up to 10^6 times as long. An additional complication is that a sediment-bound chemical is only partly available for reequilibration to the water column because some fraction becomes buried in the underlying sediments at a rate controlled by the net deposition rate in the water body. In fast-moving streams where sediments do not accumulate rapidly, most of the chemical is removed from the water body by transformation or hydrologic processes.

Volatilization Measurements and Estimates

Volatilization of chemicals from water or soil to air is an important transport process for a wide range of chemicals. Even compounds with very low vapor pressures can volatilize from water surprisingly rapidly owing to their high activity coefficients in solution.

The volatilization rate constant (k_v) can often be estimated using the Liss-Slater two-film model (Liss and Slater, 1974; Smith et al., 1981; Smith et al., 1983)

$$k_v = \frac{1}{L}\left[\frac{1}{0.7k_l^o(D_l^r)} + \frac{RT}{H_ck_g^w(D_g^r)} \right]^{-1} \qquad (4.17)$$

In this relation, k_l^o and k_g^w are the reaeration constant for oxygen in water and the gas-phase mass transfer coefficient of water vapor in air, respectively. H_c is the Henry's constant; D_l^r and D_g^r are ratios of molecular diffusion constants of the chemical and oxygen in water and the chemical and water in air, estimated for conditions in a specific type of water body; L is the depth of the water body in cm.

Equation 4.17 simplifies if $H_c > 1000$ in units of torr/molar and mass transfer is liquid-phase limited

$$k_v = \frac{k_l^o (D_l^r)0.7}{L} \qquad (4.18)$$

If $H_c << 1000$ the process becomes limited by gas-phase mass transfer and Equation 4.20 again simplifies

$$k_v = \frac{H_ck_g^w(D_g^r)}{L} \qquad (4.19)$$

Equations 4.17, 4.18, and 4.19 are complex relations which contain several environmental parameters in k_g^w, including wind velocity and water surface roughness. Because direct measurement of volatilization in natural waters is difficult to perform, these parameters are often estimated from molecular size and diffusion constants, plus some empirical term for average water surface conditions. H_c can be measured in the laboratory (EPA, 1985c) or calculated from the ratio of vapor pressure (P_L) to water solubility (S_L) for the liquid phase form of the chemical.

$$H_c = P_L/S_L \qquad (4.20)$$

Mackay and Shiu (1981) have published a critical review of H_c for many organic compounds, and it is a useful first source for reliable values for known compounds. Both kinetic and thermodynamic laboratory methods have been developed to measure H_c; Brunner et al. (1990) have discussed the scope and limitations of these methods. Smith et al. (1981) evaluated Equation 4.17 over a wide range of H_c and D_l and D_g values in a river and lake. Figure 4.1 illustrates the relation between H_c and $t_{1/2}$ for volatilization of compounds having a wide range of H_c values. In general, for $H_c > 100$ (in Torr/M), $k_v > 2 \times 10^{-7}$ s^{-1} and $t_{1/2} < 10$ days.

An example of the use of Equation 4.19 to estimate the volatilization half-life of 2,3,7,8-tetrachlorodioxin (TCDD) is based on an estimate of H_c of 1.2 torr/Molar, which in turn came from measured values of P and S for TCDD (Podoll et al., 1986). In this equation, $k_g D_g$ has the value of 2000 cm/h, and L is 2 m, which gives a value of k_v of 0.016 day^{-1} at 300 K, corresponding to a half-life of about 40 days.

From Equation 4.20, it is evident that estimating H_c reduces to estimating values of P and S from molecular structure. For many chemicals, P may be expressed as a function of normal boiling point (T_B) and melting point (T_m) of the chemical by

$$\ln P \text{ (atm)} = 10.6(1 - T_B/T) + 6.8(1 - T_m/T) \qquad (4.21)$$

where T is $< T_B$ or T_m and is usually 298 K (Mackay, 1980).
Since

$$H_c = P_L/S_L \qquad (4.20)$$

Combining Equations 4.16 and 4.21 and converting to mol/liters,

$$\ln H_c(\text{atm } M^{-1}) = 10.6(1 - T_B/T) - 4.494 + \ln K_{ow} \qquad (4.22)$$

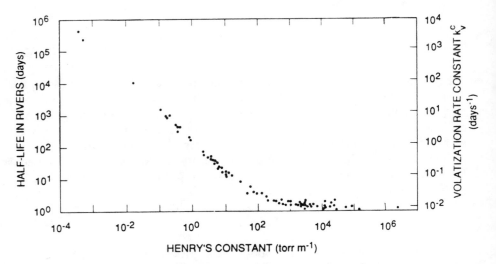

Figure 4.1. Estimated half-lives versus Henry's law constant for chemical in rivers (Smith et al., 1981).

$$H_c = 2.36 \, K_{ow} \, (1 - T_B/T) \qquad (4.23)$$

Spencer et al. (1973) have reviewed volatilization processes from soil surfaces. The overall process is complicated by variable contributions from volatilization of the chemical from the water at the surface, evaporation of water itself, and the wick effect that brings more water and dissolved chemical to the surface. Initially, volatilization of the chemical from the surface water will be rate-controlling and Equation 4.19 can be used to estimate the rate constant. As the surface chemical is volatilized, additional chemical diffuses to the surface layer and a concentration gradient develops in the soil column. Equation 4.19 no longer holds under this condition and at this time no simple laboratory measurement will reliably estimate the rate of the process in a way that can be extrapolated to the field.

Transport Processes in Cloud Water Droplets

The average cloud droplet density is about 1 cm^3/m^3 or 1 in 10^6 v/v. This means that at equilibrium, even relatively low volatility compounds will favor the vapor phase. Partitioning between cloud water and the atmosphere is characterized by complex mass transport kinetics which are still not well understood. In general, only polar low molecular weight species will favorably partition to cloud droplets, but if the species is rapidly transformed in the droplet, the droplet becomes a sink even for low solubility species such as ozone. Oxidation of atmospheric SO_2 to SO_4^{2-} (acid rain) is caused by importing SO_2, ozone, and (probably) H_2O_2 into cloud droplets from the tropospheric vapor phase, where the liquid-phase oxidation occurs several million times faster than the vapor phase oxidation of SO_2 by HO radical (Martin, 1984; Schwartz, 1984, 1988).

TRANSFORMATION PROCESSES

Introduction

Major transformation processes in the environment are listed in Table 4.1. They range from relatively uncomplicated oxidations by HO radical in the troposphere to complex redox processes in water and sediments to the very complex biotransformations by bacteria and fungi in water, soil, and sediments. This section will focus on the details of two major kinds of chemical transformations, hydrolysis and photoreactions (both direct and indirect) in water and the atmosphere, where kinetic methods for measuring key rate constants are well developed and relatively well understood in theory as well as practice. Both experimental measurement methods and empirical predictive methods will be discussed. Biotransformations will be reviewed more briefly; despite their great importance, biotransformations are still relatively poorly understood, and characterized by much more empirical measurement methods.

Hydrolysis

Hydrolysis refers specifically to introduction of HOH or OH into an organic molecule resulting in cleavage of another chemical bond. Some examples of hydrolysis include conversion of alkyl halides to alcohols, esters to acids, and epoxides to diols. Hydrolysis reactions are rapid enough in some natural waters to control the overall fate of many classes of important synthetic chemicals. Hydrolyzable chemicals include esters, alkyl halides, epoxides, amides, carbamates, alkoxysilanes, alkyl phosphonates, and alkyl phosphates (Mabey and Mill, 1978). The importance of hydrolysis for exposure modeling is that hydrolysis products, alcohols, acids, and carbonyls, are often more water soluble, and less subject to bio-uptake or volatilization than their parent compounds. A useful generalization is that hydrolysis may be important in any molecule where alkyl carbon, carbonyl carbon, or imino carbon are linked to halogen, oxygen, or nitrogen atoms or groups through σ-bonds.

There is a substantial literature on hydrolysis including rate constants, structure activity relationships (SARs), and mechanistic studies. A comprehensive review of kinetic constants for hydrolysis of organic compounds in water is found in Mabey and Mill (1978), while detailed mechanisms or pathways are discussed in reviews by Streitweiser (1962), Thornton (1964), Kirby (1972), and Lowry and Richardson (1981); regular, more specialized reviews are found in *Chemical Reviews* and *Advances in Physical Organic Chemistry*.

Kinetics and Mechanisms of Hydrolysis Reactions

Hydrolysis reactions usually are catalyzed by acids and bases, but also occur with water more slowly without benefit of catalysts. The general kinetic expression for hydrolysis of many neutral chemicals (RX) in water can be expressed as

$$d[RX]/dt = k_h[RX] = (k_A[H^+] + k'_N[H_2O] + k_B[HO^-])[RX] \quad (4.24)$$

where k_B, k_A and k'_N are process rate constants for the acid- and base-catalyzed and neutral (water) processes, respectively. The concentration of water is usually constant and is always much greater than that of the chemical RX; therefore $k'_N[H_2O]$ is also a constant, k_N. The pseudo first-order rate constant k_h is the observed rate constant for hydrolysis at a specific pH and temperature

$$k_h = k_A[H^+] + k_N + k_B[HO^-] \quad (4.25)$$

At constant pH, $[H^+]$ and $[HO^-]$ are constant, and Equation 4.24 may be rewritten as Equation 4.26 and integrated

$$d[RX]/dt = k_h[RX] \quad (4.26)$$

$$\ln(C_0/C_t) = k_h t \quad (4.27)$$

Equation 4.25 shows how pH affects the overall rate; at high or low pH, the first or last term is usually dominant, whereas k_N is often most important near pH 7. However, the detailed relationship and overall rate depend on the specific values of k_B, k_A, and k_N. Equation 4.27 is used to estimate the k_h from laboratory measurements conducted at constant pH; Equation 4.25 is then used to evaluate k_B, k_A, and k_N by solving the equation for values of k_h determined at three distinctly different pH values. If k_h is measured at pH values of 2–3 or 10–11, k_h is equal to either k_A or k_B, thus giving an easy solution for these rate constants. From k_h, the half-life of the hydrolysis reaction is

$$t_{1/2} = \ln2/k_h \tag{4.28}$$

In addition to H^+, HO^-, and H_2O, other chemical species in surface waters may act as acids or bases to promote hydrolysis. In the presence of general acid or base Z, the first-order rate expression for loss of chemical has an additional term for Z.

$$k_h = k_A[H^+] + k_N + k_B[HO^-] + k_z[Z] \tag{4.29}$$

where k_Z and $[Z]$ are the rate constant and concentration for Z. The most important effect of general acid or base catalysis is found in laboratory experiments that use buffers such as phosphate to control pH. Buffer anions may catalyze hydrolysis, leading to erroneously high values for rate constants. The simplest way to avoid or detect the problem is to use concentrations of buffers in the low millimolar range. Anionic species in natural waters also can cause general base catalysis, but these effects are usually minimal.

The graphical form of Equation 4.25, a plot of log k_h versus pH, provides an independent check on the validity of rate measurements typically performed at pH 3, 7, 11 (EPA, 1985c). Figure 4.2a-d shows plots of log k_h versus pH for compounds which undergo different combinations of acid, water, and base-promoted hydrolyses of neutral compounds. All of these plots have slopes of −1, 0, or +1 for the acid, neutral, and base-promoted processes respectively, in keeping with the log-log relation of k_h and $[H+]$. Most classes of neutral hydrolyzable compounds fit one or another of the curves as summarized in Table 4.2 (Mabey and Mill, 1978; Mill and Mabey, 1988)

The slope of the plot for log k_h vs pH is a useful criterion for distinguishing true hydrolysis from other processes that may intervene in poorly controlled systems. If k_h, measured at low, medium, and high pH values does not fit the plot, very likely some adventitious process such as volatilization, biodegradation, or even photolysis has intervened. New measurements should be conducted in sterile, closed, and darkened containers to check and ensure the validity of the data. Plots which show nonzero or nonunit slopes, but have the general shape expected for the class of compound, may be confounded by buffer catalysis.

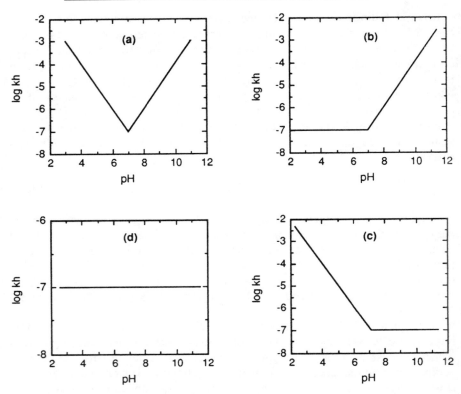

Figure 4.2. Log k_h versus pH plots (clockwise) for several different pH-rate relationships.

Ionizable compounds that form cations or anions in water at pH values between 3 and 11 may have complex pH-dependent hydrolysis rates. If, for example, the cationic or anionic form undergoes hydrolysis much faster than the unionized species, the rate of hydrolysis abruptly changes at the pH where the species ionizes. This kind of hydrolytic behavior has been reported for 1,6-

Table 4.2. Curve Types and Processes of Classes of Neutral Hydrolyzable Compounds

Compound Class	Curve Type[a]	Important Processes[b]
Esters	a	A, N, B
Alkyl halides	b	N, B[c]
Epoxides	c	A, N
Amides	b	A, N, B[d]
Phosphates	a	N, B[c]
Acetals, ketals	c	A, N
Imines	e	A, N, B

[a]See Figure 4.2a–d.
[b]A, N, B are acid, neutral, base-promoted processes.
[c]Base-promoted process is generally not important below pH 10 or 11.
[d]Acid catalysis is important only below pH 2.
[e]Complex curves; imines are protonated above pH 5.

dimethyl-1,6-dinitrosourea, N,N-dimethyl-N-nitrosourea and o-hydroxy-benzamides. Other complex pH profiles due to ionization are found for phthalamic acids and o-aminophenyl alkanoic esters (Mill and Mabey, 1988) and imines (Cordes and Jencks 1964).

Temperature Effects on Rate

Rates of hydrolysis always increase with increasing temperature. Most investigators report temperature effects in the form of the Eyring or Arrhenius equations

$$\log k = -\Delta H^{\pm}/RT + \Delta S^{\pm}/R + C \qquad (4.30)$$

$$\log k = A - E/RT \qquad (4.31)$$

where A and $\Delta S^{\pm}$ are entropy terms and E and $\Delta H^{\pm}$ are temperature coefficients. Mill and Mabey (1988) have summarized values for E or $\Delta H^{\pm}$ for several classes of compounds. Neutral hydrolyses of alkyl halides have $\Delta H^{\pm}$ values corresponding to a four-fold change in rate for each $10°C$ change near $25°C$. Acid-catalyzed hydrolysis of epoxides has a three-fold change for the same temperature change. Amides fall in this same range for both acid and base-promoted reactions, while carbamates have a wider range of rate changes.

Another, smaller, effect of increasing temperature is to increase the dissociation constant for water, (K_w) from 1.002×10^{-14} at $25°C$ to 5.474×10^{-14} at $50°C$; this change has the effect of increasing $[HO^-]$ from 1.00×10^{-7} M to 2.34×10^{-7} M. Although this effect of temperature is small, it is predictable and corrections should be applied when appropriate. The correction for changes in $[H^+]$ is not needed if the pH of the solution is measured at the temperature of the experiment; correction is needed for $[HO^-]$, however.

Hydrolysis of Alkyl Halides

Hydrolysis reactions of alkyl halides and related compounds are a subset of substitution reactions on aliphatic carbon in which group Y replaces group X

$$RX + Y^- \rightarrow RY + X^- \qquad (4.32)$$

These reactions may actually take place in several steps with formation and reaction of intermediates or in one concerted process without intermediates. In the environment, the two major substitution reactions involve neutral water or OH^- displacing on a neutral species. Pathways for aliphatic substitution reactions may proceed either by backside displacement of Y^- on the C-X bond with loss of X^- (S_N2, Reaction 4.33), or by initial loss of X^- from R followed by capture of Y^- by R^+ (S_N1, Reactions 4.34 and 4.35). The S_N1 process is important only at saturated (sp^3) carbon and only if the incipient carbenium ion (R^+) is highly stabilized by substituents. Thus, t-alkyl halides, a-haloethers, and benzyl and allyl halides react partly or largely by S_N1 processes. The S_N2 process is important for unhindered primary alkyl groups such as n-alkyl halides and sulfonates ($RX = R'CH_2X$). Many substitution reactions are some "blend" of the S_N1 and S_N2 processes (Streitweiser, 1962).

$$Y^- + RX \rightarrow YR + X^- \; (S_N2) \tag{4.33}$$

$$RX \rightarrow R^+ X^- \tag{4.34}$$

$$R^+ + Y^- \rightarrow RY \; (S_N1) \tag{4.35}$$

The nature of the leaving group X strongly influences the rate of S_N2 reactions: halides and anions of other strong acids are displaced readily in water, whereas very weakly acidic OR, OH, or NH_2 groups are unreactive with H_2O or HO^-. The order of reactivity for displacement of X by HO^- or H_2O is $I > Br \sim OSO_2R > Cl> > F$. Lowry and Richardson (1981) provide a very good overview of the subject.

Hydrolysis of Carbonyl Compounds RC(O)X

Hydrolysis of carboxylic acid derivatives such as esters, anhydrides, acyl halides, and amides has been studied extensively by several generations of physical-organic chemists and biochemists. These compounds hydrolyze by displacement reactions with H_2O or HO^-; acids also catalyze reactions of some carbonyls (Mabey and Mill, 1978).

$$RC(O)X + Y^- \rightarrow RC(O)Y + X^- \tag{4.36}$$

The detailed mechanisms for these hydrolyses follow several pathways, most of which involve initial addition of a nucleophile such as water or OH^- to the carbonyl or protonated carbonyl to form a tetrahedral intermediate, followed by loss of the other substituent to form the carboxylic acid or anion as a carbonyl analog to the S_N2 reaction.

$$\overset{\displaystyle O}{\underset{}{\|}} \qquad \overset{\displaystyle O}{\underset{\displaystyle Y}{|}} \qquad \overset{\displaystyle O}{\underset{}{\|}}$$
$$RC\text{–}X + Y \rightleftarrows RC\text{–}X \rightarrow RC\text{–}Y + X \tag{4.37}$$

$$Y = H_2O \text{ or } OH^-$$

Aliphatic substitution (S_N2) similar to that of alkyl halides also occurs in esters if the carboxylate or other anion is highly stabilized. For example, substitution in p- or sec- alkyl trifluoroacetates occurs with displacement of the trifluoroacetate moiety.

$$Y^- + R'OC(O)CF_3 \rightarrow YR' + {}^-OC(O)CF_3 \qquad (4.38)$$

Similar reactions occur with sulfonate and nitrate esters.

Base-promoted hydrolysis of esters is accelerated by electron attracting groups in R, such as halogens, whereas acid-catalyzed hydrolysis is relatively insensitive to such polar effects. The relative ordering of leaving groups in carbonyl derivatives is the same as in S_N2 reactions but the range of reactivity is compressed: halogen $> RC(O)O > > R'O > R_2'N$. Whereas hydrolysis of ethers (ROR') is very difficult except under extreme conditions of acidity or basicity, hydrolysis of most esters (RC(O)OR') is relatively rapid at pH 4 or 9 (Mabey and Mill,1978).

Laboratory Measurements of Hydrolysis Rate Constants

Laboratory methods for measuring hydrolysis rate constants, k_h, and estimating values of the specific process constants k_A, k_B and k_N are described in detail in the EPA protocols (EPA, 1985c). The loss of chemical is followed at constant pH (usually 3,7,11) and temperature (usually 25°C) in sterile, buffered solutions with low concentrations of the chemical (usually below 1 mM). The value of k_h is determined at each pH value by the slope of the regression of log (C_0/C_t) versus time (Equation 4.27), with standard errors of about ± 5–10%. For many compounds, low and high pH experiments give k_A and k_B directly because the other terms in Equation 4.25 are negligible:

$$k_h \text{ (pH 3)} = k_A(H^+) \qquad (4.39)$$

and

$$k_h \text{ (pH 11)} = k_B(HO^-) \qquad (4.40)$$

The determination of k_N from k_h is often difficult unless the compound has no acid or base-promoted reaction, in which case the pH 3 and 7 or pH 11 and 7 experiments will give similar values for k_h, indicating that k_h is identical with k_N' (see Figure 4.2 a-d).

Predictive Methods for Hydrolysis Rate Constants

Qualitative Structure-Activity Relationships (SARs)

Environmental chemists may often find that qualitative SARs are adequate for exposure modeling because many compounds are so resistant to transformation that over several days, months or years changes are too slow to have any impact on the exposure of a biological or human population. Under these circumstances, a half-life of ten, one hundred, or ten thousand years will have the same unwanted effect and estimates accurate to no more than a factor of ten will be quite adequate. Similarly, compounds that react so rapidly that 90% has disappeared in ten to sixty minutes also can be modeled adequately with an estimated value for the rate constant.

The reactivities of different classes of chemicals are conveniently expressed as half-lives ($t_{1/2}$), as defined in Equation 4.28 at pH 7 and 25°C (Mabey and Mill, 1978; Mill and Mabey 1988). These half-lives are listed in Table 4.3 for alkyl carbon and carbonyl derivatives. Hydrolysis rates at saturated carbon (RX) generally increase with branching of the carbon chain. Similarly, the leaving groups Cl, Br, and I have similar reactivities, but F is much less reactive and NHR and OR are unreactive under ordinary conditions. All of these reactions are pH independent in the region around pH 7, have no acid-catalyzed process and react rapidly with HO⁻ only at pH 10 or above (see Figure 4.2b).

Rates of hydrolysis at carbonyl carbon (RC(O)X) follow reactivity patterns similar to those for RX, but with an important difference: all leaving groups become much more reactive when attached to carbonyls than on alkyl carbon by orders of magnitude; even NHR and OR become moderately reactive in base or acid-promoted reactions (see Table 4.3). Alkyl halides with OR or NR_2

Table 4.3. Half-Lives for Hydrolysis of Organic Compounds at 25°C, pH 7[a]

Class	Half-Life, $t_{1/2}$
Alkyl Halides (X = Cl, Br)	
n-RF	> 10 yr
n-OR i-RX	2-30 days
t-RX	< 1 min
$PhCH_2X$	< 1 hr
$ROCH_2X$	< 5 min
Esters (R = n or i-R)	
RC(O)OR'	2->10 yr
$Cl_nRC(O)OR'$ (n = 1–3)	< 1 d
$C_6H_5C(O)OR$	> 10–30 yr
Amides, Carbamates	
RC(O)NHR'	> 3000 yr
$CL_nRC(O)NH_2$ (n = 1–3)	< 1 yr
$XC_6H_5NHC(O)OC_6H_5$	< 10 days
$XC_6H_5N(Me)C(O)C_6H_5$	> 5,000 yr
$R_2NC(O)OC_{10}H_9$	> 1,000 yr
$RNHC(O)OC_{10}H_9$	< 10 d
Phosphorus Esters (R = n or i-R)	
$RP(O)(OR)_2$	100–100,000 yr
$ROP(O)(OR)_2$ (R = Alkyl or Ph)	1 yr
Ethers	
ROR	> 1000 yr
$R_2COCR'_2$	10–20 d
$CH_2(OR)_2$	5000 y
$RCH(OR)_2$	< 300 da
$RC(OR)_3$	< 10 min

[a]Mabey and Mill, 1978; Mill and Mabey, 1988.

are also much more reactive than unsubstituted alkyl halides. Epoxides have a labile ether link that hydrolyzes with remarkably little influence of substituents.

Phosphorus esters $[(RO)_3PO]$ have complex dependence of hydrolysis rates on substitution in which the pK_a of the RO plays an important role.

Quantitative SARs

Organic chemists have developed quantitative SARs (QSARs) to predict rate constants for many kinds of reactions with reasonable success (Chapman and Shorter, 1972); several QSARs for environmental processes are discussed below. QSARs have been developed for many classes of hydrolyzable organic compounds, including alkyl halides, epoxides, esters, carbamates, and phosphorus esters. Most SARs are derived from either the Hammett or Taft relations (Exner, 1972; Chapman and Shorter, 1972)

$$\log k_x = \log k_o + \rho\sigma \qquad (4.41)$$

$$\log k_x = \log k_o + \rho\sigma^* + \delta E_s \qquad (4.42)$$

where k_x and k_o are rate constants for substituted and unsubstituted structures respectively, σ, σ^*, and E_s are the electronic substituent constants for substituent x in aromatic and aliphatic systems, and E_s is the steric substituent constant in aliphatic systems; ρ and σ are the measures of sensitivity of the reaction rate constant to changes electronic and steric properties respectively. The Hammett equation applies to m- or p-substituted aromatic systems where only electronic effects are important, whereas the Taft equation is useful for many classes of compounds in which steric effects at the reaction center also play an important role (Chapman and Shorter,1972).

SARs for hydrolysis sometimes relate $\log k_x$ to the pK_a of the leaving group (Brønsted equation) when loss of X is part of the rate-controlling step. Thus the ease of loss of -X is related to the acidity of its conjugate acid HX and the QSAR equation takes the form

$$\log k_x = pK_a + \log k_o \qquad (4.43)$$

Libraries of σ, σ^*, and δ constants are available for well over 500 substituents (Exner, 1978; McPhee et al., 1978; Perrin et al., 1981); where no value has been published it is sometimes possible to estimate the value from similar substituents. Hydrolysis SARs can be used to estimate values of k_A, k_B, or k_N only for related compounds which react by the same reaction path. The limited database of kinetic data usually found for a specific class of compounds, (ten values is not uncommon), limits reliable SAR-estimated values of k_B or k_A to those new compounds which have structural features most closely resembling the compounds used to form the original correlation.

SARs generally fail because (1) they are used to correlate rate constants for a reaction that has a change in reaction mechanism on a change in substitution, and (2) they are used to predict a constant for a compound with structural features not represented in the database used to develop the SAR. This latter point is quite subtle and is often difficult to evaluate. The empirical nature of current SARs clearly limits their range of applicability to those chemical structures for which statistically significant relations have been established.

SARs for some important classes of hydrolyzable compounds are listed below.

Carboxylic Acid Esters. Although aliphatic esters undergo acid, base, and neutral hydrolyses, only OH⁻-promoted hydrolysis exhibits significant variations with changes in structure of the acyl group, and the largest database for aqueous hydrolysis constants is for k_B. Drossman et al. (1988) developed several Taft and Hammett correlations for k_B for alkyl and aromatic esters. A master SAR for 125 esters of the general structures

$$RC(O)OR' \text{ or } X_1PhC(O)OPhX_2$$

follows the relation

$$\log k_B = 2.10\sigma^*_R + 2.27\sigma^*_{R'} + 2.13\sigma_{X1} + 1.25\sigma_{X2}$$
$$+ 0.81E_{sR} + 0.33E_{sR'} + 2.59 \qquad (4.44)$$

$$r = 0.979 \qquad r^2 = 0.959 \qquad n = 125$$

R and R′ represent both alkyl and aromatic (Ph) groups . The fact that ρ^* values for R and R′ are similar, but ρ values for X_2 are much smaller than X_1 may be due to the greater distance of X_2 from the reaction center. Steric effects in R or X_1Ph become more significant as the size of the companion alcohol group (R′) increases. However, the value of δ for the alcohol R′ is small.

Haloalkanes. The SAR equation for calculating $k_B(E)$ for elimination of HX from halogenated ethanes and higher homologs is (Haag et al., 1989):

$$\log k_B(E) = 2.09\Sigma\sigma^*(1\text{--}3) + 3.20f_X - 15.84 \qquad (4.45)$$

$$r = 0.99 \qquad r^2 = 0.98 \qquad n = 7$$

The halogen factor f_X accounts for halogen leaving preference, and is defined as the average of $\log (k_NX/k_NF)$ and $\log (k_BX/k_BF)$ for hydrolysis of CH_3X versus CH_3F; f_X factors are: F = 0.00; Cl = 1.33; Br = 2.60, and I = 2.02.

Epoxides. the best SAR for aliphatic epoxides is

$$\log k_A = -2.15\Sigma\sigma^*(1\text{--}4) + 0.36\Sigma E_s(1\text{--}4) + 1.02C_0 - 1.77 \qquad (4.46)$$

$$r = 0.89 \qquad r^2 = 0.80 \qquad n = 14$$

where 1–4 refers to all substitution positions on the oxirane ring and C_0 is 1 for cyclic epoxides and 0 for alicylic epoxides.

Aromatic Carbamates. Carbamates undergo acid, base, and neutral hydrolyses. Rate constants for the base-promoted reaction can be fitted to two SAR equations for k_B for carbamates (1) $X_1R_1NR_2C(O)OR_3X_2$, (2) $X_1R_1NHC(O)OR_3X_2$

With no hydrogen on nitrogen : (1)

$$\log k_B = 7.99\sigma^*_{R3} + 0.31\sigma_{X2} + 3.14\Sigma E_{sR1,R2} + 0.442 \qquad (4.47)$$

$$r^2 = 0.903 \qquad n = 18$$

With one hydrogen on nitrogen : (2)

$$\log k_B = 2.39\sigma^*_{1,2} + 0.96\sigma_{X1} + 7.97\sigma^*_3 + 2.81\sigma_{X2} - 0.275 \qquad (4.48)$$

$$r^2 = 0.973 \qquad n = 62$$

Acetals and Ketals. Cordes (1967) lists Hammett and Taft SARs for acid-catalyzed hydrolysis of acetals and ketals. One of the better SARs covers a range of k_A values of 10^7.

$$\log k_A = 3.6\sigma^* + \log k_{Ao} \qquad (4.49)$$

Hydrolysis Properties of Natural Waters

Temperature and pH

The two most important properties of natural waters that affect hydrolysis rates are temperature and pH, both of which may vary seasonally and spatially. Laboratory hydrolysis experiments are usually conducted at 25°C to facilitate intercomparison among different data sets and investigators. However, most natural waters have average spring/summer temperatures close to 15°C, while most marine systems are closer to 5°C (Mabey and Mill, 1978; Mill and Mabey, 1988). This means that rate constants and half-lives estimated from laboratory measurements need to be corrected to these average environmental conditions where rates will be slower by factors of three to fifteen. On the other hand, marine systems have nearly uniform pH values near 8.2, which means that rates of hydrolysis for several classes of compounds subject to base-promoted reactions (see Figure 4.2a-d and text) will be up to 250 times faster in seawater than in freshwater at 25°C, and perhaps 40 times faster even at 5°C.

Catalytic Anions and Metal Ions

Many natural waters contain a variety of anions and heavy metal cations that might catalyze hydrolysis of certain types of organic compounds under some circumstances. Dissolved inorganic anions such as carbonate, phosphate, and sulfides or polysulfides can catalyze hydrolysis through general base catalysis. However, Mabey and Mill (1988) and Haag and Mill (1988b) evaluated the importance of these common anions on substitution reactions of alkyl halides and concluded that only polysulfides at maximum natural abundances could compete with water in most reactions. Similarly, chloride ion at typical marine concentrations of 0.3 M has hardly any effect on even the most susceptible halogenated alkanes (Mill and Mabey, 1985).

Metal ions such as Ca^{2+}, Cu^{2+}, Co^{2+}, and Zn^{2+} which are widely distributed in natural waters are known to be hydrolysis catalysts for many classes of hydrolyzable chemicals (Hoffman, 1980). However, Drossman et al. (1991) found that the Zn-catalyzed hydrolysis of 2,4-dinitrophenyl acetate at pH 7 at an environmentally relevant Zn concentration of 1×10^{-6} M contributes less than 1% of the observed rate. Similarly, Mg- and Zn-catalyzed hydrolyses of phosphate esters (investigated by Steffans et al. (1973)) account for only 0.06 to 0.5% of the total rate at 1 mM of metal ion. Drossman et al. also measured possible catalytic effects of 0.1 mM Cu, Ca, and Zn ions on hydrolysis of several esters, amides, nitriles, and phosphates which bind metals. Only 2-ethyl picolinate with Cu^{2+} showed a significant catalytic effect, which 10 ppm added humic acid largely inhibited. They concluded from these results that most hydrolyzable molecules will not exhibit metal-ion catalysis in natural waters, where most metals are bound to humic acids. Similar results were found with halogenated alkanes in sediments (see below).

In short, rate constants measured in buffered, distilled water may be used confidently to predict rates in natural waters, after temperature and pH effects are taken into account.

Effects of Sorption on Hydrolysis Rates

The presence of bottom sediments in almost all natural waters means that many chemicals will be partly sorbed to sediment. Other chemicals applied to soils or in groundwater sediments are almost wholly sorbed. In a slurry of sediment and water the concentration of chemical in the water, C_w, is given by

$$C_w = C_T/(1 + \rho K_p) \qquad (4.50)$$

where C_T is total chemical concentration, ρ is the sediment to water ratio, and K_p is the ratio of C_s/C_w and is related to K_{OC}. If hydrolysis occurs only in the water phase (rate constant k_w)

$$k_{obs} = k_w/(1 + \rho K_p) \qquad (4.51)$$

Macalady and Wolfe (1984) showed that for several organophosphorothioates, as well as alkyl bromides (S_N2), benzyl chloride (S_N1), and hexachlorocyclopentadiene (S_N2'), water-promoted hydrolysis occurs as rapidly in the sediment as in the water phase, and k_h is unchanged by the sediment phase. At high pH, however, 2,4-dichlorophenoxyacetate esters react very slowly in the sorbed state because sediments form negatively charged carboxylate ions, which inhibit the base-promoted process.

Hydrolysis in soils and subsurface sediments (groundwater) has been investigated by Saltzman et al. (1976), El Amamy and Mill (1984), and Haag and Mill (1988a). The situation is complicated by a pronounced increase in surface acidity in clay soils with low water content. On montmorillonite and kaolinite containing between 3% and 20% water, acid-catalyzed hydrolysis of epoxides and phosphorus esters is accelerated with rate constants corresponding to a surface pH two units lower than the bulk medium pH (Saltzman et al., 1976; El Amamy and Mill, 1984). Moisture contents in excess of 25% do not cause acceleration. In real soils where some chemicals may partition to organic carbon and some to clay surface, predictions of hydrolysis rates will be uncertain. However, Haag and Mill (1988a) found that rates of hydrolysis of several halogenated alkanes in subsurface sediments were identical with those found in pure water, greatly simplifying predictions of hydrolysis rates in groundwater environments.

Photolysis

Introduction

Sunlight photolysis (or photoreaction) of organic chemicals occurs in surface waters, on soil, and in the atmosphere, sometimes rapidly enough to make photolysis the dominant environmental transformation process for some compounds. Indeed, photooxidation is the dominant loss process for more than 90% of the organic compounds in the troposphere. Not only do photoreactions control the fate of many chemicals in air and water, but these reactions often create oxidation products that are more water soluble, less volatile and less subject to bio-uptake than the parent compounds.

The key parameters to be measured or estimated for environmental photoreactions are the rate of photon absorption and the process efficiency (quantum yield) in water or air at wavelengths above 300 nm in the solar spectrum. Although it is not necessary that a compound absorb light in this region to undergo a photoreaction (see below), direct photoreaction does require absorption of a solar photon by the compound of interest.

In addition, measurement or estimation of a photoreaction rate in the environment requires a description of sunlight intensity as a function of wavelength as well as its temporal and spatial variability. These complex topics are treated in detail by Zepp and Cline (1977), Liefer (1988), Mill and Mabey (1985), Calvert and Pitts (1963), Atkinson (1986), and Finlayson-Pitts and Pitts (1986) for photolysis in surface water and in the atmosphere.

The sunlight spectrum at the earth's surface extends from 290 nm to well beyond 1500 nm. An organic compound absorbs and can be changed directly by sunlight only in those wavelength regions coinciding with UV-visible absorption bands in the spectrum of the compound. Different classes of molecular structures absorb sunlight in different wavelength regions and with different intensities, but usually below 350 nm. Figure 4.3 illustrates the overlap of the solar emission spectrum with the UV absorption spectra of several types of organic compounds that absorb in the solar region. The amount of light absorbed at wavelength λ is proportional to the absorption cross section, ϵ_λ.

$$\epsilon_\lambda = A_\lambda/[C] \times 1 \qquad (4.52)$$

A_λ is the measured absorption at wavelength λ and 1 is the pathlength of the spectral cell in cm; when concentrations are molar, ϵ_λ has units of M^{-1} cm^{-1}.

Table 4.4 lists sunlight intensities (L_λ in units of millieinsteins/cm^2 day) and ϵ_λ for p-nitroanisole (PNA), as 24-hour time-averaged intensities at the surface of natural waters within narrow wavelength regions for 40° latitude in summer. The term $L_\lambda \epsilon_\lambda$ is the rate of absorption of photons by PNA (in units of day^{-1}), and is the basis for estimating the rate of photolysis of PNA. L_λ values are available for all decadic latitudes and seasons, making it possible to estimate the light absorption rate for any compound whose spectrum is well known above 300 nm, at any location, all year, under clear sky conditions. Atmospheric equivalents of L_λ and e_λ are J_λ and σ_λ, also with units of day^{-1} for the product $J_\lambda \sigma_\lambda$ (Mill and Mabey, 1985).

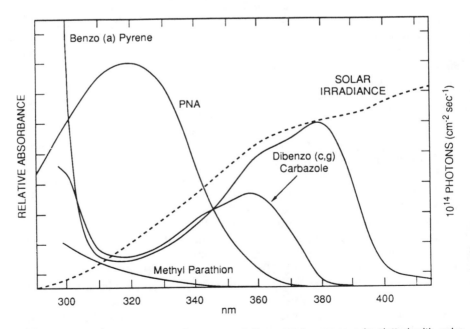

Figure 4.3. Absorption spectra for representative organic compounds plotted with solar irradiance.

Table 4.4. L_λ, ϵ_λ, and $\epsilon_\lambda L_\lambda$ for PNA

Wavelength Center, nm	L_λ, summer 40° Millieinsteins cm^{-2}d^{-1}	ϵ_λ, PNA, M^{-1} cm^{-1}	$\epsilon_\lambda L_\lambda$, d^{-1}
297.5	6.17(−5)	8076	0.498
300	2.69(−4)	8480	2.3
302.5	8.30(−4)	9030	7.5
305	1.95(−3)	9350	18.2
307.5	3.74(−3)	9790	36.6
310	6.17(−3)	10000	61.7
312.5	9.07(−3)	10260	93.0
315	1.22(−2)	10350	126
317.5	1.55(−2)	10370	161
320	1.87(−2)	10330	193
323.10	3.35(−2)	10120	339
330	1.16(−1)	9210	1070
340	1.46(−1)	7165	1050
350	1.62(−1)	4906	795
360	1.79(−1)	2968	531
370	1.91(−1)	1618	309
380	2.04(−1)	749	153
390	1.93(−1)	288	55.5
400	2.76(−1)	87	24.0
410	3.64(−1)	15	5.5
420	3.74(−1)	7	2.6
430	3.61(−1)	5	1.8
440	4.26(−1)	0	0.00

$$\Sigma \epsilon_\lambda L_\lambda = 5036 \text{ d}^{-1}$$

Because sunlight intensities, L_λ, are averaged over 24 hours and over each 3-month season, the accuracy of rate estimates will depend on how slow or rapid the photolysis process is, compared with the averaging interval. For example, a compound which photolyzes with a half-life of 1 hour will photolyze three times faster at solar noon than at 800 or 1600 hrs. However, if the half-life is one week, diurnal variations will make little difference.

Average winter sunlight intensity is only one-fourth to one-sixth of summer intensity, depending on latitude. About one-half of total solar intensity is emitted directly from the sun; the other half comes from scattered sky light so that only in heavy cloud cover does the intensity drop significantly. In high latitudes, reduced photolysis rates under persistent cloud cover can be corrected for by multiplying clear sky rate constants by the average percent reduction in light values caused by clouds.

Two Kinds of Photoreactions

Environmental photoreactions fall into two broad categories: direct and indirect. A direct photoreaction occurs when sunlight is absorbed by a chemical and the photon energy is used to form excited or radical species, which react further to form stable products. Only a small fraction of all organic compounds absorb sunlight and photolyze at significant rates. Most aliphatic compounds and simple benzene derivatives with alkyl groups or one heteroatom substituent do not absorb sunlight (Calvert and Pitts, 1963; Turro,

1967). Important exceptions are ketones and aldehydes. Generally, nitro, azo and polyhalogenated benzenes, naphthalene derivatives, polycyclic aromatics and all aromatic amines, and all ketones and aldehydes absorb sunlight somewhere between 300 and 450 nm; some polycyclic and azoaromatics (dyes) also absorb visible light beyond 700 nm. Most aromatic compounds have maxima in the 250–270 nm spectral region, with only weakly absorbing tails extending into the solar region (Calvert and Pitts, 1963).

Indirect photolysis is a more complex process because chemicals can be transformed under conditions where they absorb little or no sunlight by reactions with intermediate oxidants formed during photolysis of dissolved organic matter (humic acid, DOM) in water or soil or photolysis of ozone or NO_2 in the atmosphere. Most indirect photoreactions in water depend on relatively selective oxidants such as singlet oxygen or peroxy radicals. As a result, only phenols, furans, aromatic amines, sulfides, and nitro aromatics undergo indirect photolysis in water (Mill and Mabey, 1985). Some of these compounds also photolyze directly. By contrast, HO radical dominates atmospheric photoxidations, causing rapid oxidation of even alkanes, olefins, alcohols, and simple aromatics (Atkinson, 1986; Atkinson et al., 1987a).

Direct Photolysis: Kinetic Relations and Measurement of ϕ and k_p

Direct photolysis occurs only if light is absorbed by the chemical. The rate of a direct photoreaction (R_p) depends on the product of the rate of light absorption (R_A) and the efficiency (quantum yield, ϕ) with which the absorbed light causes a reaction [ϕ is defined as the ratio of moles of chemical transformed to moles (einsteins) of photons absorbed].

$$R_P = \phi R_A = \phi k_A[C] = k_P[C]$$

If the chemical is present in water or air at a low concentration, R_A in sunlight can be calculated from the sum of the product of ϵ_λ, at each wavelength λ where $\epsilon_\lambda \neq 0$, and sunlight intensity in the same wavelength interval (L_λ in millieinsteins/cm^2d).

$$R_A = \Sigma L_\lambda \epsilon l \, [C] = k_A[C] \tag{4.53}$$

and the photolysis rate is

$$R_p = \phi \Sigma L_\lambda \epsilon_\lambda l \, [C] = k_P[C] \tag{4.54}$$

L_λ is averaged over each 24 hour period, making it possible to integrate Equation 4.54 as a simple first-order equation.

$$\ln[C_0/C_t] = k_P t \tag{4.55}$$

where C_0, C_t, and k_P and t have the usual meaning, and since k_P is in units of days^{-1}, the half-life of the reaction (in days) is

$$t_{1/2}(\text{days}) = \ln 2/k_P \qquad (4.56)$$

Values of ϵ_λ in water are easily calculated for most compounds from their UV spectra in water or acetonitrile, and L_λ values are available from tables (Mill and Mabey, 1985; EPA 1985c; Liefer, 1988). Thus, R_A values can be calculated with a minimum of experimental work. If the value of ϕ is known, Equation 4.54 can be used with appropriate values of L_λ and ϵ_λ to estimate k_P at any other location or season.

Several methods are used to measure sunlight photolysis rate constants and ϕ in water. The EPA protocol (EPA, 1985c) calls for use of dilute solutions of a compound in water which is exposed to sunlight for several hours, days, or weeks and monitored for loss of the compound. Regression of $\ln (C_o/C_t)$ versus time will give a straight line only if rates are very fast or slow compared to diurnal variations in light intensity, and if nighttime periods are ignored. The slope of the plot is equal to k_{PS}, the photolysis rate constant in sunlight. From k_{PS} the apparent half-life of the chemical may be estimated from Equation 4.56. This method is limited to measurements conducted under clear skies in above-freezing temperatures. One can calculate ϕ from this experiment by combining the L_λ values for the location and season with ϵ_λ values for the compound.

$$\phi_s(C) = k_{PS}/(\Sigma L_\lambda \epsilon_\lambda l) \qquad (4.57)$$

Implicit in this relation is the assumption that ϕ does not change with wavelength, a reasonable assumption for many compounds. Even if ϕ_s changes with λ, this value is an average sunlight value.

A more complicated procedure to measure ϕ uses compound A with a known value of ϕ at some solar wavelength(s) (called an actinometer) to monitor the sunlight to which compound C is exposed, thus obviating the need for clear skies. A value for $\phi(C)$ is derived from the relation (Dulin and Mill, 1982)

$$\phi(C) = S\phi(A)k_{AA}/k_{AC} \qquad (4.58)$$

where S is the slope of the log-log plot of (C_o / C_t) versus (A_o / A_t); $k_{AA} = \Sigma L_\lambda \epsilon_\lambda(A)l$ and $k_{AC} = \Sigma L_\lambda \epsilon_\lambda(C)l$. If $\phi(C)$ does not change much with wavelength, the values from Equations 4.57 and 4.58 will agree. Table 4.5 shows a comparison of measurements and calculations for k_p for four compounds which photolyze in sunlight; an actinometer was used with selected wavelengths to calculate ϕ, and $t_{1/2}$ was estimated from Equation 4.56. Good agreement was found in all cases except for biacetyl, which absorbs light in several different spectral regions (hence has a large value of $\Sigma L_\lambda \epsilon_L$), but photolyzes only in one band.

Atmospheric photoprocesses can be treated in a similar manner to those in water by using bulbs containing the vapor of the chemical in air (at 1 atm), and

Table 4.5. Comparison of Calculated and Measured Half-Lives for Photolysis in Sunlight[a]

Chemical	$t_{1/2}$,[d] (calc.)	$t_{1/2}$,[d] (meas)
Dibenzothiophene	9.1[b]	6.4
Nitrobenzene	19.8[b]	19.3
p-Methoxyacetophenone	11.9[c]	12.0
2,3-Butanedione	0.028[c]	0.35

[a]Data of Mill et al. (1984)
[b]Spring, 40°L, for horizontal water surface.
[c]Summer, 40°L, for horizontal water surface.

an actinometer can be included to measure light intensity, but the methodology is limited to organics with > 0.1 torr vapor pressure ($\sim 5 \times 10^{-6}$ M). No standard protocol for this measurement is available at this time.

An example of the calculation method for k_p is shown in Table 4.4 for PNA. When $\epsilon_\lambda L_\lambda$ values are totaled, the sum ($\Sigma \epsilon_\lambda L_\lambda$) is equal to k_A from Equation 4.53: 5036 day^{-1}. The quantum yield for direct photolysis of PNA in water is 2.8×10^{-4} (Dulin and Mill, 1982), and when multiplied by 5036 day^{-1}, the product is k_p, 1.4 day^{-1} in summer and the half-life is about 12 hr. This is a photoreactive compound and the 24-hr averaged L_λ values are ill-suited for accurate estimates for this rate constant. Nonetheless, this estimate shows that PNA will disappear rapidly during most daylight conditions.

Direct Photolysis-Estimation Methods

The only estimation method available for the direct photolysis rate constant is for the theoretical maximum value of k_P. This value may be estimated for any compound from Equation 4.54 by assuming that ϕ is equal to one (100% efficiency). If the maximum calculated value is smaller than the hydrolysis, biodegradation, or volatilization rate constants, then direct photolysis cannot be an important process and need not be tested further.

Values of ϵ_λ cannot be calculated, but acceptable estimates of ϵ_λ can be made, based on the similarity of spectra for the same chromophore in other compounds. For example, ϵ_λ and λ_{max} values for most aliphatic or aromatic ketones are very similar (Calvert and Pitts, 1963). However, the near universal availability of modern UV-vis spectrometers makes acquisition of precise values of ϵ_λ a simple procedure in most cases (EPA, 1985c), from which the maximum value of k_p can be calculated for screening purposes.

No reliable estimation procedure is available for ϕ because its value depends on a competition among several activation and deactivation steps for excited state species, few of which are known with any certainty for a few compounds. Mill and Mabey (1988) have listed several ϕ values for several types of photoreactions in aerated water; Calvert and Pitts (1963) have extensive compilations for a variety of vapor phase reactions, some of which were conducted in the presence of air. Air is necessary in these measurements because oxygen quenches photoreactions and its absence from an experiment may cause the measured ϕ value to be higher than is likely to be found in the environment.

Table 4.6. Environmental Oxidants[a]

Oxidant	Average Concn.
Atmospheric Oxidants	
HO·	0.01–0.4 ppt
O_3	0.02–0.5 ppm
NO_3	<1–400 ppt
NO_2	20 ppb–0.5 ppm
Surface Water Oxidants	
$^1O_2$	1×10^{-13}M
RO_2·	1×10^{-10}M
HO·	1×10^{-16}M

[a]From Mill 1989.

And since values of ϕ may vary by up to a thousandfold, even in a series of related compounds, generalizations from these tabulations are not very useful. In the environment, ϕ is almost always less than one, often less than 0.1, and not rarely less than 0.001.

Indirect Photolysis Experimental Methods

Indirect photoreactions occur when a photon is absorbed by DOM in water or NO_2 in the troposphere to form transient oxidants such as singlet oxygen or oxyradicals, which then further react with chemicals present in the air or water. Rate laws for indirect photoreactions follow simple bimolecular kinetics. The rate expression includes the steady-state concentration of an oxidant [Ox], the concentration of the chemical, [C], and the rate constant for the reaction, k_{ox}, which is determined by the molecular structure of the oxidant and chemical, the temperature, and phase (water or air). Several different oxidants are usually generated concurrently by insolation of DOM or NO_2. Thus, the total rate of indirect photolysis (R_{PI}) in water is the sum of the second-order rate expressions for each oxidant.

$$R_{PI} = \Sigma k_{ox(i)}[Ox(i)][C] \qquad (4.59)$$

where Ox(i) is the ith oxidant and the expression is summed over all oxidants. Usually only one or two oxidants control the rate of oxidation of a specific chemical in water (Mill, 1988), whereas HO radical dominates the oxidation process in the troposphere (Atkinson 1986). The concentrations of each oxidant varies with L_λ over a 24-hr solar cycle, and also with season or latitude. Seasonally averaged values of [Ox] at 40° or 50° latitude are listed Table 4.6.

With average values of [Ox], Equation 4.59 reduces to a first-order expression

$$R_{PI} = \Sigma k_{ox(i)}[Ox(i)] [C] = k_{pI}[C] \qquad (4.60)$$

$$t_{1/2} = \ln2/k_{pI} \qquad (4.61)$$

The availability of values of [OX] (Table 4.6), together with published values of k_{ox} for HO radical (Atkinson, 1986), singlet oxygen (Monroe, 1985), and oxyradicals (Hendry et al., 1974) makes it possible to calculate rate constants for indirect photolysis (k_{pI}) for many kinds of chemicals as part of an estimation process for indirect photolysis. A simple experimental procedure is available for evaluating k_{pI}, the rate constant for indirect photolysis in water (EPA, 1988). The chemical is photolyzed in pure water solution to measure k_P, and in pure water containing a specific amount of DOM to absorb most of the UV light near 300 nm. If the chemical photolyzes more than twice as fast in the DOM water as in pure water under the same conditions in sunlight, the indirect process is important enough to warrant more careful measurements using actinometers.

Measurement of the rate constant for the indirect photolysis of a compound in the troposphere is almost as simple as in water because HO radical is the dominant oxidant and measurement of k_{HO}, coupled with a value for [HO] (Table 4.6; Singh, 1977), gives k'_{PI} directly for over 90% of atmospheric organic compounds. A laboratory method for measuring k_{HO} involves competing two chemicals, one with a known value of k_{HO} (S) and the other with an unknown value (C), for gas phase HO· radical generated by photolysis of several torr of methyl nitrite in the presence of air and NO. Loss of S and C are followed with time to give $k_{HO(C)}$ from the relation

$$k_{HO(C)} = k_{HO(S)}\{\ln[(C_0)/(C_t)]/\ln[(S_0)/(S_t)]\} \qquad (4.62)$$

To estimate the half-life of a compound in the troposphere, an average HO concentration of $\approx 1 \times 10^6$ molecules per cm³ (1.6×10^{-15} M) (Singh, 1977) is used in Equation 4.60 (k_{pI}[HO]). This method is useful for the range of reasonably volatile compounds for which more than one torr of vapor is readily obtained. Although almost all common volatile organics have been measured by more accurate methods (Atkinson, 1986), this simple competitive procedure is applicable to any unusual new volatile compounds

Very nonvolatile compounds, such as dioxins or PCBs plus the vast majority of other organics, cannot be treated this way and alternate experimental procedures have been devised, based on HO oxidation in water or fluorocarbons. The methods seems to work well for aliphatic compounds, but not as well for aromatic compounds (Dilling, et al. 1986).

Indirect Photolysis Estimation Methods

Bimolecular rate constants for oxidation by oxyradicals and singlet oxygen have been measured in numerous studies and have been compiled and used to develop accurate structure activity relationships (SARs) for photooxidations. Mill (1989) has recently summarized SARs available for photooxidants in water and air, while Atkinson (1987) has developed and published an extensive compilation of fragment constants to estimate k_{HO} for many kinds of organic

Table 4.7. Half-Lives for Organic Compounds in Indirect Photooxidations in Air and Water[a]

Class	HO·	O_3	RO_2·	$^1O_2$
		Atmosphere		
Alkanes	200 hr	$>10^6$ y	—	—
Alcohols	60 hr	$>10^6$ y	—	—
Aromatics	60 hr	$\sim 10^3$ y	—	—
Olefins	20–60 hr	< 20 y	—	—
		Surface Water		
Phenols	<10 hr	—	20 hr	5 hr
Aromatic amines	<100 hr	—	20 hr	10 hr
Furans	~ 250 hr	—	100 hr	10 hr
Hydroperoxides	~ 250 hr	—	20 hr	100 hr
Polycyclic aromatics	—	—		50 hr

[a]Based on atmospheric or aqueous concentrations from Table 4.5 and rate constants from Mill 1989.

structures. Table 4.7 lists the estimated half-lives for major classes of organics in the atmosphere and surface water toward these different photooxidants. The data clearly show that HO is the dominant tropospheric oxidant, but that only a few classes of compounds undergo indirect photooxidation in water and no one oxidant is dominant.

It is important to realize that even if k_{PI} values are measured or estimated accurately (within a factor 2 or 3), average oxidant concentrations in the environment (Table 4.4) are accurate to no better than a factor of 5 to 10 for any given location. In extreme locations, such as pristine marine waters, marine air masses or heavily polluted urban air sheds, oxidant concentrations may be 100 time smaller or larger than values listed in Table 4.6.

Estimation of indirect photoreaction constants for water is simplified by the fact that only a few classes of organics, notably phenols, aromatic amines, furans, alkyl sulfides, and some azo compounds are rapidly oxidized by oxyradicals or singlet oxygen. SARs for these reaction classes take the form of Hammett and Taft relations similar to those discussed for hydrolysis reactions (Mill 1989).

Substituted phenol and amine rate constants for oxidation by RO_2 radical are correlated by several SARs of the type

$$\log k_{ox} = \rho^+ \sigma^+ + C \tag{4.63}$$

In this equation, σ^+ is similar to σ but gives a better fit of the data, and ρ^+ has values ranging from -0.9 to -1.5. Azophenol dyes are much more rapidly oxidized as phenol anions by both radicals and singlet oxygen; the SAR accounts for this reactivity as a function of the fraction ionized, α

$$k_{ox} = 1.3 \times 10^7 (1 - \alpha) + 1.8 \times 10^8 \alpha [Ox] \tag{4.64}$$

Reactions of both singlet oxygen and oxyradicals are accounted for by this SAR (Haag and Mill 1988). Values of α can be calculated from the pK_as of the phenols, which in turn are derived using well-developed SARs (Perrin et al. 1981).

SARs for reactions of singlet oxygen with phenols, styrenes, furans, t-butylphenols, some sulfides, and aromatic amines also follow simple Hammett or Taft relations (Mill 1989).

Atkinson (1987a) has used the large corpus of accurately measured k_{HO} values to develop a fragment additivity SAR for k_{HO} for most kinds of aliphatic compounds, based on additivity of molecular structure. Reactions of HO radical with aromatic compounds are usually close to 5×10^9 M^{-1}s^{-1}, and are better correlated using a modified Hammett relationship with σ^+ (Atkinson, 1987b). Atmospheric HO radical is so reactive that a generalization that all aromatic k_{HO} constants are $\sim 5 \times 10^9$ M^{-1} s–1 is accurate enough for most aromatics within a factor of 3.

Photolysis Products in Water and Air

Pathways or products in direct photolysis depend on how absorbed photon energy is used to break bonds. UV sunlight has enough energy to rearrange or cleave carbonyl double bonds, carbon-halogen, carbon-nitrogen, some carbon-carbon bonds, and peroxide O-O bonds, but not enough to cleave most carbon-oxygen or carbon hydrogen bonds. Thus, many alkyl and aromatic compounds lose halogen, nitro, or O-atoms from nitro groups to form free radicals, while excited ketones and aldehydes rearrange C-H bonds or lose CO. Photooxidation generally involves singlet oxygen or free oxyradical reactions. By either pathway, oxygen becomes incorporated into the organic structure as carbonyl or peroxide products. Other kinds of photoproducts have been discussed in detail by Turro (1967) and de Mayo (1980).

Indirect photolyses in water and air often lead to peroxides, acids, ketones, or alcohols. Classic examples in water include formation of diacetylethylene from dimethylfuran with singlet oxygen (Zepp et al., 1977) and cumene hydroperoxide from cumene with peroxy radicals (Mill et al. 1980). HO radical oxidation in air, on the other hand, does cause deep-seated cleavage of aromatics to form simple dicarbonyls by as yet unknown pathways (Atkinson, 1986).

$$Me_2FURAN + {}^1O_2 \rightarrow MeCOC=CCOMe + H_2O_2 \qquad (4.65)$$

$$C_6H_5CHMe_2 + O_2 \rightarrow C_6H_5CMe_2OOH \qquad (4.66)$$

$$C_6H_4Me_2 + HO + O_2 \rightarrow MeCOCOMe + ? \qquad (4.67)$$

Environmental Effects on Photolysis

Temperature and pH effects. Unlike hydrolysis reactions, direct and indirect photoreactions generally show little effect of changing temperature by 20 to 40°C or pH by 3 to 6 units (Haag and Hoigné, 1986). HO radical oxidation rates increase by a factor of only two for a 40° temperature increase near 25°C (Atkinson 1986).

Surfaces. All environmental surfaces such as soil or sediments, aerosols, fly-ash, algae, and cloud drops can serve as media for several kinds of photoprocesses. The whole range of photoreactions found in solution have been observed on particulates as well. By and large, surface reactions are restricted to those compounds that sorb strongly to surfaces, either by partitioning or by site-specific sorption processes. Both singlet oxygen and free oxyradicals have been identified on isolated soil by their reactions with sorbed organics (Gohre and Miller 1986). Algae-sensitized photooxidation of anilines (Zepp et al., 1985), photooxidations of metal oxide-organic complexes (Waite, 1986) and photolysis rates and pathways of flyash-sorbed polycyclic aromatics and chlorodioxins (Fox and Olive, 1979) also have been reported. However, the heterogeneity, variability, and complexity of particulate surfaces makes it difficult to generalize about the rates and pathways to be expected.

Transition metals. Iron and manganese, often in the form of colloidal high valent oxides or organic acid complexes are ubiquitous in air and water. Photolysis of these oxides and ions leads to direct and indirect oxidation of organics with reduction of the metal ion (Waite, 1988).

$$Fe\ III + H_2O \quad \rightarrow \quad Fe\ II + HO\cdot + H^+ \qquad (4.68)$$
$$O_2 + HO\cdot + RH \quad \rightarrow \quad H_2O + RO_2 \qquad (4.69)$$
$$Fe\ III\ CO_2R \quad \rightarrow \quad Fe\ II + RCO_2\cdot \qquad (4.70)$$
$$O_2 + RCO_2\cdot \quad \rightarrow \quad CO_2 + RO_2\cdot \qquad (4.71)$$

In marine systems, Mn IV is rapidly photoreduced to the bioavailable Mn II and reoxidized biologically (Sunda et al., 1983). Iron and manganese also may catalyze oxidations in cloud droplets (Martin 1984).

Marine systems. The oceans are the largest receptors for solar protons and, although open ocean waters are strikingly clear for up to 30 meters because of the very low concentrations of DOM, all photons are eventually absorbed. Apart from the obvious decreases in rates of indirect processes caused by low DOM concentrations, this high pH (8), high salinity (3% salt) medium has few effects on direct or indirect photoreactions. Mopper and Zhou (1990) recently reported evidence that indirect photoprocesses involving HO· radical may be responsible for a significant fraction of simple carbonyls, such as pyruvate and acetaldehyde, formed in marine systems as part of the natural turnover of marine DOM, a process with a half-life of up to 20,000 years!

Biotransformations

Introduction

Biotransformation by bacteria, protozoans, and fungi are the major transformation processes for natural organic compounds in water and soil, converting complex molecules such as starches, lipids, and polypeptides to CO_2, H_2O, nitrate, and phosphate by means of enzymes evolved over millions of years. However, synthetic chemicals may not be as readily biodegraded unless their structures are closely related to naturally occurring organic structures. The recent enthusiasm for biodegradable polymers has prompted renewed interest in synthesis of polymers which structurally resemble natural polymers such as cellulose. Indeed, the widespread persistence of many synthetic organic chemicals has slowly dispelled that idea that all organic compounds are subject to microbial degradation.

Biotransformation is the least understood of environmental transformations because microorganisms have complex life cycles and can effect many kinds of chemical reactions including hydrolysis, oxidation, and reduction through a wide variety of enzymes and biochemical processes. This section presents only a brief overview of the environmental aspects of biotransformation. Recent expert reviews include those of Klecka (1985), Alexander (1980), and Ribbons and Eaton (1982).

General Features of Biotransformation

The ease or rapidity with which synthetic chemicals are biotransformed ranges from (a) immediate utilization for growth or energy by enzymes capable of transforming the chemical; (b) utilization and transformation only after a lag of hours to weeks; (c) transformation only by cometabolism with chemicals or nutrients being utilized for growth; and (d) inability to biotransform, owing to an unusual structure or inaccessibility of the structure to the microorganisms.

Bacteria, protozoans, and fungi are suspended in surface waters either freely or sorbed to suspended particles and on soil or sediment particles, often to great depths. While many organisms utilize oxygen as the ultimate electron acceptor in metabolism, anaerobic organisms use electrophilic substrates such as sulfate or nitrate; some organisms can use both oxygen and inorganic salts as oxidants. Generally aerobic processes are faster than anaerobic processes, possibly because larger numbers of aerobic organisms are typically present in surface waters; growth patterns are more rapid and energy production through ATP production is greater.

The details of microbial hydrolytic and redox processes have been reviewed extensively by the National Academy of Science (NAS 1972). Hydrolytic enzymes act on the same bonds cleaved abiotically by acid and hydroxide ion: esters, amides, carbonates, and phosphates (Mabey and Mill, 1978); some alkyl halide also are hydrolyzed but many other C-X bonds are reductively cleaved

$$R_3 \text{ C-X} + \text{e(Biol)}^- \rightarrow \text{R} \cdot + \text{X}^- + \text{(Biol)} \qquad (4.72)$$

Oxygenation and oxidation (insertion of one or two oxygen atoms into a molecule) is a common pathway for metabolism of aromatic rings and carbon hydrogen bonds in n-alkanes, analogous to reactions of hydroxyl radical with aromatics and alkanes (Atkinson, 1986).

Biotransformations of chemicals typically leads stepwise to more oxygenated products, which in turn are more rapidly oxidized, eventually resulting in "mineralization" — the conversion of organic carbon to CO_2. This sequence is uncommon in abiotic processes, except perhaps for HO radical oxidations in the atmosphere (Atkinson, 1986). Not all metabolic or biotransformation processes lead to CO_2, however. Thus, measuring the loss of the starting compound in any biotransformation rate study is of vital importance to developing reliable rate relations and structure activity relations.

Biotransformation Kinetic Relations

The Monod equation, the usual starting point for describing biotransformation kinetic relations, describes cell growth rate in terms of substrate (chemical) utilization

$$d[B]/dt = \mu[B] = \mu_{max} [B] \frac{[C]}{K_c + [C]} \qquad (4.73)$$

where [B] is the concentration of organisms (usually cells/mL), μ and μ_{max} are specific and maximum growth rates, [C] is the chemical concentration, and K_c is the chemical concentration (C') when $\mu = \mu_{max}/2$. A plot of Equation 4.73 as μ vs [C] is hyperbolic: initial large increases in μ follow from small increases in [C], but later increases in μ with additional substrate are much slower.

If C $<<$ K_c then

$$\mu = \frac{\mu_{max}[C]}{K_c} - k_b[C] \qquad (4.74)$$

Equations 4.73 and 4.74 assume that the chemical is utilized to form cell mass. If Equation 4.73 is modified to account instead for loss of C

$$d[C]/dt = \frac{\mu_{max}}{y} \times \frac{[B][C]}{K_c + [C]} \qquad (4.75)$$

Again, when [C] $<<$ K_s and [B] is nearly constant in a steady state phase,

$$d[C]/dt = \frac{\mu_{max}}{yK_c} [B][C] = k_b [B][C] = k_b' [C] \qquad (4.76)$$

Thus, biotransformations can be described by simple second- and first-order kinetic relations where the organism replaces the abiotic chemical species in the equation. Several workers have reported that with low concentrations of chemicals, good first- and second-order kinetic relations are found for some biotransformation systems (Larson, 1979; Paris et al., 1981; Smith et al., 1978). However, for many systems, the variability and unpredictability of [B], μ_{max}, and K_c in different environments has been reported and no correlations were found between microbial activity and microbial populations (Klecka, 1985). Moreover, much of the energy from biotransformation may be needed for maintenance, not growth, making the Monod assumptions invalid.

Biotransformation Test Methods

Despite limitations imposed on laboratory measurements by the often unknown and complex ecology of microorganisms in natural waters and sediments, several methods have been devised to measure the rates of biotransformation of selected chemicals. Many of the methods use ^{14}C-labeled organics to measure rates of mineralization to form $^{14}CO_2$, but this is unnecessary if the concentration of the chemical to be tested is high enough to permit the use of conventional gas measuring methods. EPA recommends an aerobic shaker flask method with unlabeled compound in which CO_2 is measured by absorbing it in barium hydroxide solution and titrating the solution with HCl. Total and dissolved organic carbon also are measured to estimate the fractional conversion to CO_2. An anaerobic test also is recommended which uses sewage sludge as the innoculum. The total pressure change due to formation of CO_2 and CH_4 is monitored by manometry. Test chemicals with known activity are used in controls to check the activity of the innoculum; test chemicals are usually easy to convert to CO_2 (EPA, 1985c).

All of the test methods suffer from the difficulties inherent in trying to replicate complex, poorly understood living systems which inevitably change on removal from natural habitats to laboratory confinement. Experiments which minimize the time between sampling the natural waters or soil and the measurements, and use low concentrations of chemicals are most likely to yield kinetic and product data meaningful for exposure analysis.

Estimation Methods

A wide variety of simple and complex estimation methods has been developed to predict biodegradability of organic chemicals. Benchmark chemicals are widely used as are well-studied reference chemicals that can be introduced into a particular microbial system to compare relative ease of biotransformation with that of a chemical of interest. Klecka (1985) has summarized some of these data. Variations in rate constants of up to 3000 times are observed for some benchmark chemicals in different aquatic systems, so this approach has limited, qualitative value only.

Other estimation methods range from using chemical or atomic structure (Paris et al., 1983; Hammond and Alexander, 1972; Dearden and Nicholson, 1986) to log K_{ow} (Yonezawa and Urushigawa, 1979), or its surrogates such as molecular surface or connectivity indices (Sabljic, 1983; Boethling, 1986), to expert systems (Boethling et al., 1989).

Some of the predictive methods are capable of ranking chemicals in order of susceptibility to biotransformation and of assigning the most likely pathways for transformation. Boethling (1986) has correlated rate constants for biodegradation with molecular connectivity indices of homologous series of esters, ethers, and acids with favorable correlation parameters ($r^2 > 0.94$). However, without experimental data using the microbial systems of interest, it is unlikely that any of the methods will give reliable predictions of rate constants in a specific microbial system.

Environmental Factors Affecting Biotransformation

Microbial ecology in natural systems has been reviewed competently by a number of workers (see, for example, Costerton and Geesey, 1979 or Grant and Long, 1981). The important factors affecting the mobility of microbial populations and biotransformation rates are adaptation, nutrients, oxygen, moisture, pH, and temperature. Unlike simple chemical systems, no single driving parameter can be singled out as critical in determining rates of biotransformation. For some chemicals, previous or long exposure of microorganisms is needed for adaptation to induce enzymes for biotransformation. Availability of macro- and micro-nutrients (C, N, P, Fe) and oxygen will often control biotransformation rates. Many biotransformations rates increase with increasing temperature within the range of 10° to 30°C (Arrhenius-type kinetics), but for some organisms and chemicals, no correlations are found between temperature and rates.

Moisture is perhaps the key parameter which influences biotransformation in terrestrial systems not only by controlling microbial population growth, but also by decreasing sorption and increasing volatilization of chemicals in soil. Laskowski et al. (1982) have noted how widely biotransformation rates of several pesticides vary in different soils, with up to an 80-fold change in one case. Undoubtedly variations in soil moisture and organic (humic) content are responsible for many of these variations. Soils with low moisture and high organic content will sequester chemicals on the clay particles preventing biotransformation (Hamaker and Goring, 1976).

Despite the great complexity of biotransformation processes, progress has been made in site-specific predictions of biotransformation rates for some pesticides in soils. Walker (1978) showed how knowledge of moisture and temperature effects in selected soil samples could be used successfully to predict loss rates in field plots.

5

Mathematical Models of Transport and Fate

Donald Mackay and Sally Paterson

INTRODUCTION

Objectives

The aim of the chemical fate and transport modeler is to gather the data that are available concerning the chemical's discharge rates and concentrations, together with information on the state of the environment; for example, its hydrology or soil characteristics; and estimate the amounts and concentrations of chemical which will be present in each medium, the rates at which the chemical is degraded and is transported from place to place, and therefore how long the chemical will reside in the environment in question. The calculation, estimation or prediction of the likely concentrations in various media requires a knowledge of the physical/chemical properties of the substance, as described in the previous section, and especially parameters such as chemical reactivity, in the form of reaction rate constants or half-lives.

Even in retrospective assessments for which measured concentrations are available, these data only give a partial snapshot of the condition of the environment at a point in time, whereas the modeler's aim is to translate this into a quantitative picture of the dynamic behavior of the contaminant in the system. It is argued that once the chemical's dominant environmental pathways and fate processes are understood quantitatively, a much clearer picture emerges of the nature of the contamination issue, the behavior of the chemical, and how conditions may be modified to reduce concentration and hence exposure.

ISBN 0-87371-875-5
© 1993 by Lewis Publishers

The Mass Balance

The assembly of this quantitative description is frequently referred to as the development of a "mass balance" model. The fundamental concept is illustrated in Figure 5.1 in which a volume in space of the environment is identified as a compartment, and a mass or material balance equation is written for this volume. The volume may be water in a section of a river, it may be a region of the atmosphere, or it may be soil to a depth such as 20 cm. It may even be an organism. It must have defined physical boundaries, and preferably be fairly homogenous in conditions within the phase envelope. The mass balance equation states that the inventory change of chemical in the phase envelope in units such as kg/yr, will equal the sum of inputs to the phase envelope, again in kg/yr, less the sum of the outputs. The input terms may include flow in air and water, direct discharges, or formation from other chemical compounds. The outputs may include flow, e.g., water from the lake, or degrading reactions, or diffusion to another compartment. It is taken as axiomatic that the mass balance equation applies; thus the modeler's task is essentially to develop expressions, equations, or quantities for each of the terms in the mass balance equation.

When assembling this mass balance, the modeler faces a number of decisions, which are often subjective in nature. The art of modeling can be viewed as the ability to make the "right" decisions when faced with a large number of alternatives.

First is defining the phase boundary or envelope. Ideally the volume within

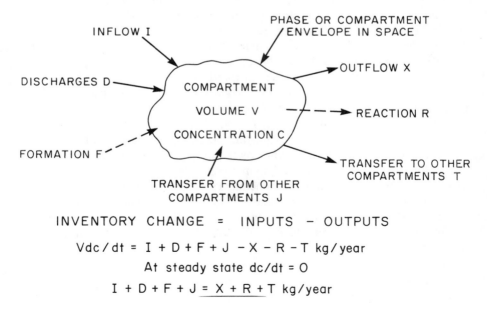

INVENTORY CHANGE = INPUTS – OUTPUTS

$$Vdc/dt = I + D + F + J - X - R - T \ \text{kg/year}$$

At steady state $dc/dt = 0$

$$I + D + F + J = X + R + T \ \text{kg/year}$$

Figure 5.1. Fundamental mass balance equations for a compartment.

the compartment should be well mixed and thus have a fairly homogenous concentration. Examples are shallow pond water or surface sediments. If a more detailed and complex model is desired, it may be preferable to segment the water column into two layers, a surface and deep layer. It may be desirable to treat particles in the water column as a separate phase, or they may be lumped in with the water column. Obviously the greater the number of compartments, the greater will be the fidelity of the model to reality, but the mathematics become more complex, more input data are needed, and the model becomes fundamentally more difficult to understand. It is also less likely to be used if it is too complex. Figure 5.2 is a simple six-compartment model of air, water, fish, soil, and bottom and suspended sediments with a chemical (trichloroethylene) introduced into three compartments. To describe the processes which apply to the chemical as it enters and leaves each of these compartments requires estimation of some 15 processes, including rates of sedimentation, run-off from soil, evaporation, reaction, and advective flows.

The second task is to define the various reaction rates by processes such as biodegradation, hydrolysis, and photolysis, as has been discussed in the previous chapter. This may be difficult because temperatures and other conditions

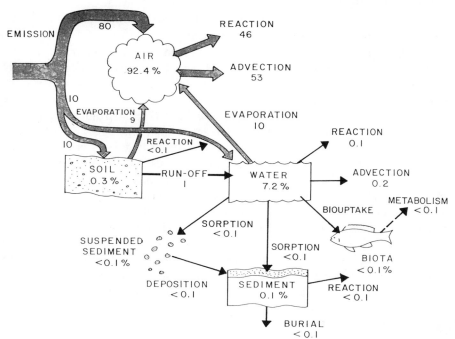

Figure 5.2. Illustrative mass balance diagram for the fate of trichloroethylene in a six-compartment system (after Mackay, 1987).

such as microbial populations, sunlight intensity, and pH may vary diurnally and seasonally. Thus it is not always clear what average conditions should be used. In some cases a distribution of values may be used. The most scientifically reliable and rigorous data are obtained by bench scale experiments involving exposure of a chemical in a beaker to a known reactive environment under highly controlled conditions. The challenge is then to translate this physical-chemical information to environmental conditions. Experiments may also be conducted in larger microcosms or mesocosms in which environmental conditions are more closely simulated, but with some loss of control over the variables which determine the reaction rate. It is also difficult to sort out the contributions of individual processes.

The third task is to define the various discharges to the system, which may be from industrial or municipal sources, spills, deliberate application of chemical (for example, pesticides) leaching from groundwater, or from use or disposal of consumer products. The total discharge rate, which is also referred to as the loading rate, is often very difficult to determine, at least for large regions in which there are multiple point and nonpoint sources, including deposition from the atmosphere. Some discharges such as pesticide applications must be treated as being dynamic in time and space. A general term that includes these dynamics as well as the discharge rate is the source term (Ch. 3). This is a critically important task because it is the magnitude and distribution of the discharge that drives the model, establishes the magnitude of the concentrations, and therefore determines the exposure.

Fourth, and finally, is the definition of transport rates between the various media. It is important to be able to assess these rates because a contaminant may be introduced into one medium such as soil, where it is relatively stable. But when conveyed into another medium, e.g., air, it becomes subject to an appreciable reaction rate. The chemical's life time in the environment may then be controlled, not by how fast it will react in soil, but by how fast it can evaporate from the soil to the atmosphere. In the early days of environmental chemistry the prevailing view was that most chemicals tended to remain in the medium into which they were discharged. For example, DDT applied to a soil would remain there indefinitely, subject only to soil degradation processes. It is now clear that most chemicals may have the capability of migrating between all environmental media, thus application of DDT to soil will result in appreciable concentrations in the atmosphere which is exposed to that soil, and that contaminated air will result in the contamination of distant areas, as in the case of DDT contamination of Arctic and Antarctic mammals, birds, and fish. Indeed, it was Rachel Carson in *Silent Spring* who first raised the issue of exposure of nontarget organisms to pesticide chemicals as a result of these intermedia transport processes.

Transport Processes

Transport processes can be classified into two general groups. First are the *nondiffusive* or advective processes, depicted in Figure 5.3, in which a chemical is conveyed from one medium to another by virtue of "piggybacking" on "carrier" material which is moving between media for reasons unrelated to the presence of the chemical. Examples are deposition of chemical from the water column to sediment by attachment to depositing particles, transport from the atmosphere to the soil or water in rainfall, dust fall, or in wet deposition (dust scavenged from the atmosphere by rain). Ingestion of contaminated food by an organism is also a process of this type. These processes are readily quantified by multiplying the concentration of chemical in the migrating medium by the flow or transport rate of the medium. For example, if rain contains 10 ng/L of PCB and is falling at a rate of 0.5 m/yr, on a region of area 5×10^6 m², the rain rate will be 2.5×10^6 m³/yr and the transport rate of PCBs will be a product of this rate and the concentration of 10 ng/L (expressed equivalently as 10 micrograms per cubic meter). The PCB deposition rate therefore will be 25×10^6 µg/yr or 25 g/yr. A calculation similar in principle to this can be done for other intermedia advective processes, including flow in water, air, and on solids, as well as ingestion in food by organisms and uptake by inhalation.

The second type of process is *diffusive* in nature in which the chemical migrates between media because it is in a state of disequilibrium. For example, considering the two-compartment environment also depicted in Figure 5.3, benzene is present in the water at a concentration of 1 mg/L and in the air at 0.1 mg/L. From the physical-chemical properties of benzene (specifically, its air/water partition coefficient or Henry's law constant which is determined by solubility and vapor pressure) it is known that whatever benzene concentrations become established at equilibrium, the ratio of air to water concentrations or the air to water partition coefficient, will be approximately 0.2 at the defined temperature. It is therefore clear that in this case the air is "undersaturated" with respect to the water. Benzene will thus migrate from water to air until it establishes concentrations of perhaps 0.8 mg/L in the water and

NONDIFFUSIVE TRANSPORT   DIFFUSIVE TRANSPORT

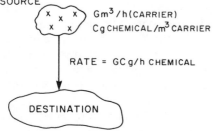
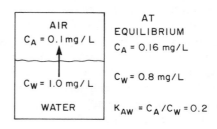

Figure 5.3.   Intermedia transport processes.

0.16 mg/L in the air. From a physical-chemical viewpoint the environment is striving to achieve a condition of equilibrium, or one in which the benzene has an equal chemical potential or fugacity in both media. It may, or may not, have time to achieve this equilibrium because of other continuous inputs to or outputs from the system, but the direction of diffusive transport is at least clear, and the rate of transport can be calculated.

It must be appreciated that this diffusion process is basically a manifestation of random mixing of benzene throughout the entire system. The benzene in the water does not "know" the concentration in the air, but the rate of loss of benzene from the water compartment to the air is proportional to the concentration of benzene in the water. Thus, that rate of loss will exceed the rate of gain from the air until such time as air and water concentrations adjust to give upward and downward rates of movement which are equal. It is common to write the diffusive transport equation in the form

$$N = kA(C_1 - C_2K_{12})$$

where N is a transport rate (mol/hr) or kg/yr, k is a transport rate coefficient (m/hr), A is the area between the media (m^2), C_1 and C_2 are the concentrations in the two media, and K_{12} is the partition coefficient such that at equilibrium, when there is no transport, K_{12} equals C_1/C_2 or C_1 equals C_2K_{12} and the term $(C_1-C_2K_{12})$ thus becomes zero. This term is usually referred to as the "driving force" for diffusion or the "departure from equilibrium."

In these diffusive rate calculations, which can be applied to other combinations of media such as water and sediment, the key quantities are obviously (a) the intermedia partition coefficients, e.g., K_{12}, which are thermodynamic quantities expressing the equilibrium condition and thus the extent of departure from it, (b) a transport rate parameter k which contains the dimensions of time and can be viewed as a velocity with which chemical moves from one medium to another, and (c) intermedia area A. A variety of methods are available for measuring, correlating, estimating, and predicting these intermedia transport coefficients. A full account is given by Thibodeaux (1979) and by Mackay (1991). The quantities involved are generally mass transfer coefficients, molecular diffusion coefficients, and diffusion path lengths or boundary layer thickness. Methods for estimating many of these quantities are conveniently reviewed in the text by Lyman et al. (1982). The general approach is usually to make measurements of intermedia transport rates, N, in controlled conditions in a laboratory in which A, C_1, C_2 and K_{12} are known, deduce k, and then correlate it with conditions such as turbulence in the atmosphere or water as reflected by wind speed or water current velocity. It is difficult to make these measurements in the environment, thus there is often uncertainty that laboratory-to-environment extrapolation is valid.

In some cases the diffusing chemical may pass through two or more layers during its migration from one medium to the other. For example, benzene evaporating from water to the atmosphere must diffuse through a near stagnant boundary layer in the water and a near stagnant boundary layer in the air. These diffusion processes occur in series and require application of the "two-resistance-in-series" concept, often referred to as the Whitman Two Film theory.

Having defined the rates, or the rate expressions, of the various terms in the mass balance equation in Figure 5.1, the remaining task is to solve the equation to obtain an expression for the desired concentrations. Here again the modeler is presented with various options.

Solutions to the Mass Balance Equation

A simple and valuable approach is to examine the steady-state condition of the system, i.e., ignore the inventory change terms on the left of the equation and write the material balance equations in the form of one or more algebraic equations. These equations then express the concentrations which would prevail in the system if the input parameters remain constant for a long period of time. Mathematical solution is accomplished by writing the set of algebraic equations and solving either by hand or matrix techniques. This solution, although somewhat artificial, is useful in that it demonstrates which processes and parameters are likely to be most important in determining the chemical's fate, concentration and hence exposure. The modeler can then devote more effort to establishing more accurate values of the critical parameters. For example, the reaction rate may prove to be so small that it is relatively unimportant compared to a transport rate by evaporation. Further effort is then justified to obtain a more accurate evaporation transport rate. The quantities in Figure 5.2 were obtained in this way, and show that for TCE the important processes are evaporation from water, and reaction and advection in the atmosphere.

The second and more rigorous approach is to solve the set of differential equations represented in specimen form by the equation in Figure 5.1. This is usually done by numerical integration techniques in which the initial or boundary conditions are first established, then the change in inventory is deduced for a specified time increment or period, and the new set of concentrations for the system calculated. The equation is applied repeatedly to the new set of conditions. Care is necessary when selecting the time step and method of integration. The output from such a calculation as illustrated in Figure 5.4 is actually the time course of concentration changes in the system as a result of inputs which may be constant, or may change with time. The difficulty with this type of solution is that results are not easily generalizable to other conditions, since each solution is specific to the initial conditions selected, and of course, to the inputs. When there is a limited number of compartments it may be possible to solve the set of differential equations analytically. This is straightforward when there are only two compartments, but when the number is three or more, the solutions become very cumbersome.

It is again in this context that the art of environmental modeling becomes

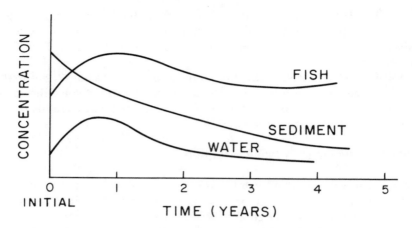

Figure 5.4.   Illustrative time course of chemical concentration changes.

apparent. Regrettably, considerable effort has been devoted to assembling complicated models of chemical migration in complex multimedia environments involving many dozens of transport and transformation processes, and hence parameters. So many parameters and assumptions are built into the model that potential users become suspicious that there may be some hidden sensitivity, error, or even a mistake. Mathematically, the problem is that the "fit" involves a multitude of adjustable and poorly quantified parameters; thus the agreement with reality may be fortuitous. It may not be obvious which are the most sensitive, fate-controlling, parameters; thus the key processes may be controlled by one, or a few, expressions or parameters which contain considerable error. Frequently the total rate of loss of chemical is known fairly accurately from field data, and can be modeled accurately, but it is the sum of several contributing processes such as degrading reactions, volatilization, and biodegradation, thus the components of this total are not known accurately. A strong case can be made that the model should be capable of being checked for intuitive reasonableness. To an experienced environmental chemist the results should be in accord with what is expected from that chemical in that environment. Often the function of the model is merely to sort the multitude of disparate processes into order of importance, and identify the key processes. A very simple "back of the envelope" model can then be assembled containing only those key processes which can give a satisfactory description of reality.

Having set the stage, we now develop and discuss a simple illustrative model containing a variety of processes, then describe some of the currently available environmental fate models.

ILLUSTRATION OF A SIMPLE MASS BALANCE MODEL

Environmental Conditions

To illustrate a simple one-compartment aquatic model we examine the fate of a chemical in a lake subject to a direct chemical discharge and which is 10 m deep and has an area of 10^6 m^2, the volume of water then being 10^7 m^3. Water flows into, and out of this pond at a rate of 1000 m^3/h, the residence time of the water thus being 10,000 hours or approximately 14 months. It is believed that there is inflow of 50 L/h of suspended sediment, deposition of 30 L/h to the bottom, and the remaining 20 L/h flows out of the system. The bottom (which we ignore here) consists of sediment. Chemical can react in the water column with a half-life of 289 days, i.e., a rate constant of 10^{-4} h^{-1}. It evaporates with an overall water-side mass transfer coefficient of 0.001 m/h.

The chemical (molecular mass 100 g/mol) has an air-water partition coefficient of 0.01, a particle-water partition coefficient K_{PW} of 5450, and a biota water partition coefficient K_{BW} or bioconcentration factor of 5000. The concentration of particles is 20 ppm and that of biota (including fish) is 5 ppm, both by volume. There is a constant discharge of 0.4 mol/h or 40 g/h, and chemical present in the inflowing water at a concentration of 10^{-4} mol/m^3; i.e., 0.01 mg/L. The aim of the model is to calculate the steady state or constant concentration in the system of water, particles, and fish, and all the loss rates.

We can undertake this calculation in two ways; first, a conventional concentration calculation, and then a fugacity calculation.

Concentration Calculation

We let the total concentration of chemical in the water (including chemical present in particles and fish) be an unknown, C_W g/m^3. We can calculate the various process rates in terms of C_W, sum them and equate them to the total input rate, then solve for C_W, and finally deduce the process rates. We use units of g/h for the mass balance calculation.

Input Rate

The discharge rate is 40 g/h. The inflow rate is the product of 1000 m^3/h and 0.01 g/m^3 (i.e., 0.01 mg/L) or 10 g/h. This gives a total input of 50 g/h.

Partitioning between Water, Particles and Fish

The total amount in the water is $V_W C_W$ g where V_W is the volume of water (10^7m^3). But this contains 20 ppm of particles, i.e., a volume of $20 \times 10^{-6} \times 10^7$ or 200 m^3 and similarly, 5 ppm of biota or 50 m^3. If the dissolved chemical concentration in water is C_D, then the concentrations in the particles will be $K_{PW} C_D$ and in the biota $K_{BW} C_D$; thus the amounts are respectively:

$10^7 C_D$ in water,
200 $K_{PW}C_D$ in particles, and
50 $K_{BW}C_D$ in biota.

Since K_{PW} is 5450 and K_{BW} is 5000 these add to give

$$C_D(10 + 1.09 + 0.25) \times 10^6 = 11.34 \times 10^6 C_D$$

But this must equal $10^7 C_W$; thus C_D is $0.882C_W$, C_P is $4800C_W$, and C_B is $4400C_W$, with 88.2% of the chemical being dissolved, 9.6% sorbed to particles, and 2.2% bioconcentrated in biota. Note that we use dimensionless partition coefficients K_{PW} and K_{BW}; i.e., ratios of (mg/L)/(mg/L). Often K_{PW} is reported and used as a ratio of (mg/kg)/(mg/L) and thus has dimensions of L/kg.

Outflow

Since the outflow rate is 1000 m³/h, the outflow rate of dissolved chemical must be $882C_W$ g/h. In addition there is outflow of 20 L/h (0.02 m³/h) of particles containing $4800C_W$ g/m³ giving $96C_W$ g/h. We assume the biota to remain in the lake.

Reaction

The reaction rate is the product of water volume, concentration, and rate constant; i.e., it is $10^7 \times C_W \times 10^{-4}$ or $1000 \, C_W$ g/h.

Deposition

The concentration on the particles is $4800C_W$ g/m³ particle. Since the particle deposition rate is 30 L/h or 0.03 m³/h, the chemical deposition rate will be $4800 \times 0.03C_W$ or $144C_W$ g/h.

Evaporation

The evaporation rate is the product of mass transfer coefficient (0.001 m/h), the water area (10^6 m²), and the concentration dissolved in water, i.e. it is $.001 \times 10^6 \times 0.882C_W$ or $882C_W$. Note that it is assumed that the air contains no chemical to create a "back pressure" or diffusion from air to water. If this were the case it would be included as another input term.

Combining the process rates, we argue that they will equal the discharge rate of 50 g/h, thus

$$50 = 882C_W + 96C_W + 1000C_W + 144C_W + 882C_W = 3004C_W$$

thus

$$C_W = 50/3004 = 0.0166 \text{ g/m}^3 \text{ or mg/L}$$

Table 5.1. Process Rates

Outflow in water	14.7 g/h	29%
Outflow in particles	1.6 g/h	3%
Reaction	16.6 g/h	34%
Deposition	2.4 g/h	5%
Evaporation	14.7 g/h	29%

The dissolved concentration is thus 0.0146 g/m³, that on the particles is 5450 times this or 80 g/m³ particle, or if the particle density is 1.5 g/cm³, approximately 53 mg/kg. This is also 0.0016 g/m³ water. The fish concentration will be 73 g/m³ fish or mg/kg fish and is equivalent to 0.0004 g/m³ water. The various process rates are shown in Table 5.1.

These total to the input of 50 g/h, thus satisfying the mass balance, the relative importance of the processes being immediately obvious. Figure 5.5 depicts this mass balance. There is now a clear picture of the chemical's fate.

Fugacity Calculation

In the fugacity approach, partitioning of chemical between phases is expressed in terms of the equilibrium criterion of fugacity, as is discussed in the final section. Fugacity can be considered a surrogate for concentration and is related to concentration through a fugacity capacity or Z value. These Z values

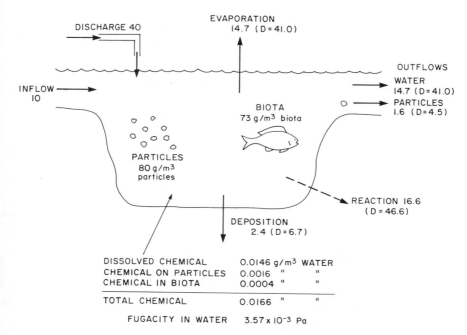

Figure 5.5.   Steady-state mass balance of a chemical in a lake, as described in the example in the text.

Table 5.2. D Values

Outflow in water	$D_1 = G_W Z_W = 1000 \times 4.1 \times 10^{-2} = 41.0$
Outflow in particles	$D_2 = G_p Z_p = 0.02 \times 223 = 4.5$
Reaction	$D_3 = V Z_{WT} k = 10^7 \times 4.66 \times 10^{-2} \times 10^{-4} = 46.6$
Deposition	$D_4 = G_D Z_p = .03 \times 223 = 6.7$
Evaporation	$D_5 = k_M A Z_W = .001 \times 10^6 \times 4.1 \times 10^{-2} = 41.0$

are calculated from a chemical's physical properties of molecular weight, vapor pressure, solubility, and octanol/water partition coefficient, as well as environmental properties such as density and fraction organic content of the phases present.

Diffusive transfer between air and water, air and soil, and water and sediment is estimated from mass transfer coefficients, interfacial areas, diffusivities, and path length. Nondiffusive processes include air and water advection, wet and dry atmospheric deposition, leaching from groundwater, sediment deposition, burial and resuspension, and runoff from soil, and are estimated from a flowrate, Z value, and fugacity. In each case the rate parameter is expressed as a "D value" such that the rate is the product of D and fugacity. Degradation in each compartment or subcompartment is included by means of a rate constant, again expressed as a D value.

A more detailed account is given by Mackay and Paterson (1982) and Mackay (1991), and an example of a more complex aquatic model of PCBs in Lake Ontario has been given by Mackay (1989).

This calculation can be repeated in fugacity format by first calculating Z values, then D values, then equating input and output rates as before. It is preferable to use units of mol/h when doing fugacity calculations.

For air, Z is assumed to be 4.1×10^{-4} mol/m³ Pa. Z for water is then calculated as Z_A/K_{AW} or 4.1×10^{-2}, and for particles Z_P as $K_{PW} Z_W$ or 223, and for fish Z_B is $K_{BW} Z_W$ or 205. The total Z value for water, particles, and biota is the sum of these Z values, weighted in proportion of their volume fractions, i.e.,

$$Z_{WT} = Z_W + 20 \times 10^{-6} Z_P + 5 \times 10^{-6} Z_B$$

or

$$4.66 \times 10^{-2}$$

The D values (in units of mol/Pa.h), are shown in Table 5.2.

The overall mass balance is then expressed in terms of the fugacity of the chemical in the water f_W. The input rate is 0.5 mol/h

$$\text{Input} = f_W D_1 + f_W D_2 + f_W D_3 + f_W D_4 + f_W D_5$$

or $\quad 0.5 = f_W \Sigma D_i = f_W \, 139.8$

∴ $\quad f_W = 3.57 \times 10^{-3}$

∴ $\quad C_W = Z_W f_W = 1.46 \times 10^{-4}$ mol/m³ or 0.0146 g/m³

$\quad C_P = Z_P f_W = 0.80$ mol/m³ or 80 g/m³

$\quad C_B = Z_B f_W = 0.73$ mol/m³ or 73 g/m³ or 73 mg/kg

The individual rates are Df, which are identical to those calculated earlier. The D values give a useful direct expression of the relative importance of each process.

Discussion

The model thus yields invaluable information on the status and fate of the chemical in this lake system. Obviously it is desirable to measure concentrations in particles, water, and biota to determine if the model assertions are correct. If (as is likely) discrepancies are observed, the model assumptions can be re-examined to test if some process has not been included, or if the assumed parameters are correct in magnitude. When reconciliation is successful, the environmental scientist is in the satisfying position of being able to claim that the system is well understood. It is then possible to explore the effect of various changes in inputs to the system.

For example, if the discharge were eliminated, the water, particle, and biotic concentrations would eventually fall to one fifth of the values calculated above because the input would now be 10 g/h instead of 50 g/h. The biota would now contain 14 mg/kg instead of 73 mg/kg. To achieve a "target" of, say, 5 mg/kg would require reduction of the input rate to about 3 g/h. The nature and magnitude of regulatory measures necessary to achieve a desired environmental quality can thus be estimated.

Another issue is that of how long it will take for such measures to become effective. This requires solution of the differential mass balance equation, which in this case is

$$V_W dC_W/dt = \text{Inputs} - \text{Outputs}$$
$$\text{or } V_W dC_W/dt = 50 - 3005 C_W$$
$$\text{or } dC_W/dt = 50 \times 10^{-7} - 3005 \times 10^{-7} C_W = A - BC_W$$

If the initial water concentration is C_{W0} at time t of zero, this equation can be solved by separation of variables to give

$$C_W = C_{WF} - (C_{WF} - C_{W0}) \exp(-Bt)$$

where C_{WF} is the final value of C_W at t of infinity, and is A/B. C_W thus changes from C_{W0} to C_{WF} with a rate constant of B or a half-time of 0.693/B hours.

Here B is $3005 \times 10^{-7} h^{-1}$; thus the half-time is 2300 hours or 96 days.

In reality the biota would respond more slowly because of the delay in biouptake or release, but the important finding is that within a year the system would be well on its way to a new steady-state condition. In fugacity terms the equivalent value of B is $\Sigma D/V_W Z_{WT}$ and the individual processes contributing to this overall rate constant is $D/V_W Z_{WT}$. It is thus apparent that water outflow and reaction have about equivalent influences on the rate or time of response, while deposition and particle outflow are relatively unimportant. Such information can be of considerable regulatory value.

It is obviously possible to include a sediment compartment, or another water compartment to give more detail. If exposure to chemical from fish consumption is important, it may be necessary to develop a separate bioconcentration or food chain model. The nature and depth of the modeling activity can be tailored to the exposure assessment needs.

ENVIRONMENTAL MODELS OF CHEMICAL FATE

Scope

In this section we describe some of the available environmental fate models which can be used for exposure estimation. No attempt is made to provide a comprehensive list; thus, this section should be viewed as merely a glimpse into the world of models. References are given to enable the reader to pursue the details of these models as desired. The primary emphasis is on multimedia models, i.e., those which attempt to describe how a chemical behaves in the entire air-water-soil-sediment-biota environment. Second are "remediation models" which attempt to describe chemical fate in the vicinity of a contaminated area such as a dump or a former industrial site. Third are a few representatives of the large number of single media models which describe chemical fate in single compartments such as lakes, soils, the atmosphere, and in food chains, and as a result of chemical spills.

Convenient reviews of available models are the text by Cohen (1986), OECD (1989), Swann and Eschenroeder (1983), Dickson et al. (1982), Jorgensen (1984), and Jorgensen and Gromiec (1989).

Multimedia Models

Fugacity Models

A series of fugacity-based models has been developed in the last decade at the University of Toronto by Mackay and co-workers (Mackay, 1979; Mackay and Paterson, 1982; Mackay et al., 1985; Mackay, 1991). They vary in complexity from a simple equilibrium distribution of a conservative chemical, to steady-state and time varying descriptions of the fate of reactive compounds.

The Level III model (Mackay et al., 1985) is most useful in estimating environmental fate of persistent organic chemicals which are continuously emitted to the environment over a prolonged period. This model treats four bulk compartments: air, water, soil, and sediment, each of which is considered to consist of subcompartments of varying fractions of air, water, and organic matter. Equilibrium is assumed to exist within, but not between, compartments. The model calculates the steady-state distribution and concentration of a chemical and its persistence in the environment. Media of accumulation and dominant pathways of transfer are determined. Figure 5.2 is an illustration of the fate of a chemical in such a system.

The algebraic steady-state model incorporates expressions for emissions, diffusive and nondiffusive transfer processes, advective flows, and degrading reactions and is readily solved algebraically.

Four mass balance equations representing the four bulk phases are solved in terms of fugacity using the principles illustrated earlier. Concentrations and amounts are estimated for the bulk phases. Fugacities of subcompartments are equated to those of the bulk phase. Thus, concentrations and amounts in subcompartments are readily calculated. The resulting predicted environmental concentrations can be combined with air inhalation and water and food ingestion rates to provide a preliminary estimate of human exposure as discussed by Mackay and Paterson (1988).

GEOTOX

GEOTOX (McKone and Layton, 1986) is a comprehensive multimedia compartmental model developed under contract from the U.S. government (U.S. Department of Energy and U.S. Army), which calculates chemical partitioning, degrading reactions, and diffusive and nondiffusive interphase transport. The concentrations estimated for various environmental compartments are subsequently combined with appropriate human inhalation and ingestion rates, and absorption factors to calculate exposure. It treats an environment representative of the southeastern United States, consisting of the following compartments: air (gas), air (particles), biomass, upper soil, lower soil, groundwater, surface water, and sediments. These media are assumed to be made up of subphases of gas, liquid, and solid. Environmental dimensions and characteristics can be adjusted to represent other regions. Chemical partitioning between compartments, interphase transport, reaction, and advective losses are described by first-order rate constants. The model can be applied to constant or time-varying chemical sources.

The soil compartment is treated as three layers: upper soil layer, lower soil layer, and groundwater zone. The upper and lower soil layers are described by parameters of depth, bulk density, porosity, water content, and fraction organic carbon. The groundwater compartment consists of solids and fluids; that is, all the pore spaces are occupied by water. Processes of adsorption, ion exchange, precipitation, colloidal infiltration, and irreversible mineralization in the groundwater are incorporated by means of appropriate sorption partitioning constants and expressions. The water phase is considered to consist of water, biota, and suspended solids in equilibrium.

Concentration in land biomass is calculated as the product of the soil concentration and plant/soil partition coefficient, i.e., an equilibrium type of expression is assumed incorporating vegetation production or growth rate. Animal biomass is not considered in the mass balance but it is treated as an exposure vector.

Data requirements include physical-chemical properties, degradation rate constants, and emission data, as well as environmental characteristics, such as fraction of land surface covered by water, and average water depth. Output is in the form of environmental concentrations, intake by various exposure pathways, as well as total intake. Relative health risk for a number of chemicals can be calculated.

SMCM (Spatial Multimedia Compartmental Model)

This model, developed by the National Center for Intermedia Transport at UCLA by Cohen and coworkers (1989, 1990), Cohen and Ryan (1985), describes the fate of chemicals in a conventional air-water-soil-sediment system under steady- or unsteady-state conditions. It has the unusual feature that it allows for concentration variation with depth in the soil and sediment, i.e., these compartments are not treated as well-mixed "boxes." This hybrid approach introduces an increased level of complexity, but yields greater fidelity. The model is user-friendly with help menus and the capability of presenting data in tabular or graphical form. Steady- and unsteady-state conditions can be treated. Particular strengths of the model are its treatment of deposition from the atmosphere and volatilization from soil.

Enpart (Environmental Partitioning Model)

Enpart (OECD, 1989) is one of a set of models developed by the U.S. Environmental Protection Agency (EPA) as a first-level screening tool for new and existing organic chemicals of possible concern. It is a fugacity-based model which estimates the steady-state equilibrium or dynamic partitioning of organic chemicals among environmental compartments. It identifies dominant pathways and data gaps, and estimates the chemicals' persistence and bioconcentration potential.

The data requirements are minimal and include properties of the chemical and some environmental parameters such as soil and sediment density, suspended sediment and biota concentrations. The output is in the form of concentration ratios between compartments rather than absolute concentrations. It is an easy-to-use approximate method intended to indicate chemicals which may require further testing.

Toxscreen

Toxscreen (Hetrick and McDowell-Boyer, 1983) is a time-dependent multimedia model, developed by the U.S. EPA to assess the potential for environmental transport and accumulation of chemicals released to the air, surface water, or soil. It is modular in concept and incorporates intermedia transfer processes. It is intended as a screening tool to assess the human exposure potential of organic chemicals.

Atmospheric dispersion is incorporated in the air compartment using a modified Gaussian plume dispersion model. Contaminant migration in soil following direct application and/or transport between other media is estimated by means of the SESOIL model. Pollutant concentrations in water bodies are estimated over a period of time using a method similar to the EXAMS approach which is described later.

Data requirements are intensive and include information on the characteristics of the environment treated and emission sources; chemical and degradation parameters; and climatological, meteorological and hydrological parameters. Files of data on climatic and soil conditions for various regions are included. The output is in the form of an estimation of the fate of a chemical over a period of time.

Environmental Exposure Potentials (EEP)

EEP (Klein et al., 1988) is a fugacity-based equilibrium multicompartment model used by member states of the European Community to determine the exposure potential of new organic chemicals. The methodology is applied to chemicals being imported or produced in quantities of 1 metric ton per year or more.

This simple model treats multiple or diffuse sources of continuous emissions. It calculates environmental partitioning, quantity in the environment, degradation, and accumulation potential in air, water, and soil, employing weighting factors to produce a "fingerprint" of environmental fate. Environmental concentrations are not calculated. Abiotic degradation is not included since this could result in transformation to another chemical.

The use patterns of a chemical are differentiated into three types: (a) intermediate, a maximum of 3% of production enters the environment; (b) processing, 10% is released; and (c) end use, whole quantity enters environment. Scores are developed for each criterion, summed and normalized to permit a standard evaluation.

Data requirements include physical-chemical properties of solubility vapor pressure and octanol/water partition coefficient, as well as classification of use pattern.

SIMPLESAL

SIMPLESAL (OECD, 1989) is a spreadsheet-based multimedia fugacity compartmental model which can be used to estimate steady-state or time-dependent concentrations of organic compounds as well as heavy metals. It determines dominant environmental pathways and processes for contaminants, and was designed for use in the Netherlands as a screening tool to predict results of various scenarios for emission control of new and existing chemicals such as benzene, cadmium, lindane, and copper. It considers compartments of air, water, suspended solids, aquatic biota, sediment, and soil.

The model incorporates processes of advective flows, diffusive and nondiffusive transfer, bioconcentration in aquatic biota, leaching from soil to groundwater, and biotic and abiotic transformation. Data requirements include dimensions and properties of, and emission rates into, various media; air and water residence times; parameters for intercompartment transfer in association with particulates; physico-chemical properties of water solubility, vapor pressure and octanol-water partition coefficient and degradation rate constants.

Multi-Phase Non-Steady State Equilibrium Model (MNSEM)

MNSEM (Yoshida et al., 1987) is a simplified kinetic model designed to predict the fate of organic chemicals under steady-state conditions of continuous loading to the Japanese environment. It calculates the chemical distribution, persistence, and concentrations in an environment representative of Japan. It also estimates recovery time after emissions are terminated.

The model environment consists of four major compartments: air, water, soil, and sediment with subcompartments of air and raindrops in air; biota and suspended solids in water; air, water, and solids in soil; and pore-water and solids in sediment. As in the fugacity models, each phase is well mixed and equilibrium is assumed to exist within, but not between, compartments. Diffusive and nondiffusive interphase transfer and degradation processes are included. Intraphase partitioning is described in terms of conventional partition coefficients.

Interphase diffusive transfer between air and soil, air and water, and sediment and water is quantified by means of mass transfer coefficients. Nondiffusive transfer processes of rainfall, surface runoff, suspended sediment deposition, and leaching from soil are characterized by rate constants. Degradation processes include hydroxy radical oxidation, photolysis, hydrolysis, and biodegradation.

Input data requirements include solubility, vapor pressure, organic carbon sorption coefficient (K_{OC}) and bioconcentration factor (BCF) and transformation rate constants. Chemical loading rates are estimated from annual production in Japan, use patterns, and physico-chemical properties. The model has been applied to the fate of chloroform, aniline, chlorobenzene, nitrobenzene, and naphthalene in Japan. Estimated concentrations are generally within an order of magnitude of measured values.

Remediation Models

AERIS (Senes Consultants, 1989) is a multimedia risk assessment model that estimates environmental concentrations and subsequently, human exposure in the vicinity of contaminated land sites. It is intended for use at sites where redevelopment is being considered. The model is novel in that it runs within a user-friendly expert system programming environment. An "intelligent" preprocessor interrogates the user about the redevelopment scenario to be assessed, assisting the user where necessary, or supplying default values.

It estimates pseudo steady-state concentrations of contaminant in compartments such as air and groundwater based on the concentration of the contaminant in the soil. These predicted environmental concentrations are used to estimate the exposure incurred by a site user. It is intended to provide a consistent approach to establishing soil guidelines and identifying cleanup objectives.

RAPS—Remedial Action Priority System

The RAPS system (Whelan et al., 1987) was developed for use by the U.S. Department of Energy to set priorities for investigation and possible cleanup of chemical and radioactive waste disposal sites. It is intended to be used in a comparative rather than predictive mode. The methodology considers four major pathways of contaminant migration: groundwater, surface water, overland, and atmospheric. Estimated concentrations in the air, soil, sediments, and water media are used to assess exposure to neighboring populations.

The methodology is not truly multimedia since it is based on use of independent modules which do not interact spatially or temporally; that is, transfer of pollutant is in one direction only. This modular framework permits updating of, or inclusion of, additional components with advancing technology.

RAPS employs user-supplied data on site and pollutant characteristics to simulate migration and fate from source to receptor by various pathways. The user must identify appropriate routes of chemical from waste site to neighboring populations through various media including air, groundwater, soil, and vegetation.

The estimated environmental concentrations form the basis of subsequent human exposure calculation and determination of the Hazard Potential Index (HPI). The modular development makes it useful for the inclusion of additional components.

Aquatic Models

Persistence

The persistence model (Roberts et al., 1981; Asher et al., 1985) was developed for the National Research Council of Canada as a screening method to estimate the fate of various organic chemicals, but especially pesticides, which are released into the aquatic environment. It considers four compartments: water, catch-all (representing suspended solids, invertebrates or other components of water system, excluding fish), sediment, and fish. It calculates both a steady-state, or fixed, solution and a time-dependent solution. Default environments for the model are a Standard Pond and a Standard Lake simulating a small, eutrophic pond and a deep, oligotrophic lake. Removal pathways include photodegradation, volatilization, and hydrolysis in water; biodegradation in fish; and microbial degradation in suspended solids and sediments.

Output is in tabular form for the steady-state or fixed solution, and in the form of tables or concentration-time curves for various compartments for the dynamic solutions. The overall persistence or "retentative capacity" of the system is also calculated.

EXAMS—Exposure Analysis Modeling System

EXAMS (Burns et al., 1981) is an interactive mass balance model developed at the U.S. EPA Research Laboratory in Athens, GA, which predicts the fate of organic contaminants in stratified surface waters as a result of continuous or intermittent releases. It is widely used by the EPA and other environmental agencies in the U.S.

The water body is subdivided into zones or segments, the mass balance of each segment being described by a differential equation. The resulting set of equations, which describes the mass balance of the entire system, incorporates comprehensive transport and transformation processes. EXAMS allows for loadings by point or nonpoint sources, dry fallout or aerial drift, atmospheric wash-out and groundwater seepage to selected segments.

The user can choose one of three operating models, depending on the complexity of the problem being studied. The modes range from a steady-state solution for continuous release of a contaminant to the dynamic solution of a time-varying source. Input data requirements are generally intensive.

QWASI (Quantitative Water Air Sediment Interaction) Model

This fugacity-based model developed by Mackay et al. (1983) treats the fate of a chemical discharge to a water-air-sediment system using Z and D values as described in the example presented earlier. It can be used as steady- or unsteady-state versions.

EXWAT

EXWAT (OECD, 1989) is a steady-state model, developed in Germany, to describe chemical fate in water bodies. It is a simple approach suitable for application to continuous single point sources. It is intended for use as a screening tool to assess comparative hazards of existing chemicals in the Rhine River. It is also a submodel of the multimedia E4CHEM system (Exposure and Ecotoxicity Estimation for Environmental Chemicals), and exposure and hazard assessment model developed in Germany for priority setting within OECD.

Inorganic Chemical Models (e.g., Metals and Phosphorus)

Modeling of inorganic compounds in the aquatic and other environments proves to be more difficult in the sense that the chemical properties and speciation tend to be unique and thus the generalizations which apply to organic chemicals do not usually apply. Bonazountas et al. (1988) have reviewed fate models for such chemicals.

An example is the work of Dolan and Bierman (1982), who developed a dynamic mass balance model of heavy metals in Saginaw Bay, Lake Huron. Extensive sampling and analysis data were collected and attempts were made to relate mass input of trace metals to observed water column concentrations.

The model requires input in the form of advection and dispersion rates in water and information on suspended sediment loadings, wind regime, depositional zones, and sediment concentrations. It incorporates a dynamic mass balance for suspended solids in the water column and sediment which is coupled with a metals mass balance through an equilibrium partition coefficient between dissolved and absorbed metal. It has been successfully applied for the site and metals studied.

Many models describing phosphorus (P) loadings in lakes have been developed in an attempt to elucidate and solve eutrophication problems. In Canada and the U.S., the focus has been on the Great Lakes, whose water quality has deteriorated as the result of rapid growth of urban populations.

One of the earliest phosphorus models was developed by Vollenweider (1969). It was an empirical, one-compartment model based on extensive data from lakes in Europe and North America. It defined the relationship between P loading, mean lake depth, and trophic state. It is steady-state and assumes removal of P is by sedimentation described by a first-order rate constant. This model was later improved (Vollenweider, 1976) to include terms for water residence time and relative residence time of P in the lake system.

There have been numerous applications of models of this type to various fresh and marine systems.

Speciation Models, e.g. MINTEQAI

MINTEQAI (Brown and Allison, 1987) is an example of an equilibrium metal speciation model applicable to metallic contaminants in surface and groundwaters. It is thus quite different in purpose from the mass balance models discussed earlier. It calculates the equilibrium aqueous speciation, adsorption, gas phase partitioning, solid phase saturation states, and precipitation-dissolution of 11 metals (arsenic, cadmium, chromium, copper, lead, mercury, nickel, selenium, silver, thallium and zinc). It contains an extensive thermodynamic base and is designed to make minimal demands on the user.

Some expertise regarding kinetic limitations at particular sites is required for proper application of the model. The resulting output is a description of the major metal species in the system.

Sediment Chronology Models

Bottom sediments serve as sinks for many metallic and hydrophobic contaminants; thus, it is possible to estimate the historic condition of a lake by examining the variation of contaminant concentration with depth of burial. Considerable effort has been devoted to deducing the likely fate of these buried chemicals when subject to diagenetic processes. These efforts usually take the form of multilayer models in which the year-by-year transport and transformation rates are estimated in slices of the buried sediment. An example of a recent comprehensive effort of this type is that of Eisenreich et al. (1989) for organochlorine chemicals, Pb and Cs in Lake Ontario. Other similar studies are available in the text edited by Hites and Eisenreich (1987).

Soil Models

Several models have been developed to describe chemical fate in soils. Notable are SESOIL (a Seasonal Soil Compartment Model, OECD, (1989)), Bonazountas and Wagner (1984), PRZM (Pesticide root zone model) by Carsel et al. (1984), PESTAN (Pesticide Analytical Model) by Enfield et al. (1982), and the "Jury" Model by Jury et al. (1983).

These models generally treat degradation, evaporation, diffusion in water and air and advective flow in water, and in some cases, in colloidal organic carbon. They are mainly applied to pesticides in an attempt to describe chemical fate and persistence, and especially transfer to groundwater. Melancorn et al. (1986) have evaluated some of these models by comparison of predictions with experimental soil core data.

Fish Uptake and Food Chain Models

Because of the importance of the human exposure route via fish consumption, considerable effort has been devoted to estimating chemical concentrations in fish. Contaminants may enter fish via the gills (bioconcentration) and, especially in the case of hydrophobic chemicals, by way of food (biomagnification). These pathways are the subject of discussion in recent examples by Thomann (1989), Clark et al. (1990), and the text by Connell (1989).

Atmospheric Models

Numerous air dispersion models with general or limited geographic applicability have been developed with the objective of deducing ground level concentrations, and hence exposures, from stack emissions. Most texts or handbooks on air pollution contain full descriptions of such models.

The general approach is to define first the emission rate or strength of one or more sources in units such as kg/day. Since most discharges to the atmosphere are from stacks, it is usually necessary to deduce a plume rise height, yielding an estimate of the height at which contaminant is released. The chemical then blows downwind and disperses horizontally and vertically, diluting steadily. A "map" of concentrations as a function of position can then be assembled, provided that information is available on the dispersion characteristics. These take the form of standard deviation terms in the Gaussian distribution equation and are functions of wind speed and atmospheric stability. Of most interest are ground level concentrations downwind of the source, since it is these concentrations which control human exposure.

Models range from simple application of Gaussian dispersion equation to complex, multisource models containing allowance for depositing particles and topographic features of the terrain. The use of such models is often written into legislation as a means of translating desired ground level concentrations into acceptable stack emission rates.

Spill Models

Finally, there are several models which can be used to deduce the fate of chemical spills. These are often used to provide guidance for spill response personnel. An example is the POSSM (PCB On-Site Spill Model) described by Brown and Silver (1986). Most regulatory agencies and many industries concerned with the marine environment (e.g., the Coast Guard) have available models which can describe the movement and fate of spills of materials such as oil, when subject to variable winds and currents. These "trajectory" models are invaluable for providing advance warning of the arrival of an oil slick. They are, of course, subject to the uncertainties inherent in weather forecasting.

CONCLUSIONS

In this chapter an attempt has been made to present and illustrate some of the underlying principles of environmental modeling. The partial list of models will, it is hoped, provide a useful entry to this large and growing literature. With the advent of faster, cheaper, more user-friendly computer systems with attractive outputs in graphical and pictorial form and a growing computer literacy, it is likely that there will continue to be extensive developments in environmental modeling. While this is certainly desirable, it is important to maintain an appreciation that the models are only as good as the quality of the input data and the expressions used for describing the various partitioning transport and transformation processes in the environment. Regrettably, there is often overconfidence that the computed results are accurate, and modelers are frequently guilty of failing to convey a full appreciation of the sensitivity of the results to errors or variation in the various parameters used to build the model. A strong case can be made that the model user should invest the time necessary to become intimately familiar with the model's idiosyncracies, its strengths and its weaknesses.

Despite these potential pitfalls, it is clear that models will play an important role in exposure assessment by estimating quantities which have not been, or cannot be measured, and in helping to synthesize information about chemical behavior into a coherent, quantitative statement of fate and thus of exposure.

6

Exposure

Glenn Suter

Exposure has been defined as contact with a chemical or physical agent (ASTM 1991, EPA 1986a). In more dynamic terms, it is the process by which an organism acquires a dose. The last stage of exposure assessment is the description and quantification of this contact between organisms and chemicals. In practical terms, this stage of risk assessment is the point of contact between the environmental chemists and engineers who estimate transport and fate, and the biologists who estimate effects. It is concerned with determining which environmental media organisms contact, and how that contact results in uptake. First, exposure assessment must consider the relevance of the concentrations in media predicted by the fate models to the environment as experienced by organisms. Second, it must consider how the behavior of organisms results in uptake of pollutants. Finally, it must use models of exposure to either equate exposure with concentration in a medium or convert concentrations in media into exposure levels.

The boundary between exposure assessment and effects assessment is fuzzy. In ecological assessments, exposure assessment usually extends to estimation of the temporal dynamics of chemical concentrations at the mouth, nasal opening, or skin of an animal, the stomates or root surfaces of a plant, or the cell wall of a microbe. This is referred to as external exposure. Because it is equivalent to the exposure measures used in ecotoxicology and is correlated with whole-organism dose, external exposure is nearly always used in ecological risk assessments. However, estimates of internal exposure are more useful in some circumstances. Internal dose can be estimated from external exposure through the use of toxicokinetic models. Internal doses may be absorbed doses (dose to the whole organism), or delivered doses (dose to particular target organs or tissues). These estimates of internal exposure may provide a better understanding of effective exposure, but in general, they cannot be related to the results of ecological toxicity tests.

Assessment of exposure begins by identifying an exposure scenario that answers the following questions: (1) given the output of the fate models (Chapter 5), which media are significantly contaminated; (2) to which contaminated media are the endpoint organisms exposed; (3) how are they exposed (i.e., routes and rates of exposure); and (4) given an initial exposure, will the behavioral response of exposed organisms modify subsequent exposure (i.e., attraction and avoidance)? A scenario for foxes on an oil field is depicted in Figure 6.1.

Given an exposure scenario, the assessor must reconcile the output of the fate models and the expressions of exposure used in the toxicity tests to the expressions of exposure suggested by the scenario. For example, if the scenario is consumption of pesticide-contaminated food, the exposure model generates deposition of pesticide per unit area, and the toxicity test results are scaled as dose per unit body mass, an exposure conversion must be performed (Kenaga 1973).

Conversions can also be performed in the effects assessment but are more often associated with the exposure assessment. It is generally easier and less expensive to modify and rerun a fate model to generate the desired estimate of exposure than to modify a test protocol and perform more tests. Similarly, it is generally more straightforward to modify the output of a fate model than the output of a toxicity test. In our pesticide example, it seems more reasonable to convert deposition to concentrations in food items and then use consumption rates to estimate dose than to convert to output of an acute lethality test (oral LD_{50}) to the equivalent output of a feeding test (LC_{50} for concentration in food) and convert that to a deposition rate, but either approach is feasible (see Chapter 7).

Ideally, the appropriate expression of exposure will be identified during the planning of an assessment so that exposure conversions can be avoided. Through such planning or by good fortune, the exposure model may reduce to the following minimal form: the medium contaminated in the toxicity test and the form in which that contamination is reported (e.g., total aqueous concentration) are equivalent to the medium and form of the exposure estimated by a fate model (e.g., solution phase concentration), and that concentration is an adequate description of exposure scenario (e.g., ventilation of fish gills by contaminated water). In the absence of matching expressions of exposure, simple exposure conversion models may be used. The use of a species's average daily food consumption rate to convert concentration in food to acute oral dose is an example. Finally, exposure simulation models might be used. An example would be use of a toxicokinetic model to estimate the concentration in food that would give a time-integrated internal dose from consumption of contaminated food equal to the time-integrated internal dose resulting from a particular acute oral dose.

The duration and temporal dynamics of exposure are as much a part of the exposure scenario as is concentration in media. However, temporal dynamics are intimately involved in both exposure and effects, and are treated in Chapter 3.

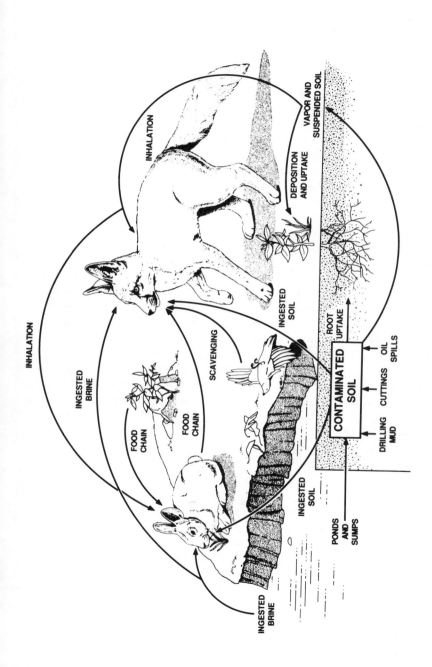

Figure 6.1. A depiction of the diverse modes of exposure of a fox on an oil field. Contaminants occur in sumps, soils, and the air. Direct routes include inhalation, ingestion of soil, and consumption of water. Indirect routes include ingestion of prey, and scavenging of oiled animals.

The discussions of exposure models and data in this chapter are rather crude, relative to those used in human exposure assessments. In fact, ecological exposures vary depending on a variety of factors not discussed here, such as stage in the reproductive cycle, seasonal changes in behavior and physiology, development and maturation, and climate. Assessors should be aware of uncertainties introduced by ignoring such factors. When risks are highly uncertain, such factors may be critical in resolving the uncertainty, in which case detailed exposure models incorporating natural history and site characteristics should be developed.

The following sections are organized in terms of the major types of exposure scenarios: respired fluids, solid substrates, ingested food and water, and multimedia exposures. Finally, the general problem of interaction between behavior and exposure, and the potential utility of kinetic models to simulate exposure are discussed.

RESPIRED FLUIDS

The simplest exposure scenarios are those involving the respired fluids; water and air. These exposures are relatively predictable because gases and aerosols in air and soluble chemicals in water are distributed homogeneously in a locally uniform medium with which organisms maintain intimate contact. Usually, the assessor can assume that the concentrations of chemicals predicted to occur in water or air are equivalent to the concentrations in aquatic and atmospheric toxicity tests in terms of their ability to impose a dose. This assumption depends on the accuracy with which the fate models predict the state of the pollutant chemical. For example, if the fate model assumes that a chemical dissolved in an effluent will simply be diluted, when in actuality much of it will sorb to suspended particles in the receiving water, the assumption that the ambient aqueous concentration and the concentration in the toxicity test will be equivalent will be highly inaccurate.

SOLID MEDIA

Exposure to contaminated sediments and soils is considerably more complex than exposure to air and water. Sediments consist of two major abiotic phases, solids and pore (i.e., interstitial) water, and soil consists of three; solids, pore water, and pore air. Benthic animals are in contact with both solid and liquid phases so they can absorb chemicals through their integuments, respire pore water or free water, and ingest sediments, sediment-associated food, or food from the water column. Plants rooted in sediments may take up material from sediments, surface water, and air. Organisms in soils are potentially subject to all of those routes of exposure plus respiration of pore air. In addition to the variety of phases, soils and sediments that support diverse communities have a greater range of physical conditions that can modify the availability of chemicals (including redox potential, cation exchange capacity, and organic carbon content) than do air or water. As a result of both of these considerations, the lethal concentrations of chemicals in whole sediment are much more variable than those in laboratory water or ambient water (Spehar and Carlson 1984, Carlson et al. 1986, Adams et al. 1983, OWRS 1989).

The most common simplifying assumption for exposure in sediment is that organisms are significantly exposed only to the chemicals dissolved in the pore water (Dickson et al. 1987, OWRS 1989). This assumption seems to be reasonably accurate for a variety of chemicals and species (i.e., chemical concentrations in sediment have highly variable toxicities, but sediments with equal concentrations in pore water are approximately equitoxic). The assumption has two practical advantages. First, it allows the assessor to use data from toxicity tests conducted in water to predict effects on organisms in sediments. Once a fate model has predicted the concentration in pore water, the assessor may assume that concentration to be toxicologically equivalent to the same concentration in free water. Second, if the pollutant will occur at more than one site, this assumption permits the assessor to predict effects for a variety of sediments. For example, if pore water concentrations are assumed to be a function of organic matter content of the sediment [as the EPA has assumed for neutral organic chemicals (OWRS 1989, DiToro et al. 1991a)], the distribution of pore water concentration can be calculated from the distribution of organic matter content in the sediments of the environment. The model is:

$$C_{sed}/C_w = f_{oc}K_{oc} \qquad (1)$$

where C_{sed} is the concentration in sediment (mg/kg), C_w is the pore water concentration (mg/L), f_{oc} is the fraction of the sediment that is organic carbon, and K_{oc} is the water/organic carbon partitioning coefficient. The EPA estimates K_{oc} by DiToro's (1985) model:

$$\log K_{oc} = 0.00028 + (0.983 \log K_{ow}) \qquad (2)$$

where K_{ow} is the octanol/water partitioning coefficient. This is effectively $K_{oc} = K_{ow}$. This and other sediment exposure models were reviewed by Landrum and Robbins (1990), who advocate more complex models.

DiToro et al. (1990) have proposed an equivalent model that uses acid volatile sulfide (AVS) concentrations to normalize sediment concentrations of metals to pore water concentrations. It is believed to be applicable to metals whose sulfides are less soluble than FeS (Cd, Cu, Hg, Ni, Pb, and Zn). Where only one metal competes for the AVS, the molar equivalent of the AVS would not be bioavailable, so the concentration in sediment (C_{st}, μmol/g) that is equitoxic to a concentration in water (C_{wt}, μmol/L) is:

$$C_{st} = AVS + K_p C_{wt} \qquad (3)$$

where K_p is the partition coefficient (L/g). This model is not applicable to fully oxidized sediments and to sediments with extremely low AVS (< 1 μmol/g). This model is more difficult to use than Equation 1 because there is no ready means to estimate K_p equivalent to Equation 2, because of the need to account for competition among metals for AVS, because AVS is not routinely measured like f_{oc}, and because AVS can be seasonally quite variable.

In cases for which the assumption that sediment exposure is equivalent to pore water exposure is not believed to hold, whole-sediment concentrations can be used as the expression of exposure and matched to the results of toxicity tests conducted with experimentally contaminated sediments. For example, if the sediment is expected to be significantly contaminated below a new outfall, sediment can be collected from the site, contaminated with the emitted chemical to obtain a concentration series, and toxicity tests can then be performed with relevant species.

Alternately, the mechanisms of exposure to sediments can be elucidated and modeled. For example, the relative importance of surface water, interstitial water, and ingested sediment as sources of hexachlorobenzene (HCB) by the deposit-feeding clam (*Macoma nasuta*) was determined by Boese et al. (1990). They fit an uptake model to data from a "clam box" which allowed estimation of the uptake rate by each route. Most of the HCB uptake was due to sediment ingestion. Landrum and Robbins (1990) discuss a similar modeling effort for an amphipod (*Pontoporeia hoyi*).

Exposure in soil has not received as much attention as in sediments. It has been suggested that the same simplifying assumption be applied to soil as to sediments, that is, that exposure is to the solution phase and solution concentrations can be calculated from equilibrium partitioning models (Baes 1982, van Gestel and Ma 1988, OWRS 1989, Connell 1990). This assumption is used in a model of organic chemical accumulation from soil by plants (Boersma et al. 1988, Lindstrom et al. 1988), and is implicit in the use of hydroponic systems for testing toxicity to plants. However, this assumption is not as easily implemented in soil because wetting and drying cycles and freeze-thaw cycles change the pH and redox conditions, thereby changing the form of the contaminant; cause shrink-swell cycles of clays which may sequester contaminants; and may cause saturation of the soil solution, thereby violating the assumption of equilibrium solid/liquid partitioning. In addition, other soil phases may be important. The vapor phase may be a critical route of exposure for burrowing mammals, and both the solid and solution phases are relevant to soil detritivores and to herbivores that ingest soil while grazing.

A more complex assumption is that contaminants occur in multiple phases in the soil, each of which can be measured by a step in a sequential extraction procedure (Bell et al. 1991, Sims and Kline 1991). The relationship of concentration in each of the phases to uptake and effects could then be determined experimentally, and a multiple regression model developed to predict uptake and effects. Although this approach seems promising, it has been only moderately successful in practice when applied to metals (Sims and Kline 1991).

Plant uptake of pollutants from soil has been estimated from field-derived plant/soil distribution coefficients (Kabata-Pendias and Pendias 1984), and from quantitative structure-activity relationships (Briggs et al. 1982, Topp et al. 1986). However, because root uptake of organic compounds is inefficient, much plant uptake of soil contaminants apparently results from volatilization and leaf uptake. QSARs for combined root and leaf uptake of organic chemicals in soils are presented by Topp et al. (1986) and Travis and Arms (1988). Similarly, uptake of organic chemicals by earthworms has been modeled empirically (Wheatley and Hardman 1968, van Gestel and Ma 1988) and on the basis of assumed partitioning from soil to soil water to earthworm lipids (Connell 1990).

Currently, these models of bioaccumulation from soil are of little use in estimating effects on the soil-exposed plants and animals because there are few data relating effects to body burdens. If water is assumed to be available in soil, sediment-water partitioning models can be used to estimate soil water concentrations which can be related to results of hydroponic toxicity tests with plants. However, the concentration in whole dry soil is the most commonly used expression of exposure of plants and soil animals, and it is matched to toxicity data from experimentally contaminated soils or from field sites (Gough et al. 1979, Kabata-Pendias and Pendias 1984, Tyler et al. 1989).

Soil ingestion rates have been measured for cattle and sheep to estimate pathways for human exposure (IAEA 1982). Similar estimates of soil ingestion rates for grazing wildlife, mammals grooming their fur, and birds and mammals consuming soil invertebrates could be used to estimate chemical consumption rates, and those might be related to toxicity data expressed as oral dose.

INGESTED WATER AND FOOD

Exposure assessment for consumption of water or food must match environmental fate data to one of two types of effects data, (a) tests of the effects of ad libitum consumption of contaminated water or food and (b) oral dosing tests. In the first case, it can be assumed that the concentration in food or water is the appropriate expression of exposure if it is first assumed that consumption in the field is equivalent to consumption in the test. The consumption is equivalent if wildlife consumption will be the same as the test animal consumption (e.g., consumption of feed by quail in a pen is approximately equal to the consumption by quail or similar granivorous birds in the field). If it is not possible to assess exposure using data from the endpoint species or from similar-sized test species, one could correct the effective dietary concentration for the relative consumption rates of the endpoint species and test species.

For example, to estimate the dietary exposure of kestrels that is equivalent to 1.0 mg/kg in the diet of chickens, one would convert the chicken diet to a

Table 6.1. Percent Water of Some Food Items That Are Representative of Wildlife Foods[a]

Food	Water Content (5)
Plant Foods	
Berries	82–87
Grains	11–12
Hays, legume	7.8–12
Hays, grass	9.3–16
Leafy vegetables	87–95
Mushroom	91
Nuts	3.0–5.3
Root crops	78–94
Animal foods	
Bivalves	80
Crustaceans	80
Chicken	71
Deer	73
Duck	54
Egg, chicken	74
Rabbit	54
Squab	58
Fishes	63–83

[a]These and other data concerning food composition are found in Spector (1956).

dose rate and then back convert to kestrel dietary concentration. One mg/kg of a chemical in the diet of a 6-kg chicken consuming 0.39 kg/day of feed equals 0.39 mg/day or a weight-scaled dose rate of 0.065 mg/kg/day. The same weight-scaled dose rate for a 0.2 kg kestrel consuming 0.01 kg/day is equivalent to a dietary concentration of 1.3 mg/kg. This example assumes dry weight basis for all food concentrations; many real cases will also require wet weight to dry weight conversions (Table 6.1).

If the available toxicity data are expressed as oral doses, it is necessary to convert the concentrations in food or water to doses. This requires estimates of the food or water consumption rates of the endpoint species. Most commonly, oral dose data are expressed as weight of chemical administered per unit body weight (e.g., mg of chemical per kg body weight). A concentration in food or drinking water (e.g., mg of chemical per kilogram of food) can be converted to a dose by multiplying by the mass of food or water consumed.

Both conversions between diet and dose and interspecies conversions of food and water consumption require size and consumption-rate data. To the extent possible, these data should come from the reports of the test data for the test animals and from measurements at the site for endpoint species, but generic literature values usually must be used for the endpoint species and sometimes must be used for the test species. Size data can be found in various faunal guides and in Clench and Leberman (1978), Norris and Johnston (1958), and Schoener (1968). Data on avian size and food requirements are presented in Kenaga (1973) and Szaro and Balda (1979). Food and water consumption rates for laboratory rodents, domestic animals, and some wildlife are presented in Spector (1956) and ECAO (1987). In the absence of

Table 6.2. Allometric Models of Food Consumption (F) in kg/day and Water Consumption (C) in L/day as Functions of Body Weight (kg) Derived from Data for Various Mammals, Based on Direct Measurements from Captive Animals[a]

Model Type	Equation	r	n
Dry diet	$F = 0.049\ W^{0.6087}$	0.95	148
	$C = 0.093\ W^{0.7584}$	0.95	148
Wet diet	$F = 0.054\ W^{0.9451}$	0.97	16
	$C = 0.090\ W^{1.2044}$	0.96	16

[a]ECAO 1987.

species-specific data, allometric regression models can be used (Tables 6.2 and 6.3). Calder and Braun (1983) presented the following models for water consumption (mL/d) as a function of weight (g) for birds:

$$C = 59\ W^{0.67}$$

and mammals:

$$C = 99\ W^{0.90}.$$

In most cases, reported toxic doses are single acute doses (administered by gavage, capsule, or bolus) and this must be compared to a dose received by drinking or feeding over some time period. Conventionally, daily consumption is assumed to be equivalent to a single acute dose. In other cases, effects of repeatedly administered oral doses are reported, and these can be compared to

Table 6.3. Allometric Models of Feeding Rates for Wildlife,[a] Based on Field Metabolic Rates[b]

Group	Equation	95% Prediction Interval[c]	n
Eutherian Mammals			
All eutherians	$F = 0.235\ W^{0.822}$	−63 to + 169%	46
Rodents	$F = 0.621\ W^{0.564}$	−64 to + 176%	33
Herbivores	$F = 0.577\ W^{0.727}$	−62 to + 161%	17
Marsupial Mammals			
All marsupials	$F = 0.492\ W^{0.673}$	−37 to + 59%	28
Herbivores	$F = 0.321\ W^{0.676}$	−46 to + 84%	12
Birds			
All birds	$F = 0.648\ W^{0.651}$	−55 to + 124%	50
Passerines	$F = 0.398\ W^{0.850}$	−31 to + 45%	26
Desert birds	$F = 1.11\ W^{0.445}$	−49 to + 95%	8
Sea birds	$F = 0.495\ W^{0.704}$	−61 to + 159%	15
Iguanid Lizards			
Herbivores	$F = 0.019\ W^{0.841}$	−59 to + 146%	5
Insectivores	$F = 0.013\ W^{0.773}$	−30 to + 43%	20

[a]From Nagy (1987).
[b]The units of F are g/d of dry matter and of W are g.
[c]The bounds of the 95% prediction intervals on mean y as percentages of the predicted mean y values.

doses resulting from consumption of chronically contaminated food or water calculated from consumption rates and contaminant concentrations in food or water. However, the doses estimated by such conversions can be inaccurate. Mallard test data show differences of 10X between LD_{50}s from 30d daily oral dosing studies and LD_{50}s estimated from dietary toxicity studies of similar durations for two pesticides and 4X for a third (Kenaga 1973). These errors are due at least in part to avoidance of contaminated food discussed below.

This discussion is based on the assumption that the toxicity is expressed as oral dose and that the uptake efficiency is equal for all forms of oral administration. If the toxicity is expressed as internal dose, some assumption must be made about uptake efficiency. For some chemicals, uptake efficiency values for rodents or humans can be found in the literature. Ideally, these should be derived by administration of realistic materials with the toxic material incorporated in a realistic manner. For example, to study the availability of dioxins in soil to mammals, Shu et al. (1988) administered to rats soil from a contaminated site, Times Beach, Missouri. In the absence of realistic uptake data, assumptions must be made based on expert judgment and analogies to similar chemicals. Of course, ad hoc uptake studies could be performed, but if the time and funds are available for such studies it would usually be preferable to go ahead and do an oral toxicity test with a realistic dosing medium.

MULTIPLE MEDIA AND ROUTES

In most ecological risk assessments it is possible to assume that one route of exposure is dominant and other routes are negligible. However, in some cases more than one route contributes significantly to the dose. In such cases, it is necessary to estimate the combined dose or to use toxicity data from tests that include all significant routes. For aquatic organisms, the need to assess exposure by multiple routes is limited to chemicals that have significant dietary exposure due to their persistence, affinity for organic materials, and low depuration rates. Synthetic chemicals with these properties are largely excluded from the environment by regulatory agencies, but some, such as PCBs and dioxins, are still commonly addressed because of incidental production (e.g., in incinerators) or reintroduction into the environment (e.g., dredging), and in retrospective risk assessments (Chapter 10). Organic chemicals with octanol-water partition coefficients greater than 4.3 (Oliver and Niimi 1983) to 5 (Bruggeman et al. 1981) are likely to have significant food-chain input to accumulation by fish. Such chemicals occur at higher concentrations in organisms at higher trophic levels, a phenomenon termed biomagnification. Food-chain input to bioaccumulation is seldom a concern in assessments of inorganic chemicals, but mercury is generally recognized to biomagnify through food chains in the methyl form, and recently selenium has been recognized to biomagnify through aquatic food chains (Lemly 1985). These cases can be addressed by models that combine food chain bioaccumulation with direct bioconcentration (e.g., Thomann 1981).

Cases of potentially significant multiple-route exposures are more easily identified for terrestrial environments. A bird in a field or forest that is sprayed with pesticides can receive an inhalation exposure, a dermal exposure through the feet and legs, and oral exposures through preening and ingestion of contaminated food. Similarly, wildlife on a waste site or oil field drink from liquid waste sumps, contact spilled oil or wastes resulting in dermal and oral exposures, consume potentially contaminated food, and ingest waste-contaminated soil by burrowing, traversing land farms, etc. and then grooming their paws and pelts (Figure 6.1). Avian pesticide exposures are currently assessed by assuming that dietary exposure dominates (Urban and Cook 1986), but more complete avian exposure models are being developed for pesticide regulation. Garten and Trabalka (1983) reviewed terrestrial food-chain data and concluded that only organic chemicals with K_{ow} values greater than 3.5 significantly bioaccumulated from food in mammals and birds. The exception was methyl mercury, which does not bioaccumulate primarily by a hydrophobic mechanism.

The most complex exposures take place where contaminants occur in both terrestrial and aquatic media, and organisms come into contact with both environments. Figure 6.2 depicts an exposure model developed for a complex of waste sites that includes contaminated soil, water, and sediments. Mink, which are highly sensitive to some chemicals, take a diversity of prey from both environments, and even great blue herons take some terrestrial prey.

In the absence of methods to model multiple routes of exposure and appropriate toxicity data, exposure by multiple routes must be simulated by toxicity tests in microcosms, mesocosms, or field sites. For aquatic contaminants, the need to estimate exposure by individual routes can be avoided by observing effects resulting from doses to entire mesocosms. For pesticides, terrestrial field tests can serve the same purpose (Chapter 9). The obvious advantage of this approach is realism; the test organisms, rather than an exposure model, integrate the various routes of exposure. The disadvantage is lack of generality; in current practice, the assessor must assume that exposure in the mesocosm or field test site is equal to exposure at all sites where the contaminant will occur. However, it is conceptually feasible to use the results of field tests to develop realistic contaminant-specific exposure models that could then be adapted to other sites, rather than assuming that all sites are the same.

BEHAVIORAL MODIFICATION OF EXPOSURE

The behavioral responses of organisms to toxicants may modify subsequent exposure. The most commonly reported example is avoidance of contaminated food or media. However, toxicants may also attract organisms, may reduce

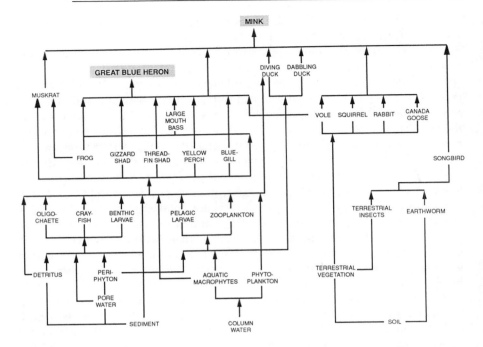

Figure 6.2.   A diagrammatic representation of the routes of exposure of two endpoint species, mink and great blue heron, to contaminated terrestrial and aquatic environments (from Macintosh et al. 1992).

feeding, thereby reducing the oral dose, or may cause sessile aquatic organisms to lose contact with the substrate and drift out of the area.

Because behavioral data are lacking for most chemicals, it is generally assumed that behavior does not modify exposure. The simplest way to incorporate behavioral data is to assume that if a clear avoidance response has been demonstrated in a test, exposure will be avoided in the field, where possible. If the contamination is local and brief, avoidance can prevent the occurrence of effects on mobile organisms. If organisms discriminate and avoid contaminated food such as treated seeds, effects may be avoided even if the contamination is long-term or widespread. However, avoidance of chronically contaminated media or natural foods results in habitat or food loss which can seriously affect populations. For example, the most serious effect on fish of the experimental contamination of Shaylor Run with copper was loss of habitat due to avoidance (Geckler et al. 1976). The importance of habitat and food limitations relative to other limitations on populations is an important and complex topic in ecology, but for purposes of assessment, it would not be overly conservative to assume that populations of exposed organisms that are not exploited by humans are limited by the availability of habitat or food, and the loss due to long-term avoidance is equal to the density of the population times the area in which media or food are so contaminated as to cause avoidance. If the population of interest is limited by human exploitation, it may generally be assumed that organisms avoiding pollution will find alternate habitat.

Unfortunately, avoidance responses are often not as clear as this assessment approach suggests. In fact, avoidance consists of two distinct reactions. The simplest is repellency or sensory aversion, the direct response to a substance that is repugnant to an organism. It occurs rapidly and its induction is a simple process of detection and reaction. The other type of avoidance is conditioned aversion, the learned avoidance of materials that cause illness. This reaction takes longer (12–24 hr for pesticides and birds — Bennett 1989) and requires that the organism (1) associate the contaminated material with the illness, (2) survive long enough to develop the aversion, and (3) discriminate the contaminated material from uncontaminated materials.

Most avoidance studies for water pollutants assume repellency and are conducted during acute exposures. However, some studies suggest that chemicals that are repellent at acutely lethal concentrations are not avoided at concentrations that cause sublethal effects or chronic lethality (e.g., Hellawell 1986). Also, after extended exposures habituation may occur. Therefore, repellency may not protect organisms from effects occurring after long exposures to low concentrations.

Both repellency and conditioned aversion depend on the unimpaired functioning of the organism's sensory and motor functions. The toxic effects of a contaminant may diminish or eliminate the ability of organisms to sense and avoid subsequent exposures. Doane et al. (1984) found that bluegill sunfish avoided low concentrations of jet fuel but not lethal concentrations. Therefore, repellency occurs only in a range of exposure concentrations between the lower concentration that is not repellent and an upper concentration that interferes with avoidance.

Toxicological interference with avoidance is particularly likely for conditioned aversions in which organisms must associate illness with a source and may be required to discriminate contaminated materials using clues other than repellency. The simple discrimination trials used to determine the potential for avoidance may exaggerate the ability of organisms to make complex discriminations among degrees of contamination in the field. Bennett (1989) found that bobwhite quail avoided methyl parathion at a tenth of the LC_{50} when presented with one container of treated food and one of untreated, but when presented with five containers of each they could only discriminate and avoid concentrations of half the LC_{50} or greater, and when presented with nine contaminated containers and one uncontaminated container, they could not discriminate concentrations greater than the LC_{50}. In addition, prior exposures to the contaminant may interfere with subsequent avoidance (Bussiere et al. 1989).

It is not clear how well avoidance and preference behaviors extrapolate between taxa. Hose et al. (1981) found that the concentration of chlorine that is avoided by fish is relatively consistent, and Smith and Bailey (1990) found that striped bass, steelhead trout, and chinook salmon were all attracted to refinery effluents. However, Hansen et al. (1974) found large inconsistencies in PCB avoidance by fish, Hellawell (1986) found that sublethal toxic concentrations of ammonia attracted stickelbacks but not green sunfish, and Gray (1990) found that fathead minnows, but not rainbow trout, avoided the water-soluble fraction of a synthetic oil.

The degree of protection conferred by avoidance is limited by the mobility of the animals relative to the area contaminated and the rate of onset of toxic effects. Avoidance confers no protection on sessile organisms, except for those such as bivalve mollusks that may close themselves off in response to some contaminants. Even highly mobile species such as fish and birds have life stages that are incapable of avoidance such as eggs, larvae, altricial young, and incubating adult birds. Avoidance may also be constrained by territoriality.

Because of these limitations and complications, the assumption that avoidance will protect organisms from toxic exposures should be applied with caution. Some of the diversity of results of exposures to chemicals that can induce avoidance are illustrated in Figure 6.3. Compilations of avoidance data can be found in Schafer et al. (1983), Schafer and Bowles (1985), Hara et al. (1983), Rand (1985), and Smith and Bailey (1989).

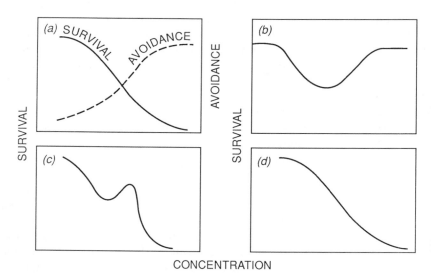

Figure 6.3.   Models of the influence of avoidance on concentration/response relationships: (a) separate concentration/avoidance and concentration/mortality functions; (b) a function resulting from simple joint action of mortality and avoidance; (c) a function resulting from suppression of avoidance at high concentrations; and (d) a function resulting from prevention of avoidance by territoriality, rapid onset of effects, inability to find refugia, etc. (Modified from Gray 1990).

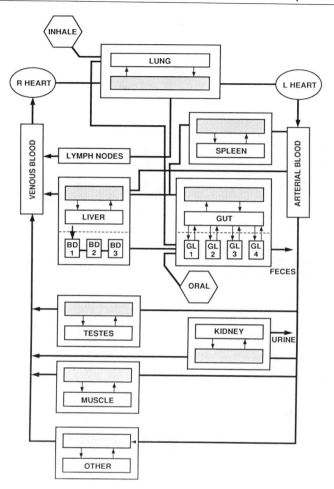

Figure 6.4. Diagram of mammalian toxicokinetics (redrawn from Menzel 1987). Gray boxes represent the organ's blood volume.

TOXICOKINETIC MODELS

Mathematical simulation models that estimate the internal dose of toxic chemicals from estimates of the external dose are called toxicokinetic models. Human toxicokinetic models are increasingly mechanistic, simulating the uptake, internal transport, partitioning among internal compartments, transformation, metabolism, and depuration of chemicals (Figure 6.4). They are derived from the pharmacokinetic models developed by the drug industry to estimate human doses from animal doses, and to estimate the appropriate dose for individuals of different sizes, ages, and sexes. They are increasingly used in health risk assessments to estimate effective doses to humans from animal data. Toxicokinetic models for nonhuman organisms other than laboratory rodents are relatively poorly developed and are never likely to be as precise and detailed in their representation of compartments and processes. However, human models serve as a useful guide for development of kinetic models for other species. Reviews of these models include Bischoff (1980), Dedrick and Bischoff (1980), Ramsey and Gehring (1980), Gibaldi and Perrier (1982), O'Flaherty (1986), Welling (1986), and Menzel (1987).

Toxicokinetic models have two potential uses in exposure assessment. The first is to estimate an internal dose that could be compared to effects data expressed as a function of internal dose. Such effects data might be obtained by analyzing the body burdens measured in organisms that have died or experienced other effects, or by administering internal doses by injection and noting the resulting effects. This approach is commonly used in health risk assessments (e.g., Baskin and Falco 1989), but is seldom used in ecological risk assessments because ecotoxicological effects data are very rarely expressed in terms of internal doses. The second potential use is as a means of converting the irregular temporal dynamics of exposure to match the exposures used in toxicity tests. That is, they can be used to estimate the exposure concentration in a toxicity test with a particular duration that would cause an internal dose equal to the internal dose resulting from the pattern of external exposure predicted by fate models. This use amounts to creating a duration scale for the integration of exposure and effects (Chapter 3). Use of toxicokinetic models to extrapolate between species and life stages and to develop toxicodynamic models is discussed in Chapter 7.

The obvious disadvantage of using toxicokinetic models rather than matching exposure in toxicity tests to the predicted pattern of exposure is that the models require considerable understanding of how organisms handle and respond to the chemical of interest. Like other mechanistic models, they can fail catastrophically if a critical process is not included or if the functional form is incorrect. Also, because of the large number of parameters and functions required to simulate the uptake from food and media, partitioning among organs, and various routes of metabolism and excretion, the cumulative uncertainty in such models can be quite large if parameters are poorly specified.

The obvious response to the problem of having too little knowledge to fully simulate the kinetics of a particular chemical in an organism is to make simplifying assumptions. This is most commonly done by treating the organism as a single compartment and accumulating the chemical into that compartment using linear uptake and depuration kinetics, and assuming that when the body burden reaches some concentration, death or some other effect results. The model is

$$\frac{dC_i}{dt} = r_u C_a - r_d C_i \tag{6.1}$$

where C_i and C_a are the concentrations in the organism and the ambient medium, respectively, and r_u and r_d are the corresponding uptake and depuration rates. The model is commonly used to estimate bioaccumulation by fish for health risk assessments, and can be similarly used in models for estimating risks to piscivorous wildlife. However, in exposure assessments for effects on fish the C_i that is solved for is the concentration causing an effect, rather than the equilibrium concentration in edible parts. The parameters of the model can be estimated either from QSARs or by measuring uptake and loss rates in laboratory studies (e.g., Kara and Hayton 1984, Connolly 1985, McKim et al. 1986). Even models as simple as this can fit ("explain") data from a variety of chemicals reasonably well and can be used to suggest where more complex models are needed (McKim et al. 1986).

Models with two internal compartments (an active compartment and an inactive storage compartment) are receiving increasing attention in aquatic toxicology. Kara and Hayton (1984) used a two-compartment model to simulate the kinetics of di-2-ethylhexyl phthalate in sheepshead minnows. Parameters were estimated by least squares fit to experimental data for the concentrations in water and in fish, and the concentration of metabolites.

Barber et al. (1988) have developed a biophysical toxicokinetic model for fish that can be used in the absence of data on kinetic properties of nonpolar organic chemicals. They assumed that kinetics result from passive transfer between an ambient aqueous phase, an internal aqueous phase, and an internal lipid phase; that uptake and depuration occur only through the gills; and that chemicals are not metabolized or actively excreted. The limitation of the model to thermodynamic processes makes it easy to parameterize, but raises concerns about model error relative to empirically-based toxicokinetic models.

Deviations of kinetics from simple models have become apparent in mammalian pharmacology and toxicology, where there is considerably greater experience. Examples include capacity-limited uptake and elimination, saturation of target binding sites, induction or inhibition of enzymes, and alteration of blood flow to target or metabolic tissues (O'Flaherty 1986). In addition, the application of simple toxicokinetic and toxicodynamic models to toxic effects relies on the assumption that active biological processes such as active uptake rates, metabolic rates, damage repair rates, and rates of replacement of inactivated enzymes are not affected by toxic injury. Human pharmacologists and toxicologists are increasingly using physiologically based toxicokinetic models that explicitly represent the interaction of chemicals with specific external and internal compartments and processes, as portrayed in Figure 6.4.

Barron (1990) advocates mechanistic physiologically-based toxicokinetic models that are specific to classes of chemicals. He argues that fish bioaccumulation models based on partitioning between lipid and aqueous phases are inadequate for chemicals with low K_{ow} or large molecular size, and cannot account for differences in rates of contaminant metabolism. His diagram of a model structure for a chemical that is taken up by the gills and metabolized in the liver is shown in Figure 6.5. A partial implementation of this approach that

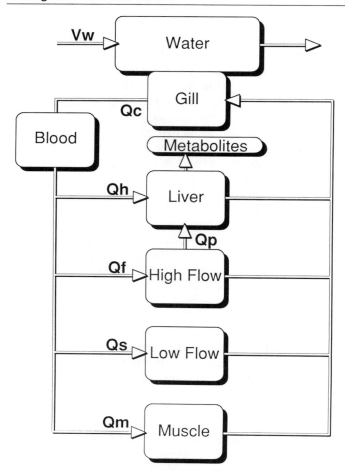

Figure 6.5. A diagram of fish toxicokinetics for organic chemicals that are taken up by the gills and metabolized by the liver (redrawn from Barron 1990).

demonstrates the importance of metabolism as a model variable for one class of chemicals (carboxyl esters) is presented by Barron et al. (1990).

In assessments of air pollutant effects on plants, models analogous to pharmacokinetic models can be used to estimate the amount of pollutant that reaches the vegetation canopy, individual leaves, and tissues within a leaf (Taylor et al. 1988). Unlike the empirical and system-dynamic approaches that dominate kinetic modeling of chemicals in animals, these models are based on physical theory. The most common class of models is analog resistance models, an analogy to Ohm's law for resistance in series:

$$J = C/R \qquad (6.2)$$

where J is the deposition rate of the pollutant, C is chemical potential of the pollutant, and R is the resistances of the canopy and various components of the leaves to movement of the gas. An analog resistance model of a vegetated landscape is presented diagrammatically in Figure 6.6. A variety of models have been developed, ranging from models of the interaction of air with structured plant canopies (Hosker and Lindberg 1982, Baldocchi 1988), to

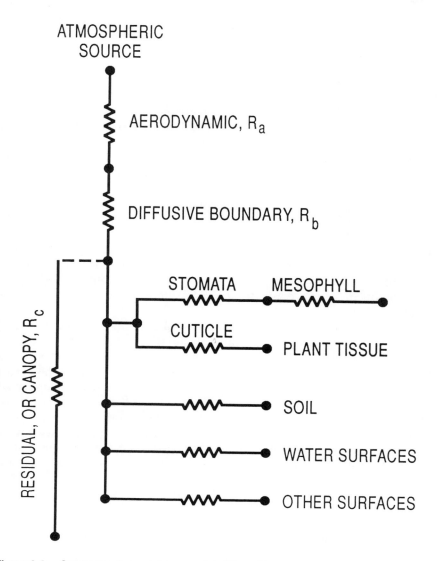

Figure 6.6.   General analog resistance model of deposition of gaseous pollutants to terrestrial landscapes (Taylor et al., 1988).

three-dimensional models of gas movement within leaves (Parkhurst 1986). Because of the direct path from environmental medium (air) to the target tissues (primarily leaf mesophyll), the detail and completeness of air/plant kinetics models exceed that of any other class of ecological exposure models.

Plant uptake from soil and distribution from tissues has primarily been treated empirically. However, a mechanistic model of plant toxicokinetics has been developed by Boersma et al. (1988) that assumes that uptake of pollutants by roots and distribution in plant tissues is based on passive movement with water.

EXPOSURE OF POPULATIONS AND ECOSYSTEMS

Assessment of exposure is nearly always conducted at the level of the individual organism, as in the discussion above. This approach is appropriate if the effects on populations and ecosystems are assumed to be a function of effects on the component organisms (see Chapters 8 and 9). In that case, exposure can be assumed to be equal for all members of a population or community, or some distribution of individual exposures can be derived. For example, Breck et al. (1986 and 1988) modeled the exposure of hypothetical populations of fish in a lake to a pulsed input of pollutants. The movement of fish was represented by a time-varying random walk biased by home range fidelity, habitat preference, or food availability. Fish were assumed to vary in their sensitivity to effects and degree of avoidance. Models of this sort could be used to estimate the proportion of a population that would be lost due to lethal combinations of concentration and duration in a heterogeneous habitat.

Alternatively, exposure of an entire ecosystem might be assessed as a unit. The most conspicuous recent example is the assessment of the ecological effects of acid deposition. In that example, an ecosystem's external exposure is the deposition rate, and internal exposure is usually expressed as the pH of water in soil, streams, and other ecosystem compartments. This treatment of exposure at the ecosystem level is necessary because of the need to link atmospheric sources with the whole-system exposure processes that cause whole-system effects such as loss of soil nutrients and buffering capacity. Other ecosystem-level risk assessments will need to consider whether analogous measures of whole-system exposure are more appropriate than measures of organism exposure.

PART III

EFFECTS ASSESSMENT

In this section we discuss methods for estimating the type and magnitude of ecological effects that result from exposure to pollutant chemicals. The discussion of effects is organized hierarchically. The lowest organizational level considered is the individual organism. Although effects on organs, tissues, cells, or even individual molecules can be determined, they have no practical significance except in helping to predict or interpret effects on organisms. Although physiological and histological responses are potentially precise and rapid screening tools, their use in toxicity testing has been limited by (1) the lack of data showing that they are good predictors of whole organism responses, (2) the difficulty in identifying tests that are neither so general that toxic effects cannot be distinguished from natural stress nor so specific to a single mode of action that they can be used only with well-characterized chemicals or effluents, and (3) the difficulty of interpreting the results (Mehrle and Mayer 1985, Neff 1985). A physiological response may be a sign of negative toxic effects (distress) or merely a sign of adaptation to the chemical (eustress). However, these problems do not interfere with the use of suborganismal characteristics to diagnose the cause of observed organismal or higher level effects (Chapter 10). Physiological measurements may also be used to develop mechanistic models of organismal responses (Chapter 7). At the other extreme, we do not consider prediction of effects on organizational levels higher than the individual ecosystem even though risk analysis could be applied at regional, continental, or planetary scale problems such as CO_2-induced climatic change or stratospheric ozone depletion (Chapter 11).

Organism-Level Effects

Glenn Suter

. . . but time and chance happen to them all.

— Ecclesiastes

The individual organism has been the primary focus of ecological toxicology. This is because (1) ecological toxicology is derived from human-oriented mammalian toxicology, which is concerned with the fate of individuals; (2) for macroorganisms, higher-level processes require too much space and physical structure to be readily observed in the laboratory; and (3) organismal responses are generally more easily observed and more easily interpreted than those of lower or higher levels. In sum, individual organisms have the optimum scale and organizational characteristics to be studied by individual human organisms. Because testing is easily performed on organisms and because test data for this level is abundant, there has been much more testing of effects than modeling of effects on individual organisms.

TESTS FOR EFFECTS AT THE ORGANISM LEVEL

Of the many types of tests that are conducted at the organism level, we will emphasize those that are standard. This is not because they are the best possible tests; rather, they are the only ones for which data are readily available and that can be readily conducted by testing laboratories, and therefore they are the only ones that will be available for development and implementation of assessment models and procedures. Nonstandard tests may be highly useful in specific situations. However, in standard assessment schemes such as that for the U.S. National Ambient Water Quality Criteria, they are usually left out of computations and are relegated to a table of "other results" (Stephan et al. 1985). Standard test methods are published by the Environmental Protection Agency (EPA) (1982b and various reports and *Federal Register* notices), Organization for Economic Cooperation and Development (OECD 1981, 1984), American Society for Testing and Materials (ASTM 1991), American Public Health Association (APHA 1985), and various other government agencies and standard setting organizations.

Aquatic Algae

Test endpoints for effects on algae have been less well standardized and their relevance to the field are less clear than for animals (Lewis 1990). Reported responses included mortality, growth, CO_2 fixation, cell numbers, chlorophyll content, and others. Durations have been various, and a variety of statistical expressions derived from both hypothesis testing and curve fitting have been used. There is now some agreement on the use of 96-h EC_{50} values for some measure of productivity. However, there is still no agreement on whether the appropriate measure is weight, number of cells, chlorophyll, or carbon assimilation, and whether the benchmark should be based on the final value, the time-integrated value, or the maximum rate of increase. The EPA calls for the use of final cell weight, cell number, or an equivalent indirect measurement, whereas the OECD calls for the use of the maximum growth rate based on cell number (EPA 1982b and OECD 1981, 1984). The appropriate indicator depends on both toxicological and ecological considerations. Different toxicants may reduce the final cell concentration, reduce the growth rate, or both (Figure 7.1, Adams et al. 1985, p. 337). The nature of control of productivity may determine which of these effects is more important in the assesed ecosystem. If planktonic algae are limited by nutrient availability, equilibrium biomass or cell numbers may be more relevant. However, if algae are limited by herbivory, the ability of a population to replace losses (i.e., maximum growth rate) may be more relevant. In fact, the same algal population in nature may experience both types of control at different times of the year.

Because the life cycles of microalgae in a rapidly growing culture are much shorter than both test durations and most effluent releases, the laboratory test results may be considered to be population-level responses. However, it should be remembered that algal communities are generally nutrient limited, and, over the course of chronic exposures, populations of resistant algal species will tend to replace those of sensitive species. The implications of these changes in community composition depend on the effects of the algae on water quality and their palatability to herbivores.

Good reviews of algal toxicity testing is provided by Calamari et al. (1985), Adams et al. (1985), and Lewis (1990). Toxicity data for aquatic algae have been compiled in the EPA's AQUIRE database, and in the EPA's water quality criteria support documents (EPA 1980, 1985a).

Aquatic Vascular Plants

Relatively few tests of effects on aquatic vascular plants have been conducted, but they are becoming more common. The most popular of these is the 96-h EC_{50} for growth of duckweed (*Lemna spp.*) (Walbridge 1977). This test examines a rapid and ecologically significant response, development of new fronds, which is likely to be of ecological importance in many circumstances. However, the amount of data on responses of aquatic vascular plants is still

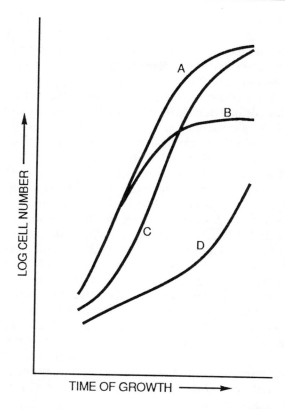

Figure 7.1. Possible algal patterns of response to increasing concentrations of a chemical. Curve (A) is typical of untreated algal cultures and curves (B-D) represent types of responses observed with chemicals having different modes of action (redrawn from Adams et al. 1985).

relatively small and includes few species, so the utility of the data for risk analysis is uncertain.

Fish

Toxicity data for fish are exceptionally abundant and the methods are well standardized (EPA 1982b, OECD 1981, OECD 1984, ASTM 1991, Peltier and Weber 1985, Weber et al. 1988, 1989). The standard tests for fish are acute lethality tests, life cycle tests, and the various early life stage tests. Test conditions may be flow-through (test solution is constantly renewed), static (the original solution is used for the entire test), or static-renewal (test solution is replaced periodically). Results from flow-through test data are generally the most useful for analysis because the exposure concentration is nearly constant. Conversely, static test data are generally the least desirable because concentrations may decline considerably over the course of the test. However, static tests may be better to simulate a spill or other pulse input of a pollutant to a poorly mixed body of water, and may also simulate a planktonic organism drifting with a pollutant plume. Although toxicity generally declines as pollutants degrade in a static system, oxidation products may be more toxic than the parent compound (Eisler 1970, Mill 1988).

The most commonly used test endpoint for fish is the 96-h LC_{50} for adult or juvenile (post-larval) individuals. It is "acute" in the sense of severe, but the duration is fairly long relative to most pulsed exposures. For many chemicals, 96 h corresponds to the time-independent or incipient LC_{50} (Ruesink and Smith 1975). However, Eisler (1970) said that 96 hours is simply the longest test duration that does not interfere with the weekend. Because it does not protect early life stages and implies mass mortality of late life stages, it can be thought of as an endpoint for conspicuous fish kills (large numbers of large dead fish). Although the median response was chosen for the test endpoint because of its small variance relative to other levels of mortality, a correction factor must be applied if the assessor is interested in preventing low-level mortality, a process that adds considerable uncertainty.

Another problem with LC_{50}s is that in most cases only the response at 96 h is reported. Many assessments involve transient events, in which case the time to mortality is more important than the percent mortality. However, despite the recommendations of Sprague (1970), Alabaster and Lloyd (1982), ASTM (1991), and others, the time course of mortality is seldom reported. In defense of the 96-h LC_{50}, it has been argued that it is only meant to be used for comparative purposes, and not for assessing effects. However, assessments are conducted and criteria are set using LC_{50}s because they are available, and better numbers are generally not.

Toxicity tests for fish that are termed chronic include (1) full life cycle tests that go from egg to egg and may require two years, (2) partial life cycle tests that go from adults to juveniles, (3) early life stage tests that go from eggs to juveniles, and (4) five to eight day "short-term chronic" tests involving eggs and larvae or just larvae. Although chronic EC_{01} values have been suggested as alternate chronic test endpoints by the EPA (Weber et al. 1989), the standard endpoint for chronic tests on fish has been the MATC, more recently termed the chronic value by the EPA (Weber et al. 1989, Stephan et al. 1985). MATCs have all of the considerable faults of endpoints that are derived by statistical hypothesis testing (Chapter 3). In this context, it is important to reiterate that assessments based on MATCs do not provide a consistent level of protection, and the industry that performs the poorest tests will, on average, be the least stringently regulated.

The most generally useful test endpoint for assessing effects on fish would be a set of concentration-duration-response surfaces, one each for mortality of each of the life stages that will be exposed, and for growth and fecundity (Figure 3.9). Individual effects thresholds or expected levels of effect could then be selected for each assessment, depending on the life stages that will be exposed and the duration of the exposure.

If all life stages will be exposed to a relatively constant concentration of the toxicant, then a global endpoint, one that integrates the individual measured effects (Javitz, 1982), may be preferred as an expression of test results. The simplest such endpoint is the standing crop of fish at the end of the test. More commonly, the weight of young per initial female (or initial egg, in the case of early life stage tests) is calculated as

$$\text{Wt./Fem.} = S_1 S_2 \ldots S_n MW \qquad (7.1)$$

where S_x is the survivorship of life stage x, M is fecundity, and W is the average weight in the final cohort (e.g., Mount and Stephan 1967, McKim et al. 1976, Eaton et al. 1978).

The main advantage of global test endpoints is that they combine a diversity of measured responses, some of which have little intuitive significance, into a parameter that has the form of a population-level response. Global responses may be more sensitive than individual responses when a number of small toxic effects are combined into one large global response. Conversely, sensitivity can be reduced if toxic effects are combined with hormetic or pseudohormetic effects or (if hypothesis testing is used) with highly variable effects.

Aquatic Invertebrates

The most commonly tested aquatic invertebrates are cladoceran crustaceans of the genera *Daphnia* and *Ceriodaphnia*. The standard *Daphnia* tests are the 48-h EC_{50} for immobilization or death and the 21-d test for mortality and reproduction. The standard *Ceriodaphnia* tests are a 48-h EC_{50} and a 7-d test for mortality and fecundity. The most common salt water test invertebrates are the mysid crustacean *Mysidopsis bahia* and the shrimp *Penaeus duorarum* which are used in 96-h LC_{50} tests and 28-d tests of mortality, reproduction, and growth. The oyster (*Crassostrea sp.*) embryo-larval 48-h LC_{50} is also standardized and is now in common use. A variety of other aquatic invertebrates are used in 48- or 96-h lethality tests, including insects (e.g., *Chironomous sp.*) and annelids (e.g., *Neanthes arenaceodentata*).

The standard test endpoint for the 1 to 4 week tests is a NOEC or MATC (chronic value) based on hypothesis testing statistics and the 48- and 96-h lethality test endpoints are obtained by curve fitting. The comments on these test endpoints and alternate endpoints made above are also applicable here. The test endpoint could be based on the number or weight of young per initial female. In addition, because multiple generations of daphnids are produced in the 1-to 4-wk tests, population parameters such as the intrinsic rate of increase (r) can be calculated (Chapter 8). These endpoints represent growth of a population founded by a single individual.

Terrestrial Plants

Existing toxicity data for terrestrial plants is highly diverse and nonstandard (Fletcher et al. 1985). Effects on plants are measured and statistically analyzed in a variety of ways, during various stages of their life cycles, and including various organs. Plants can be exposed to toxins through the stomates, leaf surfaces, or roots. Measures such as yield, growth, or numbers of particular organs that directly express productivity are more useful in risk analysis than those such as visible injury and changes in gas exchange rates that do not correlate with production. Because available phytotoxicity data are not well standardized, it is important for the analyst to carefully consider the relevance of the duration and route of exposure in the test to the exposure being assessed.

The most common type of phytotoxicity test is the seedling growth test. This type of test can be conducted in soil or hydroponic systems and can be adapted to test chemicals in air, sprays, soil, or irrigation water. There is little agreement on either test durations or test endpoints, but the EPA (1982) recommends the determination of EC_{10} and EC_{50} values for weight and height after 14 days. Tests for effects on seed germination and hypocotyl elongation have been used as quicker and less-expensive phytotoxicity tests, as well as indicators of effects on those particular life stages (EPA 1982b). However, the relationships of these short-term responses to other plant responses have not been established. A definitive test would include the entire life cycle from seed germination to germination of daughter seeds, but such tests are rarely performed. A partial life-cycle test using *Arabadopsis* has been developed by the EPA (Ratsch et al. 1986).

Phytotoxicology for toxic effects of air pollutants and for other pollutants (pesticides, herbicides, and industrial chemicals) has developed almost independently. While phytotoxicologists who work in the context of registration of pesticides and industrial chemicals have addressed thousands of chemicals, air pollution phytotoxicologists have largely confined their attention to four criteria air pollutants (sulfur dioxide, nitrogen oxides, ozone, and particulates). As a result, air pollution phytotoxicologists have been able to devote considerable attention to factors that are relatively neglected in other fields of environmental toxicology, including effects of physical conditions, differences between populations and between genotypes, species interactions, and mechanisms of action. Unfortunately, most of the resulting data have relatively low utility in environmental analysis. A major problem is the lack of standardization of test methods, which prevents quantitative comparisons of results or extrapolation of results to other situations. In addition, most of these data have been generated in academic laboratories, and there has been no process of methods standardization to settle differences between the interests of academicians and regulators. Consequently, many of the studies use concentrations, durations, or response measurements that are inappropriate for either assessing new pollutant sources or setting criteria (Sigal 1988).

Heck et al. (1979) provide a good review of methods for testing effects of air pollutants on plants including laboratory, greenhouse, and field methods. Good reviews of the air pollution literature are found in Smith (1990), Winner et al. (1985), Heck et al. (1988), MacKenzie and El-Ashry (1989), and recent issues of the EPA's air quality criteria support documents (EPA 1982c, 1982d, 1984). Phytotoxicity data for nongaseous organic chemicals have been compiled in the EPA's PHYTOTOX data set. Phytotoxicity data for metals are compiled by Lepp (1981), Kabata-Pendias and Pendias (1984), Adriano (1986), Pahlsson (1989), and Tyler et al. (1989).

Terrestrial Invertebrates

The only terrestrial invertebrate that is routinely used in ecological assessment is the honeybee. Topical lethality tests are required for registering pesticides to which bees are likely to be exposed (Urban and Cook 1986). Earthworms have been advocated as test organisms, but no standard method has been generally adopted (Roberts and Dorough 1985). Recently, the EPA (Greene et al. 1988) has proposed a 14-day earthworm lethality test to assess toxicity of soils at waste sites, based on the method of Goats and Edwards (1982) which was developed for the OECD.

Terrestrial Vertebrates

The most common test endpoint available for assessing effects on wildlife is the acute, oral, median lethal dose (LD_{50}) for laboratory rodents. Avian toxicologists have followed the mammalian example by relying largely on acute LD_{50}s for adults (e.g., Hudson et al. 1984), but subacute median lethal dietary toxicities for young birds (LC_{50}s) have become common (e.g., Hill et al. 1975, Hill and Camardese 1986) and have been adopted by the EPA (1982b) and ASTM (1991). Although no wild mammal species is used in routine testing, the EPA may require LD_{50}s or LC_{50}s for mammals in the areas likely to be affected (OPP 1982, Urban and Cook 1986). These test endpoints are applicable to short-term exposures such as result from application of nonpersistent pesticides. In assessments of such cases, the concentration in food has been the primary expression of exposure; therefore, oral LC_{50}s are directly applicable, whereas intake must be estimated to calculate doses before LD_{50}s can be used (Chapter 6). In a few cases, notably when the exposure results from direct consumption of granular pesticides and coated seeds or from cleaning pelt or plumage, an oral LD_{50} is more directly applicable. Because the relative sensitivities of adults and young and the effects of exposure duration are less well known for birds than fish, the comparability and utility of these endpoints are uncertain.

The other standard wildlife test endpoint is the threshold for statistically significant effects in the avian reproduction test (EPA 1982b, ASTM 1991). This test resembles the MATC for chronic and subchronic effects on fish in that the endpoint is usually derived by applying hypothesis testing statistics to an array of measured parameters. Like the MATC, it would be more useful for assessment if curve fitting were used to establish a consistent level of effect, and if a global parameter (such as the weight of young per female) were calculated along with the measured responses (fecundity, survival, growth, and eggshell thickness). The duration of exposure in this test (6–10 weeks) should approximate a chronic adult exposure to a discrete release of any but the most persistent and bioaccumulated chemicals; however, because the young are not exposed, this is not a life-cycle test.

Because some pesticides can significantly affect birds, field tests or large pen tests may be required for pesticide registration (OPP 1982, Urban and Cook 1986, Fite et al. 1988). Unfortunately, the results of most of these tests are proprietary, and so are available only to the manufacturer and the regulatory agency.

Few data are available for assessing the toxic effects of nonpesticide chemicals and effluents on wildlife. It is generally necessary to resort to the use of the health literature for such assessments. LD_{50} values for domestic mice (*Mus musculus*) and rats (*Ratus ratus*) are the most common, relatively consistent test endpoints for terrestrial vertebrates and must be at least as relevant to wild rodents as to humans. Ecological assessors have made little use of mammalian toxicity data because it has been generally assumed that if humans are protected, other mammals will be also. However, differences in mode of exposure and sensitivity can cause major effects in other mammals in areas where humans are apparently unaffected; fluorosis and selenosis in ungulates (NRC 1971, Shupe et al. 1980, Eisler 1985) are examples (see also Box 1.1). Although birds are thought to be more sensitive to toxicants than mammals, data on avian toxicity is largely limited to pesticides. Tests for effects on reptiles and amphibians are rare and are largely confined to aquatic larval amphibians. [See the oral LD_{50} test results for bullfrogs in Hudson et al. (1984) and tests of amphibian larvae in Mayer and Ellersieck (1986) and Birge et al. (1981).] Compilations of wildlife toxicity data for pesticides and repellents include those of Hill et al. (1975), Schafer et al. (1982, 1983), Hudson et al. (1984), Schafer and Bowles (1985), and Hill and Camardese (1986). Newman (1979, 1980) has reviewed the relatively small body of data concerning air pollution effects on wildlife. Effects of metals on mammals have been reviewed by Venugopal and Lucky (1978).

MODELS FOR EFFECTS AT THE ORGANISM LEVEL

Although mathematical and statistical models have been much less important than toxicity testing in assessment of organism-level effects, the number and variety of models are still greater for assessing effects of chemicals on individual organisms than for populations and ecosystems. They include dose-response models and similar statistical models of test results, structure-activity relationships for predicting toxic doses from characteristics of the chemical, toxicodynamic models, and models for extrapolating toxicity data between species, life stages, exposure durations, and exposure conditions.

Models of Test Results

Toxicity tests are complex phenomena involving a variety of responses by numerous organisms over a period of time. This complexity is managed by choosing a model of toxic effects that determines what observations will be recorded and the manner in which those observations will be summarized and reported (the test endpoint). Ideally, the model should be chosen to represent those aspects of the dimensions of toxic effects (concentration, duration, severity, and proportion responding) that are important to the assessment for which the test is intended (Chapter 3). More commonly, standard models are applied and they are used to calculate single numeric test endpoints rather than to describe dynamics.

The most common type of model of test results is the dose-response (or equivalent concentration-response) function. The pattern of response with increasing dose is typically S-shaped, so the most commonly used functions are the S-shaped probit and the logit. Both functions can be expressed as linear functions of a logarithm of dose (D):

$$P_t = a + b \ln(D) \tag{7.2}$$

where P_t is proportion responding transformed to probits or logits and a and b are fitted constants. Transformed values can be obtained from tables in texts of biostatistics or from statistical software packages. The probit function assumes that the distribution with respect to log dose of the sensitivity of individuals is normal; a probit is the normal equivalent deviate plus five (Hewlett and Plackett 1979). The logistic function, unlike the probit, is easily calculated due to its relatively simple algebraic form:

$$P = (a + bD)/(1 + a + bD) \tag{7.3}$$

where P is proportion responding.
   The Weibull function:

$$P = 1 - \exp(-aD^b) \tag{7.4}$$

has been advocated for use in aquatic toxicology (Christensen 1984, Christensen and Nyholm 1984) and has been used for gaseous pollutant effects on plants (Rawlings et al. 1988). It is related to the single-hit, multi-hit, and multi-stage models of cancer induction and implies the assumption that an effect occurs when a single receptor has been exposed to a dose unit of the chemical.

These three functions assume no threshold but that is usually not a concern in ecological risk assessments because very low effect levels are not estimated. These three functions also assume a dichotomous response, such as mortality, of recognizable (i.e., countable) individuals. However, continuous responses and nondichotomous quantal responses can also be fit to these functions by representing the response as a proportion of the control response. Because there is no consensus on appropriate dose-response functions for nonbinary responses, functions may be chosen simply on the basis of fit to the data. Descriptions of dose-response models and their use can be found in Ashton (1972), Altshuler (1981), Brown (1978), Finney (1964, 1971), Hewlett and Plackett (1979), Kooijman (1981, 1983a, 1983b), and Stephan (1977).

Because algal and other microbial populations grow and decline within the span of conventional laboratory toxicity tests, they display concentration-response dynamics that are more complex than those of macroorganisms. This complexity can be avoided in many cases by limiting these tests to the period of maximum growth and fitting simple functions like the exponential. This can be difficult to do in algal or other microbial growth tests where nutrient or substrate limitations may exert their effects quite early. In addition, this simplification eliminates the possibility that toxic effects on maximum population can be observed. Another solution is to develop separate numeric endpoints (EC_{50}s) for different aspects of algal population dynamics. However, it is preferable to use a model that reflects all aspects of population dynamics. Li (1984) added terms to the logistic function so that it could reflect both the growth and decline of algal cultures over time.

In addition, many test chemicals may stimulate growth of algae or microbes at low concentrations. Bahner and Oglesby (1982) developed an algal response function that includes stimulation of growth as well as inhibition. Kooijman (1983a) recognized three effects of chemicals on algae (mortality, decreased carrying capacity, and decreased reproduction rate), and prepared models for each effect and for all effects. Such functions are potentially valuable in describing a variety of population and ecosystem processes that have natural temporal dynamics similar to the growth and decline of an algal culture, or that may be both promoted and inhibited by different concentrations. The practical problem with such functions is that more fitted parameters (there are four in Li's function, six in Bahner and Ogelsby's, and nine in Kooijman's) require more toxicity data than the conventional two parameter functions. Otherwise there will not be sufficient degrees of freedom for the regression statistics, and the full range of effects will not be revealed.

Complex dose-response dynamics can also occur in conventional organism-level toxicity tests. The most common cause is hormesis, improved performance relative to controls of organisms receiving small doses of nonnutrient chemicals (Stebbing 1982). Lucky and Venugopal (1977) estimated that "50 percent of all potentially harmful agents will be stimulatory in minute doses" and that hormesis generally occurs at a tenth to a thousandth of toxic doses. Hormesis may be a real effect of the chemical that is due to stimulation of the organism. It may also be due to processes that are apparently irrelevant such as reduced aggression in the presence of a narcotizing test chemical, or to processes that have questionable applicability such as the increase in food for larval fish that results from microbial growth on the test chemical or solvent. Paradoxical dose-response dynamics (increase, then decrease, then increase in effects as dose increases) may also occur due to induction of metabolic enzymes, or to decreases in uptake or response due to narcotizing effects (Filov et al. 1979). The assessor must decide whether deviations from a monotonic increase in effects are real and whether they reflect phenomena that are relevant to the field. The common practice of ignoring hormesis and paradoxical effects in the name of conservatism may not be justified in all cases.

Concentration-duration functions are considerably less common. Green (1965) suggested using a linearized hyperbolic function:

$$LD_{50} = a + b(t^{-1}) \qquad (7.5)$$

because its intercept represents the effective dose for exposures of indefinitely long durations. Some authors have used existing data to generate concentration-duration models. For example, Wang and Hanson (1984) found chlorine LC_{50}s from 16 species of aquatic invertebrates with multiple exposure durations of 96 hours and less. They fit six functions to each set of data and chose the function that provided the best fit for each species. In most cases, the best function was either log linear with respect to duration or log-log linear.

Three-dimensional models of toxicity test results are uncommon in environmental toxicology, and four-dimensional models are not known to occur in the literature. The three-dimensional models are typically expansions of the standard dose-response models. For example, the probit plane model:

$$P_p = a + b \ln(D) + c \ln(t) \qquad (7.6)$$

where P_p is the proportion responding in probits and t is time, has been advocated by Finney (1971) and Hewlett and Plackett (1979) for modeling dose-duration-response relationships.

Structure-Activity Relationships for Effects

Structure-activity relationships (SARs) are models for predicting the properties of a chemical from its structure or from some physicochemical property that is more easily determined than the estimated property. SARs originated in the fields of drug and pesticide design where they have attained considerable sophistication (Martin 1978, Topliss 1983). They can be used in ecological risk assessments to estimate toxic effects when toxicity data are not available, as is often the case when registering a new industrial chemical (DiCarlo et al. 1985) or reviewing existing chemicals (NRC 1981a, Klein et al. 1988, Walker and Brink 1988). They can be divided into qualitative and quantitative SARs. Qualitative SARs consist of either (1) identifying a chemical that is sufficiently similar to the chemical of interest to serve as a surrogate, or (2) identifying a substructure of the chemical that is believed to be responsible for toxicity. The chemical of interest is assumed to have the same mode of toxic action as the surrogate chemical or as other chemicals possessing the substructure, and is expected to have roughly the same quantitative toxicity. These qualitative approaches depend heavily on the expertise and experience of the assessors.

Quantitative structure-activity relationships (QSARs) are statistical models that are nearly always obtained by regressing values of a common test endpoint for a series of chemicals against one or more quantifiable properties of the chemicals. The properties may be (1) structural characteristics of the molecules such as the number of carbon atoms or presence of a particular substituent; (2) derived structural descriptors such as molecular connectivity indices and topological indices; (3) inherent physicochemical properties including electronic properties (dipole moments, molar refractivity, acid and base dissociation constants, and Hammett constants) and steric properties (Taft constants and molecular volume); or (4) physicochemical properties that relate to environmental behavior such as water solubility, partitioning coefficients, and vapor pressure. The relative utility of these various properties is a subject of debate.

There are also multiple approaches to definition of the chemical series to which the QSAR applies. Ideally the series would be homologous in that all members would have the same mode of toxic action (toxicodynamics) and the same modes of uptake, partitioning, metabolism, and excretion (toxicokinetics) (Rekker 1985). The best-described example is the nonionizing or baseline narcotics (Konemann 1981, Veith et al. 1983, Lipnick 1985). These organic chemicals apparently have no specific mode of action, but physically disrupt cell membranes, enzymes, or other organic structures to which they adsorb. The kinetics of these chemicals is dominated by partitioning so that the most hydrophobic chemicals are the most toxic relative to their concentrations in water (up to the point at which transfer rates become limiting). When there is no information on toxicodynamics or kinetics, an alternate approach used to define a chemical series is to simply adopt conventional chemical categories such as organochlorides or alcohols. Finally, if not enough chemicals have been tested for a particular test endpoint to partition them into homologous or chemically consistent classes, or if there is no reliable basis for identifying classes, a heterogeneous series of chemicals is used (e.g., Zaroogian et al. 1985). In any case, it is essential that the chemical series to which a QSAR applies be clearly defined so that a user of the model knows whether it applies to his chemical.

A number of different approaches can be used to select independent variables for QSARs. The simplest is to identify a variable that is believed to dominate the toxic process as hydrophobicity does for the aquatic effects of narcotic chemicals. A more complicated version of this approach is to identify the variable associated with each individual process in the induction of toxic effects and use that variables in a multivariate model. For example, Lipnick's (1985) model for proelectrophilic chemicals is:

$$\log (1/LC_{50}) = A \log P + B E^\circ + C \log k + D \qquad (7.7)$$

where P is the octanol/water partitioning coefficient, E^o is the redox potential, k is the reaction rate of the activated chemical with biological macromolecules, and A, B, and C are fitted constants. This model corresponds to the conceptual model of effects induction shown in Figure 7.2. A third approach is to use basic chemistry to identify molecular descriptors that capture the important differences among members of a chemical series, assuming that if you have captured the essence of the molecular structures, the toxicology will follow. Finally, stepwise regression can be used to select variables from a list of variables without prejudging their importance. This last approach is less desirable because of the risk of chance correlation, particularly if the number of toxicity data is not much greater than the number of potential variables.

This discussion should indicate to the reader the diversity and dynamism of QSAR research. The number of QSARs in the literature is very large. Table 7.1 presents the QSARs that are used by the EPA Office of Toxic Substances. This list constitutes the nearest thing to a standard QSAR set for ecological toxicology, but the reader should be aware that the models are not all of the highest quality.

Rather than developing empirical QSARS, it is potentially possible to mathematically simulate the uptake, transfer, metabolism, storage, elimination, and effects of a chemical using toxicodynamic models with physicochemical parameters as input. Toxicodynamic models differ from the more commonly used toxicokinetic models (Chapter 6) in that they include effects. The mathematical modeling of effects has been limited by the fact that the sites, targets, and modes of action of most chemicals are not known. DNA-damaging chemicals, particularly radionuclides, are a major exception, but these effects have not been an important consideration in most ecological risk assessments. Toxicodynamic models are discussed at greater length below.

Effects Extrapolation Models

We use the term effects extrapolation to refer to the process of predicting some effect of interest Y such as a threshold for lethality in indefinitely long exposures from some measured effect X such as a 96-h LC_{50}. Effects extrapolation models are analogous to SARs except that the toxic effect of interest is estimated from a measured toxic effect for another species, life-stage, or test type, rather than from physical or chemical properties of the toxicant. To sustain the analogy, these models might be termed Activity-Activity Relationships (AARs).

The practice of extrapolating effects data has been controversial. For example, Tucker and Heagele (1971) recommended that "extrapolation of toxicity data from one species to another should be avoided." Wiemeyer and Sparling (1991) stated that "the response of one species to a given pesticide should not be used to predict the sensitivity of another species to the same pesticide," and Dobson (1985) stated that "extrapolation of imperfectly understood data to the field situation is worthless." Clearly, these positions are

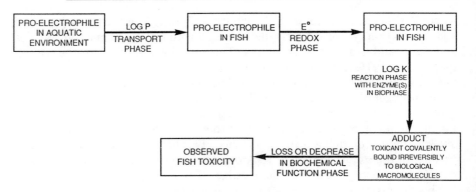

Figure 7.2. Diagram of the processes underlying Lipnick's (1985) QSAR for effects of proelectrophilic chemicals.

impractical because we cannot test all species, much less all life stages and conditions. Decisions must be made even when our understanding is imperfect. However, it is also unreasonable to assume that any laboratory test result is a perfect representation of responses by different organisms under different conditions. As Peakall and Tucker (1985) concluded, "the degree of extrapolation that one is prepared to make depends on the degree of uncertainty of risk estimation that one is prepared to accept." In this section we consider the different methods of effects extrapolation and the ways in which they have been used to address the experimental variables that contribute to the differences in test results.

Two classes of models have been used for effects extrapolation in ecological assessments. The first class includes additive and multiplicative factor models:

$$Y = a + X \text{ (additive)} \tag{7.8}$$

$$Y = bX \text{ (multiplicative)} \tag{7.9}$$

The second class is statistical models, the most common of which is the linear regression model:

$$Y = a + bX \tag{7.10}$$

Although these models are mathematically similar, their use in ecological assessment has been quite different. Factor models are deterministic and tend to rely on expert judgment guided by informal data analyses, whereas formal statistical models are probabilistic and rely more on the data than on the expertise of the assessor.

Table 7.1. QSARs for Toxic Effects Used by the EPA Office of Toxic Substances[a]

Endpoint	Equation	Source
Neutral Organics		
1. Fish 96–hr LC_{50} (Freshwater)	$\log LC_{50} = -0.94 \log P + 0.94 [\log (0.000068\ P) + 1] - 1.25$	Veith et al. 1983
2. Fish 14-d LC_{50} (Freshwater)	$\log (1/LC_{50}) = 0.871 \log P - 4.87$	Konemann 1981
3. Fish 96-hr LC_{50} (Marine)	$\log (1/LC_{50}) = 0.73 \log P - 3.69$	Zaroogian et al. 1985
4. Daphnid 48–hr LC_{50}	$\log (1/LC_{50}) = 0.91 \log P - 4.72$	Hermens et al. 1984
5. Mysid 96-hr LC_{50}	$\log (1/LC_{50}) = 1.25 \log P - 4.83$	Zaroogian et al. 1985
6. Daphnid 16-d LC_{50}	$\log (1/LC_{50}) = 0.64 \log P - 3.27$	Hermens et al. 1984
7. Daphnid 16-d EC_{50}	$\log (1/EC_{50}) = 0.72 \log P - 3.05$	Hermens et al. 1984
8. Green Algae 3-hr EC_{50}	$\log (1/EC_{50}) = 8.865 - 1.0446 \log P$	Nabholz and Johnson
Acrylates		
9. Fish 96-hr LC_{50} (Freshwater)	$\log LC_{50} = -1.46 - 0.18 \log P$	Nabholz and Platz
10. Green algae 8-d NOEC (Growth)	$\log NOEC = -0.66 - 0.34 \log P$	Nabholz and Platz
11. Blue green algae 8-d NOEC (Growth)	$\log NOEC = -1.25 - 0.48 \log P$	Nabholz and Platz
12. Protozoan 48-hr NOEC (Growth)	$\log NOEC = -0.81 - 0.26 \log P$	Nabholz and Platz
Methacrylates		
13. Fish 96-hr LC_{50} (Freshwater)	$\log LC_{50} = 0.552 - 0.715 \log P$	Nabholz and Platz
Aldehydes		
14. Fish 96-hr LC_{50} (Freshwater)	$\log LC_{50} = 0.4487 \log P - 3.314$	Clements and Nabholz
Anilines		
15. Fish 14-d LC_{50} (Freshwater)	$\log (1/LC_{50}) = 0.988 \log P - 4.02$	Hermens et al. 1984

Table 7.1. Continued

Endpoint	Equation	Source
Benzotriazoles		
16. Fish 96-hr LC_{50} (Freshwater)	$\log LC_{50} = 3.366 - 0.587 \log P$	Nabholz
17. Daphnid 48-hr LC_{50}	use Equation 4	Nabholz
18. Green algae 96-hr EC_{50}	$\log LC_{50} = 3.366 - 0.587 \log P$	Nabholz
Carbamates		
19. Sea urchin 48-hr NOEC (Development)	$\log NOEC = 3.509 - 0.7174 \log P$	Nabholz and Johnson
Esters		
20. Fish 96-hr LC_{50} (Freshwater)	$\log LC_{50} = -0.535 \log P - 2.75$	Veith et al. 1984
Esters, Phosphate		
21. Fish 96-hr LC_{50} (Freshwater)	$\log LC_{50} = 2.9305 - 0.5178 \log P$	Nabholz
Esters, Phthalate		
22. Fish 96-hr LC_{50} (Freshwater)	use Equation 20	none
23. Daphnid 48-hr LC_{50}	use Equation 4	Nabholz
24. Daphnid 21-hr NOEC (Reproduction)	use Equation 7	Nabholz
Phenols, Chlorinated		
25. Fish 96-hr LC_{50} (Freshwater)	$\log (1/LC_{50}) = 0.75 \log P - 3.51$ [at pH 6.1]	Konemann and Musch 1981
26. Fish 96-hr LC_{50} (Freshwater)	$\log (1/LC_{50}) = 0.58 \log P - 3.20$ [at pH 7.3]	Konemann and Musch 1981
27. Fish 96-hr LC_{50} (Freshwater)	$\log (1/LC_{50}) = 0.46 \log P - 3.04$ [at pH 7.8]	Konemann and Musch 1981
Phenols, Substituted		
28. Fish 96-hr LC_{50} (Freshwater)	$\log (1/LC_{50}) = 0.73 \log P - 0.61$ [at pH 6.0]	Saarikowski and Viluksela 1982
29. Fish 96-hr LC_{50} (Freshwater)	$\log (1/LC_{50}) = 0.61 \log P - 0.38$ [at pH 7.0]	Saarikowski and Viluksela 1982

30. Fish 96-hr LC$_{50}$ (Freshwater)	log (1/LC$_{50}$) = 0.52 log P – 0.31 [at pH 8.0]	Saarikowski et al. 1982
Polycationic Polymers		
31. Fish 96-hr LC$_{50}$ (Freshwater)	log LC$_{50}$ = 1.3076 – 0.534 (% amine N)	Nabholz
32. Daphnid 48-hr LC$_{50}$	log LC$_{50}$ = 1.3116 – 0.7606 (no. of + charges/g mol. wt.) tabular	Nabholz
Ethomeen Surfactants		
33. Aquatic acute toxicity	tabular	Nabholz
Linear Alkyl Benzene Sulfonates		
34. Fish and Daphnid LC$_{50}$	log LC$_{50}$ = [(ave. no. C – 16)^2 – 10.643]/12.935	Nabholz
35. Green algae 96-hr EC$_{50}$	tabular	Nabholz
Nonionic Surfactants		
36. Fish 96-hr LC$_{50}$	tabular	Nabholz
37. Daphnid 48-hr LC$_{50}$	tabular	Nabholz
Quaternary Ammonium Surfactants		
38. Fish acute LC$_{50}$ (Freshwater)	tabular	Nabholz
39. Daphnid acute LC$_{50}$	tabular	Nabholz
40. Snail acute LC$_{50}$	tabular	Nabholz
Ureas, Substituted		
41. Algae 4-hr EC$_{50}$ (Photosynthesis)	log(1/EC$_{50}$) = 1.29 log P – 2.867	Hansch 1969

P = octanol water partitioning coefficient.
Units are moles/L for Equations 1 and 14, millimoles/L for Equations 9–13 and 28–30, mg/L for Equations 31–40, and micromoles/L for all others.
Sources without dates are internal EPA documents authored by the individuals cited.
[a]Clements, 1988.

Factor Models

Factor models are the most common method of extrapolating environmental effects data and are the only models applied to effects data in conventional hazard assessments. The factors are variously referred to as safety factors, uncertainty factors, or correction factors, depending on whether the goal is to add a margin of safety, account for a recognized source of uncertainty, or correct for proportional differences between types of data. Traditionally, a single number was used that incorporated all of the assessor's knowledge and beliefs about the relationship between the test result and the anticipated effect in the field (Mount 1977). More recently, it has become common to use multiplicative strings of factors, each of which accounts for a different correction or source of uncertainty (e.g., OWRS 1985). These multiplicative chains imply an assumption that everything will go wrong at once. For example, the most sensitive life stage of the most sensitive species will be exposed to the most concentrated effluent at low-flow conditions while debilitated by stress, and the actual response is at the limit of our range of uncertainty. If carried out consistently, this approach would be extremely conservative. In actual applications, only a fraction of the possible uncertainties and corrections are included, so that the product of the factors is not unacceptably large. Reviews of the use of factors in aquatic assessments can be found in Mount (1977) and Tebo (1986).

Three methods are commonly used to derive factors. The first is expert judgment. Expert judgment is often the only available basis for the decision, and in some cases it is more accurate than available formal models. However, it has the general disadvantage of making the assessment opaque to the decisionmaker and others who may review the decision. More important, it is a biased source of uncertainty factors because experts consistently overestimate the accuracy of their estimates (Fischoff et al. 1981). The second method is to use the ratio of observed values of the two classes of data that the factor is intended to extrapolate between. These classes may be discrete data types such as acute and chronic data or they may be data that characterize a distribution such as the LC_{50} values for the most and least sensitive species tested with the same chemical. For example, Chapman (1983) concluded that taxonomic differences could have a 18,000X effect on toxic effects of a chemical because the largest ratio of the highest to lowest LC_{50} in a large data set was 18,000. The third method is to use the difference between observed values of the two classes of data that the factor is intended to extrapolate between (e.g., the difference between the LC_{50}s for the most and least sensitive species).

It should be obvious that the use of differences and ratios to estimate factors is equivalent to assuming that there is a linear relationship between Y and X, but in the first case the slope is assumed to be 1, and in the second case the line is forced through the origin. That is, $Y = a + X$ and $Y = bX$, respectively. This observation leads us to the topic of formal statistical extrapolation models.

Statistical Extrapolation Models

The simplest statistical models to extrapolate toxicity data are probability density functions. These models assume that a toxicological parameter is characterized by a normal, log normal, binomial, or other probability density function, and individual future observations are drawn from that distribution. The functions are fitted to existing data to estimate the parameters of the functions. The parameterized functions can then be used to estimate percentiles of the distribution. The purpose of these models is to extrapolate from a sample of observations to the full population. Their most common use is to estimate distributions of sensitivity of a community of species with respect to exposure concentrations from test data for a few species (Box 2.2).

Regression models assume that two classes of data, such as LC_{50} values for rainbow trout and cutthroat trout, are not simply randomly drawn from some probability distribution, but are correlated, and therefore each can serve as a predictor of the other. Most regressions used for effects extrapolation are purely empirical. That is, they are simply statistical expressions of an observed relationship and are intended to provide a best estimate of the predicted variable rather than to explain the mechanism that underlies the relationship. An early example of this type of model in ecological toxicology is the regressions of acute toxic responses of pairs of standard test species by Kenaga (1978). As discussed above, linear regression models can be forced to produce multiplicative and additive factors by eliminating either the slope or intercept. However, in most cases we have no a priori reason to be sure that the slope is 1 or the intercept is 0. Although all regression models for effects extrapolation known to the author are linear, log linear, or log-log linear, there is no reason that nonlinear models could not be used.

A discussion of regression models for ecological assessment is presented in Box 7.1 for those who are interested in developing such models. One somewhat technical point that will be discussed here is the proper expression of the quality of a regression model. The most commonly used expressions are the correlation coefficient (r) or its square (r^2). These values express the ability of a model to explain the data and are appropriate for comparing different models that are being applied to the same set of data. However, r and r^2 are not appropriate for comparing models applied to different data sets and do not indicate how good the model is as a predictor, although they are often used that way (e.g., Kenaga 1979b). As Peakall and Tucker (1985) point out in their critique of Maki (1979b), a high r value can hide a great deal of uncertainty about the relationship of individual x and y values. The value of a regression model to the assessor depends on the accuracy with which it predicts a new y value, given a new x value. This accuracy is estimated by the variance of the individual points about the line. Confidence intervals calculated from this variance are termed prediction intervals, and the prediction interval for y at mean x can be used to express the goodness of an extrapolation model.

Examples of the use of prediction intervals can be found in Call et al. (1985), Suter et al. (1987b), and Holcombe et al. (1988).

Box 7.1 Regression Models

Most regression analyses are performed using ordinary least squares (OLS) regression algorithms, without consideration of whether the underlying assumptions are met. However, alternate regression models exist and the choice of an appropriate regression model for a particular problem is a complex decision that deserves more attention (Ricker 1975). The toxicity data used in extrapolation models have the following characteristics:

1. the observed values of X and Y are subject to measurement error and inherent variability (stochasticity),

2. X is not a controlled variable (like settings on a thermostat),

3. values assumed by X and Y are open-ended and non-normally distributed.

These characteristics suggest that an OLS model would be inappropriate, and an errors-in-variables model should be used. Since we can estimate the value of the ratio of the point variances of Y and X from variances of replicate tests, a functional model provides maximum likelihood estimators of the regression parameters.

The estimators of slope (β) and intercept (α) are:

$$b = \{\Sigma y^2 - \lambda\Sigma x^2 + [(\Sigma y^2 - \lambda\Sigma x^2)^2 + 4\lambda(\Sigma xy)^2]^{1/2}\}/2\Sigma xy \qquad (B7.1)$$

and

$$a = \bar{y} - b\bar{x} \qquad (B7.2)$$

where $x = X_i - \bar{X}$ and $y = Y_i - \bar{Y}$ for $i = 1 \ldots n$. The variance of a predicted Y-value for a given X-value ($X = X_0$) is given in Mandel (1984) as:

$$\text{var}(Y|X_0) = s_e^2\{1 + 1/n + [1 + (b^2/\lambda)]^2[(X_0 - \bar{X})^2/\Sigma u^2]\} \qquad (B7.3)$$

where

$$s_e^2 = (b^2\Sigma x^2 - 2b\Sigma xy + \Sigma y^2)/(n - 2),$$

and

$$\Sigma u^2 = \Sigma x^2 + 2b/\lambda\Sigma xy + (b/\lambda)^2\Sigma y^2.$$

For ease in using the models generated by this method, the variance formula is reduced to:

$$\mathrm{var}(Y|X_0) = F_1 + F_2(X_0 - \bar{X})^2. \tag{B7.4}$$

When using errors-in-variables models, the inverse regression (X from Y) can be calculated as $x = (y-a)/b$. Variance for this inverse regression reduces to:

$$\mathrm{var}(X|Y_0) = G_1 + G_2(Y_0 - \bar{Y})^2. \tag{B7.5}$$

The means of X and Y and the F and G factors are presented in Tables 7.4, 7.7, and 7.8.

Toxicodynamic Models for Effects Extrapolation

Mathematical simulation models of the response of individual organisms to exposure to toxicants are termed toxicodynamic models. They are largely derived from the toxicokinetic models that are used to estimate internal exposure (Chapter 6). Toxicokinetic can be used to simulate toxic effects if one is willing to assume that the effect of interest occurs at a particular internal concentration or activity, and if there is some means for estimating that concentration or activity. Alternately, the process by which the chemical induces the effect can be simulated using a true toxicodynamic model. The possible advantages of these simulation models are (1) they potentially offer a more precise prediction of effects than models that relate effects to ambient concentrations; (2) by integrating variation in concentration over time, they provide a means of dealing with complex temporal dynamics; and (3) by explicitly considering the effects of size of different body compartments, differences in metabolic rates, enzyme complements, etc., these models can simulate variance in effects among life stages, sexes, and species. The obvious disadvantage of toxicodynamic models is that they require all of the understanding of how organisms take up, distribute, metabolize, and excrete chemicals of a toxicokinetic model, plus understanding of how the effects are induced.

A continuous response that is proportional to internal concentration might be represented as an internal concentration-response function (O'Flaherty 1986). The internal concentration causing an effect can be determined by analysis of the body burden at the time of death or other observable response (Mount et al. 1966). However, many chemicals cause persistent cumulative damage or for other reasons cause effects that are not proportional to internal concentration. For such chemicals, the dynamics of damage and the response to damage must be explicitly considered (Figure 7.3). Connolly (1985) presented a model of the inhibition of cholinesterase activity by diazinon in fathead minnows in response to an internal concentration that he estimates from a kinetic model. He did not make the next step from cholinesterase inhibition to death or other overt effects. Such toxicodynamic models should

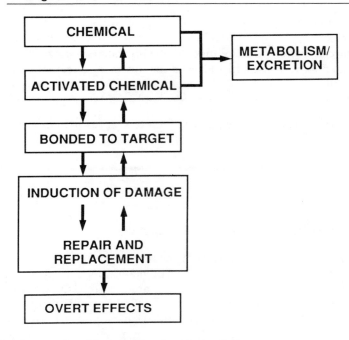

Figure 7.3. A general model of organism-level toxic effects.

be able to address anomalous patterns of response such as the post-exposure mortality observed following episodic exposure of rainbow trout to Cd (Pascoe and Shazili 1986), but they are beyond the current state of the ecotoxicological art.

VARIABLES

In this section we examine how differences in species, test conditions, exposure durations, and other variables influence the magnitude of test endpoints. This review draws on previous reviews of experimental variables in toxicity tests (Sprague 1970, 1985; Tucker and Leitzke 1979; Alabaster and Lloyd 1982; Mayer and Ellersieck 1986), but it is concerned with conceptual and methodological issues, and does not attempt an exhaustive survey of relevant research.

Taxonomic Differences

The most obvious issue to be considered in effects extrapolation is differences in sensitivity among species. These taxonomic extrapolations can be either specific or generic. Specific taxonomic extrapolations estimate the sensitivity of one species from the measured response of another species, as in estimation of human sensitivity from test data on rats. This type of extrapolation occurs in ecological risk assessment when effects must be estimated for a valued species such as the Atlantic salmon or bald eagle, and data are available for test species such as the fathead minnow or Japanese quail. Generic taxonomic extrapolations use test data for one or a few species to estimate the sensitivity distribution of the members of the biotic community that will be exposed to a pollutant. Generic extrapolation has received more attention both because of the desire to protect communities, and because protection of the most sensitive member of a community can be assumed to confer protection on individual valued species.

Generic Taxonomic Extrapolations

Most of the research relevant to taxonomic extrapolations has involved efforts to identify sensitive species. If such species can be identified, then one might assume test endpoints for that species to be protective of the biotic community. The model is, $Y = X$, where Y is the response of the most sensitive species in the exposed community, and X is the response of the sensitive test species. Although this model is not explicitly stated, it is the implicit rationale behind rankings of sensitivity of test species. The consensus that has emerged from these exercises is that on average (and particularly for insecticides), arthropods are more sensitive that fish, fish are more sensitive than amphibian larvae, and salmonids are the most sensitive fish (Macek and McAllister 1970, Birge et al. 1986, Tucker and Leitzke 1979, Mayer and Ellersieck 1986). A recent analysis of the data from the national ambient water quality criteria for aquatic life found that daphnids and salmonids were frequently sensitive freshwater animals, but that daphnids could be 200X more resistant than the most sensitive species, and salmonids could be 44,000X more resistant (Host et al. in press). Mysid and penaeid crustaceans have been found to be the most frequently sensitive salt water species (Suter and Rosen 1988, Host et al. in press).

Kenaga and Moolenaar (1979) argued that algae and other aquatic plants were nearly always less sensitive than aquatic animals. This conclusion is not supported by other studies (e.g., Patrick et al. 1968, Walsh et al. 1982), possibly because Kenaga and Moolenaar limited their algal tests to the green alga *Chlorella*. Maloney and Palmer (1956) found in a study of 30 algal strains that on average diatoms were most sensitive, green algae were least sensitive, and blue-green algae were intermediate.

One result of attempts to rank species by sensitivity was the discovery that there is no consistently most sensitive aquatic animal (Patrick et al. 1986, Kenaga 1978, Mayer and Ellersieck 1986, Kooijman 1987), alga (Blanck et al. 1984), or bird (Tucker and Heagele 1971, Hill et al. 1975). This finding has led some investigators to attempt to identify clusters of species that would include the most sensitive species, or would at least come close. Kimerle et al. (1983) looked at combinations of four commonly tested aquatic animals (*Daphnia*, fathead minnow, rainbow trout, and bluegill) as predictors of the most sensitive tested species from among LC_{50}s for 82 chemicals. They found that when one of the three fish species was paired with *Daphnia*, the lower LC_{50} for the pair was within an order of magnitude of the lowest LC_{50} for 90% of the chemicals. With their average data set of 9.5 species/chemical, any combination of two species would include the most sensitive species by chance alone 21% of the time. Sloof et al. (1983) and Sloof and Canton (1983) tested a standard species set consisting of an alga (*Selenastrum capricornutum*), a fish (*Poecilia reticulata*), and *Daphnia magna*, and found that it failed to get within a factor of 10 of the most sensitive species about 25% of the time for sets of both "(sub)acute" and "(semi)chronic" tests.

Although no sensitive species or small group of species invariably contains the most sensitive test species for all chemicals in a data set, the idea of a sensitive test species dies hard. One answer to the problem is to say that observed cases of greater sensitivity are inconsequential. In other words, no species is significantly more sensitive than the sensitive test species. This solution raises the question of what is a significant difference in sensitivity. The conventional answer is to use statistical significance as a criterion (Schafer and Brunton 1979), but this confuses statistical significance with biological significance. Another answer is to declare that differences in sensitivity among species are insignificant unless they are larger than differences among tests of a species-chemical combination. For example, Kenaga (1978) argued that differences in sensitivity of less than one order of magnitude were insignificant because rat LD_{50}s for the same chemical (but different formulations and modes of administration) varied as much as 2.5 orders of magnitude. Similarly, Fogels and Sprague (1977) argued that differences of an order of magnitude are insignificant, based on nearly order of magnitude variation among laboratories in test results for fish. Because most LC_{50} values for fish species tested with the same chemical fell within a one order of magnitude range, Fogels and Sprague concluded that all fish were about equally sensitive. The problem with this criterion is that it confuses variability among species when tested in a single laboratory with a single protocol with variability among laboratories and protocols. Also, because test protocols are becoming more standardized, variance among laboratories has declined, so any particular difference in sensitivity is becoming more significant by this criterion. Finally, it is clear that order-of-magnitude differences in concentration can be important in the field and so cannot be safely ignored.

Because no one species or small set of species reliably contains the most sensitive species and it cannot be demonstrated that the cases of greater sensitivity are necessarily inconsequential, it is prudent to at least use a correction factor for variance in sensitivity among species. Early examples were based on the ranges of sensitivity observed by experienced toxicologists. For example, Lloyd (1979) suggested a factor of 5X for native fish, 50X for aquatic invertebrates, and 100X for algae. More recently, analyses of sensitivity ranges in defined data sets have been published. Using data set for the U.S. National Water Quality Criteria, Chapman (1983) found that LC_{50}s for individual chemicals differed by as much as 18,000X among animal species. Mayer and Ellersieck's (1986) analysis of data for 82 chemicals tested with 6 or more species indicated that the average range of LC_{50}s was 256X and the maximum was 166,000X. Sloof et al. (1983) found that in "(sub)acute" tests of 22 species with 15 chemicals, the maximum difference was 9,000X and in "(semi)chronic" tests of 11 species and 8 chemicals the range was 10,000X (Sloof and Canton 1983). The use of observed ranges of sensitivity as correction factors for taxonomic variance depends on the assumption that the relative sensitivity of the test species and the most sensitive species in the community is equal to the relative sensitivity of the most and least sensitive species in the data set. This is likely to be a very conservative assumption.

The EPA Office of Water Regulations and Standards (OWRS 1985) has based their uncertainty factor for taxonomic variance on the results of Kimerle et al. (1983), discussed earlier. If an invertebrate and a vertebrate have been tested, then a factor of 10 is used for "species sensitivity." Although the logic of this factor is not explained, it seems to be based on three assumptions: that any invertebrate is as sensitive as *Daphnia*, that any vertebrate is as sensitive as the three fish used by Kimerle et al. (1983), and that protecting a small number of test species 90% of the time is adequate. This approach combines the use of a most sensitive species group with the concept of factors based on ranges of sensitivity discussed above.

Given that there are 769 species of fish in North America (only 6% of which have been tested for even one of the priority pollutants—Seegert et al. 1985) and hundreds of thousands of other aquatic species, the expectation that the limited set of species that have been tested with any set of chemicals contains either the most sensitive aquatic species or a species approximately as sensitive as the most sensitive species, is rather hopeful. To avoid this assumption, one way is to assume that the sensitivity of species follows some probability distribution (Mount 1982) and to define the concentration that affects the most sensitive species (or a sufficiently sensitive species) as the lower Xth percentile concentration. This approach has been used in calculating the U.S. National Water Quality Criteria (Erickson and Stephan 1985, Stephan et al. 1985). For each chemical, a log triangular distribution is fit to LC_{50} values for at least eight species from at least eight different families. The Final Acute Value (FAV) is the lower 5% concentration, so the intent is to protect 95% of species. Similarly, Kooijman (1987) and van Straalen and Denneman (1989) fit log logistic distributions to toxicity data for aquatic and soil species, respectively (Figure 7.4). The choice of distribution makes little difference, but the choice of protection level (percent of species), the confidence level (certainty that no more than the prescribed percentage of species will be affected), and the uncertainties included in the estimation of confidence are quite important (Box 2.2).

Kooijman (1987) differs from the other authors in protecting all of the species in a community rather than some percentage of species. This approach implies that communities with more species require a lower criterion. One

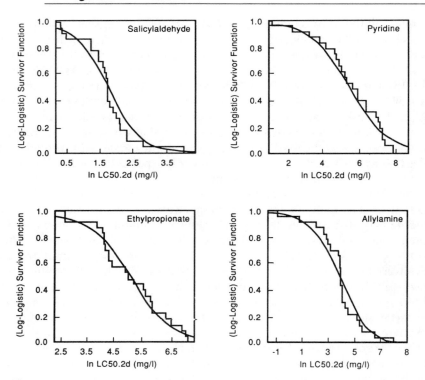

Figure 7.4. Observed survivor functions (proportion of species that have log LC_{50}s greater than the concentration X) for different toxic compounds, compared to the fitted logistic functions (redrawn from Kooijman 1987).

might argue to the contrary that more diverse communities require less restrictive criteria because they are buffered by functional redundancy (Van Straalen and Denneman 1989).

The use of sensitivity distributions assumes that test species are randomly drawn from the biotic community. It has been argued that this is not the case because test species are selected for their ease of maintenance in the laboratory, but that argument is probably incorrect for two reasons. First, there is no evidence that tolerance of laboratory conditions is necessarily correlated with tolerance of chemicals. Second, tolerance of laboratory conditions is a variable that increases with greater understanding of proper laboratory techniques, but sensitivity to chemicals does not change. At least two standard aquatic test species, *Daphnia magna* and rainbow trout, were once considered difficult to maintain in the laboratory.

If pollution damage is easily observed in the field, it is possible to establish the actual distributions of sensitivity of an entire community under ambient exposures. McLaughlin conducted a field inventory of degree of visible injury to establish the distribution of sensitivity to SO_2 injury of plant species around a coal-fired power plant (Figure 7.5).

If there are too few data to confidently estimate the parameters of the distribution of sensitivity, it would be useful to look at the distribution across chemicals of the parameters of sensitivity distributions. For example, if the distribution across chemicals of the ratios of the *Daphnia* LC_{50}s to percentiles of the species sensitivity distributions are known, a single *Daphnia* LC_{50} can be used to estimate a criterion concentration or to estimate the probability that a species with unknown sensitivity will be affected at a prescribed concentration. The state of Michigan requires that chemicals be tested for acute lethality with *Daphnia magna* and either rainbow trout or fathead minnow. To extrapolate to the FAV (the fifth percentile of the species sensitivity distribution), they plotted ratios of the lowest LC_{50} of the *Daphnia*-fish pair on log normal probability paper and picked off the ratios that would underestimate the FAV for 80% of chemicals (Figure 7.6). The values were 7 and 9 for fathead minnow and rainbow trout, respectively (Giesy and Graney 1989). Stephan and Erickson (n.d.) have used this approach to develop factors that, when applied to the lowest LC_{50} for an aquatic animal, estimate the lower 5% confidence bound on the FAV (Table 7.2). Two sets of factors are required, one for sets of LC_{50}s that include daphnids, which are frequently more sensi-

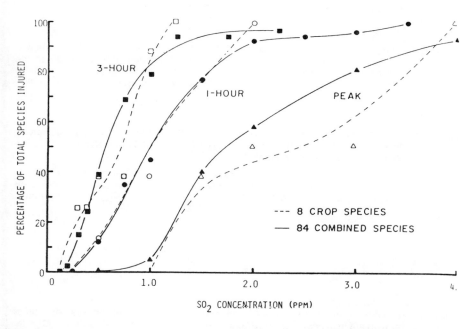

Figure 7.5. Percentage of plant species on which visible injury was detected as a function of peak, 1-h average, and 3-h average SO_2 concentrations near a coal-fired power plant (from McLaughlin and Taylor 1985).

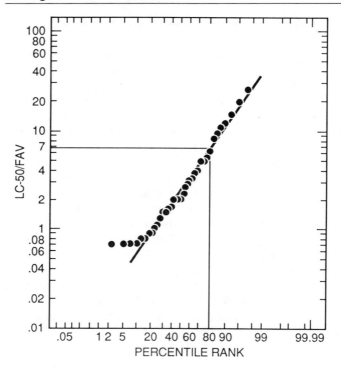

Figure 7.6. Probability plot of the ratio of the LC_{50} for the most sensitive species of the pair (*Daphnia magna* and rainbow trout) to the estimated fifth percentile of the distribution of species (final acute value) from the national ambient water quality criteria, as a function of percentile rank (redrawn from Giesy and Graney 1989).

Table 7.2. Factors for Estimation of the Lower 5th Percentile of the Distribution of Species Sensitivities from Sets of 1–7 LC_{50}s with 95% Confidence That the Value Will Not Be an Overestimate[a]

Number of LC_{50}s[b]	Factor for Data Sets That Include an LC_{50}[b] for a Daphnid[c]	Factors for Data Sets That Do Not Include an LC_{50}[b] for a Daphnid[c]
1	94	10,000
2	58	1,700
3	51	420
4	42	180
5	31	110
6	22	64
7	13	40

[a]Erickson and Stephan n.d.
[b]All LC_{50}s should be from different families and otherwise should meet the requirements of data used to calculate final acute values (Stephan et al. 1985, EPA 1985a). Multiple LC_{50}s for a species should be geometrically averaged to generate a species mean LC_{50}. Species mean LC_{50}s for congeneric species should be geometrically averaged and those genus mean LC_{50}s should be used in the calculations.
[c]Daphnids includes members of the genera *Daphnia*, *Ceriodaphnia*, and *Simocephalus*.

Table 7.3. Average Log Deviation of Standard Test Species LC_{50}s from the Mean Test Species for Fish (Freshwater Osteichthys)[a]

Species	Mean Log Deviation
Rainbow trout	−0.196
Fathead minnow	−0.054
Golden orf	0.056
Bluegill sunfish	−0.204
Zebra fish	0.213

[a]Volmer et al. 1988.

tive than average, and one for sets with no daphnids. Volmer et al. (1988) calculated a weighted mean variance for log LC_{50}s of fish (0.232) and mean deviations of LC_{50}s for standard tests species from the mean LC_{50} for all fish (Table 7.3). These can be used to estimate the mean and percentiles of the log normal distribution of fish sensitivities.

The Office of Pesticide Programs of the EPA takes an approach to protecting sensitive species that is logically distinct from those discussed above (Urban and Cook 1986). They assume that differences in sensitivity among species are equivalent to differences in sensitivity among individuals in a test population. Thus, their safety factor for differences among wildlife taxa is $LD_{50}/LD_{0.01}$. Based on an average slope of dose-response functions, this factor is five.

Specific Taxonomic Extrapolations

The simplest approach to specific taxonomic extrapolation (predicting the response of an untested species from data on a tested species) is to use the ratio of the responses of the two species to other chemicals for which they have both been tested. Schafer et al. (1983) determined average ratios of LD_{50}s for species of birds (e.g., redwing blackbird/Japanese quail = 1.4). Peakall and Tucker (1985) plotted ratios of LC_{50}s as histograms (Figure 7.7). Although these authors used the distributions of ratios to argue that the responses of one species could not be used to accurately predict the responses of another species, the distribution could also be used to make probabilistic estimates of response. That is, they could be used to generate statements of the form "given a Japanese quail LC_{50} of x, the expected redwing blackbird LC_{50} is y/x and redwing LC_{50}s would be lower than z/x for less than 5% of chemicals" (where y = 1.4 and z is the fifth percentile of the distribution of ratios).

The obvious statistical technique for relating the response of one species to that of another is regression analysis (Figure 7.8). Kenaga (1978, 1979) regressed LC_{50}s and LD_{50}s for all combinations of eight terrestrial and aquatic species using data for 75 pesticides. He concluded that both species of a highly correlated pair need not be tested, and that regression-derived equations could be used to predict responses of untested species. Since then, several publications have appeared regressing pairs of aquatic species against each other,

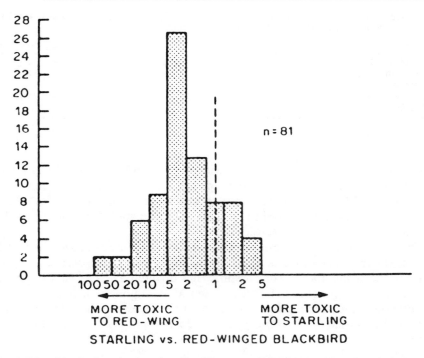

Figure 7.7. Distribution of ratios of starling LD_{50}s to red-winged blackbird LD_{50}s from Peakall and Tucker (1985).

including Maki (1979), Doherty (1983), LeBlanc (1984), Thurston et al. (1985), Sloof et al. (1986), Mayer and Ellersieck (1986), and Mayer et al. (1987).

Some authors have reported that taxonomic regressions are influenced by the types of chemicals tested. LeBlanc (1984) and Mayer and Ellersieck (1986) obtained better correlations between fish and invertebrates with nonpesticide chemicals than with pesticides. Thurston et al. (1985) and Suter et al. (1986) noted that acute lethality of narcotic chemicals to fish is highly predictable, and that relative sensitivities of species could differ among modes of action.

Fewer attempts have been made to apply interspecific regression analysis to wildlife species. Hill et al. (1975) correlated dietary LC_{50}s for four species of birds, and Hudson et al. (1979) regressed mallard duck against rat oral LD_{50}s. Sigal and Suter (1989) regressed mallard and pheasant against rat LD_{50}s for organophosphate pesticides as a means of assessing the risks of organophosphate nerve agents to birds. Interspecies regressions have also been used in human health-oriented mammalian toxicology; Craig and Enslein (1981) showed a consistent log-linear relationship between acute lethality in laboratory rats and mice; and Crouch (1983) showed a good relationship for carcinogenicity in those species.

Regression between pairs of species allow one to predict responses for only

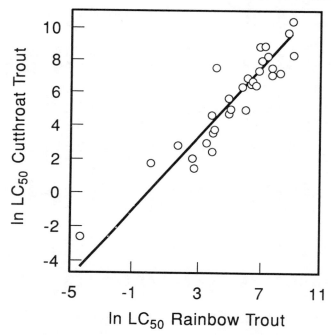

Figure 7.8. Regression of cutthroat trout LC_{50} values against those for the congeneric species rainbow trout. The model is presented in Table 7.4, line 1. The data are from Mayer and Ellersieck (1986).

the few species that have been frequently tested. In risk assessments, it is often necessary to consider effects on species that are not standard test species. Suter et al. (1983) devised a system to address this problem, based on taxonomic relationships between the tested species and the species of interest. Regressions were performed between all pairs of species that occur in a common genus, all pairs of genera within common families, families within orders, etc. Extrapolations are made between taxa having the next higher taxonomic level in common. For example, extrapolations from fathead minnow to largemouth bass constitute an extrapolation between Cypriniformes and Perciformes. A set of hierarchical taxonomic extrapolations for freshwater and salt water fish and arthropods is presented in Table 7.4.

This use of the taxonomic hierarchy is based on the assumption that a species can be represented by the members of a taxon to which it belongs, but that the uncertainty associated with that representation increases with increasing taxonomic distance. In other words, the evolutionary processes that result in increasing differences in the morphological and biochemical traits used by taxonomers, also result in increasing differences in the characteristics that control uptake and response to chemicals. This assumption is the basis for preferring mammals over nonmammals and primates over nonprimate mammals in testing for toxic effects on humans. The generalization holds on aver-

Table 7.4. Equations for Extrapolating Between Taxa of Freshwater and Marine Fish and Arthropods; Units are log μg/L[a]

No.[b]	X taxon	Y taxon	Slope	Intercept	n	$\bar{X}$[c]	F_1[c]	F_2[c]	$\bar{Y}$[c]	G_1[c]	G_2[c]	PI[d]
Species within Genera												
1F	Oncorhynchus clarki	O. mykiss	0.98	0.04	18	2.47	0.24	0.0095	2.45	0.25	0.010	0.96
2F	O. kisutch	O. clarki	0.95	0.15	6	2.34	0.075	0.0065	2.38	0.083	0.0080	0.54
3F	O. kisutch	O. mykiss	1.01	-0.05	21	2.67	0.21	0.0024	2.64	0.20	0.0023	0.89
4F	O. tschawytscha	O. mykiss	1.02	0.01	6	3.24	0.13	0.0092	3.32	0.12	0.004	0.96
5F	S. salar	S. trutta	1.0	0.09	7	2.53	0.13	0.0073	2.65	0.125	0.0069	0.70
6F	Ictalurus melas	I. punctatus	1.0	-0.11	12	2.23	0.11	0.0035	2.13	0.11	0.0034	0.66
7F	Lepomis cyanellus	L. macrochirus	1.1	-0.62	14	2.39	0.17	0.0060	1.99	0.14	0.0043	0.80
8S	F. heteroclitus	Fundulus majalis	1.1	-0.32	12	1.67	0.15	0.0094	1.56	0.12	0.0060	0.76
9F	Daphnia magna	D. pulex	0.81	0.26	9	0.68	0.59	0.069	0.81	0.90	0.16	1.51
10F	Gamarus fasciatus	G. lacustris	0.84	-0.06	11	1.32	0.15	0.015	1.05	0.21	0.029	0.76
Genera within Families												
11F	Oncorhynchus	Salmo	1.0	-0.23	58	2.52	0.092	0.0009	2.42	0.082	0.0007	0.59
12F	Oncorhynchus	Salvelinus	1.1	-0.43	45	2.86	0.15	0.0025	2.77	0.12	0.0016	0.76
13F	Salmo	Salvelinus	1.1	-0.31	23	2.65	0.11	0.0027	2.60	0.093	0.0019	0.66
14F	Carassius	Cyprinus	1.0	-0.47	8	3.04	0.09	0.0064	2.73	0.08	0.0052	0.58
15F	Carassius	Pimephales	1.0	-0.27	19	2.79	0.17	0.0037	2.61	0.16	0.0032	0.82
16F	Cyprinus	Pimephales	0.93	0.24	10	2.90	0.17	0.011	2.95	0.20	0.014	0.82
17F	Lepomis	Micropterus	1.0	-0.20	30	2.33	0.22	0.0025	2.24	0.20	0.0021	0.92
18F	Lepomis	Pomoxis	0.82	-0.01	8	1.28	0.23	0.0086	1.04	0.34	0.019	0.94
19S	Cyprinodon	Fundulus	0.96	0.21	12	1.33	0.17	0.011	1.48	0.19	0.013	0.82
11F	Oncorhynchus	Salmo	1.0	-0.23	58	2.52	0.092	0.0009	2.42	0.082	0.0007	0.59
12F	Oncorhynchus	Salvelinus	1.1	-0.43	45	2.86	0.15	0.0025	2.77	0.12	0.0016	0.76
13F	Salmo	Salvelinus	1.1	-0.31	23	2.65	0.11	0.0027	2.60	0.093	0.0019	0.66
14F	Carassius	Cyprinus	1.0	-0.47	8	3.04	0.09	0.0064	2.73	0.08	0.0052	0.58
15F	Carassius	Pimephales	1.0	-0.27	19	2.79	0.17	0.0037	2.61	0.16	0.0032	0.82
16F	Cyprinus	Pimephales	0.93	0.24	10	2.90	0.17	0.011	2.95	0.20	0.014	0.82
17F	Lepomis	Micropterus	1.0	-0.20	30	2.33	0.22	0.0025	2.24	0.20	0.0021	0.92
20F	Daphnia	Simocephalus	0.92	0.35	51	1.48	0.16	0.0012	1.71	0.19	0.0017	0.78
21F	Pteronarcella	Pteronarcys	1.03	-0.05	8	1.34	0.14	0.012	1.33	0.14	0.011	0.75
Families within Orders												
22F	Centrarchidae	Percidae	0.95	-0.02	47	1.96	0.27	0.0022	1.85	0.29	0.00	1.01
23F	Centrarchidae	Cichlidae	0.93	0.40	6	0.90	0.081	0.042	1.29	0.51	1.67	0.56
24F	Percidae	Cichlidae	1.4	0.15	5	1.42	0.33	0.13	2.19	0.16	0.03	1.12
25F	Salmonidae	Esocidae	1.4	-0.49	11	1.05	0.23	0.13	0.99	0.12	0.03	0.94

Code												
26S	Atherinidae	Cyprinodontidae	0.90	0.50	32	1.42	0.18	0.0027	1.77	0.22	0.0042	0.83
27S	Mugilidae	Labridae	0.82	0.70	12	1.33	0.79	0.041	1.79	1.17	0.090	1.74
28S	Cyprinodontidae	Poecillidae	0.75	0.19	12	0.15	0.073	0.016	0.31	0.13	0.049	0.53
29F	Perlidae	Pternarcylidae	1.11	0.21	11	0.17	0.40	0.18	0.39	0.32	0.12	1.24
30F	Perlodidae	Pternarcylidae	0.75	0.54	9	1.12	0.22	0.015	1.39	0.39	0.046	0.92
31S	Cradidae	Crangonidae	1.33	-2.25	5	4.29	0.23	0.24	3.46	0.13	0.08	0.94
32S	Crangonidae	Palaemonidae	0.98	0.46	21	1.29	0.31	0.01	1.74	0.32	0.01	1.09
33S	Crangonidae	Pandalidae	0.81	0.60	10	4.26	0.10	0.03	4.13	0.15	0.06	0.61
34S	Crangonidae	Portunidae	1.18	-0.33	9	4.00	0.09	0.03	4.41	0.06	0.01	0.58
35S	Palaemonidae	Portunidae	1.14	-1.19	11	1.32	0.37	0.02	0.68	0.18	0.004	1.18
36S	Palaemonidae	Penaeidae	0.71	1.33	5	3.68	0.84	0.03	3.94	1.67	0.04	3.94
37S	Pandalidae	Portunidae	0.97	0.59	7	4.04	0.06	0.03	4.51	0.06	0.04	0.48
Orders within Classes												
38F	Salmoniformes	Cypriniformes	0.87	0.90	225	2.32	0.45	0.0008	2.92	0.59	0.0014	1.31
39F	Salmoniformes	Siluriformes	0.85	0.87	203	2.35	0.66	0.0017	2.86	0.91	0.0033	1.59
40F	Salmoniformes	Perciformes	0.94	0.33	443	2.34	0.31	0.0003	2.53	0.35	0.0004	1.09
41F	Cypriniformes	Siluriformes	0.93	0.23	111	2.59	0.28	0.0011	2.63	0.33	0.0015	1.04
42F	Cypriniformes	Perciformes	0.99	0.39	219	2.66	0.59	0.0012	2.24	0.61	0.0013	1.51
43F	Siluriformes	Perciformes	1.1	0.74	190	2.67	0.82	0.0025	2.15	0.71	0.0019	1.78
44S	Anguiliformes	Tetraodontiformes	0.89	1.09	12	1.40	0.30	0.016	2.34	0.38	0.025	1.08
45S	Anguiliformes	Perciformes	0.96	0.21	34	1.11	0.51	0.012	1.28	0.55	0.014	1.40
46S	Gasterosteiformes	Gasterosteiformes	1.0	0.52	8	0.85	0.39	0.097	1.40	0.36	0.082	1.22
47S	Atheriniformes	Atheriniformes	1.0	0.06	46	1.31	0.23	0.0034	1.43	0.21	0.0028	0.94
48S	Cypriniformes	Cypriniformes	0.82	1.93	7	2.26	1.88	0.036	3.79	2.77	0.078	2.69
49S	Tetraodontiformes	Tetraodontiformes	0.88	1.00	46	2.26	0.29	0.0040	2.26	0.38	0.0067	1.06
51S	Perciformes	Perciformes	0.92	0.10	148	1.27	0.51	0.0019	1.27	0.61	0.0027	1.41
52S	Gasterosteiformes	Gasterosteiformes	0.94	0.49	36	0.92	0.24	0.0090	1.36	0.28	0.011	1.35
53S	Tetraodontiformes	Tetraodontiformes	1.12	0.31	8	1.40	0.42	0.099	1.88	0.33	0.063	1.27
54S	Perciformes	Perciformes	1.15	0.67	33	1.47	0.62	0.036	1.03	0.47	0.020	1.55
55S	Tetraodontiformes	Tetraodontiformes	0.91	0.93	34	1.28	0.52	0.013	2.11	0.62	0.018	1.41
56S	Perciformes	Perciformes	0.46	0.35	6	3.14	0.18	0.0081	1.10	0.86	0.17	0.84
57S	Cypriniformes	Cypriniformes	1.20	0.76	7	2.98	0.0059	0.0008	2.80	0.0041	0.0004	0.15
58F	Cladocera	Ostracoda	0.62	0.79	22	1.05	0.96	0.040	1.44	2.53	0.28	1.92
59F	Cladocera	Amphipoda	0.91	0.27	105	1.14	0.63	0.0033	1.31	0.76	0.0048	1.56
60F	Ostracoda	Isopoda	2.05	1.10	7	1.26	1.23	0.60	1.49	0.29	0.034	2.17
61F	Ostracoda	Amphipoda	2.30	2.74	14	1.62	2.06	0.33	0.99	0.39	0.012	2.82
62F	Isopoda	Amphipoda	0.45	0.22	20	1.92	0.92	0.037	0.66	4.45	0.87	1.88

Table 7.4. Continued

No.[b]	X taxon	Y taxon	Slope	Intercept	n	X̄[c]	F_1^c	F_2^c	Ȳ[c]	G_1^c	G_2^c	PI[d]
63F	Isopoda	Decapoda	1.85	2.31	5	2.00	4.42	2.09	1.39	1.29	0.18	4.12
64F	Amphipoda	Decapoda	1.67	0.65	14	0.89	2.73	0.25	2.14	0.98	0.032	3.24
65F	Plecoptera	Odonata	0.53	0.60	13	0.55	0.61	0.10	0.89	2.16	1.24	1.53
66F	Plecoptera	Diptera	2.46	0.77	18	0.18	3.15	1.68	1.22	0.52	0.046	3.48
67F	Ostracoda	Decapoda	1.37	1.05	9	1.86	1.33	0.12	1.51	0.71	0.035	2.27
68F	Ephemeroptera	Odonata	0.28	0.38	5	1.10	0.90	0.12	0.69	11.75	19.85	1.86
69S	Amphipoda	Decapoda	1.23	0.69	50	1.93	1.29	0.02	1.68	0.82	0.01	2.22
70S	Calanoidea	Decapoda	3.05	3.12	14	1.82	2.08	1.27	2.42	0.22	0.01	2.83
71S	Decapoda	Harpacticoidea	1.18	1.06	23	3.38	2.59	0.09	3.88	1.84	0.05	3.16
72S	Decapoda	Mysidae	0.35	0.16	10	1.38	0.15	0.01	0.65	1.18	0.49	0.75
All fish from standard species, within media												
73F	P. promelas	Osteichthes	1.01	−0.30	354	2.77	0.45	0.0006	—	—	—	1.31
74F	L. macrochirus	Osteichthes	0.96	0.17	500	2.52	0.49	0.0005	—	—	—	1.37
75F	O. mykiss	Osteichthes	0.99	0.29	480	2.42	0.38	0.0004	—	—	—	1.20
76S	C. variegatus	Osteichthes	0.97	0.03	51	1.25	0.58	0.0085	—	—	—	1.49

[a]From Suter et al. 1986 and 1988, and Suter and Rosen 1988, modified to reflect changes in salmonid taxonomy.
[b]F = freshwater, S = salt water.
[c]The means of X and Y and the factors F_x and G_x are used to calculate the variance as described in the text (Box 7.1).
[d]PI, the 95% prediction interval at the mean, is log Ȳ ± the number in this column.

Table 7.5. Summary of Freshwater and Marine Taxonomic Extrapolations for LC$_{50}$s

Taxonomic Level	n[a]	Mean of 95% Prediction Intervals[b]
Species		
Marine fish	1	0.75
Freshwater fish	6	0.78
Freshwater arthropods	2	1.10
Generas		
Marine fish	1	0.82
Freshwater fish	8	0.75
Freshwater arthropods	2	0.78
Families		
Marine fish	3	1.00
Freshwater fish	4	0.97
Marine crustaceans	7	0.94
Freshwater arthropods	3	1.37
Orders		
Marine fish	13	1.27
Freshwater fish	10	1.35
Marine crustaceans	4	2.38
Freshwater arthropods	10	2.06
Classes		
Freshwater vertebrates	1	3.84
Freshwater anthropods	1	2.26

[a]n = the number of pairs of taxa at that taxonomic level.
[b]Weighted by the number of points (chemicals) in each regression.

age for fresh water and salt water animals and for terrestrial plants exposed to organic chemicals (Tables 7.5 and 7.6). Extrapolations between taxa within the same family (i.e., between congeneric species and between confamilial general) can be made with fair certainty, but extrapolations between orders, classes, or phyla tend to be highly uncertain. This generalization is supported for aquatic animals by Macek and McAlister (1970), LeBlanc (1984), Sloof et al. (1986), and Mayer and Ellersieck (1986).

The generalization that similarity of toxic response is related to taxonomic similarity will not hold if the traits that determine sensitivity are extremely evolutionary labile or conservative. That is, if the traits that control responses

Table 7.6. Summary of Taxonomic Extrapolations for Vascular Plant EC$_{50}$s[a]

Taxonomic Level	n[b]	r^{2}[c]
Species	15	0.87
Genera	5	0.56
Families	4	0.13
Orders	6	0.08

[a]From Fletcher et al. (1990).
[b]n = the number of pairs of taxa at that taxonomic level.
[c]The mean across pairs of taxa of coefficient of determination for regressions of log transformed EC$_{50}$ values.

to a chemical are common to all members of a higher taxonomic group, there will be no taxonomic gradient of predictability within the group. However, if those traits are highly modified by the adaptions to the specific environments and ecological roles that distinguish species and subspecies of that taxonomic group, closely related species will be as dissimilar as distantly related species. In other words, there will be no gradient of sensitivity across levels of a higher taxon if there is no gradient of similarity of the traits controlling sensitivity. For example, arid land plants are more resistant to gaseous pollutants than are those that grow in humid climates (see below). Because adaptation to aridity or humidity is an evolutionary labile trait in plants, one would expect that sensitivity to gaseous pollutants would not show clear taxonomic pattern in plants, and that appears to be the case. We may speculate that the apparent lack of taxonomic consistency in the relative sensitivity of birds to pesticides (Tucker and Haegele 1971) is due to evolutionary conservatism; all birds are about equally evolutionarily unprepared to resist these xenobiotic toxicants.

Regression analysis can also be used to address the general problem of predicting whole taxa on the basis of the response of a test species. Suter et al. (1986, 1987b), using a data set containing tests of 66 fish species, regressed against each of three standard test fish all other species that had been tested with a common chemical. They found that other fish species fell within ± 1.31, ± 1.37, and ± 1.20 log units of the regression lines 95% of the time for fathead minnow, bluegill, and rainbow trout, respectively (Table 7.4). The parameters of the regression equations indicate that fathead minnows are a little less sensitive than the average fish, whereas rainbow trout are more sensitive. Holcombe et al. (1988) used the same technique, regressing LC_{50}s and EC_{50}s for 44 aquatic animal species and 1 alga, against fathead minnow LC_{50}s. They concluded that responses of most species fell within ± 1 order of magnitude of the line (Figure 7.9). They noted, as had Thurston et al. (1985) in regressions of pairs of species, that deviations from the line tended to increase with increasing toxicity, suggesting that more toxic chemicals have more selective modes of action.

Dose-scaling is a technique for interspecies extrapolation that is conventional in human health risk assessment, but it has received much less use in ecological assessments. Dose-scaling consists of converting dose to a scale that can be considered to be common to all species so that test data from all species can be treated as equivalent. The most common scale is dose per unit mass (mg/kg). Use of this scale suggests that effects occur at a particular concentration in the organism. However, because weight is linearly related to more than 100 morphological and physiological parameters, other interpretations are possible (Calabrese 1984). The second most common scale is dose per unit surface area (mg/m²). Because heat loss is related to surface area, and the demands of thermoregulation drive metabolism in homeotherms, a variety of physiological parameters including O_2 consumption, blood volume, renal function, and water requirements are approximately constant when scaled to surface area (Calabrese 1984). Neither weight nor surface area has proved to

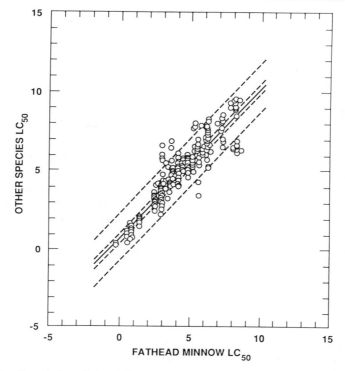

Figure 7.9. Regression of log LC_{50} values for all fresh water fish, amphibians and invertebrates against the log LC_{50} for fathead minnow (redrawn from Holcomb et al. 1988). The fitted line is: Other species LC_{50} = 0.42 + 0.91 (Fathead Minnow LC_{50}), with r^2 = 0.86 and n = 309. The inner and outer sets of dashed lines are the 95% confidence interval and 95% prediction interval, respectively.

be consistently preferable (Crouch 1983). Because some pertinent processes such as DNA repair are related to life expectancy and response rates such as the rate of cancer induction are proportional to life expectancy, life expectancy can also be used as a scaling factor (Calabrese 1984). Therefore, tests that are run for most of a rodent's life span (100–130 weeks) have been treated as representative of a human lifetime exposure.

Dose scaling is commonly used in wildlife toxicology where doses are routinely scaled to body weight (mg/kg). Kenaga (1973) suggested that doses to birds be scaled as mg/kg/day. Exposure could then be calculated on the basis of food consumption rates, with smaller birds eating proportionately more than larger birds.

A common variant of dose-scaling used in human health assessment is allometric regression. It consists of plotting the proportion responding against scaled dose for all of the species that have been tested with a particular chemical, and fitting a curve to the points. For example, in the Soviet Union lethal concentrations for humans have been estimated by determining the LD_{50} for at least three species of mammals and fitting the data to the equation

$$\log LD_{50} = a + b \log M \tag{7.11}$$

where M is mass (Krasovskij 1976).

Allometric regression has also been used by Patin (1982) for acute toxic effects on marine organisms. He regressed $\log LC_{50}$ against log body length for oil, PCBs, and five heavy metals with organisms ranging from microbes to fish (Figure 7.10). High correlation coefficients (0.73 to 0.91) resulted, despite (1) using size estimated in some cases, (2) using test durations of 2 to 4 days, and (3) using tests with a variety of conditions and endpoints. Patin attributed this result primarily to the influence of surface-volume ratios on uptake from water, and secondarily to the effect of size on metabolic rate and the length of life cycles.

Lipnick (1985) suggested that allometric regression for fish be based on the dimensions of fish gills and water pumping rate. The suggestion is based on the hypothesis that mortality is induced by narcotic chemicals at a constant internal concentration for all fish. Differences in sensitivity to acute lethality then are due to differences in kinetics, which he assumes are primarily due to differences in uptake rate.

Although interspecies extrapolations have been primarily based on taxonomy and physical scale, other factors contribute to differences between species

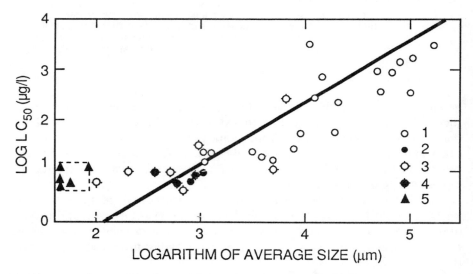

Figure 7.10. Regression of log of LC_{50} values for various marine organisms and mercury against average size of the organisms. Points of types 1 and 2 are adult organisms, 3 and 4 are early life stages, and 5 are algal EC_{50}s. The algal points, which are bounded by a dashed line, were excluded from the regression. (redrawn from Patin 1982).

and may prove useful. One is the diet of animals. Herbivores are generally more resistant to toxicants than predators because they must deal with plant toxins, and the mixed function oxidase system that oxidizes plant toxins also degrades pesticides and other organic chemicals (Plapp 1981, Mullin et al. 1982). Among herbivores, polyphages are more resistant than oligophages, which are more resistant than monophages (Croft and Morse 1979). Again, this is presumed to be because organisms that must deal with the widest variety of plant toxins possess the greatest repertoire of detoxifying enzymes. Extrapolation models that predict the sensitivity of trophic groups would be particularly useful in ecosystem-level assessments because the diversity of sensitivity within a trophic group determines the degree to which functional redundancy buffers an ecosystem against toxic effects (Chapter 9).

Guild theory has also been suggested as a means to organize species into groups that respond similarly to environmental impacts (Severinghaus 1981). Guilds are groups of organisms that use a common resource in a similar way (Root 1967, Root 1973, Cummins 1974). Members of a guild should respond similarly, not only because they have the same diet, but also because they are found in the same areas and behave in similar ways, they should have similar exposures to chemicals. For example, predatory insects not only have fewer detoxification enzymes than herbivores but also are more active than herbivorous insects, so they are more likely to come into contact with contaminated surfaces and are less resistant to the resulting dose. In addition, neurotoxic chemicals are more effective against more active species (Plapp 1981).

Using guild-like groupings, Kenaga (1973) suggested that birds should be categorized by food habits into consumers of vertebrates, invertebrates, nectar, seeds, or both invertebrates and seeds. For greater consistency in exposure, these categories might be further subdivided. Seed eaters that hull seeds such as finches might be separated from those that do not, such as doves, and consumers of foliage invertebrates separated from consumers of soil and litter invertebrates.

Sharing a common environment may have other implications for relative sensitivity of species to pollutants. If a pollutant resembles a natural environmental variable then the extent of adaptation of a species to high levels of the variable will serve to predict the response of the species to the pollutant. In evolutionary terms, species may be preadapted to pollution. The most conspicuous example is the dominance of aquatic communities exposed to high loadings of organic pollutants by species that are adapted to low dissolved oxygen concentrations. Resh and Unzicker (1975) used the distribution of midge species along a pollution gradient as evidence to attack the idea that congeneric species have similar pollution tolerance. However, because congeneric species of midges are likely to differ in their tolerance for low dissolved oxygen, the differences in response of congeneric species to organic waste is not surprising. Rather, it appears to be an example of how evolutionary lability of a trait that controls pollution sensitivity can mask the influence of taxonomic similarity.

An example of preadaptation to pollutants is the low transpiration rate of arid-land plants. The transpiration rate is an indicator of the rate of uptake of pollutants from the air through stomates and, because transpiration controls the rate of water uptake from the soil, it is also an indicator of uptake of pollutants from the soil by the roots. As a result, plants adapted to arid environments may be generally more resistant to both air and soil pollution than those adapted to humid environments (McLaughlin and Taylor 1981, McFarlane et al. 1987).

Keller (1984) suggested various factors that appear to affect the sensitivity of plants to air pollutants. Height increases sensitivity by getting the leaves into higher air velocities and causing the plants to filter more air per unit cover. Longevity makes a perennial plant more subject than an annual to cumulative effects. For pollutants such as SO_2 and NO_x that provide nutrient elements, rapid growth and its associated nutrient requirements will protect a rapidly growing annual, but a tree that is growing slowly and internally cycling most of its nutrients will experience toxic effects (see also Roberts 1984).

Neuhold (1987) attempted to derive a model for predicting the sensitivity of untested species of fish based on natural history. He hypothesized that sensitivity to pollutants would be determined by whether the species was adapted to a variable environment and therefore had high fecundity and low survivorship (r selected) or was adapted to a stable environment and therefore had low fecundity and high survivorship (K selected). He chose maximum age, maximum weight, and maximum fecundity as the variables to represent these adaptive strategies. Unfortunately, for only 15% of chemicals were the models statistically significant with the correct signs on the parameters for the three variables (i.e., the signs required by the hypothesis).

Interspecies extrapolations might also be developed using as predictive variables biochemical traits that control the internal dose or toxic dynamics of a chemical. This type of model might be similar to the dose-scaling models discussed earlier but use biochemical traits such as mixed function oxidase (MFO) inducible activity instead of gross morphological traits. Techniques for assaying detoxification enzymes in animals have recently improved and become more widely used, and the environmental variables that affect enzyme levels are being elucidated (Jimenez et al. 1988). Detoxification enzymes that are responsible for variation in sensitivity in plants have also been identified (Sekiya et al. 1982). Some physiological traits that have no obvious relation to dose or response may serve as predictors of sensitivity. For example, CAM plants may be more resistant than C_3 plants to SO_2 (Olszyk et al. 1987).

Multivariate statistical models might be developed for interspecies extrapolation that combine taxonomic, morphological, and other traits to predict the response of a particular species or of a most sensitive species. Multivariate models are commonly used in QSARs and multivariate hybrids of QSAR and interspecies extrapolation have been developed. Enslein et al. (1988), for example, developed models that combine chemical structure and rat LD_{50}s to predict *Daphnia* EC_{50}s.

Toxicokinetic and dynamic simulation models that incorporate the critical differences in morphology and physiology between the taxa are a potential alternative to purely empirical, statistical methods for these extrapolations. These differences include differences in (1) uptake and depuration rates, (2) biotransformation rate, (3) biotransformation products, (4) response of the target, and (5) whole organism response to effects on the target. The last two differences have been most difficult to address. Menzel (1987) has suggested that differences in the sensitivity of species to a particular internal dose can be corrected for in pharmacokinetic models by using the relative sensitivity of cell cultures drawn from the two species to calculate a "ratio of inherent sensitivity". However, this approach still does not account for differences in whole organism response to cellular effects. For example, Macek and McAllister (1970) noted that differences in susceptibility of fish to cholinesterase inhibitors appeared to be due to differences in tolerance of low brain cholinesterase levels rather than in activity of the toxicant.

Even for human health assessments where the knowledge base is large, simulation models for taxonomic extrapolation of toxic effects remain relatively poorly accepted. Dedrick and Bischoff (1980) argued that because mammals share a common arrangement and structure of organs, partitioning among tissues can be extrapolated between species on the basis of allometric scaling of parameters. The common exception is toxicant metabolism, which might be estimated from metabolic products and metabolic rates in cell cultures. Clearly, mathematical simulation of the differences between species will work best when the species are well characterized, like rats and humans. The difficulties are greater in ecological assessments where extrapolations are made across very large taxonomic distances, and differences in metabolism and sensitivity to tissue doses are likely to be large and poorly characterized. However, as McKim et al. (1986) point out, these models potentially provide a mechanism for using information from human-oriented toxicology in ecological assessments.

Most discussions of taxonomic extrapolation have addressed only one life stage and test condition, implicitly ignoring the possibility that intertaxa relationships may not be constant. This is a result of the lack of studies addressing in a consistent manner both multiple species and multiple life stages or conditions. However, Stene and Lonning (1984) found that the sensitivity of fish eggs was much more variable among species than the sensitivity of larvae. They attributed this result to the large differences in structure and lipid content of eggs. The mode of exposure may also affect the relative sensitivity of species. The relative sensitivity of mallards and gallinaceous birds is frequently the opposite for 30-day repeated doses of what it was for a single oral dose (Tucker and Lietzke 1979). These findings suggest that the interaction of taxonomy with other factors needs more attention in taxonomic extrapolations.

Life Stage and Size

As discussed in the previous section, the size of organisms is an important determinate of toxic response, and scaling for body mass and surface is used to extrapolate between mammalian species. Size is also an important variable among members of a species but is covariate with changes in morphology and physiology associated with development and maturation. Growth and development have a number of effects on sensitivity to toxicants including (1) small organisms have larger chemical absorbing surfaces per unit volume, (2) young, growing organisms and small homeothermic organisms eat more per unit mass, (3) young and small organisms have higher respiratory ventilation rates, and (4) young organisms have incompletely developed organs and metabolic capacities (Tucker and Lietzke 1979).

In general, eggs are the least sensitive life stage of fish (Danil'chenko 1977, Stene and Lonning 1984, Mayer and Ellersieck 1986, Suter et al. 1986, 1987b). It is also generally true that larval fish are more sensitive than adults to lethal effects (Pickering and Henderson 1966, Mayer and Ellersieck 1986, Suter et al. 1986, 1987b). However, one adult process, the production of viable eggs, is generally the most sensitive life stage of fish in life cycle tests (Suter et al. 1987b). Aquatic invertebrates generally become less sensitive with maturity (Mayer and Ellersieck 1986). Chapman (1983) suggested that life stage made a factor of 30 difference in results of acute lethality tests of aquatic fish and invertebrates.

Tests of age effects in birds have been more systematic. Hudson et al. (1972) determined oral LD_{50}s (mg/kg basis) for 36 hour, 7 day, 30 day, and 6 month old mallards for 14 pesticides. They found no consistent pattern of sensitivity with age. In most cases, the extreme values differed by less than a factor of 3, but in one case they differed by a factor of 7. Hill and Camardese (1986) determined dietary LC_{50}s for 9 pesticides to Japanese quail over a period of 3 weeks post-hatching. They found a 2.5-fold increase in LC_{50} on average, which is equivalent to a 1.05 per day rate of increase. However, that adjustment was not useful when applied to independent data. Currently, age appears to be a relatively small and nonsystematic source of variance in avian toxicity.

The primary concern of most of these studies has been determining the effects of variance in size and age on the consistency of test results or finding a most sensitive stage to use in testing. However, some extrapolation models have been developed for aquatic assessments. In toxicity tests of individual species, weight can be used as a parameter of the exposure-response model, or allometric regression can be used to normalize individuals for size (Tsai and Chang 1984). Such models can be used to extrapolate from the organisms used in tests to the distribution of organism sizes found in the field. Regression models for extrapolating between life stages of fish based on data from life cycle tests are presented in Table 7.7. The variances of the regressions were large relative to taxonomic extrapolations of acute toxicity data. Mayer and Ellersieck (1986) developed regression equations for effects of weight on LC_{50}s

Table 7.7. Equations for Extrapolating Between Quartile Effective Concentrations for Different Life Stages of Fish (units are log μg/L).

No.	X variable[a]	Y variable[a]	Slope	Intercept	n	$\bar{X}$[b]	F_1[b]	F_2[b]	$\bar{Y}$[b]	G_1[b]	G_2[b]	PI[c]
1.	MORT2 EC25	MORT1 EC25	1.27	-0.31	13	1.41	0.21	0.016	1.48	0.13	0.0063	0.90
2.	MORT2 EC25	HATCH EC25	1.00	0.25	27	2.35	0.20	0.0024	2.62	0.20	0.0024	0.88
3.	HATCH EC25	MORT1 EC25	0.78	-0.14	6	0.86	0.30	0.026	0.53	0.49	0.070	1.1
4.	EGGS EC25	HATCH EC25	0.84	0.71	6	1.05	0.18	0.015	1.60	0.26	0.030	0.84
5.	EGGS EC25	MORT1 EC25	0.94	0.52	22	0.72	0.12	0.004	1.19	0.14	0.0056	0.69
6.	MORT2 EC25	EGGS EC25	1.30	-0.93	17	1.63	0.46	0.027	1.19	0.27	0.0095	1.3
7.	WEIGHT EC24	HATCH EC25	1.12	0.088	15	2.11	0.23	0.0058	2.44	0.19	0.0038	0.94
8.	WEIGHT EC25	MORT2 EC25	1.04	-0.080	46	1.93	0.12	0.0011	1.94	0.11	0.0009	0.67

[a] MORT1 EC25 = the concentration estimated to cause a 25% increase in mortality of parental (postlarval) fish, MORT2 EC25 = the concentration estimated to cause a 25% increase in mortality of larval fish, HATCH EC25 = the concentration estimated to cause a 25% decrease in normal hatches of eggs, EGGS EC25 = the concentration estimated to cause a 25% decrease in production of viable eggs per surviving female, WEIGHT EC25 = the concentration estimated to cause a 25% decrease in the weight of early juvenile fish.

[b] The mean of X and Y and the factors F_x and G_x are used to calculate the variance as described in the text (Box 7–1).

[c] The 95% prediction interval at the mean is log $\bar{Y}$ ± the number in this column.

for ten combinations of six chemicals and five species of fish. Slopes ranged from 0.0167 to 2.3758, but were similar across species for individual chemicals. Mayer and Ellersieck's results suggest that extrapolation models for weight of fish are worth pursuing. However, because the numbers of points and the ranges of weights are small in all but one case, these particular models have limited utility as assessment tools.

Exposure Duration

Ideally, the influence of the temporal dynamics of exposure on effects would be determined in the toxicity tests. However, in most cases only the results for the ends of tests are available, and extrapolations must be made to the durations of interest. As was discussed in Chapter 3, temporal dynamics are most commonly treated categorically, with each category corresponding to a class of toxicity tests. In certain circumstances, the assumption of negligible temporal effects on toxicity within temporal categories may be justified. Bailey et al. (1984) found that in static acute tests with bluegill the 24-to 96-hour LC_{50}s did not significantly differ for nine of ten assorted chemicals, and did not significantly differ for two chemicals in flow-through tests. Such results can occur due to a genuine absence of increasing toxicity over the period of interest (e.g., the organism achieves equilibrium with the polluted media within 24 hours and damage does not accumulate), or because loss of the chemical from the exposure medium offsets the effects of longer exposures.

If results of tests for a temporal category are unavailable, they must be estimated by extrapolation. The simplest and commonest model is the ratio of test endpoints from tests that differ in duration, which is referred to as a chronicity ratio or, when standard acute and chronic tests are used, an acute-chronic ratio. However, such ratios are often complicated by differences in the test other than duration, and in such cases they may reveal little about temporal dynamics.

When comparable temporal data are available for a set of chemicals, regression analysis can be used to estimate general temporal relationships. Bailey et al. (1984) developed equations to estimate factors to convert 96-h LC_{50}s for bluegill to shorter durations, one for 1–8 hours and one for 24–72 hours. The linear regressions provided good correlations, but only after removing acrylonitrile from the data set because of its extraordinarily high chronicity.

Another approach to temporal extrapolation is to assume a concentration-duration function. The simplest assumption is that the product of exposure concentration and time is a constant for a given effect. Given any toxic endpoint concentration (c) and associated duration (t), a constant can be derived from the formula

$$ct = k \tag{7.12}$$

and then for any desired time, the estimated effective concentration is

$$c = k/t \qquad (7.13)$$

This is equivalent to Green's (1965) hyperbolic function for calculating an LD_{50} for exposures of infinite duration (Equation 6), but with intercept set to zero. Because there is no threshold, any concentration is lethal if the exposure is long enough. This property makes the model overly conservative when applied to long exposures.

Tucker and Leitzke (1979) advocated the use of this formula (Equation 7.13) and stated that for the range of 24 to 96 hours, k was usually constant for fish and pesticides, although they cited some exceptions. Similarly, Anderson et al. (1981) reported that in various exposures of shrimp to petroleum, the product of concentration and time was constant. The model of toxic spill effects for CERCLA type A damage assessments assumed this model, originally calculating lethal concentrations as $[(96h\ LC_{50}/4) \times d]$, where d is the exposure duration in days (DOI 1986). Later the model was changed to a function based on the average slope of observed concentration-duration functions for fish $LC_{50}s$ (DOI 1987).

Kenaga's (1973, 1977) proposal to scale doses to wildlife as mg/kg/day amounts to an assumption that the toxic dose is linearly proportional to time. That is, if mg/kg/day is a constant, then the effective dose in mg/kg = kt, and the effective dose for any one duration can be calculated from any other. In other words, the effective dose must be increased if it is spread over more time, and any dose is safe if spread out over enough time.

Of course, if toxicity data is available for multiple exposure durations, Green's hyperbolic function (Equation 6), a probit plane model (Equation 7), or any other appropriate model can be fit to the data for purposes of interpolation or extrapolation.

These temporal extrapolation models are intended to deal with constant exposures of various durations, but exposure concentrations are often pulsed or irregular. If it is not possible to associate these exposures with a constant exposure in a test system or to perform a pulsed exposure toxicity test, then it is necessary to extrapolate from constant exposures to pulsed or irregular exposures. One solution is to treat fluctuating exposure as a source of uncertainty that has a 2X or less effect on toxicity in both acute and chronic aquatic tests relative to constant exposures with the same average concentration (Chapman 1983). Another simple solution is to integrate the area under the time-concentration curve and assume that this value is equivalent to the product of a constant concentration times the duration of the exposure, which is then assumed to be constant for a particular effect, as discussed above for constant exposures. This approach was advocated by Larson (1978) for intermittent chlorination of aqueous effluents.

Toxicokinetic and dynamic models offer the potential to simulate variable exposures, but values for the model parameters are seldom available. One solution is to fit functions with the form of kinetic or dynamic models to concentration-duration data from toxicity tests with constant exposures. Mancini (1983) rearranged the integral of the simple linear, two compartment (internal and external) bioaccumulation model to obtain a negative exponential function, which he fit to concentration-time data for fish experiencing a particular percent mortality in continuous exposures. The fitted parameters were the depuration rate and the ratio of the internal concentration to the uptake rate, which he termed equivalent dose. He then used these parameters in the model to simulate the effects of time-varying exposures of varying duration, but each with a constant concentration. In each period, the toxicant concentration accumulated in the prior period dies away and new toxicant is accumulated. Wang and Hanson (1984) applied the same approach to pulsed chlorine exposures but added a formulation for triangular pulses. Breck (1988) showed that a damage-repair model reduces to a probit plane or logit plane (depending on the assumed distribution of sensitivity) if the repair rate was negligible relative to the damage rate. He used this model to estimate proportional mortality in response to time-varying exposures to physical factors such as pH and heat that do not accumulate in the organism.

The difference between these models and those that represent the temporal dynamics of toxicity tests with constant exposures is simply the interpretation that is put on the parameters. By interpreting the statistical parameters fitted to test data as kinetic or dynamic parameters, toxicokinetic and dynamic processes can be estimated. This approach differs from toxicodynamic simulation modeling as discussed above in that the parameters are not estimated from direct evidence concerning the processes that they represent. However, it is always possible to estimate the kinetic and dynamic parameters directly, and then mathematically simulate the onset of effects for any temporal pattern of exposure (e.g., McKim et al. 1986).

Various temporal aspects of exposure, including diurnal and seasonal differences in the consequences of exposure, can be accounted for by modifying the exposure metric. In the National Crop Loss Assessment Network (NCLAN), ozone dose to plants has been expressed as the seasonal average hourly concentration. Rawlings et al. (1988) accounted for the greater effectiveness of concentrations during ozone peaks and of concentrations occurring during periods of high plant activity by weighting the hourly concentrations for these factors. That is,

$$\text{Weighted Dose} = \Sigma(w_i x_i)/\Sigma w_i \qquad (7.14)$$

where w_i is the weights and x_i is the hourly ozone concentration. Time of day and solar radiation were used as surrogates for plant activity, and peaks of ozone exposure were expressed as powers of concentration (i.e., $w_i = x^p$). Weighted dose-response functions were then fit to the effects data.

Acclimation and Adaptation

Acclimation is an increase in tolerance resulting from physiological and biochemical adjustments by an organism to environmental conditions, including pollutant exposure. It has been shown for a variety of chemicals and species that preexposures to chemicals can increase their resistance in toxicity tests. However, reviews of the evidence concerning acclimation by aquatic organisms by Chapman (1983, 1985) and Weis and Weis (1989) indicate that preexposure may result in acclimation, may have no effect, or may reduce tolerance of the chemical, presumably by causing accumulation of the chemical or of damage. These reviews found no predictable patterns among species or chemicals. In addition, acclimation expressed as resistance to acute lethal effects may not imply resistance to other effects such as reduced reproduction (Chapman 1985). Finally, acclimation may require more time than is typically allowed in tests; exposure to boron caused an extended algistatic period before exponential algal growth was re-established (Stockner and Antia 1976). Chapman (1983, 1985) concluded that acclimation could not be predicted and only causes about a 2X improvement in resistance to metals.

Adaptation is increased tolerance of pollutant exposure or other environmental conditions resulting from acquisition of heritable traits. Adaptation to toxic chemicals can occur due to changes in the target cells or molecules, or due to changes in the kinetics of the chemical, particularly in metabolic capability. Adaptation cannot be distinguished from acclimation without breeding experiments, but it has been demonstrated to occur in the field for fish (Angus 1983, Weis and Weis 1989), aquatic invertebrates (Klerks and Levinton 1988), plants (Roose et al. 1982, Shaw 1990, Taylor 1978), and insect pests exposed to insecticides (Plapp 1981, Brattsten et al. 1986). Adaptation is, in general, unpredictable, but the slope of the dose-response curve might be used as an indicator of the selectable variance in susceptibility. Adaptation has generally been ignored in ecological assessments because effects on a population that create enough selective pressure to induce adaptation is assumed to be unacceptable, and because it is believed to take too long to be considered ameliorative. However, selection may proceed quite rapidly in microbial populations, and even in higher organisms, recovery from selective pressure may be rapid. Plapp (1981) argued that because there are relatively few mechanisms for resistance, organisms in a polluted environment may be preadapted to new pollutants. However, adaptation to mercury by mummichogs decreased tolerance of lead and had no effect on tolerance of PCBs, as well as reducing the fitness of the fish (Weis and Weis 1989). In any case, adaptation should be considered because it can have a large effect on sensitivity. Acclimation and adaptation to insecticides by fish can result in a 1000X increase in resistance and often results in a 100X increase (Chapman 1983).

Mode of Exposure

Sometimes it is necessary to use data from a test with one mode of exposure to estimate effects resulting from another mode of exposure. The most common version of this extrapolation is using mammalian or avian toxicity data from acute oral doses to estimate effects from consumption of contaminated food or water (Chapter 6). The obvious solution is to convert concentrations in food or water to dose by multiplying by the amount of food or water consumed. This conversion depends on a number of assumptions including, (1) the chemical is not repellent, (2) the chemical is equally available in all media to gut uptake, and (3) either the dietary exposure is acute, like the dose, or the chemical or its effects accumulate without significant loss or repair. DDT, with its high bioaccumulation potential and low metabolic rate, is an example of a chemical that fits this model (Tucker and Lietzke 1979). Kenaga (1977) argued that these modes of exposure should not be treated as equivalent in assessments, based on comparison of LC_{50} and LD_{50} determinations for mallard ducks. McCann et al. (1981) compared LC_{50}s estimated from LD_{50}s from dietary toxicity tests, all with laboratory rats. They found that 35% of pesticide hazard decisions would be different for the estimated and measured LC_{50} values. In the case of TEPP, which rapidly disappears from food, the difference was 100X.

Another approach is to correlate test results for the different mode of exposure. Filov et al. (1979) correlated per os (oral), intraperitoneal, subcutaneous, and cutaneous doses for rats, mice, and rabbits, obtaining correlation coefficients greater than 0.8 in all cases. This approach needs to be explored for ecological assessment. Kenaga (1979a) found that for 69 pesticides, rat chronic dietary no effect levels were not significantly correlated with acute oral LD_{50}s, but were correlated with bobwhite dietary LC_{50}s, suggesting that route of exposure may be more important than rather large taxonomic differences. There is also some evidence that LC_{50}/LD_{50} relationships are not consistent across wildlife species (Tucker and Lietzke 1979).

The most common difference in mode of exposure encountered in aquatic assessment is that between static and flow-through tests. The EPA (1979) compared 24 pairs of static and flow-through tests for 12 criteria chemicals and concluded that static LC_{50} values could be multiplied by 0.71 to estimate the flow-through LC_{50}; the 0.71 value was the geometric mean of the flow-through/static ratios. Mayer and Ellersieck (1986) examined 123 pairs of static and flow-through tests with 41 chemicals. They found a range of ratios of 0.12 to 8.33, with a mean of 0.51. They did not recommend using 0.51 as a correction factor because of the large range of ratios. However, the distribution of ratios could be used to estimate the probability that a flow-through LC_{50} would be as low as some critical value, given a static LC_{50}. Alternatively, flow-through LC_{50}s could be regressed against static LC_{50}s.

Severity

Because most test results are for rather severe effects (i.e., mortality), much attention has been paid to the problem of estimating low severity effects from severe effects. Unfortunately, lower severity has usually been considered in conjunction with longer duration and increased number of life stages in an attempt to estimate a generically safe concentration. In aquatic toxicology, a variety of factors have been proposed to estimate these safe concentrations from acute LC_{50}s. Henderson (1957) defined the application factor as the ratio of the chronic threshold concentration to the acute LC_{50}, and proposed a value of 0.1. The 1972 water quality criteria suggested factors of 0.1 or 0.01, depending on the persistence and bioaccumulation potential of the chemical (NRC 1972). Canton and Sloof (1978, cited in Peakall and Tucker 1985), showed that in 37.7% of 77 cases, the application factor was greater than 0.1, and in 20% of cases it was less than 0.01, with 5.2% less than 0.001. Kenaga (1979b, 1982) showed similarly broad ranges of application factors. The state of Michigan fit a log normal distribution to a set of application factors, and chose as a generic factor 0.022, the 80th percentile of the distribution (Giesy and Graney 1989). Stephan and Erickson (n.d.) proposed an acute/chronic ratio of 40 (application factor = 0.025), which is approximately the 95th percentile of final acute/chronic ratios in the EPA's national water quality criteria.

Mount and Stephan (1967) proposed that, rather than using factors for all chemicals and species, the application factor should be determined once for each chemical and then applied to other species. However, Mount (1977 — based on an unpublished study by Andrew et al. 1977) and Kenaga (1979b) concluded that application factors were not even consistent across species for a single chemical. Nevertheless, chemical-specific application factors are used to calculate water quality criteria (Stephan et al. 1985).

Regression of chronic thresholds against acute LC_{50}s gives somewhat more accurate results than generic factors, because application factors tend to decline as LC_{50}s increase. Andrew et al. (1977) regressed MATCs from fish life cycle tests against 96-hr LC_{50}s for 31 combinations of species and chemicals, obtaining the equation:

$$\log(\text{MATC} \times 10^5) = 4.78 + 0.675 \log(LC_{50} \times 10^2) \qquad (7.15)$$

with an $r = 0.679$, where concentrations are mg/L. Suter et al. (1983, 1986) presented acute/chronic equations for fish to predict MATCs and chronic EC_{25}s from 96-h LC_{50}s (Table 7.8). Sloof et al. (1986) developed an equation based on 164 tests of fish and daphnids:

$$\log \text{NOEC} = 0.95 \log LC_{50} - 1.28 \qquad (7.16)$$

where units are μg/L.

Genuine comparisons of severity are much less common than these hybrid

Table 7.8. Equations for Extrapolating from 96-hr LC_{50}s for Fish to MATCs and Quartile Effects Concentrations for Specific Life Stages (units are log µg/L).

No.	Y Variable[a]	Condition[b]	Slope	Intercept	n	$\bar{X}$[c]	F_1[c]	F_2[c]	PI[d]
1.	MATC	FW, LC	0.90	-1.16	55	2.75	0.51	0.01	1.4
2.	MATC	FW, ALL	1.07	-1.51	98	3.13	0.59	0.008	1.5
3.	MATC	FW, ALL, N	0.90	-0.42	23	3.87	0.09	0.002	0.6
4.	MATC	FW, ALL, M	0.73	-0.70	25	3.25	0.37	0.02	1.2
5.	MATC	SW, ALL	0.98	-0.60	41	1.80	0.42	0.01	1.3
6.	HATCH EC_{25}	AW, ALL	1.1	-1.3	28	3.2	0.83	0.013	1.8
7.	MORT1 EC_{25}	AW, ALL	0.87	-0.87	28	2.2	0.53	0.016	1.4
8.	EGGS EC_{25}	AW, ALL	1.4	-3.0	36	2.6	1.2	0.050	2.2
9.	WEIGHT EC_{25}	AW, ALL	1.0	-0.99	50	2.9	0.60	0.0048	1.5
10.	MORT2 EC_{25}	AW, ALL	1.0	-0.86	77	2.7	0.55	0.036	1.4
11.	WEIGHT/EGG EC_{25}	AW, ALL	0.90	-0.86	34	2.6	0.62	0.0055	1.6
12.	Daphnia MATC	LC	1.11	-1.30	57	2.73	0.48	0.01	1.35
13.	Daphnia MATC	LC, M	0.96	-1.08	27	2.44	0.63	0.02	1.56

[a]MORT1 EC_{25} = the concentration estimated to cause a 25% increase in mortality of parental (postlarval) fish, MORT2 EC_{25} = the concentration estimated to cause a 25% increase in mortality of larval fish, HATCH EC_{25} = the concentration estimated to cause a 25% decrease in normal hatches of eggs, EGGS EC_{25} = the concentration estimated to cause a 25% decrease in production of viable eggs per surviving female, WEIGHT EC_{25} = the concentration estimated to cause a 25% decrease in the weight of early juvenile fish.

[b]FW = freshwater, SW = saltwater, AW = all waters, LC = life cycle or partial life cycle tests, All = LC + early life stage tests, N = narcotic chemicals, M = metals.

[c]The mean of X and Y and the factors F_x and G_x are used to calculate the variance as described in the text (Box 7.1).

[d]The 95% prediction interval at the mean is log $\bar{Y}$ ± the number in this column.

extrapolations. Tucker and Lietzke (1979) concluded after their review of insecticide effects on vertebrate wildlife and fish that "no sublethal effect reported in the current literature tends to occur at a level less than a quarter or a sixth of the level producing a comparable percent mortality in equivalent tests." The important qualifier here is "equivalent tests," meaning the same life stage, duration, and mode of exposure. Mayer et al. (1986) concluded similarly that statistically significant effects on nonlethal endpoints occurred at concentrations within a factor of 5 of a lethal endpoint approximately 95% of the time. However, this conclusion is dependent on the use of statistical significance instead of biological significance to define the low severity effects level (Suter et al. 1987).

Because it is readily observed, visible injury of plants has been the most commonly reported effect of air pollution. However, the more severe response in terms of human utility is reduction in yield. There appears to be little or no correlation of either yield or long-term growth with visible injury (Roberts 1984).

Proportion Responding

Unlike extrapolation from 50% mortality to some less severe effect, extrapolating to a lower proportion responding is reasonably straightforward. If the exposure-response function is available or can be calculated, and response is expressed as proportion responding, then it is simply a matter of solving the function for 0.01, 0.10, or other desired proportion. Although the variance on the estimate is larger at these lower effects levels, they are still small enough that the choice of function is not a major source of uncertainty as it is in human health risk assessments where 10^{-6} effects levels are estimated.

In an unfortunately large number of cases, LC_{50} and LD_{50} values are reported without the information needed to calculate other proportional responses. In such cases, a slope must be assumed. The EPA's Office of Pesticide Programs determined that, for terrestrial and aquatic species, one-fifth of the LD_{50} or LC_{50} corresponds to a 0.1% mortality ($LC_{0.1}$) in the average test and 10% mortality in the test with the lowest slope (Urban and Cook 1986). The EPA's Office of Water Regulation and Standards (OWRS 1985) recommends using a factor of 0.3 to estimate LC_{01} from LC_{50} values for aqueous effluents because only 9% of effluent tests gave lower LC_{01}/LC_{50} ratios for aquatic species (Figure 7.11). Interestingly, this factor exactly equals one of the first published safety factors for ecological assessment, which was intended to estimate zero lethality concentrations from LC_{50}s (Hart et al. 1945, cited in Tebo 1986).

Conditions

Physical/chemical conditions including temperature, pH, hardness, and salinity can greatly influence the results of toxicity tests. Reviews of tempera-

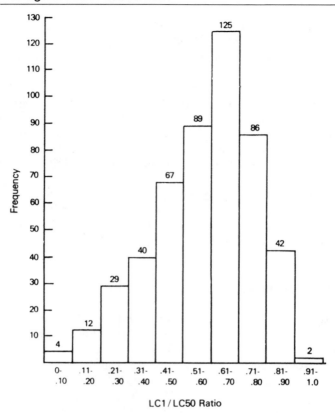

Figure 7.11. Frequency distribution of LC$_1$ to LC$_{50}$ ratios for effluent toxicity tests (from OWRS 1985).

ture effects in aquatic toxicology by Cairns et al. (1975a,b) and Mayer and Ellersieck (1986) indicate that temperature and toxicity are positively correlated for most chemicals and species. For most organic and inorganic chemicals, the increase in toxicity per 10°C increase in temperature is 2X to 4X, which is the same range as the temperature coefficient (Q_{10}) for respiration. The major exceptions are the organochloride insecticides DDT, DDD, and methoxychlor, for which LC$_{50}$s are negatively correlated with temperature. Sprague (1985) argued that temperature effects were not predictable and should be treated as a small source of error in acute lethality tests (primarily affecting time to death rather than the lethal concentration) and may have a negligible effect in chronic exposures. Mayer and Ellersieck (1986) derived 26 regression equations to predict temperature-specific 96-h fish LC$_{50}$ values for 19 pesticides (Table 7.9). The parameters of the equation tend to be similar among species for a chemical and, to a lesser extent, among chemicals within a pesticide class. They concluded that for most chemicals having positive tem-

Table 7.9. Linear Regression Models for Effect of Temperature on Acute Static Toxicity (96-hr LC_{50}) of Organic Chemicals to Fishes[a]

Chemical Species	Temperature Range (°C)	LC_{50}s (n)	LC_{50} Range (μg/L)	Intercept	Slope	r^2 (%)
Aminocarb						
Brook trout	7–17	3	2600–9400	4.36	−0.056	100
Yellow perch	7–22	4	1700–11700	4.43	−0.052	94
Azinphos-methyl						
Yellow perch	7–22	3	2.4–40	2.17	−0.082	100
Carbaryl						
Brook trout	7–17	3	680–2100	3.98	−0.064	92
Yellow perch	7–22	4	1200–13900	4.60	−0.068	98
Chlordecone						
Redear sunfish	7–29	5	29–140	2.36	−0.030	100
Chlorpyriphos						
Rainbow trout	2–18	4	1.0–51	1.94	−0.100	96
DDT						
Bluegill	7–29	5	1.6–5.8	0.028	0.028	89
Dimethrin						
Bluegill	12–22	3	21–85	0.652	0.060	94
Diuron						
Bluegill	7–29	5	2800–9500	4.220	−0.022	74
Endothall, Aquathol K						
Bluegill	7–22	4	343–1740[b]	6.668	−0.047	86
Endrin						
Bluegill	7–29	5	0.19–0.73	0.101	−0.026	90
Malathion						
Bluegill	7–29	5	20–87	2.215	−0.028	92
Merphos						
Bluegill	12–22	3	8000–32000	5.104	−0.055	100
Methomyl						
Bluegill	12–27	9	430–2000	3.642	−0.039	75
Methoxchlor						
Brook trout	7–17	12	7.0–30	0.762	−0.028	42

[a]From Mayer and Ellerseick 1986.

perature coefficients, the effects of a 10°C change in temperature could be estimated as:

$$\log LC_{50_{t \pm 10}} = \log LC_{50_t} \pm 0.4956 \qquad (7.17)$$

but for organophosphate pesticides, the following formula should be used:

$$\log LC_{50_{t \pm 10}} = \log LC_{50_t} \pm 0.7113 \qquad (7.18)$$

The 95% confidence intervals on the parameters are 0.3775 to 0.6137 and 0.5836 to 0.8390, respectively.

Temperature effects are less well studied for terrestrial organisms. Increased temperature, with constant atmospheric humidity and soil moisture, causes increased stomatal opening and increased uptake of air pollutants by plants. However, Norby and Kozlowski (1981a, 1981b) showed that temperatures after exposure as well as during exposure can affect response. In some species of woody plants, increased resistance to the effects of SO_2 with increasing temperature was sufficient to compensate for increased uptake. No regularity in temperature effects were identified that would serve as a basis for assessment.

Hydrogen ion activity (pH) influences the aquatic toxicity of chemicals primarily by ionizing or neutralizing them; neutral forms are more readily taken up by gills and thus are more toxic (Maren et al. 1968). The influences of pH and temperature on ammonia toxicity are incorporated into site-specific water quality criterion concentrations for ammonia (EPA 1985a). Mayer and Ellersieck (1986) applied analysis of covariance to 100 pH tests of 49 chemicals and obtained slopes significantly different from zero in 23 tests of 10 chemicals (Table 7.10). Parameters of the equations are reasonably consistent among species of fish. The influence of pH on these chemicals is large, and is explainable after the fact in terms of ionization, neutralization, or hydrolysis. However, methods are not available to predict its influence on new chemicals.

Water hardness significantly influences the toxicity of metals primarily by reducing the concentration of free metal ions, and also apparently by increased resistance through unknown mechanism (Pascoe et al. 1986). The U.S. national water quality criteria for some metals are presented as functions of hardness rather than fixed concentrations (EPA 1985). As would be anticipated, soaps are 10–100 times less toxic in hard water than soft water, but hardness has less than a 2X effect on detergent toxicity (Henderson et al. 1959, Hokanson and Smith 1971). Hardness has relatively little effect on the toxicity of organic chemicals, and the reported relationships between toxicity of organic chemicals and hardness is probably due in nearly all cases to differences in pH (Pickering and Henderson 1966, Mayer and Ellersieck 1986).

Salinity has been shown to influence aquatic toxicity, particularly of metals (e.g., Engel and Fowler 1979). However, comparisons of toxic effects of chemicals on freshwater and saltwater species show no consistent differences in sensitivity when tested in appropriate water, even for metals and other ionizable chemicals (EPA 1979, Klapow and Lewis 1979, Suter and Rosen 1988). These results suggest that the influence of salinity will be important primarily in estuaries, where salinity variations are large. Sprague (1985) suggests that euryhaline species are most resistant by as much as 10X at isosmotic salinities (about one third seawater), due to minimization of osmotic stress.

Temperature, pH, hardness, and salinity are only the best studied of the properties that modify aquatic toxicity. Others include dissolved oxygen level, redox potential, photon flux, alkalinity, dissolved organic matter, and suspended particulate matter. One solution to the problem of accounting for the effects of all of these factors is to assume that they only modify the form of the

Table 7.10. Linear Regression Models for Effect of pH on Acute Static Toxicity (96-hr LC_{50}) of Organic Chemicals to Fishes[a]

Chemical Species	pH Range	LC_{50}s (n)	LC_{50} Range (µg/L)	Intercept	Slope[b]	r^2 (%)
Aminocarb						
Rainbow trout	7.5–9.5	3	5000–21000	6.89	−0.31	84
Bluegill	6.5–9.5	4	1000–10000	6.45	−0.33	84
Yellow perch	6.5–9.0	4	425–7000	7.50	−0.50	69
Benomyl						
Rainbow trout	6.5–8.5	3	160–880	−0.49	0.37	82
Bluegill	6.5–8.5	3	1200–6400	0.56	0.35	71
Carbaryl						
Cutthroat trout	6.5–8.5	3	970–7100	6.36	−0.36	–
Yellow perch	6.5–9.0	4	350–4200	7.58	−0.54	80
2,4-D dodecyl/tetra decyl amine salt						
Fathead minnow	6.5–8.5	6	1900–8400	5.92	−0.29	86
Dinoseb						
Cutthroat trout	6.5–8.5	3	41–1350	−3.78	0.76	96
Lake trout	6.5–8.5	3	32–1440	−4.43	0.83	91
Diquat						
Bluegill	6.5–9.5	4	115000–498000	7.38	−0.23	92

Table 7.10. Continued

Chemical Species	pH Range	LC$_{50}$s (n)	LC$_{50}$ Range (µg/L)	Intercept	Slope[b]	r² (%)
Glyphosate						
Rainbow trout	6.5–9.5	4	1400–7600	5.27	-0.23	66
Mexacarbamate						
Coho salmon	6.5–9.0	3	4000–23000	7.06	-0.37	73
Cutthroat trout	6.0–9.0	3	630–15800	6.25	-0.34	46
Bluegill	6.5–9.5	5	600–22900	8.60	-0.57	86
Phosmet						
Rainbow trout	6.5–9.5	5	105–4700	-2.18	0.57	85
Bluegill	6.5–8.5	3	22–640	-3.70	0.73	97
Trichlorophon						
Cutthroat trout	6.5–8.5	3	375–4750	7.57	-0.55	99
Rainbow trout	6.5–9.0	12	210–11400	6.64	-0.43	49
Atlantic salmon	6.5–8.5	4	300–4400	7.54	-0.58	81
Brook trout	6.5–9.0	7	240–9200	8.64	-0.65	85
Bluegill[b]	6.5–9.0	12	1720–50000	8.04	-0.49	56

[a]From Mayer and Ellersieck 1986.
[b]All are significantly different from 0 (alpha < 0.05).

chemicals. In that case, assessors could isolate and functionally define the effects of the physical/chemical conditions on the form of the chemicals, and mathematically model the form or mixture of forms to which the organisms are exposed (i.e., treat it as strictly an exposure assessment problem). This is a formidable job in itself but it would also require that the toxicity of each chemical form and the combined effects of mixtures of forms be identified. The problem is manageable for chemicals with only two forms and few physical/chemical influences like ammonia (Thurston et al. 1983), but is currently impractical for polyvalent metals and organics with complex ionization, hydrolysis, and sorption dynamics. Most current environmental exposure models distinguish only dissolved and sorbed phases. An alternative approach is to identify a few critical physical/chemical variables and develop response surfaces for their combined influence on toxicity for major pollutant chemicals and important taxa. This approach is currently being pursued at the EPA Gulf Breeze Laboratory. When a physical/chemical variable is routinely associated with a pollutant, as with elevated temperature and chlorine in cooling water, that relationship can be worth working out in detail (Hall et al. 1983 —

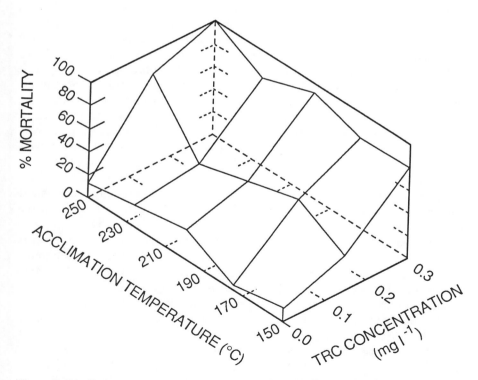

Figure 7.12.   Response surface for percent mortality of striped bass larvae exposed to total residual chlorine (TRC) at various temperatures (Hall et al. 1983).

Fig. 7.12). Finally, where an identifiable effluent at a particular site is being assessed, the site water can be used as dilution water in toxicity tests.

Physical/chemical conditions have not been an issue in wildlife toxicology, but soil chemistry and atmospheric conditions are major considerations in phytotoxicology. As discussed previously, transpiration drives the uptake of soil water and of associated chemicals, and the opening of stomata and changes in internal leaf conductance associated with transpiration determine the rate of entry of pollutant gases (McFarlane et al. 1987, McLaughlin and Taylor 1981). These relationships imply that plants will be more susceptible to toxic chemicals when well watered and when humidity is high. In addition, soil chemistry, including pH, clay content, and organic matter content, can affect the availability of metals and other chemicals (Kabata-Pendias and Pendias 1984, Van Straalen and Denneman 1989). However, Fletcher et al. (1990) found that the differences between greenhouse and field conditions made, on average, less than a factor of two difference in EC_{50}s for chemicals applied to foliage.

Organism Condition

The relative condition of organisms in the laboratory and field has been a topic of considerable conjecture and debate. Some investigators believe that organisms in the laboratory are stressed and malnourished; others believe they are unnaturally well fed, healthy, and unstressed (e.g., Buikema et al. 1980, Price and Swift 1985). The only aspect of this problem that has received much attention is diet. The quality of nutrition can obviously affect an organism's ability to generate toxicant-metabolizing enzymes, repair injury, and otherwise resist toxic effects. In addition, a well-fed organism has more fat in which to sequester hydrophobic chemicals. On the other hand, more rapidly growing fish were found to take up lead faster and experience more severe effects (Hodson et al. 1982). Malnourished animals can be highly susceptible. For example, ducks fed corn and treated with lead suffered 50% mortality in four days while ducks fed a complete diet had no mortality in 50 days at the same lead dose (Roscoe and Nielsen 1979). Reviews of the effects of diet on the outcome of laboratory tests (Tucker and Lietzke 1979, Dave 1981, Chapman 1983, Mayer and Ellersieck 1986) indicate that differences in laboratory diets have typically caused 2-6X differences in effects. This range is undoubtedly diminishing as the nutrition of test species is better understood and as more test protocols include defined diets. However, the variance in nutritional state in natural populations and its effects on sensitivity are unknown.

The stress of confinement has also been hypothesized to be a confounding variable in toxicity testing. The commonly used test species have attained their popularity in large part because they tolerate captivity, so for most tests this is not an issue. However, species used in ad hoc testing programs may be seriously stressed by confinement. Walton et al. (1981) found that the effects of captivity on the condition and reproductive development of Atlantic cunners was greater, relative to the source population, than the effects of oil treatment (0.3 to 0.8 ppm) relative to controls.

Behavioral interactions may also affect an organism's resistance. Resistance to toxicants declines following harassment of starlings by humans (Tucker and Lietzke 1979), harassment of bluegill sunfish by conspecifics (Sparks et al. 1972), and the interactions of male/female pairs of pigeons (Dobson 1985).

Interactions among host organisms, parasites, pathogens, and pollutants can be complex (Suter 1987). Depending on the species combination and timing, parasitism and disease may make a host organism more susceptible to pollutants, pollutants may make an organism more susceptible to disease, or pollutants may have a prophylactic effect if the parasite or pathogen is more susceptible than the host. Even in a fairly narrowly defined combination such as gaseous pollutants, coniferous trees, and fungal pathogens it is not possible to generalize about the signs of the pathogen/pollutant interactions (Suter 1987). This issue has not been studied well enough to even allow estimation of the degree of variance introduced by parasites and pathogens. In general, laboratory test organisms are kept free of disease and parasites, and if observable infection occurs, tests are usually terminated. In contrast, most macroorganisms in the field have some parasite burden and disease outbreaks are not uncommon.

The number of other factors that could affect the susceptibility of individual organisms is indefinitely large, and the interactions among them are complex. Porter et al. (1984) examined the effects of an immunosuppressant, a virus, reduced food, reduced water, a plant hormone, and a PCB mixture on the reduction of survival and weight of young produced by deer mice. All factors had an effect, either alone or in combination; some combinations produced severe effects at ostensibly safe concentrations of PCBs; and the plant hormone reduced growth when water was reduced. In addition, stress can produce anomalous effects. Poor diet can cause *Daphnia* to produce fewer but heavier and more resistant young (Cowgill et al. 1985). Finally, the influence of stressors on susceptibility is toxicant-specific. Alexander and Clark (1978) found that starvation and temperature stress both significantly reduced the time to death of rainbow trout exposed to phenol; only starvation significantly changed the effects of sodium pentachlorophenate, and neither factor significantly changed the effects of sodium azide, copper sulphate, or disodium sulphate. These results suggest that the interaction of stress and toxicity depends on the relationship between the mechanism of action of the toxicant, and the mechanism of the stress. Therefore, even if there is a general adaptation syndrome in response to diverse nonspecific stresses that leads in a regular fashion to death or recovery (Selye 1950), that does not imply a regular relation between nonspecific stress and the specific effects of toxicants.

Even if the nature of the relationship between stress and effects of a particular chemical or class of chemicals is specified in the laboratory, that would not provide a basis for assessment because the degree to which organisms are stressed in the field is unknown. Even in human health risk assessments, where

Table 7.11. Categories of Responses to Mixtures of Chemicals[a]

	Similar Action	Dissimilar Action
Non-interactive	simple similar (concentration addition)	independent (response addition)
Interactive	complex similar	dependent

[a]From Hewlett and Plackett 1952, 1979.

the frequency of disease, malnutrition, and other stressors are relatively well defined, there are no generally accepted techniques for treating these factors in risk models (Clayson et al. 1985).

One possible approach is to examine the circumstantial evidence supplied by field monitoring and experiments. Field experiments suggest that in general, when physical/chemical conditions are equivalent, organisms in the field respond at about the same exposure levels as organisms in the laboratory, including aquatic fish and invertebrates (Crossland 1982, Hansen and Garton 1982) and birds (McEwen et al. 1971). However, in at least two studies, fish in outdoor channels and ponds were more sensitive than those in the laboratory (Hedtke and Arthur 1985, Boyle et al. 1985), and wild-caught fathead minnows were twice as sensitive as laboratory-reared fish to endrin (Dave 1981). The frequently observed conjunction of bacterial infections and high pollution levels in fish kills (Wedemeyer 1970, Snieszko 1974) suggests that periodic associations of stress factors may be a significant issue in assessment. Models of the frequency and consequences of co-occurring risk factors are common in engineering risk analysis, but much more research is needed before such models could be developed for effects on organisms in nature.

MODELS OF CHEMICAL MIXTURES

Most of the research and testing in environmental toxicology has studied individual chemicals, but chemicals often occur together as complex materials such as oils and mixed formulations of agricultural chemicals, as complex effluents, or as chemicals from various sources mixed in an environmental medium. The occurrence of chemicals in mixtures influences toxicity in two ways. First, by jointly inducing effects, chemical mixtures can cause a toxic effect that is qualitatively or quantitatively different from any of the components acting alone. Second, the effects of one chemical may influence the kinetics of uptake, metabolism, and excretion of other chemicals. Examples include the coating of fish gills by thick mucus when exposed to zinc, and the damage to nephridia caused by cadmium-metalothionine complexes. The metabolic kinetics of a chemical may also be affected by other chemicals that induce or inhibit enzymes, or that simply reduce the physiological capacities of an organism.

The most commonly used classification of joint toxic action is that of Plackett and Hewlett (1952) (Table 7.11). They divided the joint actions of pairs of chemicals into four classes based on whether they had similar joint action and on whether interactions occurred. In this system, chemicals are called similar if their sites and modes of action are the same. Interactions include effects of one chemical on the kinetics of the other, modification of the site of action, or production of an effect by the joint action of chemicals when one or both of the chemicals is not active alone. Interactions may be potentiating (also termed synergistic or enhancing) or antagonistic (also termed inhibiting or attenuating). Terminology is used inconsistently in this field, and some authors make other distinctions among these terms, as well as using terms like supra-, hyper-, super-, infra-, sub-, hypo-, and hyper-additive. Simple similar action is commonly referred to as concentration or dose addition and independent action is referred to as response addition.

Various versions of the concentration addition model have been formulated for quantal responses, based on sensitivity distributions like those used in dose-response models (Finney 1971, Hewlett and Plackett 1952, Kodell and Pounds 1985). If a set of chemicals differs only in their potency, then mortality or any other quantal response can be modeled using Finney's (1971) formulation:

$$Y = a + b \log (C_1 + r_2C_2 + \ldots + r_nC_n) \tag{7.19}$$

where
   Y is the probit of the response to the mixture,
   C_i is the concentration of chemical i,
   r_i is the potency of chemical i relative to chemical 1, and
   a and b are fitted intercept and slope parameters.

For a particular level of effect such as an LC_{50}, concentration addition reduces to:

$$\Sigma (C_i/LC_{50_i}) = 1 \tag{7.20}$$

The ratio C_i/LC_{50_i} is termed the toxic unit. Chemicals that have the same mode of action [i.e., are fit by the same quantitative structure activity relationship] would be expected to follow this model. In addition, concentration additivity has commonly been treated as the nominal case for joint toxic action of chemicals because many mixtures of chemicals approximately fit the model (Alabaster and Lloyd 1982).

The response addition model assumes that the site or mode of action of the chemicals are different and the chemicals do not interact, so that an individual organism dies of the effects of the chemical to which it is most sensitive relative to the chemical's concentration. That is, for two chemicals:

$$
\begin{aligned}
P(C_1 + C_2) &= P(C_1) + [1 - P(C_1)]P(C2) \\
&= P(C_1) + P(C_2) - P(C_1)*P(C_2)
\end{aligned}
\tag{7.21}
$$

where $P(C_1)$ is the probability of a quantal response given concentration C_i. This is simply the joint probability of independent events, so the formulation for more than two chemicals can be derived analogously. If the probabilities of the effects of the two chemicals acting individually are small or if the probabilities are assumed to be perfectly negatively correlated (i.e., organisms that are least tolerant of one chemical are most tolerant of the other so that $r = -1$), then the product of the probabilities can be dropped and the probabilities of single chemical responses can simply be added. The other extreme version of response additivity is perfect correlation of sensitivity to the chemicals ($r = 1$), so that all organisms die of the most toxic chemical and the other chemicals do not contribute. To determine whether test results for a pair of chemicals fit this model, the probit or other suitable concentration-response function $[F(C_i)]$ can be fit to dose-response data for each chemical and the goodness of fit of a response additivity model:

$$P(C_1 + C_2) = F_1(C_1) + F_2(C_2) - F_1(C_1)*F_2(C_2) \qquad (7.22)$$

to test data for mixtures is determined (e.g., Kodell and Pounds 1985). The single-hit and linear dose-response functions give identical results in the concentration addition and response addition models (Kodell 1987).

The possibilities for interactive joint action are revealed by examination of an isobologram (Figure 7.13), a diagram of the possible combinations of two chemicals that can produce a particular effect. Isoboles are lines of equal effect. The conventional interpretation is that isoboles that fall into the areas marked potentiation and antagonism indicate interaction between the chemicals. There are no standard biometrical models to fit these cases.

There are at least three ways to assess the effects of mixtures. First, if an effluent or other complex toxic material already exists, is available to the assessor, and has a reasonably stable composition, the most realistic approach is to test the toxicity of the material. This approach has gained considerable popularity among aquatic toxicologists (Bergman et al. 1986), and provides the basis for regulation of aqueous effluents under the EPA's policy of "water quality-based toxics control" (OWRS 1985). The results can be used in assessments like test results for single chemicals. If the toxicity test results suggest that the material is hazardous, however, these tests do not indicate what chemicals in the mixture are responsible for the toxicity and need to be removed by treatment of the effluent or modification of the product. Chemical analysis of the material may reveal component chemicals that occur at high concentrations relative to their known effective concentrations. Otherwise the material must be fractionated, the fractions tested, and the most toxic fractions analyzed and possibly refractionated and tested, until one or more treatable toxic fractions are identified (Rubin et al. 1976, Parkhurst 1986, Doi and Grothe 1988). The EPA has developed a standard approach to this process termed "toxicity identification evaluation" for acute effects on aquatic animals (Mount 1989, Mount and Anderson-Carnahan 1988, 1989).

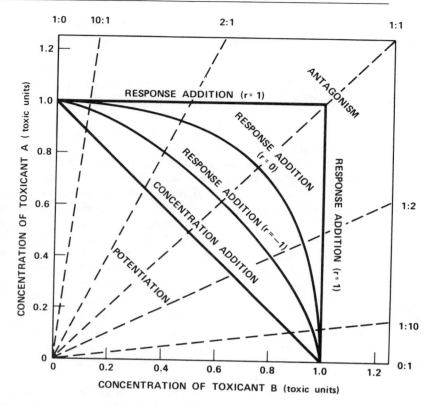

Figure 7.13. Isobologram (isobole diagram) for a quantal response to mixture of two toxic substances. Dashed lines represent constant ratios of concentrations of the two toxicants. Solid lines are isoboles (lines of equal response) and represent the set of concentrations of A and B at which the endpoint quantal response (e.g., LC_{50}) occurs. They suggest the type of interaction of the two toxicants.

Second, when the material of interest is not available, or when the material has relatively few components and the assessor is interested in the effects of variation in their proportions, then synthetic mixtures can be tested. This situation may also apply to complex mixtures if the toxicity is known to be due to a few components. Alabaster and Lloyd (1982) argued that chemicals occurring at less than 1% of their LC_{50} for fish may generally be ignored (because effects of some chemicals in chronic exposures can occur at concentrations less than one percent of the LC_{50}, this rule-of-thumb should only be applied to acute exposures). The advantage of this approach is that the nature of the joint toxicity can be elucidated and used to predict the effects of variability in the material or to optimize the composition of a product. For example, a herbicide formulation might be designed to maximize toxicity to plants relative to animals. Tests of synthetic mixtures may also be used to confirm that fractionation studies have revealed the sources of observed effects (Meyer et al. 1985).

Results of these tests of chemicals mixed in varying portions must be fit to a model, preferably a simple concentration or response additivity model. Results of such studies are relatively common in the aquatic toxicology literature (e.g., Hermens et al. 1984a, 1984b, Broderius and Kahl 1984, Spehar et al. 1978), in mammalian toxicology for human health effects (e.g., Landau and Wellington 1987, Vouk et al. 1987), and in the air pollution phytotoxicity literature (e.g., Shinn et al. 1976).

Third, in some cases the assessor cannot obtain test data for a mixture of interest. In those cases, a mode of joint action must be assumed. For example, to assess the effects of a mixture of chemicals in a predicted synfuels effluent, Suter et al. (1985) used the concentration addition model to estimate the potential for acutely lethal effects on fish of a predicted synfuels effluent. Because the LC_{50}s for the various chemicals came from various fish species, taxonomic extrapolation equations (Table 7.4) were used to convert them all to a common unit, LC_{50}s for largemouth bass. The use of the concentration addition model was justified by the finding of Lloyd (1986) that "there is no evidence of synergism (i.e., more than additive action) between the common pollutants; at toxic concentrations the joint action is additive and at concentrations below those considered 'safe' there is circumstantial evidence for less-than-additive joint action." However, this conclusion is based primarily on studies of acute lethality, and several studies indicate that the mode of joint toxicity of a mixture depends on the duration of the exposure and on the response being measured (Hermens et al. 1984a, Hermens et al. 1984b, Hermanutz et al. 1985, Spehar et al. 1978).

Calabrese (1991) provides an excellent review of the toxicology of mixtures for humans and their surrogate species and for fish. Rao et al.(1988) provide a recent review of effects of air pollutant mixtures on plants.

RISK CHARACTERIZATION FOR ORGANISM-LEVEL EFFECTS

This section explains how the various endpoints, factors, statistical models, and mathematical models can be used to estimate risks to individual organisms exposed to toxicants.

Factors and Quotients

Factors are used in hazard assessment and in the screening levels of risk analysis to identify hazards by what has been termed the quotient method (Barnthouse et al. 1982, Urban and Cook 1986). If the quotient of the expected environmental concentration (EEC), divided by a toxicological endpoint concentration (TEC), is less than some lower-bound safety or uncertainty factor or product of factors ($EEC/TEC < F_L$), the release is considered safe and further assessments are not conducted. If quotients exceed some upper-bound factor or product of factors ($EEC/TEC > F_U$), it is considered to be unsafe and further assessment is not conducted. Quotients between F_L and F_U, however indicate uncertainty about safety and imply the need for further assessment. In many cases, iterative assessment is not possible and a single factor is used (i.e., if $EEC/TEC < F$, the release is considered safe, otherwise it is not).

An alternative formulation is:

$$f_1 f_2 \ldots f_n(EEC) < F_1 F_2 \ldots F_n(TEC) \tag{7.23}$$

where f_x is any factor influencing the expected environmental concentration and F_x is any factor influencing the toxicological endpoint concentration. If the inequality holds, the release is considered safe.

Often, rather than formulating factors in advance, the quotient is calculated, and a judgment is made as to whether the implied safety factor is large enough. This post hoc version of the quotient method is indicated by statements of the form: "since the calculated ground-level concentration of SO_2 is only twenty percent of the lowest concentration known to cause toxic effects, no significant impacts are expected." Such post hoc judgments should be avoided.

Probabilistic Methods Based on Extrapolation

The inevitable uncertainty concerning the magnitude of effects is best represented by probability density functions. These functions provide a more complete portrayal of the assessor's understanding than do factors, and they avoid the conservatism associated with the use of extreme values to calculate factors. In addition, they allow the assessor to calculate risks as probabilities of occurrence. If we say that a value such as the sensitivity of fish species in acute lethality tests has a log normal distribution with respect to concentration, that is equivalent to saying that the probability (risk) of acute lethal effects on a particular species at a particular concentration is the cumulative probability density at that concentration. The uncertainty associated with any measurement or model can be used to estimate risk in this way if they are believed to control the probability of occurrence of effects. For example, the models fit to test data have associated uncertainty, and the lower 95% confidence bound on the function can be used as an estimate of a dose with a low probability of causing the prescribed magnitude of the measured effect (Dourson 1986).

Statistical models used to extrapolate test data have associated probability density functions that describe the uncertainty associated with estimation of effects on the endpoint organisms and conditions. Kenaga and Lamb (1981) suggested that the confidence limits on regressions between species and test methods could be used as safety factors in hazard assessments. Suter et al. (1983, 1986) showed that the probability density functions for values estimated by regression models for the steps in a lab to field extrapolation could be combined and used to estimate risk. They called this risk assessment method Analysis of Extrapolation Error (AEE).

AEE can best be introduced by presenting an example graphically. Assume that we wish to estimate the probability that the expected environmental concentration of a chemical will exceed a level that causes significant reductions in the production of juvenile brook trout (*Salvelinus fontinalis*) and that the only datum available for the chemical is an LC_{50} for rainbow trout (*Oncorhynchus mykiss*). In that case we must extrapolate between the genera *Oncorhynchus* and *Salvelinus*, which are both in the family salmonidae, and between the LC_{50} and an early life-stage effect (a 25% reduction in the weight of young per initial egg in a life cycle or early life stage test). The relationship between the two genera is illustrated in Figure 7.14. Each of the points represents an individual chemical for which a member of both genera has been tested using a common protocol and with the results expressed as 96-h LC_{50}s. The relationship between LC_{50}s and weight/egg EC_{25} is shown in Figure 7.15. The points in the figure represent different species-chemical combinations for which both a 96-h acute lethality test and a standard life-cycle or early life stage test have been conducted in the same laboratory for a particular fish species-chemical

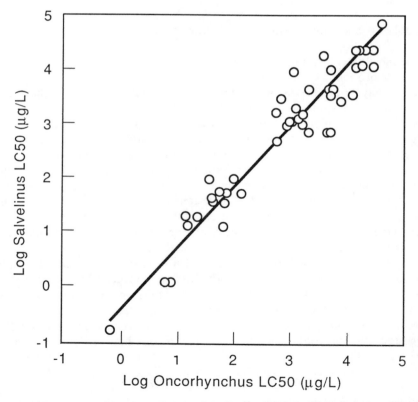

Figure 7.14. Logarithm of LC_{50} values for *Salvelinus* regressed against *Oncorhynchus* using an errors-in-variables regression. (See line 12, Table 7.4)

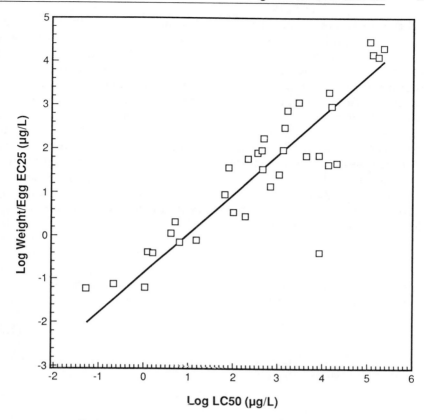

Figure 7.15.   Logarithms of EC_{25} values for weight of juvenile fish per initial egg from life cycle or partial life cycle tests plotted against logarithms of 96-h LC_{50} values determined for the same species and chemical in the same laboratory. The line is derived by an errors-in-variables regression; the parameters are presented in line 11 of Table 7.8.

combination and from which a LC_{50} and a EC_{25} for weight/egg could be calculated. If we use the rainbow trout LC_{50} as the x value in the Figure 7.14 relationship, we can estimate a brook trout LC_{50} and an associated variance that can be used in the Figure 7.15 relationship to estimate a brook trout weight/egg EC_{25} and its associated variance. The estimated EC_{25} and its total variance can be represented as a probability density function, as in Figure 2.4. The risk that the EC_{25} will in fact be exceeded is the probability that a realization of the EC_{25}, chosen at random from that probability density function, will be less than a similarly chosen value from the probability density function for the expected environmental concentration.

AEE consists of five steps: (1) define the assessment endpoint (e.g., the probability of causing reduction in brook trout productivity, given an early life stage exposure) in terms of a test endpoint that can serve as a surrogate for the assessment endpoint (e.g., the probability of exceeding the brook trout weight/egg EC_{25}); (2) identify the existing datum for the chemical of interest that is most closely related to the assessment endpoint (e.g., a rainbow trout 96-h LC_{50}); (3) break the relationship between the datum and the surrogate assessment endpoint into logical steps for which an extrapolation model exists or can be developed (e.g., rainbow trout to brook trout and LC_{50} to EC_{25}); (4) calculate the distribution parameters of the surrogate assessment endpoint extrapolated from the datum; and (5) calculate the risk that the expected environmental concentration (EEC) will exceed the endpoint concentration. Step 1 depends on the assessment situation and on the assessor's and decision-maker's conceptualization of environmental values. Steps 1, 2, and 3, however, are severely constrained by the state of the science of environmental toxicology as reflected in the available test endpoints for the organisms in question.

In this method, risk is defined as:

$$\text{Risk} = \text{Prob(EEC} > \text{SEC)} \tag{7.24}$$

where SEC is the surrogate endpoint concentration. If we assume that the EEC and SEC are independent and log-normally distributed, then

$$\text{Risk} = \text{Prob(log SEC} - \text{log EEC} < 0) \tag{7.25}$$

$$= \text{Prob}[Z < [0 - (m_s - m_e) / (s_s^2 + s_e^2)^{1/2}] \tag{7.26}$$

$$= d_z[(m_e - m_s) / (s_s^2 + s_e^2)^{1/2} \tag{7.27}$$

where (m_s, s_s^2) and (m_e, s_e^2) are the mean and variance of the log SEC and log EEC, respectively, and

$$Z = [\text{log SEC} - \text{log EEC}) - (m_s - m_e)] / (s_s^2 + s_e^2)^{1/2} \tag{7.28}$$

a standard normal random variable with d_z as its cumulative distribution function. If it is assumed that the EEC is constant and certain, the risk calculations reduces to

$$\text{Risk} = \text{Prob}\{Z < [(\text{log EEC} - m_s) / s_s]\} \tag{7.29}$$

$$= d_z[(\text{log EEC} - m_s) / s_s] \tag{7.30}$$

Given this definition, risk depends on the definitions of the EEC and SEC and their associated variances (i.e., on m_e, m_s, s_e^2, and s_s^2). For the SEC, the mean and variance can be estimated by fitting statistical models to toxicity data and by statistical extrapolation of the test endpoints. The appropriate variance is the variance of the individual observations in the regression equations or equivalent variances (e.g., the variance of a species sensitivity distribution).

When multiple extrapolations are used, as in the example above, the total variance is the sum of the variances of each step. For a series of errors-in-variables regression models, the variances can not simply be added because that would result in double counting. If we are extrapolating from X to Y to Z, the variance of Y occurs in the regressions of Y on X and Z on Y. To avoid this, we use the following formula for the variance of a predicted Z value for a given X ($X = X_0$):

$$\text{var } (Z|X) = \text{var}(Z|Y_0) + b^2\text{var}(Y|X_0) \tag{7.31}$$

where b is the slope of the regression of Z on Y.

In our example, assume that the rainbow trout LC_{50} is 5300 $\mu g/L$ for the chemical of concern. Substituting the log of that LC_{50} into the *Oncorhynchus-Salvelinus* extrapolation model (Table 7.4, Equation 12F) gives an estimated log brook trout LC_{50} of 3.49; using Equation B7.4, the variance is 0.093. Substituting 3.49 into Equation 11 from Table 7.8, gives an estimate of 2.28, for the log of the brook trout EC_{25}, with a variance for this extrapolation of 0.62. Using Equation 7.31, the total variance for the double extrapolation is $0.62 + 0.81(0.093) = 0.69$.

If the log of the expected environmental concentration (EEC) is 2.0 with a variance of 0.5, then the probability that a realization of the brook trout EC_{25} for weight/egg is less than a realization of the EEC is determined from Equation 7.28, by calculating

$$(2.0 - 2.28) / (0.5 + 0.69)^{1/2} = -0.26$$

The cumulative probability for this Z value (obtained from a Z table) is 0.40. Thus, the risk that the ambient concentration would cause as much as a 25% reduction in the production of early juvenile brook trout is 0.40, or we are 60% certain that a 25% reduction would not occur. More simply, it is a toss-up.

The AEE approach is not limited to toxicity extrapolation models derived by regression. Dose-response models, quantitative structure-activity relationships, species sensitivity distributions, or any other model that generates a statistical distribution of effects could be used (Volmer et al. 1988, Nendza et al. 1990). The basic idea of (1) defining the effects assessment as an extrapolation or series of extrapolations, (2) developing a statistical model of each extrapolation, (3) calculating the point estimates and cumulative variances to generate a probability distribution for the test endpoint that serves as a surrogate for the assessment endpoint, and (4) calculating the probability of exceeding the endpoint, given a distribution of exposure concentrations. AEE can accommodate a variety of extrapolation methods. These might include distributions estimated by expert judgment when no data are available. For example, a judgment by a panel that, given an oral LD_{50} of Y for bobwhite quail, the LD_{50} for bald eagles must lie between X and Z, establishes a uniform probability distribution for the eagle LD_{50} with bounds X and Z.

As an example of AEE without regression, consider the estimation of LC_{50}s for other fish from the rainbow trout LC_{50} of 5300 µg/L using the models developed by Volmer et al. (1988) for cases in which there are insufficient data for a particular chemical to estimate relative sensitivity distributions. In this case, we estimate the mean log LC_{50} for osteichthys by adding 0.196 (from Table 7.3) to the log rainbow trout LC_{50} to obtain 3.92. The weighted average variance of log osteichthys LC_{50}s from Volmer et al. (1988) is 0.232. If we assume that all other sources of variance are negligible relative to the variance among species, then the mean and variance of fish LC_{50}s for this chemical are 3.92 and 0.232. If the log of the environmental concentration is assumed to be fixed at 3.0 (1,000 µg/L), then from Equation 7.30 the z statistic is -1.9, and the risk is 0.028, or roughly 3%. This result can be interpreted two ways: (1) On average, there is a 3% likelihood that any particular fish species will experience 50% mortality when exposed at 1,000 µg/L for at least 96 h; (2) On average, 3% of species will experience 50% mortality at that exposure.

The same problem can be solved using Equation 75F from Table 7.4. From that model, the mean log LC_{50} for osteichthys is 3.98 and the variance is 0.38. From Equation 7.30, the z statistic and risk at 1,000 µg/L are -1.6 and 5%, respectively. In this case, the two models, using different statistical approaches and data sets, give nearly identical estimates of the risk.

These examples have assumed that the distributions of toxicity data are acceptable estimates of the true distributions of species and life stage sensitivities in nature. If this assumption is not accepted, the extrapolation models could be elaborated to include the uncertainty due to deriving the models from a sample of the population of species and chemicals. Methods have been developed to incorporate this uncertainty in species sensitivity distributions (Box 2.2). The factors in Table 7.3 incorporate this uncertainty to obtain estimates of a concentration that would protect 95% of aquatic vertebrate and invertebrate species 95% of the time. If we use the same hypothetical rainbow trout LC_{50} of 5,300 µg/L and divide by 10,000 we get 0.53 µg/L. This is more than three orders of magnitude lower than best estimates of the fifth percentile (50th percentile estimates of the fifth percentile of the sensitivity distribution). This illustrates the extreme conservatism necessary to ensure high confidence in a highly uncertain estimate.

The assessor who does not believe that the test endpoint estimated by extrapolation is an adequate surrogate for the assessment endpoint may readily incorporate that subjective uncertainty by adding an increment of variance before calculating the risk. It is important to clearly document such judgments, including who made them and on what basis. It is also important to make such judgments prior to the calculation of endpoint values and risks, so as to avoid the temptation to fiddle with the conclusion.

Suter et al. (1986) tested the ability of AEE to predict from fathead minnow LC_{50}s the results of standard chronic toxicity tests (MATCs) for fish other than fathead minnows. AEE estimates were compared to fathead minnow MATCs, fathead minnow MATCs with correction factors (CF = true LC_{50} / fathead minnow LC_{50}), and fathead minnow LC_{50}s with acute/chronic factors of 20 and 100. The AEE consisted of an appropriate taxonomic extrapolation for each case and an LC_{50} to MATC regression. The test data set consisted of 41 MATCs for 10 chemicals and 13 species. The MATCs from the extrapolation models were closer to the true MATC about as often as the fathead minnow MATCs or the fathead MATCs with application factors. The LC_{50}s with factors performed poorly. In other words, a statistical extrapolation model using only LC_{50}s from fathead minnows could predict MATCs for a particular species of fish about as well as a measured MATC for fathead minnows or a measured LC_{50} for the species of interest with factors. All of the methods were in error by more than a factor of 10 in more than 10% of cases, but AEE has the additional advantage that it calculates an estimate of its associated uncertainty. In this test set, the 95% prediction intervals on the predicted MATCs included the true MATC in 98% of the cases, suggesting the reasonableness of the variance estimates.

The main advantage of the AEE method is that it clearly distinguishes, quantifies, and displays both the extrapolations that must be made from the toxicity data to relate it to the assessment endpoint and the uncertainties associated with the process of extrapolation. In contrast, the quotient method with factors treats uncertainties and corrections as equivalent and does not systematically account for either one. Compared with mathematical simulation models, AEE has the advantages of using the test endpoints that constitute the de facto assessment endpoints of current regulatory assessment schemes and of not requiring quantification of the mechanisms that result in toxic effects.

One disadvantage of analysis of extrapolation error is that it requires knowledge of basic statistics, whereas the quotient method requires only arithmetic. Another disadvantage is that the development of formal statistical models requires standard test endpoints, and fields of environmental toxicology, such as air pollution effects on plants, that have no standard tests are not amenable to this method. Finally, AEE like other empirical methods, assumes that existing data sets are representative of future toxicity data. Because of this assumption, it is important to pay close attention to the potential biases in available data sets and be aware of which sources of variability (e.g., water chemistry, interlaboratory variability, or different strains of the test species) are represented in the data set and which are implicit in the assessment (e.g., should data from laboratories of unknown reliability be used, and should the results of the assessment apply to a variety of sites). In some cases, the extrapolations can be inappropriately precise as the result of using a highly standardized data set. For example, studies of the acute effects of narcotic chemicals in Lake Superior water on the Duluth population of fathead minnows are used in QSARs that generate predicted LC_{50}s that are more precise than replicate tests in different laboratories using different waters and fish populations (Veith et al. 1983). More often, some sources of variance in the data sets are extraneous to the assessment (e.g., variance in test protocols), but cannot be avoided because more appropriate data sets are not available. In those cases, the extraneous variance is simply part of the uncertainty associated with performing assessments with limited knowledge.

Although the AEE method was developed to provide estimates of risk, it has various other potential uses. The regression and error propagation portions can be used to estimate toxic effects parameters for population and ecosystem models and to generate the parameter distributions used in Monte Carlo simulations. This use is described in Chapter 8. Another potential use is in designing testing programs. Decisions about the need for additional testing of a chemical could be made on the basis of the expected reduction in the total uncertainty concerning the true value of the endpoint, the expected reduction in risk, or the probability that the test will cause a change in a regulatory decision. Finally, in addition to making decisions for testing individual chemicals, AEE could be used to elucidate the implications of the decision rules in tiered testing schemes or to devise new decision rules.

Simulation Modeling of Individual Responses

As with other risk assessment techniques, toxicodynamic models and any other simulation models for responses of individual organisms can be used to estimate the probability that effects will occur as a result of exposure. This is done by error analysis using Monte Carlo simulation to propagate the uncertainties associated with the model parameters through the model to generate a probability distribution of outcomes (Chapter 2). Because most simulation modeling has been directed to predicting effects at population and ecosystem levels where test data are scarce, the use of error analysis techniques to generate risk estimates is discussed in those chapters (8 and 9).

SUMMARY

Methods for characterizing effects on individual organisms include (1) use of factors to adjust quotients for uncertainties and extrapolations (the quotient method), (2) use of statistical models to estimate risks that environmental exposures will exceed assessment endpoints (analysis of extrapolation error), and (3) use of probabilistic simulation models of toxicodynamics to estimate risks that environmental exposures will exceed assessment endpoints (uncertainty analysis). The first method has been in use for decades and is the easiest to perform but has severe technical limitations. Analysis of extrapolation error has the advantage of providing estimates of the exposure resulting in the endpoint response, of the uncertainty associated with that estimate, and of the resulting risk. However, it is a new technique and has not been widely used. Simulation of the response of individual organisms is rare in ecological risk assessment and most future simulation modeling efforts are likely to emphasize the population and ecosystem levels. However, most toxicity testing will continue to be done at the organism level. Therefore, it will be necessary to continue to develop models of organismal response both to interpret the test results and to provide a link to models of higher level responses.

8

Population-Level Effects

Lawrence W. Barnthouse

If the only problem facing assessors was to identify limits on contaminant loadings or other stress levels that would protect all organisms from harm, then this book could have ended with the previous chapter. However, as noted earlier in Chapter 2, it is often impractical to completely eliminate toxicity from waste streams, to prevent spills, breakdowns of treatment systems and other accidents, or to eliminate toxicity to nontarget organisms of agrochemicals. If toxic effects are unavoidable, then regulatory agencies must assess the amount of mortality (or growth reduction, etc.) that can be tolerated by the exposed species.

The toxicity tests and extrapolation models discussed in Chapter 7 are insufficient for addressing these problems. From a population viewpoint, the death or impairment of an individual organism is meaningless, because organisms die after brief lives (on a human time scale) and few organisms achieve their full reproductive potential or maximum growth. The questions of interest in risk assessment relate to effects on the abundance, production, and persistence of populations and ecosystems. Responses of these "higher" levels of organization cannot be predicted from toxicity tests alone. The response of a fish population, for example, to a contaminant exposure will depend on the spatial pattern of exposure as related to the distribution of individuals in time and space, on the magnitude of other impacts being imposed (including especially exploitation by man), and the inherent capacity of the population to "compensate" or to evolve in response to exposure.

Despite general agreement that population and ecosystem-level effects are of paramount importance in ecological risk assessment (NRC 1981b, Moriarty 1988, Levin and Kimball 1984, Westman 1985), few workers have attempted to assess these effects. Toxicity testing is of limited value because the few organisms amenable to toxicity testing at the population level (e.g., microbes, *Daphnia* spp. and *Triboleum* spp.) are unrepresentative of most endpoint populations and are of little assessment interest in their own right.

247

Controlled laboratory studies will continue to be the predominant source of information on the ecological effects of toxic chemicals. The objective of this chapter is to show how individual-level toxicological data can be integrated with existing ecological theory to provide useful assessments of population-level effects of toxic contaminants.

The problem of quantifying population-level responses to death or impairment of individual organisms is not new. Fish and wildlife managers have struggled for decades with the problem of defining the effects of harvesting, habitat modification, and disease on the abundance and stability of exploited populations. The methods used by resource managers provide a useful frame of reference for assessments of other sources of stress, including toxic chemicals. Fisheries-derived models have for the last 20 years been used in assessments of effects of power plant cooling systems on fish populations (Barnthouse et al. 1986). Provided we accept an analogy between fishing and other stresses such as toxic contaminants, these models form a useful starting point for population-level risk assessments. Later, we will discuss problems for which the analogy with fishing fails, and suggest some new approaches that are on the forefront of current population biology.

This chapter provides some basic definitions, briefly describes some of the models that are especially relevant to assessing ecological effects of toxic chemicals, and discusses some applications. Readers interested in a more in-depth treatment of the principles of population biology are advised to consult one or more of the excellent textbooks and reviews that are available. Most modern undergraduate-level ecology textbooks (e.g., Krebs 1985, Colinvaux 1986) explain the fundamentals of population analysis. However, many of the most readable and thorough accounts are older. Especially noteworthy and still generally available are Andrewartha and Birch (1954) and Slobodkin (1961). The texts of choice for fish populations are Ricker (1975) and Gulland (1977); for other vertebrates Caughley (1977). For the mathematically inclined, Caswell (1989) provides a thorough exposition of the theory of matrix population models. Readers interested in developing computer simulation models of populations should consult Swartzman and Kaluzny (1987).

Note that this chapter emphasizes models relevant to management applications. These models are usually age, size, and sex-structured. Much of theoretical population biology is based on logistic growth or on Lotka-Volterra dynamics. Although providing valuable theoretical insights into population and community dynamics, the extreme simplifications involved limit their applicability to specific problems of management interest. Unlike other chapters in this section, this chapter does not include a discussion of toxicity tests. Conventional laboratory toxicity tests, which provide the input to the models presented here, are discussed in Chapter 7. Ecosystem-level tests, which often include measures of effects on population processes, are discussed in the next chapter (9).

BASIC CONCEPTS AND DEFINITIONS

The fundamental objective of population biology is to infer characteristics of groups of organisms (the populations) from the characteristics of individuals. The population characteristics typically of interest include total numbers or biomass, the rate of population growth or decline, and the age, size, sex, or genotypic composition of the population. Managers of exploited populations may be interested in the number of organisms available to be harvested; conservationists may be interested in the probability of extinction of a population of a given size. These population characteristics are simply collective expressions of the state and fate of the individual organisms: their reproduction growth, and death summed over every organism. The individual characteristics are, in turn, governed by (1) innate processes such as development and senescence, (2) the effects of the physical environment, (3) interactions with other organisms, and (4) deliberate and unintentional actions by man.

Population-Level Assessment Endpoints

Population studies address a variety of endpoints of interest in risk assessment. The most basic of these are the endpoints traditionally used in management of exploited populations: total population density or biomass, age or size distribution of the harvestable stock, and yield available to man. Models relate these endpoints to management actions such as bag limits, creel limits, or length of fishing season. More recently, due to concerns about conservation of endangered species, population biologists have formulated new models, containing environmental and demographic stochasticity, to address endpoints related to persistence of populations in variable environments: frequency or probability of extinction within a given time period or expected time to extinction, as functions of population size or size of habitat required.

General Implications of Life History for Population-Level Risk Assessment

Are some species inherently more vulnerable to environmental stress because of their life history? Are some life stages more important than others to the survival of a population? General aspects of population biology were a major topic of theoretical research during the 1960s and early 1970s (see review by Stearns 1977); however, little attention has been paid to the the consequences of life history for risk assessment. Both theoretical analysis and management experience have shown that long-lived vertebrates such as large mammals, predatory birds, and whales are more sensitive to mortality imposed on adults than are short-lived, highly fecund organisms such as quail and anchovies. Conversely, short-lived species are often vulnerable to short-term catastrophes that affect critical life stages. Populations in which survival or reproduction are strongly related to density should on theoretical grounds be less vulnerable than populations with a low degree of density-dependence (e.g., due to intensive exploitation by man). Qualitatively, it seems clear that the response of a population to a toxic contaminant is influenced by the preexisting pattern of natural environmental variability, the age-specific survival and reproduction of the organisms, and the intensity and duration of contaminant exposure. Quantitative conclusions require case-by-case studies of populations of interest.

Representation and Propagation of Uncertainty

Chapter 2 discusses a variety of sources of uncertainty of interest in risk assessment. Potentially important sources for population analysis include (1) environmental variability in time and space, (2) variations in sensitivity among individuals and life stages, and (3) stochastic birth and death processes.

Inter-individual and inter-life-stage variability are discussed in Chapter 7. Stochastic birth and death result from the fact that each individual organism has an indeterminate life span, even if the average life span for the population can be very precisely estimated. In practice, random birth and death processes are important only in small populations (e.g., 50 individuals or less). Even for small populations, Goodman (1987) has shown that this source of uncertainty is quite small compared to environmental variability.

Temporal environmental variability has been relatively thoroughly discussed in the scientific literature and is readily incorporated in population models. Both periodic and stochastic variation have been studied. The principal mathematical tools available include time series analysis and stochastic modeling. These approaches may be used to quantify environmental variability (e.g., to estimate a periodic function of some important driving variable such as temperature or rainfall), or to estimate probability distributions for temporally varying population parameters (e.g., mortality rates).

Spatial variability is more difficult to address using conventional techniques for population analysis. Classical population modeling techniques assume that each organism in a population is exposed to the same environment. A significant theoretical literature exists for population dynamics in patchy environments (Levins 1968, Slatkin 1974, Levin 1976), and some of the results have been found relevant for designing biotic reserves (Gilpin 1987). Individual-based models, discussed below, provide a new technique of potential use in spatial representation of population dynamics.

Density Dependence

Population regulation has been a fundamental problem in population biology since its inception. What prevents populations with high reproductive rates from increasing without bounds? How do fish and wildlife populations persist in the face of intensive exploitation by man? Simply stated, when numbers are high, mortality increases and reproduction decreases, and when numbers are low, mortality decreases and reproduction increases. The qualitative observations of density-dependent mortality and growth are many (McFadden 1977). Empirical and theoretical studies (e.g., Tanner 1967, Royama 1977, McFadden 1977) have documented that some form of density-dependence is universal in nature. The list of potentially important density-dependent mechanisms is large (McFadden 1977); however, most of these mechanisms are exceedingly difficult to identify or measure in the field.

Although density-dependence affects responses of populations to management actions, it is not always necessary to build density-dependence into predictive models. Fish and wildlife managers have been reasonably successful with density-independent models, provided that the prediction horizon is short and the population changes being modeled are relatively small. For long-term predictions, explicit incorporation of density-dependence is usually necessary.

APPROACHES TO POPULATION ANALYSIS

A variety of approaches to population analysis has been developed over the last several decades. This section provides an overview of the principal methods, with an emphasis on their conceptual relationships and past applications. Specific case studies involving toxic contaminants are presented in the section entitled "Applications to Toxic Chemicals."

Reproductive Potential

The simplest approach to population analysis is quantification of reproductive potential. The theory in its present form was developed by Lotka (1924), Fisher (1930), and Cole (1954). Prior to the 1960s, reproductive potential was the only approach to population dynamics used outside fisheries management. The approach requires only a compilation of (1) the fraction of organisms surviving from one age to the next, and (2) the average number of offspring of an organism of a given age. Define l_x as the fraction of organisms surviving from birth to age x, and m_x as the average number of offspring produced by an organism of age x. Suppose that the maximum age of organisms in the population is n years. If l_x and m_x are constant, these parameters uniquely determine the relationship between reproduction, mortality, longevity, and population growth. This relationship is mathematically given by the equation:

$$\sum_{x=1}^{n} e^{-rx} l_x m_x = 1 \qquad (8.1)$$

The parameter r, the coefficient required to make the left side of Equation 8.1 sum to 1, has been termed the "Malthusian parameter," the "intrinsic rate of natural increase," or the "geometric rate of increase." In this chapter it will be termed the instantaneous rate of population change. If r is greater than 0, the population will increase indefinitely. If r is less than 0 the population will decrease to extinction, and if r is exactly 0 it will remain unchanged. It can be shown that if undisturbed, the age composition of this population will converge to a stable age distribution in which the fraction of organisms in each age class is the same from each generation to the next. Once this state is achieved, both the population as a whole and the fraction of organisms in each age class will grow (or shrink) exponentially with time:

$$N_t = N_0 e^{rt} \qquad (8.2)$$

where

$$N_t = \text{population size at time t, and}$$
$$N_0 = \text{population size at time 0.}$$

Equations 8.1 and 8.2 are often expressed in alternative forms:

$$\sum_{x=1}^{n} \lambda^{-x} l_x m_x = 1 \qquad (8.3)$$

$$N_t = N_0 \lambda^t \qquad (8.4)$$

where
$\lambda = e^r = $ the finite rate of population change.

The value of r and changes in r that might relate to the decline or extinction of a population are important for population management. Many authors have investigated the sensitivity of r to changes in fecundity or mortality. Mertz (1971) showed that because of its very low reproductive rate, the California condor population was extremely vulnerable to increased mortality of adults due to hunting by man. Mertz also concluded that management actions designed to increase reproductive success in this population were unlikely to improve the prospects for recovery. Unfortunately, Mertz's predictions proved correct and the California condor is now extinct in the wild. More recently, numerous authors have used reproductive potential methods to assess the viability of the northern spotted owl (USDA 1986, Dawson et al. 1987, Marcot and Holthausen 1987, Lande 1988).

In ecotoxicology, the reproductive potential approach has found a niche in the interpretation of chronic toxicity tests using *Daphnia* and *Ceriodaphnia* (e.g., Gentile et al. 1983, Daniels and Allan 1981, Rago and Dorazio 1984, Meyer et al. 1986, Barbour et al. 1989). Measurements of daily survival and reproduction obtained from these tests are sufficient to obtain estimates of r. Changes in r resulting from exposure to toxicants can be used as an index of chronic effects. Although the calculated values of r cannot be directly extrapolated to the field, this approach to test data interpretation has the advantage of combining information on survival and reproduction into a single index.

Projection Matrices

Age-structured or stage-structured projection matrices are closely related to the reproductive potential model. The two approaches were developed at essentially the same time by the same people. However, whereas reproductive potential analysis has been a principal tool of population biology since the late 1940s, matrix models were not used extensively until the 1960s, and did not begin to appear in texbooks until the 1970s. Many undergraduate-level ecology textbooks still do not mention them. Caswell (1989) attributed the lack of attention to (1) unfamiliarity of biologists with matrix algebra, which was not widely taught until quite recently, and (2) lack of access to computers, which greatly facilitate the numerical calculations.

The simplest matrix model is the linear "Leslie matrix" (Leslie 1945). It contains exactly the same information as the reproductive potential model, but the information is expressed in matrix form. The change in abundance of each population in time can be represented by the matrix equation

$$N(t) = LN(t-1) \qquad (8.5)$$

where N(t) and N(t-1) are vectors containing the numbers of organisms in each age class ($N_0, \ldots N_k$) and L is the matrix defined by

$$L = \begin{matrix} s_0 f_1 & s_1 f_2 & s_2 f_3 & \cdots & s_{k-1} f_k & 0 \\ s_0 & 0 & 0 & \cdots & 0 & 0 \\ 0 & s_1 & 0 & \cdots & 0 & 0 \\ 0 & 0 & s_2 & \cdots & 0 & 0 \\ 0 & 0 & 0 & \cdots & s_k & 0 \end{matrix} \qquad (8.6)$$

where

$$s_k = \text{age-specific probability of surviving from one time interval to the next and}$$
$$f_k = \text{average fecundity of an organism of age k.}$$

The matrix analog to Equation 8.4 is:

$$N(t) = L^t N(0) \qquad (8.7)$$

where

$$N(0) = \text{age distribution vector at time 0 and}$$
$$L^t = \text{the matrix } L \text{ raised to the power } t.$$

Leslie (1945) showed that any population growing according to Equation 8.7 will converge to a stable age distribution, after which it will grow according to:

$$N(t) = \lambda^t N(0) \qquad (8.8)$$

The parameter λ, which in matrix algebra is termed the "dominant eigenvalue" of the matrix L, is the same finite rate of population change that appears in Equations 8.3 and 8.4.

For a population that is *not* at a stable age distribution, Equation 8 can be used to predict the abundance and age distribution of the population over the next few time intervals. Otherwise, as long as the coefficients of the matrix L are viewed as constant parameters, the population projection approach is essentially equivalent to the reproductive potential approach and has little added value for risk assessment. The real power of the matrix representation of population dynamics is its flexibility. The coefficients of L can be viewed not as constants, but as random variables or functions of environmental parameters. Alternatively, a matrix of life stages and stage transitions (as opposed to ages and age transitions) can be constructed.

Both of these modifications permit entirely new types of analyses, many of substantial value for risk assessment. For example, entirely new endpoints can be addressed. In the strictly linear and deterministic models discussed so far, the only endpoint that can be addressed is the future trend of abundance: will the population increase or decline in the future, and how sensitive is the rate of increase or decline to changes in life history parameters? Modification of the matrix approach in the ways described below permits assessment of any endpoint for which an operational definition can be formulated.

Caswell (1989) presents a detailed discussion of the mathematics of stage-classified models. In addition to representing survival and reproduction rates, the coefficients of a stage-classified model can represent probabilities of transition from one size class or life stage to the next. Ecological examples include Sinko and Streifer (1967, 1969), Taylor (1979), Law (1983), Law and Edley (1990), De Roos et al. (1992), and many other papers cited by Caswell. These models appear especially useful for plants and invertebrates, which have complex life cycles in which population dynamics are more strongly influenced by size and developmental stage than by age. Emlen and Pikitch (1989) used stage-classified models of generalized vertebrate populations to analyze the sensitivity of different types of vertebrate life cycles to mortality imposed on different stages. With this exception, stage-classified models have found few management applications outside of entomology.

The remainder of this section deals strictly with modifications of Equation 8.6. It should be emphasized that because Equation 8.6 is a discrete-time model, its principal applicability is to populations in which reproduction occurs during brief seasonal or annual periods. It is difficult to apply to populations in which reproduction occurs continuously. De Roos et al. (1992) have recently generalized the Leslie matrix to obtain a continuous-time age/stage-structured model that can accommodate continuous reproduction. All applications of this approach to date have been to *Daphnia* populations.

One straightforward modification of L is to make the coefficients random variables. Stochastic Leslie matrices are generalizations of Equations 8.5 to 8.8, in which one or more of the matrix coefficients, for example, the survival rate of newborn organisms (s_0), is assumed to be a random variable. The purpose of this modification is to account for the fact that most natural populations exhibit substantial fluctuations in abundance in response to variable environmental conditions (see Chapter 2). Quantification of variability in survival and reproduction using a stochastic matrix model permits quantification of the potential variability in future population sizes, including risks of population extinction. In recent years a significant theoretical literature on stochastic matrix models has developed (see Caswell 1989). Much of the theoretical literature on these models deals with stochastic analogs to the stable age distributions and growth rates associated with deterministic models (Cohen 1976, 1979, Tuljapurkar and Orzack 1980). Investigators have also used these models to estimate (1) the influence of population parameters (both mean values and variances) on the probability of extinction of populations with different initial sizes and age distributions and (2) the average time required for populations to become extinct (Tuljapurkar and Orzack 1980, Lande and Orzack 1988).

The theoretical literature on stochastic matrix models emphasizes the use of analytical mathematics to obtain general results. In applications to specific populations, however, numerical simulations using Monte Carlo methods (Chapter 2) are usually adequate and sometimes are preferable. Applications to fish populations have been described by Goodyear and Christensen (1984) and Barnthouse et al. (1990). An application to the Everglades kite was described by Nichols et al. (1980). A generic software system for developing stochastic matrix models has been developed by Ferson et al. (1987). Special software isn't really needed anyway! Equations 8.3 and 8.5 can be readily implemented on a spreadsheet, and software for performing Monte Carlo analysis of spreadsheets is now widely available.

Another common modification of the basic matrix model is to make the coefficients density-dependent; i.e., to make one or more of the vital rates a function of the number of individuals either in the entire population or in some of the age classes. The purpose of incorporating density-dependence is to account for the fact that, in spite of fluctuations in abundance, populations in nature are always bounded within limits.

Inspection of Equations 8.2, 8.4, and 8.8 shows why explicit incorporation of density-dependence in population models is often necessary. For the deterministic case, unless r is *exactly* 0 (λ *exactly* 1), the population must either grow without bounds or decline to zero. Inclusion of variability in life history parameters does not alter this behavior. Given enough time, any density-independent model population will either grow to infinite size or go extinct. Most models intended to simulate population behavior over more than one generation incorporate density-dependence. Two common functions for density-dependent survival are the Beverton and Holt (1957) function:

$$s = c_1/(1 + c_2 n) \qquad (8.9)$$

where c_1, c_2 = constant parameters, and

$$n = \text{population size;}$$

and the Ricker (1954) function:

$$s = \alpha e^{-\beta n} \tag{8.10}$$

where α, β = constant parameters.

Density dependence in models creates as many problems as it solves. Depending on the particular function and parameter values chosen, many different population behaviors are possible. If $N(t)$ converges to a single age distribution vector N^* from a range of initial vectors $N(0)$, the model is said to be *stable*. If $N(t)$ does not converge, the model is said to be *unstable*. A model may be stable for some values of $N(0)$ and unstable for others. If it is unstable, the population may oscillate within well-defined limits, may grow without bound, or may decline to zero. Density-dependent models have occupied the attention of theoretical biologists for decades, but relatively few general (i.e., applicable to more than a few functional forms) results have been obtained. Caswell (1989) reviewed the literature on density-dependent matrix models.

The problems with application of these models to real populations are that the functional forms used do not generally reflect real biological processes and the parameters are not directly measureable. There is no way to determine a priori which functional form is appropriate for a given population and, having chosen a functional form, usually no way to directly measure the critical parameters. Instead, values are obtained empirically from time series of population data (Cushing and Harris 1973, Vaughan 1987, Lawler 1988, Barnthouse et al. 1990).

Predictions about future behavior or response to stress can be extremely sensitive to variations in functional forms and parameter values. Because these are so difficult to obtain for real populations, management applications of density-dependent models have had mixed success. Much of the existing literature relevant to ecological risk assessment relates to effects of thermal power plants on fish populations. Cohen et al. (1983) and Heyde and Cohen (1985) analyzed a stochastic model of the Chesapeake Bay striped bass population. Hess et al. (1975), Christensen et al. (1977), DeAngelis et al. (1977), Vaughan (1981), and Levin and Goodyear (1980) described density-dependent matrix models of fish populations. Goodyear and Christensen (1984) analyzed a model that incorporated both density dependence and stochastic variation. Barnthouse et al. (1990) used a similar approach to develop models of contaminant impacts on fish populations. The successes and failures of these efforts have been reviewed by Barnthouse et al. (1986). Long-term projections of even the best-supported models are highly uncertain. Density-dependent models can still be used for comparative purposes or for short-term projections, but it is not clear whether anything is gained over a purely density-independent model.

Aggregated Models

Readers of the literature on population biology will be familiar with other kinds of models, such as the logistic model and stock-recruitment models. These approaches to population modeling differ from the models discussed above in that all organisms are aggregated either into one or two components such as "population size" or "parents and offspring." The logistic model has a long history in population biology and readers interested in it can consult Slobodkin (1961), which is still the most readable general introduction. Because of its extreme simplicity, most applied population biologists view the logistic model and its variants as being of little practical value. One particular form of the logistic, known as the Shaeffer surplus production model, is used in fisheries management. Good discussions of surplus production models and of the various forms of stock-recruitment models developed by fish population biologists can be found in Gulland (1977).

Because these models can be viewed as special cases of age-structured models in which there are only one or two age classes, we do not provide a detailed discussion of them in this chapter. All of the conclusions drawn above for age-structured models apply to these others as well.

Individual-Based Models

Ultimately, the health of a population is no more than a collective expression of the health of the individual organisms. The models discussed above are at best abstractions that capture the general (we hope the essential) features of the biology of the organisms. Some, like the reproductive potential model or the density-dependent Leslie matrix, are basically bookkeeping devices with which the deaths and births occurring during a given time are tabulated, and the biological mechanisms responsible for reproduction and mortality are ignored. All organisms within a given class (however defined) are assumed to be indistinguishable. Clearly, all organisms are *not* indistinguishable, and variations between individuals can have substantial influences on the responses of populations to anthropogenic stresses or management actions.

Recognition of these problems has led to interest in "individual-based" models, i.e., models in which population dynamics are represented in terms of the physiological, behavioral, or other properties of the individual organisms. Either by brute force simulation or by elegant mathematics the properties of the population as a whole are derived from the characteristics of the individuals (Figure 8.1). Individual-based models have made important contributions to understanding successional patterns in forests (Huston and Smith 1987), comparing the structure and development of different forest types (Shugart 1984), and predicting the effects of environmental stress on forest composition (Dale and Gardner 1987). A small number of applications to fish populations have also been described (Sperber et al. 1977, Adams and DeAngelis 1987, DeAngelis et al. 1990). Huston et al. (1987) elegantly argued that for most

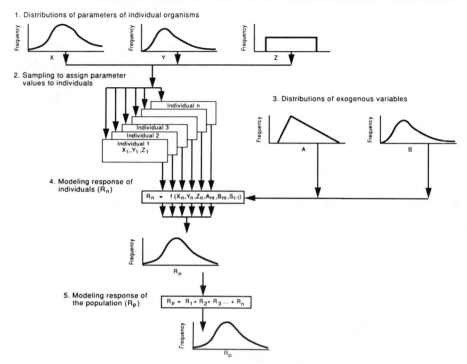

Figure 8.1. A schematic representation of individual-based population modeling. X_n, Y_n, and Z_n are characteristics of an individual organism n such as size and leaf area. A_{nt}, and B_{nt} are characteristics of the environment experienced by individual n at time t such as temperature, pollutant concentrations, and prey availability. S_{t-1} is the state of the organism at the previous time step. R_n is response of individuals such as death or maturation. R_p is response of the population such as abundance or harvestable biomass.

theoretical problems in modern ecology, the individual-based approach is inherently superior.

There are two broad approaches to developing individual-based models, one of which emphasizes Monte Carlo simulation and the other of which emphasizes analytical solutions to equations. The subtleties of the approaches and criteria for choosing one over the other have recently been discussed by Caswell and John (1992), and by DeAngelis and Rose (1992). Elegant examples of the analytical approach have recently been published by McCauley et al. (1990), and by Hallam et al. (1990). The principal advantage of the analytical approach is that results obtained are general, easy to verify, and easy to understand. The level of detail, however, must be compromised to achieve analytical tractability. In practice, the published biological applications of analytical individual-based models all deal with relatively simple organisms such as *Daphnia*.

In the simulation approach, the population is analyzed by using computers to simulate the daily activity of each individual organism. The number of processes and parameters that can be included is unlimited. The above-mentioned work on forest stand succession was all performed using the simulation approach. DeAngelis et al. (1991) simulated the influence of seasonal temperature patterns, male size distribution, and nest density on the growth and survival of young-of-the-year smallmouth bass. The model used by the authors accounts for temperature and size-dependent spawning, nest defense, energetics, foraging behavior, and size-dependent cannibalism. Models that simulate hundreds of individuals for a year can run in a few minutes on today's personal computers. The principal limitation on the use of the simulation approach is not computing power or cost, but the quantity and complexity of the data required.

The individual-based approach is now being widely applied in biological research. Application in risk assessment can be expected in the near future.

APPLICATIONS TO TOXIC CHEMICALS

Most of the work discussed above was related either to theoretical problems, to management of exploited populations, or to conservation of endangered species. Recently, the number of applications to toxicological problems has been growing.

Reproductive potential models are now frequently used to interpret life-cycle toxicity tests with cladocera. Daniels and Allan (1981) and Gentile et al. (1983) used r as a measure of population-level effects of contaminant exposure. Others who have used this technique include Rago and Dorazio (1984), and Meyer et al. (1986). The paper by Meyer et al. (1986) is especially noteworthy because these authors provided a statistical method for calculating confidence intervals on r. Barnthouse et al. (1987) used reproductive potential to explore hypothetical effects of toxic contaminants on fish populations. Other studies have employed projection matrices. In all of these studies, contaminants have been treated as a source of mortality analogous to fishing or hunting.

Impacts of Carbofuran Poisoning on Raptors

The U.S. Environmental Protection Agency (OPP 1989) performed a population-level assessment to support its proposal to cancel registration of granular carbofuran. This pesticide is widely used throughout the United States for control of soil and foliar pests on agricultural crops, particularly corn. Although it has many desirable properties (e.g., short half-life, low bioaccumulation potential, and noncarcinogenicity), carbofuran is acutely neurotoxic to birds. A single granule contains sufficient pesticide to kill a small bird. Kills are often observed in field tests of carbofuran and many instances of mortality in actual use have been documented (OPP 1989). Secondary poisoning of scavengers, insectivores, and raptors has been observed.

Secondary poisoning of bald eagles has been suspected. EPA used a stochastic matrix projection model originally described by Grier (1980) to "illustrate" potential effects of pesticide-related mortality on the bald eagle population along the lower James River, Virginia. Estimates of annual age-specific survival rates were available from the literature. Reproduction per nesting female was assumed to be stochastic, with number of young ranging from 0 to 4. Starting with 9 breeding pairs and 40 immature birds of different ages, EPA simulated population histories over 10 years. Consistent with the results for California condor (Mertz 1971) and spotted owl (Lande and Orzack 1988), EPA found that the population projections were highly sensitive to mortality imposed on adults. Modest levels of pesticide-induced mortality (10% per year) were sufficient to cause declines in the average abundance. The EPA's study could easily have been extended to calculation of the likelihood of extinction as a function of the rate of pesticide-induced mortality.

Effects of Pesticide Exposure on Quail

Tipton et al. (1980) used a population projection matrix to link cholinesterase inhibition levels to population changes. Cholinesterase inhibition is a physiological marker of acute exposure to neurotoxins: the higher the inhibition, the greater the likelihood of mortality. Cholinesterase inhibition can be measured in the field and used as an index of exposure and probable effects on birds experimentally exposed in field trials. The matrix developed by Tipton et al. is not a Leslie matrix, but a generalized stage-based model. The stages are not developmental stages, but pathological stages: 0–100% cholinesterase inhibition, pesticide-induced death, natural death. Starting with an initial population of birds in various states, the matrix elements represent the probability of each bird moving to another stage during a given time interval t and $t+1$. The transition matrix T is given by

$$T = \begin{matrix} 0.23 & 0.0 & 0.0 & 0.0 & 0.0 & 0.0 \\ 0.7 & 0.04 & 0.0 & 0.0 & 0.0 & 0.0 \\ 0.0 & 0.79 & 0.1 & 0.0 & 0.0 & 0.0 \\ 0.0 & 0.0 & 0.48 & 0.01 & 0.0 & 0.0 \\ 0.05 & 0.15 & 0.49 & 0.97 & 1.0 & 0.0 \\ 0.02 & 0.02 & 0.02 & 0.02 & 0.0 & 1.0 \end{matrix} \qquad (8.11)$$

The matrix projection was performed according to:

$$N(t) = T^tN(0) \qquad (8.12)$$

The rows and columns of T correspond (from top to bottom and left to right) to 0%, 25%, 50%, and 100% cholinesterase inhibition, death due to pesticide poisoning, and natural death. The elements of T represent weekly probabilities of transition between states. For example, a bird that is healthy (0% inhibition) during week t has a 23% chance of remaining healthy until week $t+1$, a 70% chance of having its cholinesterase activity reduced by 25%, a 5% chance of dying from pesticide exposure during week t, and a 2% chance of dying of natural causes during week t.

The matrix T is informative. Simple inspection of T shows that a healthy bird has only a 5% risk of dying in the first week of exposure. However, there is a 70% risk that its health (as measured by cholinesterase inhibition) will decline. Each subsequent week of exposure increases the probable level of cholinesterase inhibition and the risk of death due to pesticide exposure.

Tipton et al. (1980) projected populations over an eight-week exposure period at two hypothetical pesticide treatment levels. At the higher treatment level, all exposed birds would be killed by the end of the treatment interval. As presented, the model does not allow for a probability of recovery. Recovery could be readily incorporated by allowing nonzero values in the upper right-hand part of the matrix. If functional relationships between pesticide exposure and cholinesterase inhibition could be established, the model could be further generalized by making the elements functions of pesticide exposure rather than constants.

Recovery of Seabird Populations from Oil Spills

Samuels and Ladino (1983) simulated the effects of hypothetical oil spills on seabird populations in the mid-Atlantic region of the United States. The study was performed as part of an assessment of the environmental impacts of oil field development on the mid-Atlantic outer continental shelf. The authors employed a density-dependent matrix projection model in which the number of young produced per breeding bird was assumed to be inversely related to the total adult population size. The authors developed models of herring gull and common tern breeding colonies from published life table data, and then used the models to evaluate recovery from oil spill incidents. Toxicity data were not used. Instead, various fractions of the populations were assumed to be killed by exposure to spilled oil. Recovery was examined as a function of the magnitude and distribution of oil-related mortality. Alternatives examined included:

(a) equal mortality to chicks only
(b) equal mortality to juveniles only
(c) equal mortality to breeding adults only
(d) equal mortality to all age classes
(e) mortality to chicks as a function of adult mortality.

Percentages of vulnerable ages assumed to be killed ranged from 25% to 95%. Patterns of response varied markedly between scenarios but were similar for both species. In accord with theoretical expectations, the long-lived, low-reproductive tern population was found to recover much more slowly than the herring gull population, for any given level of mortality. Recovery times increased in length systematically from scenario (a), in which only chicks are affected, to scenarios (d) and (e), in which birds of all ages are affected. For even a severe spill in which 95% of herring gull chicks were killed but no other ages were affected, the population would recover within five years. For a moderate spill in which 25% of all age classes were killed (scenario d), the tern population would require nearly 20 years to recover.

The authors coupled the seabird population models to an oil spill model that predicted frequencies of oil spill events, by season, over projected 30-year production lifetime for a typical lease tract. Spill frequencies were calculated (a) for individual lease sites, considering two alternative methods of delivering the oil to shore (pipeline and tanker), and (b) for cumulative transport of oil through the mid-Atlantic region, including tanker transport of imported oil. Impacts of the spills were calculated in terms of percentage of a 30-year leasing period in which the populations were projected to be below 10%, 25%, 50%, and 75% of baseline population size.

Samuels and Ladino's study illustrates a principal use of ecological models in risk assessment: expression of risks in terms of endpoints relevant to decisionmaking. Reductions in seabird populations are an important assessment endpoint in assessments of impacts of offshore oil field development. The authors were able to evaluate the relative risks to seabird populations of alternative oil field development strategies (tankers vs. pipelines) and to compare risks of oil field development to existing similar risks associated with tanker transport through the region. Reed et al. (1989) used a similar approach to assess risks to the Pribiloff Island fur seal population of hypothetical oil spills in the Bering Sea.

Risks of Toxic Chemicals to Fish Populations

Barnthouse et al. (1987, 1989, 1990) developed models that directly link toxicity test data to fish population models, and then used the combined models to evaluate the ecological implications of toxicity test data. This section describes the population models used and relates them to the types discussed previously. The next section describes the extrapolation models used to link test data to the parameters of the population models.

Two different approaches were used. First, standard survival and reproduction data were used to calculate an index of reproductive potential. The index was defined (Barnthouse et al. 1987) as the expected contribution of a female recruit (a one-year-old fish) to future generations of recruits, taking into account (1) her annual probability of survival (s_i), probability of being sexually mature (m_i), and age-specific fecundity (f_i), and (2) the probability that a spawned egg will hatch and survive to age 1 (s_0). The reproductive potential of a one-year-old female recruit is given by

$$P = s_o \sum_{i=1}^{n} s_i f_i m_i \qquad (8.13)$$

The reproductive potential index P is frequently used in assessments of the impacts of power plant cooling systems on fish populations (Barnthouse et al. 1986), and has also been used in an assessment of the impact of acid deposition on lakes (Baker et al. 1990). Although Equation 8.12 contains the same information as Equations 8.1 and 8.6, reproductive potential indices are not used to make numerical estimates of future abundance or age composition. Instead, they are used as relative measures of the impact, expressed as a fractional reduction in reproductive potential (R_s)

$$R_s = (P - P_s)/P \qquad (8.14)$$

where P is the reproductive potential index in the absence of stress and P_s is the reproductive potential index in the presence of stress. Stress, in this context, refers to a change in survival or fecundity. For stress due to toxic contaminants that can affect either survival or fecundity, Barnthouse et al. (1989) calculated P_s as

$$P_s = s_o(1 - C_m) \sum_{i=1}^{n} s_i(1 - C_r)^{i-1}f_iC_jm_i \qquad (8.15)$$

where C_m is the probability of contaminant-induced mortality during the first year of life, C_r is the probability of contaminant-induced mortality for one-year-old and older fish, assumed equal for all age classes, and C_f is the fractional reduction in fecundity due to contaminant exposure, assumed equal for all reproducing age classes.

The reproductive potential approach, like the density-independent Leslie matrix, cannot account for natural environmental variability or density dependence. To explore the influence of these processes on responses of fish populations to toxic contaminants, Barnthouse et al. (1990) developed density-dependent, stochastic matrix projection models for two especially well-studied populations: the Gulf of Mexico menhaden population and the Chesapeake Bay striped bass population. The models employed conventional projection matrices, but with the survival coefficient for young-of-the-year fish (s_0) containing both density dependent (using Equation 8.10) and randomly varying components. Estimates of the coefficients were obtained from published abundance, age structure, and mortality statistics for these two populations.

Survival of young-of-the-year fish (s_0) was calculated using:

$$s_0 = e^{-\alpha + R_i\sigma - 0.5\sigma^2 - \beta N_0} \qquad (8.16)$$

where α is the expected annual instantaneous rate of density-independent mortality, σ is the standard deviation of α, R_i is a unit random normal deviate, and β is the coefficient of density-dependence. The standard deviation term σ is included to eliminate a bias in stochastic matrix projections that was discovered by Goodyear and Christensen (1984).

Effects of contaminants on young-of-the-year are incorporated by replacing α in Equation 8.15 with

$$\alpha' = \alpha - ln \ (1 - C_m) \qquad (8.17)$$

where C_m is the fraction of young-of-the-year expected to die from effects of contaminant exposure.

Effects of contaminant exposure on fecundity were incorporated by multiplying each age-specific fecundity rate (f_i) in the population matrix by a fecundity reduction factor (C_f). Effects of contaminants on survival of 1-year-old and older fish were not considered, because all fish that are susceptible to lethal effects in their post-larval life stages are assumed to die in the first year. However, lethal effects on older fish could have been incorporated simply by multiplying the age specific survival rates (s_i) by similar coefficients.

Barnthouse et al. (1987, 1988, 1990) used the above models for a variety of purposes. Extrapolated population responses were compared to MATCs (Chapter 7) derived from the same data sets, demonstrating that MATCs are not equivalent to no-effects thresholds (Barnthouse et al. 1987, 1988). Comparisons between uncertainties associated with different test endpoints showed that fecundity responses are substantially more variable and introduce more uncertainty than are mortality responses (Barnthouse et al. 1988). Comparisons of uncertainty introduced in different extrapolation steps and toxicity test types were used to quantify the relative value for risk assessment of different toxicity testing strategies (Barnthouse et al. 1990). Comparisons between responses of menhaden and striped bass populations showed that life history-related uncertainties are relatively small, compared to toxicity-related uncertainties. These results are useful principally for predictive toxicity risk assessment and standard-setting. The methods and data sets used were not intended for site- or population-specific prediction, and cannot validly be used for that purpose.

LINKING TOXICITY DATA TO POPULATION MODELS

Quantifying the effects of contaminant exposure using the above model requires (1) quantifying the relationship between exposure and the response coefficients C_m and C_f, and (2) projecting the matrix through time to determine the resulting changes in abundance or other population characteristics. The authors used the procedures described in Chapter 7 of this book to quantify exposure-response relationships that explicitly incorporated three types of uncertainty in lab-to-field extrapolations: test variability, species-to-species uncertainty, and acute-to-chronic uncertainty.

Estimating Concentration-Response Functions

The concentration-response function used was the logistic model

$$P = e^{a+BX}/(1 + e^{a+BX}) \tag{8.18}$$

Where P is the fractional response of the exposed population, X is the exposure concentration, and a and B are fitted parameters with no direct biological interpretation. When fitted to concentration-response data, the logistic function has a sigmoid shape similar to the probit model.

Concentration-response data sets can be fitted to Equation 8.18 using nonlinear least-squares regression. The data required are (1) the number of replicates tested at each concentration (including the controls), (2) the number of organisms in each replicate, and (3) the number of organisms dying in each replicate (including the controls). Test concentrations are entered as $\log_{10}$ (concentration in $\mu g/l$), so that the units represent orders of magnitudes of concentrations. The fraction of organisms dying in each replicate is corrected for control mortality using Abbott's formula (Abbott 1925). The SAS procedure NLIN (SAS Institute 1985) or one of several other nonlinear regression programs can be used to produce least-squares estimates of a and B, along with a variance-covariance matrix. Uncertainty concerning the shape and position of the concentration-response function, as reflected in the variances and covariances of a and B, can be represented graphically as a confidence band surrounding the fitted function.

Chapter 7 explained procedures for extrapolating toxicity test data from tested to untested species, and from acute to chronic responses. These extrapolations are necessary because the species commonly used in toxicity tests are rarely of interest in actual risk assessments and because full life cycle tests are seldom performed. Relatively few toxicity tests of any kind have been performed using striped bass, and none have been performed using menhaden. Barnthouse et al. (1987) described a method for using the extrapolation procedures described in Chapter 7 to estimate the parameters of the logistic concentration-response model.

The logistic model can be reformulated as

$$X_P = \{ln[P/(1 - P)] - a\}/B \tag{8.19}$$

where X_P is the concentration producing a fractional response equal to P.

If a and B are specified, then X_P can be directly calculated from Equation 8.18. Alternatively, if X_P and B are specified, a can be calculated from:

$$a = ln[P/(1 - P) - BX_P] \tag{8.20}$$

Equation 8.19 implies that the complete concentration-response function can be obtained by specifying either a and B or B, and the concentration associated with a single response level such as the LC_{25}. The parameter B specifies the curvature of the logistic function and is independent of the position of the curve on the concentration axis. If two logistic functions hve different LC_{25}s but the same curvature, their B parameters will be equal.

Estimates of LC_{25}s for menhaden or striped bass, for any tested chemical, can be readily calculated using the procedures described in Chapter 7. No procedures exist for extrapolating values of B between species or test types. Barnthouse et al. (1987) applied the logistic model to 77 chronic concentration-response data sets and found that (1) the estimated B values were distributed approximately lognormally, (2) the distributions of B for different life stages were approximately the same, and (3) B values were not correlated with LC_{25}s (i.e., when plotted on a log scale, the curvatures of the logistic functions were independent of their positions on the concentration axis). In the absence of any systematic variations in B, estimates of B for any particular concentration-response function can be obtained simply from drawing a random value from the observed lognormal distribution. Barnthouse et al. (1987) showed that uncertainties in test data extrapolation, as measured by the variance estimates of extrapolated LC_{25}s, are generally much larger than uncertainties generated by randomly selecting values of B from the distribution of values observed in toxicity tests.

Linkage to Population Parameters

Barnthouse et al. (1987, 1988, 1990) used the above procedure to estimate the parameters a and B of the logistic concentration-response functions (with associated variances) for chronic mortality to three life stages (eggs, larvae, and juveniles/adults). Barnthouse et al. (1990) also used this method to estimate concentration-response functions for reduction in female fecundity. Barnthouse et al. (1988) used a quadratic regression model, rather than the logistic model, to estimate concentration-response functions for fecundity reduction.

Estimates of life-stage-specific concentration-response functions were then used to generate statistical distributions of values of C_m and C_f, the estimated reductions in survival and fecundity caused by contaminant exposure. The response functions for eggs, larvae, and juveniles were combined to estimate C_m:

$$C_m = 1 - (1 - C_e)(1 - C_l)(1 - c_j) \qquad (8.21)$$

where C_e, C_l, and C_j are, respectively, the fraction of eggs, larvae, and juveniles expected to die from exposure to toxic contaminants. This procedure assumes that although natural mortality varies between years, contaminant-induced mortality is constant for any exposure concentration. The reduction in fecundity (C_f) estimated from the concentration-response data is assumed to apply similarly to each age class of mature females.

The response parameters C_m and C_f provide the coupling between toxicity test data and population responses as they appear both in the reproductive potential Equation 8.14 and as modifiers of the age-specific survival and fecundity rates (i.e., the elements of the Leslie matrix).

Results

Figure 8.2 shows a concentration-response function and uncertainty band for the reduction in female reproductive potential of brook trout (*Salvelinus fontinalis*) exposed to methylmercuric chloride (data from McKim et al. 1976).

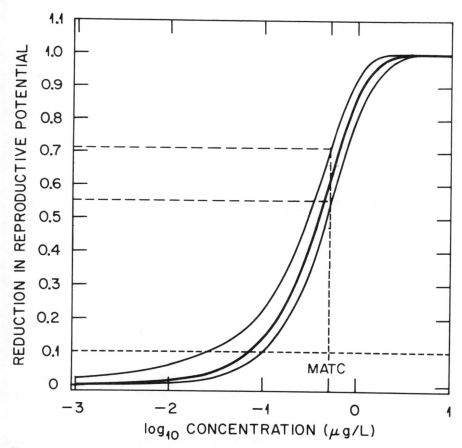

Figure 8.2.   Fitted concentration-response function and uncertainty band for the reduction in female reproductive potential of brook trout (*Salvelinus fontinalis*) exposed to methylmercuric chloride. The lower dashed line denotes the 10% effects level (EC_{10}). The two upper dashed lines denote the 90% confidence band for the effects level associated with the Maximum Acceptable Toxicant Concentration (MATC, Ch. 7). Data from McKim et al. (1976); figure from Barnthouse et al. (1987).

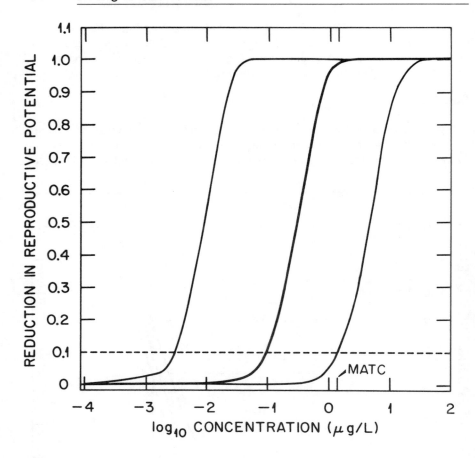

Figure 8.3. Synthetic concentration-response function and uncertainty band for the reduction in female reproductive potential of rainbow trout (*Onchorhynchus mykiss*) exposed to methylmercuric chloride (from Barnthouse et al. 1987).

The lower dashed line denotes the 10% effects level (EC_{10}). Figure 8.2 shows that conventional methods of summarizing toxicity test data can sometimes provide misleading information about the population-level implications of the data. The two upper dashed lines on Figure 8.2 denote the 90% confidence band for the effects level associated with the Maximum Acceptable Toxicant Concentration (MATC, Chapter 7). As noted in Chapter 7, the MATC has been interpreted as an effects threshold at or below which no adverse effects of contaminant exposure should occur. Figure 8.2 shows that this interpretation is often incorrect. The median value of the EC_{10} is 0.07 μg/L, and the prediction interval is approximately 0.03 to 0.1 μ/L. By comparison, the brook trout MATC for methylmercury, 0.53 μg/L, corresponds to a 60% to 78% reduction in reproductive potential. This level of reduction in reproductive potential is equivalent to very heavy fishing pressure.

In Figure 8.2 the concentration-response function and confidence band were calculated by fitting the logistic dose-response model directly to test data. No interspecies or interlife stage extrapolations were required. Figure 8.3 shows an analogous function and confidence band for rainbow trout (*Onchorhynchus mykiss*) exposed to methylmercuric chloride. Full life-cycle data for rainbow trout are not available for methylmercuric chloride. Figure 8.3 was constructed by extrapolating from an acute LC_{50} for rainbow trout to a chronic LC_{25} using the procedures described in Chapter 7. The median responses in Figure 8.3 are similar to the median responses in Figure 8.2, but the prediction intervals are much wider. The prediction interval for the EC_{10} in Figure 7.3 ranges from 0.03 to 1.2 μg/L. The rainbow trout MATC for methylmercuric chloride (1.2 μg/L) extrapolated from brook trout (Chapter 7) corresponds to a 10% to 100% reduction in rainbow trout reproductive potential.

As noted earlier, the reproductive potential index can, with appropriate expert judgment, be used as an index of population-level effects of changes in mortality and reproduction. It does not, however, directly correspond to changes in yield, abundance, or risk of extinction. Matrix projection models can be used to directly quantify these changes.

Barnthouse et al. (1990) exposed the model menhaden and striped bass model populations described above to a hypothetical toxic contaminant that affects only young-of-the-year survival. They selected three levels of "expected" contaminant-induced mortality: 8%, 20%, and 30%. Uncertainty concerning the true toxicity of the contaminant was modeled by taking the expected level as the mode of a lognormal distribution and setting the variance of this distribution such that (1) for the "medium uncertainty" case, the 5th and 95th percentiles of the distribution would differ by approximately a factor of two, and (2) for the "high uncertainty" case, the 5th and 95th percentiles would differ by a factor of 10. Factor-of-two precision is the approximate precision achieved in estimating impacts of power plants on year-class abundance of fish populations (Barnthouse et al., 1984), and probably represents an upper limit on the precision obtainable in assessments of contaminant effects on young-of-the-year fish. Experience with extrapolating toxicity test results to population-level effects suggests that order-of-magnitude (or greater) uncertainty in predictions of contaminant mortality is typical. For each low uncertainty case, the lognormal standard deviation was assumed to be zero.

For each population, 100×100-year simulations were performed with each simulation using a random contaminant-induced mortality rate drawn from the appropriate lognormal distribution. Each simulation used a unique time series of random density-independent natural mortality rates. A "low uncertainty" case was included in which the variance of the contaminant effects distribution was set to zero and the population simulations reflect only natural variation. For each case, the percent of populations going extinct during the simulation period was tabulated.

Figure 8.4 compares the responses of the model striped bass and menhaden populations to additional mortality. Figure 8.4 shows that the estimated risks of extinction at each of the three mortality levels was influenced both by the life history of the species and by uncertainty concerning the true level of contaminant-induced mortality. No extinctions were observed for either species at the lowest mortality level. For menhaden, no extinctions were observed in any of the "low uncertainty" cases, regardless of mortality level. In only one of the cases examined, involving both high expected mortality and high uncertainty, was there extinction of more than 5% of the model menhaden populations. Extinctions were much more frequent in the striped bass population runs. Extinctions were observed for all three uncertainty cases at the highest mortality level. Extinction percentages greater than 5% were observed in three cases and, for the highest mortality and uncertainty levels, more than 25% of the model populations became extinct.

Both model populations were exposed to exactly the same patterns of pollutant mortality. Therefore, any differences in their response reflect differences in the life histories of the species modeled. Thus, provided that the existing information on these populations is reasonably accurate, these results show that the Gulf menhaden population has a greater capacity to tolerate

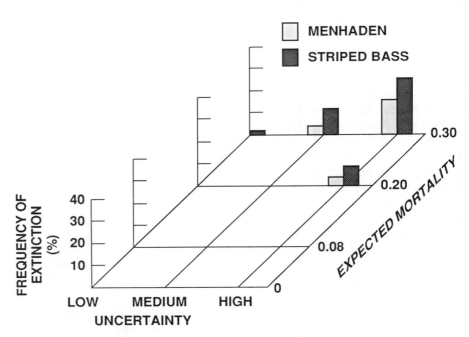

Figure 8.4. Frequencies of extinction of model menhaden (*Brevoortia patronus*) and striped bass (*Morone saxatilis*) populations subjected to three levels of expected mortality due to a hypothetical toxic contaminant and three levels of uncertainty concerning the true level of mortality (from Barnthouse et al. 1990).

pollutant-induced mortality to young-of-the-year than does the Chesapeake Bay striped bass population.

In theory, increasing the rate of fishing should increase the sensitivity of a population to any given contaminant exposure. This increase in sensitivity would be reflected in decreases in the values of the 5th percentile, median, and 95th percentile population-level EC_{25}s. Conversely, decreasing the rate of fishing should increase these values. Because the Cheseapeake Bay striped bass population was apparently overexploited and declining, this population should be more sensitive to changes in fishing mortality than the more stable Gulf menhaden population.

Barnthouse et al. (1990) tested these predictions by performing simulations in which the baseline fishing mortality rates for the two populations were either doubled or halved. Extrapolation from an LC_{50} for a standard test species to the life-cycle responses of the modeled species was used for the sensitivity analysis because for the great majority of chemicals, acute toxicity studies on standard test species such as fathead minnow or sheepshead minnow are the only data available for risk assessment. Results of the sensitivity analysis were in accord with expectation. Compared to the baseline, the median, 5th percentile, and 95th percentile EC_{10}s decreased when fishing mortality was doubled and increased when it was halved. The changes were greater for striped bass than for menhaden. For striped bass, the median EC_{10}s for the 2x and 0.5x cases differed by an order of magnitude, as compared to a factor of 6 for menhaden.

Examined by themselves these differences appear substantial. However, compared to the width of typical uncertainty bands for extrapolation from LC_{50}s for standard test species (approximately a factor of 300), they are relatively modest. Barnthouse et al. (1990) concluded that for a site-specific assessment, where full or partial-life-cycle chronic data for species of interest could be obtained, inter-specific differences in life history and exploitation intensity might be important. For typical screening-level assessments [e.g., premanufacturing notification (PMN) assessments under TSCA], uncertainties related life history and exploitation would be negligible compared to toxicological uncertainty resulting from the use of QSARs and short-term toxicity tests to predict long-term population responses.

A word of caution about extrapolation is in order at this point. Models of the kind described in this chapter are relatively easy to construct but substantial expert judgment is required to interpret them correctly. As with any other kind of modeling, extrapolations outside the range of actual observations must be regarded with some skepticism. Figures 8.2 and 8.3, constructed using the logistic dose-response function, all show responses declining to zero at low concentrations. This behavior represents "common sense" and is also a mathematical consequence of the form of the logistic, probit, and other similar models. Figure 8.5, from Barnthouse et al. (1988) shows a quite different behavior. Instead of converging, the confidence band in Figure 8.5d *diverges* at low concentrations. The reason for this is that Barnthouse et al. (1988) used

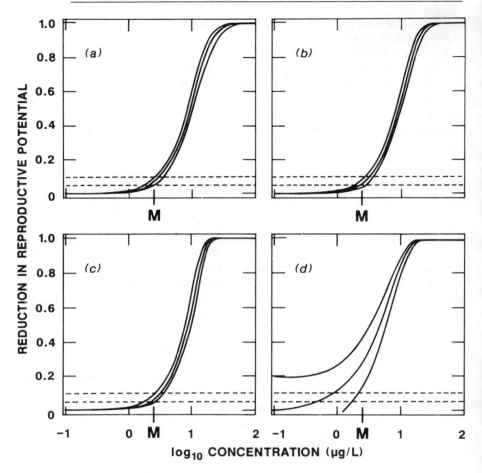

Figure 8.5. Estimated reduction in reproductive potential of a female striped bass as a function of exposure to trifluralin, based on data of Parrish et al. (1978) on chronic toxicity to sheepshead minnow (*Cyprinodon variegatus*). Panel a is based on data for embryolarval mortality only. Panel b includes embryolarval and juvenile/adult mortality and assumes that adults that survive their first spawning are unaffected thereafter. Panel c includes the same data, but assumes that mortality is repeated annually throughout the entire life cycle. Panel d includes, in addition, data on reductions in fecundity. The MATC estimated from the original data set is denoted as M on the concentration axis (from Barnthouse et al. 1988).

a quadratic model rather than the logistic model to define concentration-response functions for fecundity. The quadratic model is *not* constrained to converge to zero at low concentrations and confidence bands around quadratic functions fitted to data *diverge* outside the range of observations. The region of divergence in Figure 8.5d is the region of extrapolation below the lowest test concentration.

Within the range of test concentrations, the logistic and quadratic models yield similar results. Whether one or the other is used to extrapolate to low concentrations is dependent on the philosophical inclinations of the user rather than on objective reality. Common sense dictates that below some concentration no biological effects should occur, and therefore a converging model is appropriate. Strict empiricism postulates that the data themselves provide all that is knowable. If the data are noisy they reveal little or nothing about responses beyond the range of observations, and convergence should not be assumed.

RELEVANCE OF POPULATION MODELING TO MAJOR ENVIRONMENTAL PROBLEMS

The examples discussed in this chapter cover a fairly wide range of applications of population modeling in risk assessment. Population modeling has always been the principal quantitative underpinning of fisheries management, and has been extended in a straightforward way to assessments of impacts of power plants on aquatic resources (Barnthouse et al. 1988). Life table techniques and matrix projection models are commonly used in management of small populations, especially of endangered species. Studies of the spotted owl (Salwasser 1986, Dawson et al. 1987, Lande and Orzack 1988) provide the best-known current examples. Stochastic extinction models (Goodman 1987) and individual-based models (DeAngelis and Gross, 1992) are likely to be used with increasing frequency for this purpose within the next few years.

The principal focus of this book is on toxicological risk assessments. A principal purpose of this chapter is to show that techniques developed and used in classical resource management and impact assessment can be used fruitfully in assessments of ecological risks of toxic chemicals. Applications to problems involving toxic contaminants is not only feasible in principle, it has already been done. When coupled to contaminant fate and transport models, population models can and are being used to predict consequences of catastrophic events, especially oil spills. Natural Resource Damage Assessments (NRDAs) under CERCLA require that damages resulting from contaminated sites as well as from oil spills should be quantified so that liabilities can be properly assigned. Because many of the resources of concern in NRDAs are economically or aesthetically valued populations (birds, gamefish), population models provide an attractive tool for quantifying ecological effects in terms suitable for economic evaluation. The work of Samuels and Ladino (1983) on seabirds and of Reed et al. (1989) on fur seals demonstrate the power of this approach.

When coupled to toxicity test data, population models can be used in other kinds of predictive toxicological assessments. Tipton et al. (1980) demonstrated the use of a matrix projection model to quantify the relationship between pesticide exposure, cholinesterase inhibition, and survival of quail in pesticide field tests. Barnthouse et al. (1990) used field-calibrated models of fish populations to evaluate the utility of standard toxicity test designs for predicting population-level consequences of contaminant exposure.

Further uses of models in retrospective risk assessments to identify causes of observed population change and evaluate management options are considered in Chapter 10. There are situations in which consideration of population dynamics alone is insufficient for an adequate ecological risk assessment because ecosystem-level responses are more important. However, there are no insurmountable scientific barriers to using population models wherever they are relevant.

9

Ecosystem-Level Effects

Glenn Suter and Steve Bartell

Assessments of ecotoxicological effects are performed at the ecosystem level for two reasons. First, the assessment endpoints may be ecosystem properties such as probability of eutrophication or loss of diversity. In those cases, the methods used must be able to estimate the responses of those ecosystem properties. Second, ecosystem-level methods may support an assessment of lower-level endpoints. For example, an assessment of toxic effects on fish population production may employ aquatic ecosystem models to account for indirect effects on fish due to direct effects on lower-level food web organisms. The particular strengths of ecosystem-level methods can complement the particular weaknesses of lower-level approaches in assessments that use multiple lines of evidence to enhance their credibility (Chapter 2). Given that, ecosystem-level tests and models clearly have a place in ecological risk assessment.

Four types of effects may occur in ecosystem-level tests and models that do not occur at the organism or population level: (1) effects on a population's ability to interact with populations of other species, such as effects on ability to avoid predators; (2) indirect effects on a population due to effects on the populations with which it interacts, such as reduction in the abundance of a predator due to toxic effects on prey; (3) changes in structural properties of an ecosystem, such as the number of species or trophic levels; and (4) changes in the functional properties of an ecosystem, such as primary production. The last two categories are true ecosystem-level effects, and the first two are population-level effects that are observed only in an ecosystem context.

Despite these arguments that ecosystem-level tests and models have a role in assessments of environmental effects, the nature of that role has been the subject of inordinate controversy. Some ecologists have argued that ecological assessments of chemicals must emphasize effects at the ecosystem level,* and that assessments based on traditional organism-level toxicology are necessarily false and incomplete (O'Neill and Waide 1981, NRC 1981b, Kimball and Levin 1985). Although some ecological toxicologists have attempted to rebut that premise (e.g., Suter 1981a, Smies 1983, Mount 1985, Giesy 1985), most have acknowledged the importance of ecosystems while continuing to rely on organism-level toxicology because of practical problems with ecosystem-level methods (Box 9.1). However, all parties agree that it is desirable that risk assessments assure, in some sense, the value of ecosystems.

It is desirable to avoid the extreme positions that only ecosystem-level methods are legitimate or that ecosystem-level methods are inherently impractical and unnecessary. The ecosystem is simply one level in a conceptual hierarchy which may be central to an assessment or may be peripheral, depending on the circumstances and goals of the assessment.

Some of the controversy arises because those who customarily work at one level of organization commonly develop habits of thought that are not appropriate to other levels of organization. Therefore, they need some acculturation before they can judge the methods and results used at other levels. Those who are unfamiliar with ecosystem-level concepts and methods should devote some time to considering the development (McIntosh 1985) and current status of ecosystem science (Allen and Star 1982, O'Neill et al. 1986, Reiners 1986). These ecosystem concepts are briefly discussed in the following section. Succeeding sections consider ecosystem-level tests, and mathematical models of ecosystem-level effects.

---

*We do not discuss communities separately from ecosystems. Although the methods and models of community ecology are quite divergent from those of ecosystem ecology, the distinction is not important in ecological toxicology. The laboratory and field systems used to test community ecological parameters (e.g., species number) and ecosystem ecological parameters (e.g., net production) are the same.

Box 9.1. Impediments to Ecosystem Level Assessment

Despite the obvious incompleteness of any assessment scheme that does not include ecosystem-level methods, ecosystem-level effects testing and modeling are rarely used in ecological assessments because of a number of practical considerations:

1. Although some ecosystem-level tests are less expensive than some organism-level tests, ecosystem-level toxicity tests are generally more expensive and lengthy.

2. Few ecosystem-level test designs are standardized and there is no agreement about which are the best designs.

3. Relatively few testing laboratories are experienced in running ecosystem-level tests.

4. There is no consensus about what should be measured in an ecosystem level test, or how the test endpoint should be expressed.

5. Although an organism-level test clearly contains all of the necessary components in the right proportions (one liver, two kidneys, etc.), it is not clear when an ecosystem-level test is sufficiently complex, and the relative proportions of components in ecosystem-level tests may be unnatural.

6. The effects of variation in test conditions on the outcome of ecosystem-level tests is generally unknown.

7. The relationship between effects observed in ecosystem-level tests and those occurring in the range of actual exposed ecosystems is generally unknown.

8. Toxic effects are difficult to detect in ecosystem-level tests because replication tends to be low, variance among replicates tends to be high, and replicate systems tend to diverge over the time periods required for the effects of interest to occur (LaPoint and Perry 1989).

9. Ecosystem-level endpoints cannot respond to lower levels of exposure to a toxicant than the most sensitive organisms and populations, and are generally much less responsive. Because of the considerable functional redundancy in ecosystem processes, significant toxic effects can occur in sensitive populations without affecting ecosystem functional properties (Hill and Weigert 1980, Odum 1985, Schindler et al. 1985). As a result, population-level responses have been used as a standard against which to judge the sensitivity of ecosystem-level test endpoints (Pontasch et al. 1989a).

10. Because monitoring all species is laborious and requires tremendous expertise, even organism- and population-level properties tend to be taxonomically aggregated in ecosystem-level tests (Cairns 1986). This aggregation tends to obscure population-level effects, thereby making ecosystem-level tests less sensitive.

11. Parameterization of ecosystem-level risk models requires that toxic effects on numerous species be estimated.

These points do not negate the argument made in the introduction that there are important responses at the ecosystem level that are not adequately addressed by organism-level or population-level studies. Rather, they serve to point out that the adoption of ecosystem-level assessment techniques has been inhibited by significant practical and conceptual problems. In other words, environmental toxicologists and regulators have had practical reasons for not adopting ecosystem level approaches and have not been simply ignorant or perverse. However, many of these points are less compelling when the goals of the assessment are clearly defined. For example, it does not matter that organism-level properties are more sensitive if the goal is to assess effects on ecosystem properties, and the lower cost of organism-level tests is irrelevant if they cannot be extrapolated to the assessment endpoint.

ECOSYSTEM CONCEPTS

The concept of "ecosystem" continues to evolve in relation to fundamental changes in our understanding of biogeochemical systems. From earlier descriptions emphasizing balance, continuity, and determinism in structure and function (e.g., Forbes 1887, Odum 1971), "ecosystem" has come to imply asymmetries, scale-dependent and hierarchical structure, event-driven and chaotic dynamics (e.g., Rosen 1975, Allen and Starr 1982, O'Neill et al. 1986). Ecosystems have eluded formal definitions as entities for observation and study. This evolution in thought, while intellectually healthy and necessary, may forestall consensus concerning the important attributes of ecosystems for risk analysis and assessment. Such conceptual elusiveness has correspondingly slowed the development of generally accepted and applied methods for assessing effects or estimating risks at the ecosystem-level.

One reason why generally accepted methods for ecosystem assessments are lacking is the difficulty in objectively defining ecosystem boundaries. Ecosystems cannot be objectively located, identified, or counted. Boundaries are defined commonly by habitat (e.g., forest, lake, grassland, tundra, wetlands), despite attempts at formally defining system boundaries as locations (in time or space) across which energy transfers are minimal (Margalef 1968, Allen and Starr 1982). Note that this latter definition applies equally well to "bounding" individual organisms. Application of this definition in isolating parts of nature (e.g., microcosms and mesocosms) may (1) further the standardization of microcosms and mesocosms for assessment purposes, (2) reduce high variance associated with arbitrary isolations, and (3) provide a basis for scaling rules for interpolation, extrapolation, and cross system comparisons.

Lacking generally agreed upon criteria for defining ecosystems, one is left with several perspectives of nature that have, nonetheless, proved useful in recording observations and increasing the quantitative understanding of the world, as we measure it. These viewpoints emphasize either energy flow, material cycling, or a naturalist's outlook on ecosystems. "Ecosystem" has become a convenient shorthand for these perspectives. These viewpoints are, of course, interrelated, and all of potential use in risk analysis. A synopsis of these perspectives follows, along with a discussion of their corresponding relevance to risk analysis.

An Energy Perspective

The capture, distribution, transformation, and dissipation of biochemical potential energy remain as focal points for the fundamental study of ecosystems (Odum 1971, Klopatek et al. 1981, Prigogine 1982, O'Neill and Waide 1981, Bartell et al. 1988). Waide (1988) outlined a general theory of ecosystems framed in terms of nonequilibrium thermodynamics:

> ". . . the ecosystem may be viewed as a configuration of matter and energy that persists in far-from-equilibrium states by dissipating biologically elaborated free energy gradients, thereby forming organic structures out of inorganic elements mobilized from the physicochemical environment. Thus, the ecosystem may be viewed as a functional biogeochemical system . . . Energy in solar radiation is converted into chemical bond energy by autotrophic photosynthesis. This chemical bond energy represents a free energy gradient which is dissipated in order to build persistent high-energy organic structures out of low-energy inorganic compounds acquired from the surrounding geochemical matrix. Upon death, organic structures are decomposed, with energy being dissipated as heat and elements returned to the geochemical matrix in a state of low chemical potential."

In ecological risk analysis, two components of system energetics bear importance. One, energetics might provide, from first principles, some objective means for assessing change in the functioning of a particular ecosystem in response to stress. Two, meaningful cross-system comparisons of response to stress might be made in terms of comparative ecosystem energetics.

Assuming acceptable boundary definitions, the state of an ecosystem can be defined on the basis of energetics. Each ecosystem has some potential for biological productivity, determined primarily by the fundamental source of energy, usually solar energy. Any ecological structure exists as some integration of this potential, constrained by the availability of essential material resources. The normal rate of energy capture can provide a quantitative assessment of ecosystem state. Thus, ecosystem functional state might be defined in terms of gross primary production integrated over some space-time scale. The material structures capturing the energy input or resulting from the allocation of this realized potential are not of immediate concern. The assessment is purely one of function.

Ecosystem production (P) and respiration (R) have been offered as measures of natural system function (Odum 1969), as well as system response to stress (Odum 1985). By definition, ecosystem in approximate steady-state will exhibit P/R ratios of 1. If a system-level stress increases total respiration, decreases system production, or both, the ratio will decrease. Thus, changes in the P/R ratio might be used to estimate ecosystem response to perturbation. Similarly, changes in P/R might be used as an endpoint in ecosystem risk analysis (e.g., O'Neill et al. 1982).

Between production and respiration lie all the food web transfers of energy within the system. Thus, deviations from the normal pattern of energy flow through an ecosystem might also be used as ecosystem-level endpoints. The development of methods for ecosystem risk analysis will undoubtedly benefit from increased technical capabilities in measuring energy flows through ecosystems.

Holistic ecosystem response to disturbance might be measured remotely as changes in the energy signature or reflectance measured at some specified wavelength. For example, Pierce and Congalton (1988) combined thematic mapper data and meteorological measures to quantify the allocation of energy, as indicated by mapping latent heat flux, in two forest cover types. Remote sensing offers the potential for comprehensive and continuous monitoring. The disadvantages lie in the cost of data acquisition and analysis. Furthermore, changes in the energy signature of a system remain difficult to assess in terms of processes or mechanisms underlying the measured response. Nonetheless, future improvements in remote sensing capabilities (e.g., EOS, see Baker 1990) may facilitate the implementation of routine ecosystem-level measures for ecological risk analysis.

Despite this promise, problems in system identification may continue to retard the development of formal tools for rigorous analysis of ecosystem energetics. As a possible solution, Margalef (1968) suggested that ecosystem boundaries might be operationally identified as locations where energy flows were near zero or negligible in comparison to energy flows internal to such boundaries; for example, coastlines, forest edges, thermoclines, or watersheds. This definition is consistent with hierarchical descriptions of ecosystems, wherein strongly coupled (i.e., energy flow) components are isolated as subsystems that are weakly connected to other subsystems (Allen and Starr 1982, O'Neill and Waide 1981, O'Neill et al. 1986).

A Material Cycling Perspective

Ecosystems are material cycling systems (Pomeroy 1970, Vitousek and Reiners 1975, Likens et al. 1977). The dissipation of energy in ecosystems provides in part for the transport, decomposition, and recycling of materials bound up in the biomass, living or dead, of system components. The amount of energy captured by primary producers in ecosystems is often constrained by the availability of essential raw materials, including macronutrients (e.g., phosphorus, nitrogen, calcium) and trace nutrients (e.g., iron, manganese, molybdenum).

Patterns of material cycling in ecosystems have changed in response to perturbation (Likens et al. 1977, O'Neill and Reichle 1980). For example, excessive losses of nitrogen have been measured for forested ecosystems exposed to perturbations in the form of logging and insect outbreaks. To describe patterns of material flow through ecosystems, Finn (1976) derived a measure of nutrient cycling efficiency, the Cycling Index (CI), based on an econometric input-output analysis of ecosystem compartmental nutrient flows. Bartell (1978) found nonlinear changes in Finn's CI calculated for phosphorus cycling in modeled pelagic systems subjected to increased intensity of size-selective predation by fish on zooplankton. Using these kinds of analytical methods, changes in nutrient flux along system pathways might be used to assess ecosystem response to stress. Similarly, nutrient cycling efficiency might serve as an endpoint for ecosystem risk analysis.

A Naturalist Perspective

It may be possible to develop ecosystem-level methods for risk analysis or assessment based on the perspective of naturalists and community ecologists, who value species composition as a fundamentally important attribute of ecosystems (Rapport et al. 1985). To naturalists, the flora and fauna are valued in their own right and not necessarily in relation to their contribution to productivity, nutrient cycling, or other ecosystem functions. The loss of valued native species or the establishment of unwanted exotic species would be viewed as a major ecosystem degradation. Closely tied to this outlook is the concept and measure of species diversity (Margalef 1968). Diversity has been shown to generally decline in response to disturbance for a variety of different stressors and ecosystems (e.g., Table 1 in Rapport et al. 1985).

In addition to the basic appeal to naturalists and lay people, a population-oriented approach to ecosystem risk analysis has some practical advantages. If sensitive species can be identified that also serve a major function in the ecosystem, then a population-level approach may serve to protect ecosystems. For example, microcrustaceans occupy a pivotal role in aquatic food web structures, serving as the conduits of primary production to other trophic guilds (Burns 1989). Although no taxon is consistently most sensitive (Chapter 7), the daphnid Cladocera are sensitive to many toxic chemicals. Thus, these zooplankters appear to be useful as an ecosystem level indicator taxon.

Ecosystem Stability

Ecosystem stability is another measurable property that might be used to develop methods for risk analysis or assessment. The successful development of methods based on system stability requires careful consideration of several factors detailed in Waide (1988): One, the stability properties of ecosystem components must be evaluated against a backdrop of natural system variability in system structure and function. Two, proper comparison of stability across different ecosystems requires consideration of the characteristic spatial-temporal scales of the systems being compared (e.g., forests and planktonic systems). Assessments of system change referenced solely to human scales can produce misconceptions of system stability. Three, potential relationships between ecosystem structure and function and system stability (e.g., the complexity-stability hypothesis) suggest that ecosystem stability cannot be predicted simply from the properties of its constituent species or populations. Continued investigation of relations between ecosystem structure and stability are necessary to develop effective models and measures for risk assessment.

Despite the need for additional research, at least two components of ecosystem stability might be usefully applied in ecosystem risk analysis. These components are ecosystem resistance and resilience (Waide 1988). Assuming appropriate scaling, resistance is inversely proportional to the change from normal ranges of system behavior in response to some perturbation. Resilience measures the rate of return to predisturbance state after the perturbation. Resistance and resilience can be measured for single or multiple system descriptors.

Measures of resistance and resilience are fundamental to considerations of ecosystem risk and recovery (see also Chapter 3). Resistance is inversely related to risk; the lower the resistance to some disturbance, the greater the risk. Resilience is directly related to ecosystem recovery; more resilient systems portend a greater probability of recovery. Just as ecosystem risk is the conditional probability of some deleterious response to disturbance, ecosystem recovery pertains to whether the disturbed system will return to predisturbance conditions, and if so, when. Thus, recovery complements risk and similarly requires estimation in a probabilistic framework. These relationships between risk/recovery and resistance/resilience indicate that the development of capabilities in ecosystem risk assessment should not proceed in isolation from the theoretical issues of ecosystem stability (Bartell et al. 1992).

TESTS FOR EFFECTS AT THE ECOSYSTEM LEVEL

Tests of toxic effects at the ecosystem level include tests of interactions between species, microcosm tests, mesocosm tests, and field tests. Tests of species interactions are included here because they can be thought of as a fundamental unit of ecosystem dynamics, However, they fall between population-level tests and true ecosystem-level tests. All categorizations of nature are artificial and result in untidy categories.

Tests for Effects on Species Interactions

Species interactions, including predation, herbivory, parasitism, and competition, provide the functional connection between population dynamics and ecosystem processes. Although species interactions have been an important subject of study in ecology, toxic effects on species interactions have been relatively neglected. Potential species interaction tests have been reviewed by Giddings (1981) and Suter (1981b and 1987).

Species interaction tests offer two potential advantages. First, they may be highly sensitive. Interactions, particularly those with pathogens and parasites, can stress organisms, making them more sensitive to toxicants. In other words, the interaction results in a more sensitive organism level response (Chapter 7). In addition, interactions may involve sensitive behaviors or physiological responses that are characteristic of the interaction. Examples include avoidance of predation (Sullivan et al. 1978) and maintenance of the lichen symbiosis between algae and fungi (Ferry 1982). Second, the ecological and economic importance of certain species interactions, such as the predatory behavior of biocontrol agents, make it desirable that toxic effects on species interactions be considered in assessments.

These advantages are mitigated by a number of factors. First, except for studies of air pollution effects on interactions between plants and fungi and plants and herbivorous insects, there is little evidence from the field that pollutants affect species interactions (Suter 1987). Although it is commonly observed that when a toxicant reduces the abundance of a species, the abundance of species with which it interacts are affected, there is little evidence that the effects are due to direct effects on the rate or nature of the interaction. This absence of evidence is likely to be due to a lack of appropriate studies. Second, there are literally millions of species interactions that are potentially susceptible to toxic effects. It is clearly not possible to test all of them and it does not appear to be possible to generalize about toxic effects within broad categories of species interactions. For example, mosquito fish exposed to chemicals that cause hyperactivity can be either more or less susceptible to predation, depending on the species of predatory fish (Goodyear 1972, Herting and Witt 1967). The difference seems to be due to differences in behavior between ambush and pursuit predators. Finally, species interaction tests require the maintenance of cultures of two or more species and the maintenance during the test of conditions in which they will interact normally. As a result, these tests can require more time, space, and labor than a test of either individual species.

Given these considerations, the most useful species interaction tests are likely to be those that are easily maintained in the laboratory and have considerable ecological or economic importance. An example is the symbiosis between legumes and the nitrogen-fixing bacteria of the genus *Rhizobium*. The only standard species interaction test is a test for effects of chemicals on this symbiosis developed by Garten (1985) and adopted by the EPA (1987). It constitutes a good phytotoxicity test which, for a little additional effort, yields information on the effects of chemicals on a highly important species interaction.

Microcosms

Microcosms are laboratory systems that are intended to physically simulate an ecosystem or a major subsystem of an ecosystem. They represent an attempt to create systems that display ecosystem properties while permitting control of conditions and replication of treatments at reasonable cost. A variety of microcosms have been used for toxicity testing at numerous laboratories, routine use of microcosms has been advocated by scientists in academia, industry, and government, and reviews of microcosms have been regularly published (Gillett and Witt 1979, Giesy 1980, Giddings 1981, Suter 1981b, Gillett 1989, Gearing 1989). Despite this support, microcosms are still primarily a research tool and are not yet routinely used in environmental toxicology and assessment.

An impediment to adoption of microcosms is the failure of microcosm proponents to lay aside advocacy of their own favorite system and answer three questions. (1) What are the relative utilities of excised and assembled microcosms for answering specific assessment questions? (2) Can microcosms simulate the responses of entire classes of ecosystems, or are they better used to simulate a particular site? (3) Are microcosms best used to measure whole-system responses, measure the responses of individual organisms and populations in a relatively realistic context, or serve as a source of data and mechanistic understanding for mathematically modeling ecological effects? Because of the great number and diversity of microcosms, the following discussion is formulated in terms of these general questions concerning the utility of microcosms. The examples are largely limited to microcosms with standard EPA protocols.

Excised Versus Assembled Microcosms

The most basic methodological distinction among microcosms is that between those systems that are assembled from specified component species and materials and those that are excised from the environment. The only assembled microcosm that has received any degree of use in ecological toxicology and has a standard protocol (EPA 1987, ASTM 1991) is the standard aquatic microcosm developed by Taub (1969). It consists of ten species of algae, five zooplankters (cladoceran, amphipod, ostracod, protozoan, and rotifer), and a bacterium (plus any other bacteria introduced with the zooplankters) in a defined aqueous medium with a sterile sand sediment, all contained in gallon jar under fluorescent lights. Analogous terrestrial systems are the various assembled agroecosystem microcosms including a crop seedling in a container of soil (Lichtenstein et al. 1977), soil, crop seedlings and invertebrates in a large jar (Metcalf et al. 1971), and crops and invertebrates in soil in a chamber with controlled air and water flow (Nash et al. 1977, Gile and Gillett 1979).

The principal advantage of such assembled systems is that they can be standardized. Similar results can be obtained from different laboratories, it is possible to clearly define the adequacy of a microcosm test for quality assurance, and results with different chemicals can be compared with some assurance that differences in response are due to differences in toxicity rather than differences in the microcosms. In addition, the limited and constant array of species makes it more likely that the cause of observed responses can be determined, which makes it possible to model the ecosystem-level interactions for extrapolation to the field. The chief disadvantage is the relative physical and biological simplicity of assembled microcosms. To the extent that ecosystem responses are a function of the complexity of natural ecosystems, the relative simplicity of assembled microcosms diminishes their utility.

Excised microcosms are segments of ecosystems that have been removed from the environment as a unit or a few units (e.g., sediment and water) and placed in containers in the laboratory. They contain a natural assemblage of biota and natural media, and often retain much of the natural structure of the soil or sediment. As a result, excised microcosms are likely to be more realistic than assembled microcosms, lending some credibility to their results. However, because they cannot be as completely standardized as assembled microcosms, excised microcosms are less amenable to quality control or to comparison of results among laboratories and chemicals. Because the variability among replicates is likely to be higher, toxic effects may be more difficult to detect in excised microcosms. Because their components are less completely defined, their results are more difficult to interpret.

Two excised terrestrial microcosms have standardized protocols. The simplest is the soil microbial community test, which consists of a container of fresh, coarsely sieved surface soil with associated biota (Suter and Sharples 1984, EPA 1987). It is similar to test and experimental systems used for decades by soil microbiologists (e.g., Stotzky 1965), and this body of research suggests that it is a reasonably representative model of soil microbial community responses in the field. The second is the soil core microcosm (VanVoris et al. 1985, EPA 1987, ASTM 1991). It consists of a soil core (17 cm by 80 cm) taken from an agricultural field or a natural ecosystem with plants small enough to be maintained in the core. The core is contained in a plastic cylinder which fits in the coring tool so that there is minimal disturbance of the soil during extraction, and the core rests on a Buchner funnel for collection of leachate during the test.

Three excised aquatic microcosms have standard protocols: (1) The simplest is mixed flask culture (EPA 1987). It consists of a mixed culture of microbes and microinvertebrates derived from one or more natural communities and held in the laboratory in one-liter flasks until a microcosm-acclimated community has developed; (2) The naturally derived pond microcosm consists of water, sediment, macrophytes, and associated biota obtained from a shallow pond or the littoral zone of a lake or slow-moving river and contained in an aquarium (Giddings 1986, EPA 1987); (3) The site-specific aquatic microcosm attempts to maintain a microcosm community that is similar to the one from which it is derived (Perez et al. 1984, EPA 1987). It consists of a large tank of ambient water, a sediment core suspended in the water, and associated biota. The size of the sediment core is limited to create an appropriate ratio of water volume to sediment surface, the water is stirred to simulate natural mixing, water flow over the sediment surface is controlled, and the temperature is varied to match the ambient water temperature.

Generic Versus Site-Specific Microcosms

Although the distinction between site-specific and generic microcosms is often a matter of the way microcosms are used, some microcosms clearly relate to no particular ecosystem and others go to lengths to simulate specific sites. For example, Taub's standard aquatic microcosm resembles no lentic ecosystem but shares some properties with all lentic ecosystems. On the other hand, Perez and others have gone to great lengths to match the structural and functional properties of particular natural ecosystems.

Most uses of ecosystem-level test data in assessments are based on the assumption that microcosms and mesocosms physically simulate the ecosystems to be protected, rather than simply helping to understand how effects are induced. Under this assumption, the utility of these tests for assessment depends on one's concept of ecosystems to be simulated. The advocate of generic microcosms could argue that we cannot perform tests of chemicals in all ecosystems so we must assume that the response of ecosystems to pollutants is relatively constant, at least within broad classes of ecosystems. In other words, just as a fish is a fish is a fish (Mount 1982), an ecosystem is an ecosystem is an ecosystem. If this premise is accepted, then a relatively simple microcosm that contains the major components of a class of ecosystems could capture the important responses of those ecosystems. Advocates of site-specific microcosms argue that ecosystem-level responses depend heavily on the details of food-chain structure, the physical and chemical properties of the media, and the species present. As a result, "a chemical introduced into similar ecosystems may not produce the same type of change in each" (NRC 1981). Therefore, microcosms are best applied to site-specific assessments, and, if they are to be applied to generic assessments, it is better to have simulated one real ecosystem than no particular ecosystem. In large part, the resolution of this apparent conflict depends on the clearer definition of the goals of the ecological risk assessments performed for the various regulatory and commercial purposes.

An alternative to assuming that microcosms simulate ecosystems is using microcosms to reveal mechanisms of response which are then incorporated into conceptual or mathematical models of responses of field ecosystems. The question then becomes which types of microcosm best reveal the response mechanisms that are thought to be important for a particular combination of chemical and ecosystem type, rather than which is a better model ecosystem. Extrapolation from microcosms is discussed further below.

Appropriate Microcosm Endpoints

Most of the parameters that have been thought to be important or sensitive responses to chemicals can be measured in microcosms. As a result, microcosm studies tend to measure far more responses than organism- or population-level tests. They can be divided into four general categories: (1)

organism-level parameters, (2) abundance of component organisms grouped into species, higher taxa, or functional groups, (3) community parameters such as number of species and various diversity indices, and (4) ecosystem functional parameters. The responses measured in the EPA-standardized microcosms are presented in Table 9.1. Only one organism-level parameter is required and none of the microcosm protocols call for measurement of community parameters. A mixture of abundance and ecosystem functional responses are used in all aquatic microcosms but the two terrestrial microcosms emphasize ecosystem functions.

The exclusion of community responses is readily explained. Community responses are derived from abundance determinations, are less sensitive than the component abundances, and have less information content than the component abundances. The abundance responses provide an indication of the combined direct and indirect effects of the test chemical on species or groups of species but do not indicate which types of effects are responsible for the changes in abundance. Such conclusions about mechanisms must be inferred from single species toxicity data and knowledge of interactions; that is, from some sort of model of the microcosm community.

Ecosystem functional responses are, in effect, the chemical and energetic input and output of a black box containing the biota and physical media. The biota is treated as an energy and material processing system. Like community parameters, ecosystem functional responses are less sensitive than responses of the component populations or organisms, but ecosystem processes are generally felt to have importance independent of the fate of individual organisms and populations. For example, the retention of mineral nutrients by the soil is desirable, independent of changes in the soil biota. In addition, it is impractical to assess effects on a significant fraction of the soil populations involved in nutrient retention, so the fact that some populations must be affected before nutrient retention declines is immaterial.

Mesocosms

Mesocosms are outdoor experimental systems that are delimited and to some extent enclosed. They offer more realism than microcosms due to their larger size and more natural physical conditions but can still provide replication, control of chemical exposure, and some control of the biotic components. They have been used less than microcosms because they are less convenient, more expensive, and more subject to catastrophic agents such as extreme weather and vandals. Even more than microcosms, they have been primarily limited to research, but "large-scale field pens" have been used to test effects of pesticides on wildlife (EPA 1978a) and pond mesocosm tests have recently become a common requirement for pesticide registration in the U.S. (Touart 1987, Voshell 1989).

Like microcosms, mesocosms may be assembled or may be delimited portions of existing ecosystems. Two types of assembled mesocosms are in use,

Table 9.1. Responses To Be Measured in the EPA's (1987) Microcosm Protocols

Response Parameter	Standard Aquatic Microcosm	Mixed Flask	Pond Microcosm	Site-Specific Microcosm	Soil Microbial Community	Soil Core
Organism-Level						
Visible plant injury						X
Population and Community Abundance						
Algal species	X			X		
Phytoplankton		X	X	X		
Macroinvertebrate species	X			X		
Zooplankton		X	X	X		
Protozoa, total	X					
Rotifers, total	X					
Benthic fauna types			X	X		
Ecosystem Function						
Net production	X	X	X			X[a]
Net respiration	X	X	X		X	
Chlorophyll	X	X	X			
pH	X	X	X			
Nutrient concentrations	X				X	
Glucose decomposition		X				
Alkalinity			X			
Conductivity			X			
Dissolved oxygen			X			
Nutrient loss						X

[a]Monocots and dicots are separated.

artificial ponds and outdoor artificial streams. Pond mesocosms are lined holes in the ground of standard size and shape to which sediment or soil, lake or pond water, and biota are added (Giddings et al. 1984, DeNoyelles and Kettle 1985). Artificial streams are ditches or troughs to which gravel or other sediment, and biota are added. Water for the streams may be recirculated or drawn from a natural source and directed to flow through once. Examples of artificial stream studies include Sigmon et al. (1977) and Stout and Cooper (1983).

Delimited mesocosms are created by enclosing, or at least defining the edges of, portions of an ecosystem that will receive some set of treatments. They are highly diverse in design and operation because of the difficulty of delimiting portions of ecosystems in such a way as to allow sufficient control of the ecosystem components and of the level of treatment while minimizing distortions of the system. In aquatic ecosystems these are primarily bags and limnocorrals. The bags are large plastic containers, closed at the bottom, filled with ambient water and associated biota, and usually held open above the surface by some arrangement of floats (Grice and Reeve 1982, Gamble et al. 1977) (Figure 9.1). Limnocorrals are plastic cylinders, the tops of which are held above the surface by floats and which are either anchored at the bottom in the

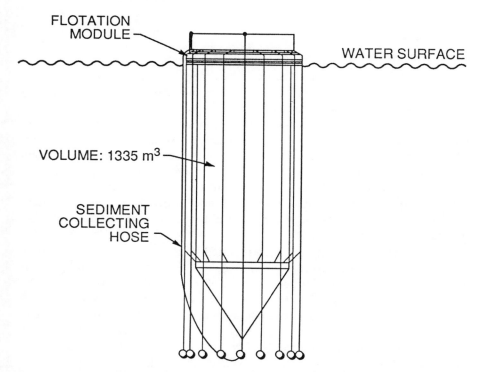

Figure 9.1. A mesocosm consisting of a plastic cylindrical bag held open at the surface by a ring of floats (redrawn from Grice and Reeve 1982).

sediment, or are held open at the bottom below the euphotic zone or at some other suitably deep level (Grice and Reeve 1982, Baccini et al. 1979). The simplest terrestrial mesocosm is a plot which may be enclosed by a wall or fence to control movement of animals or surface water flow (Barrett 1968, Anderson and Barrett 1982, Goodman 1982). Plots may also be completely enclosed for fumigation with treated air (Treshow and Stewart 1973, Hill et al. 1974). A more sophisticated test system is the open-top chamber, a self-supporting plastic cylinder with a plenum around the bottom of the interior for dispensing treated air (Figure 10.3) (Heagele et al. 1979). Alternately, plots can be fumigated by arrays of tubes (Lee and Lewis 1978, Miller et al. 1977). As shown in Figure 10.3, a systems of sprinklers and moving roofs can be used to control deposition of acidified or otherwise contaminated rain.

As with microcosms, the relative utility of assembled versus delimited mesocosms, generic versus site-specific mesocosms, and various test endpoints have not been resolved. Because relatively few mesocosm tests are conducted, there has been little discussion of these issues as they pertain to mesocosms. The greater realism of mesocosms and their ability to include more species and processes, as well as their greater expense means that the best approaches to testing in mesocosms are not necessarily the same as those for microcosms.

Field Tests

Field tests are tests of the effects of a treatment in an unconfined ecosystem conducted at approximately the same scale as the actual contamination. They are the most difficult and expensive type of test, but, if properly conducted, they offer unimpeachable realism. Some authors have argued that microcosms are fatally unrealistic, and only whole-ecosystem studies give reliable results (Perry and Troelstrup 1988). However, the responses in field tests are so complex and diverse and so influenced by weather, movement of organisms, and other uncontrolled variables, that their results may be difficult to understand and generalize. Also, they are often unreplicated or poorly replicated and have few treatment levels because of expense and because of the difficulty of finding sufficiently similar sites.

The only field tests that have been required by routine regulation have been the natural pond tests formerly required by the EPA, Office of Pesticide Programs and the terrestrial field tests for effects on wildlife still required by the EPA for certain pesticides (Fite et al. 1988). Field testing has also been used for verification of laboratory tests and associated assessment models (Geckler et al. 1976) and for studies of major large-scale environmental problems such as acid rain (Schindler et al. 1985). If conducted on some fraction of chemicals, field tests might be used to verify the quality of ecological risk assessments.

Assessments Based on Ecosystem-Level Test Data

Relatively little attention has been paid to the problem of determining how to use the results of ecosystem-level tests in risk assessments. This is in part because assessors rarely encounter such test data and in part because those who conduct such tests often have research goals rather than assessment goals. Mount (1985) argued that ecosystem-level tests have been designed to reveal problems, not to solve them. The quotient method has been used to incorporate ecosystem-level test data in ecological assessments. However, if the relationship between ecosystem-level test endpoints and assessment endpoints can be defined, analysis of extrapolation error (AEE) could be applied. Whether establishing factors for the quotient method or statistical extrapolation models for AEE, it is necessary to establish a conceptual model of the relationship of the test endpoint (a property of the test system) to the assessment endpoint (a property of the real world) (Chapter 2).

When ecosystem-level test data have been used in assessments, the most common extrapolation model has been the assumption that the tests are representative of responses in all potentially exposed ecosystems. For example, the EPA's Office of Toxic Substances has applied an assessment factor of 1 to field test data to estimate concentrations "which may cause adverse environmental effects" (Environmental Effects Branch 1984). This model of the relationship between ecosystem-level test data and field effects assumes that (1) the test endpoints are the same as the assessment endpoints, or at least respond at the same level of exposure, and (2) all exposed ecosystems respond in the same way as the test system.

The first assumption begs the question, what properties of ecosystems need to be preserved (i.e., what are good ecosystem-level assessment endpoints?). One commonly employed approach to selecting endpoints for ecosystem-level testing and assessment is to measure a diversity of organism-, population-, community-, and ecosystem-level responses and search for sensitive responses within this array of measurements. This approach suggests that any deviation in the state of an ecosystem is unacceptable. An alternate approach is to identify ecosystem-level assessment endpoints a priori and measure test endpoints that correspond to or are predictive of the assessment endpoints. Because practices such as forestry, farming, stocking, fishing, and suppression of fire and forest insects imply that undisturbed ecosystem properties are not society's goal, the latter approach is more logical. However, it requires that ecological risk assessors and risk managers be more decisive and open about environmental values.

The second assumption, that all exposed ecosystems respond in the same way as the test system, is not likely to be true. The responses of different ecosystems to a particular chemical are qualitatively and quantitatively quite variable (NRC 1981b, Harras and Taub 1985). The assumption may be saved if only one ecosystem will be exposed to the pollutant, and the test system is excised from the ecosystem to be protected or is designed to simulate it.

However, even subtle differences between the test system and ecosystem may prove to be critical. For example, when Pratt et al. (1988) created microbial communities by suspending foam blocks in a lake and then exposed them to equal concentrations of total residual chlorine in flow-through laboratory tanks and in large bags in the lake, neither the absolute sensitivities nor relative sensitivities of their structural and functional measurement endpoints corresponded. Clearly, most ecosystem level tests differ from the ecosystems that they are designed to simulate in nonsubtle ways such as scale, wall effects, and lack of large organisms. These differences between ecosystem-level test systems and ecosystems of interest are greater than, but not conceptually different from, the differences between organisms in the laboratory and organisms in the field. However, we know far more about the factors that influence the sensitivity of organisms and the uncertainties that they engender than about the factors that determine the sensitivity of ecosystems and their associated uncertainties.

If the assumption of ecosystem equality is rejected, then some approach must be developed for extrapolating the results of ecosystem-level tests to the ecosystems that will be exposed. If a most-sensitive test ecosystem could be identified, then the extrapolation model could be: the test endpoint concentration is less than the concentration that would cause significant effects in real exposed ecosystems. Although no such system has been established, it is likely on theoretical grounds that a system with long food chains and few species in any guild or functional group would be particularly sensitive. This approach is likely to fail because a most-sensitive ecosystem-level test would require most-sensitive component species, as well as a most sensitive structure. As discussed in Chapter 7, species do not have consistent relative sensitivities to different chemicals. As a result, the assessor must always deal with the probability of the pollutant occurring in a more sensitive ecosystem.

One simple way to incorporate differences in the responses of ecosystems would be to assume that ecosystem sensitivity is a random variate with some distribution function, as is done with individual species (Box 2.1 and Chapter 7). The distributions of ecosystem-level effective concentrations for common toxic chemicals could be used as estimators of the uncertainty associated with the assumption that all ecosystems respond at the same exposure level. However, an attempt by Suter (unpublished) to estimate such distributions was abandoned because the differences in the design of ecosystem toxicity studies were so great as to completely obscure differences in ecosystem susceptibility. There are no standard measures of effect, pattern of dosing, durations of exposure, frequency of effects measurement, etc. In addition, most ecosystem-level tests have too few exposure levels to define effects levels with reasonable precision. Finally, poorly defined concentrations combined with differences in physical conditions make it impossible to distinguish differences in sensitivity from differences in exposure.

A more conservative assumption is that the ecosystem-level test system is representative (with some error) of a well-defined class of ecosystems. For this purpose, it is not sufficient to assume that conventional biogeographic or ecological classifications are applicable. Rather, the classes must be based on characteristics that control ecosystem-level toxic response. We have already mentioned the length of food chains and the number of species in trophic levels, guilds, or other functional groupings as potential predictors of ecosystem sensitivity. The nature of control in the system is also likely to be important. Harrass and Taub (1985) argue that the differences in response to copper of lake ecosystems are explicable in terms of control of phytoplankton by nutrient limitations, light limitations, or grazing. Because trophic structure, functional diversity, and ecosystem control changes seasonally and over the course of succession, the ecotoxicological classification of an ecosystem may change over time.

Once classification criteria have been developed, it will be necessary to develop procedures for using the resulting classifications in assessments. The procedures must include (1) a means of creating broad or narrow ecosystem classes to suit the goals of a particular assessment, (2) a means of assigning particular microcosms, mesocosms, and field tests to ecosystem classes, and (3) a means of estimating the uncertainty associated with those assignments.

A general approach to extrapolating between ecosystems is to derive a mechanistic model of effects induction that fits the tested system and then apply it to other ecosystems by changing the values of the parameters to fit those ecosystems. This may be done by statistical fitting of parameters (Box 9.2), or more conventionally by measuring and estimating parameters that are judged to be important. However, this method and any other credible method for extrapolating toxicological responses between ecosystems is likely to require much more understanding of the causes of responses than is typically available. Because the physical and chemical processes that control the fate of chemicals in ecosystems are better understood and more easily described for particular ecosystems than the biological processes that control ecosystem-level responses, risk assessments commonly employ ecosystem-level fate models and organism or population-level effects models. For example, assessments of acid precipitation effects employ ecosystem characteristics such as watershed size, soil chemistry, and water chemistry to estimate lake acidification and aluminum leaching but effects on fisheries are estimated from organism-level test data (Christensen et al. 1988).

Box 9.2. A Proposed Method for Extrapolating Between Ecosystems

Reckhow (1983) has developed a theoretical approach to the problem of extrapolating responses between classes of ecosystems. It requires that the assessor first define a model of the response to be extrapolated; that is, an exposure-response model for some response that corresponds to the ecosystem-level assessment endpoint. Second, the assessor must define a model that describes the tested systems and the systems to which the assessor wishes to apply the exposure-response model for the purpose of determining their similarity. Reckhow proposed that this similarity model should be derived by a multivariate analysis of the variables that are believed to causally control the exposure-response relationship. Estimation of these variables for both sets of systems permits quantification of the similarity of the systems based on comparison of the centroids and covariance matrices of the variables (see Green 1978). Multivariate statistical tests can then be used to estimate the probability that the ecosystems of interest belong to the same state space as the ecosystems used to develop the exposure response model and the power of the hypothesis test to discriminate among the ecosystems. If, taking into consideration both type I and type II error, the assessor decides that the two sets of ecosystems are not the same, he can (1) expand the variance of the exposure-response model by an amount indicated by the difference between the two sets of systems, or (2) respecify the model so that it is applicable to the systems of interest. This approach has not been applied to effects assessment and will require more development.

In sum, there are a number of potential techniques for using ecosystem-level test data in risk assessments but none have been satisfactorily developed. The problems are different for test systems that emphasize effects on the component organisms and species than for those that emphasize ecosystem processes (Van Leeuwen 1990). The abundances of individual species are generally more sensitive to toxicants than community parameters or ecosystem processes, but it is not clear how much is gained when species responses are measured in an ecosystem-level test. When the nature of the single-species response is not the same in an ecosystem-level test as in an organism-level test, the discrepancy is often explained from knowledge of the system's trophic structure (Hansen and Garten 1982). For example, many chemicals are more toxic to aquatic herbivorous invertebrates than to algae, so that ecosystem-level tests result in an increase in algae biomass or production even though there is some degree of toxicity to algae. This pattern of response has been cited as evidence that ecosystem-level testing is needed (Kimball and Levin 1985, Harras and Taub

1985, Gearing 1989), but it is hardly unpredictable. The simplest ecosystem model would suffice to predict that response from the conventional test data plus the quite different response that could occur if the phytoplankton were not limited by herbivory. This is not to say that genuinely surprising and sensitive responses never occur in microcosms or mesocosms, but, if they occur, there is no established method to assess the implications of those surprising responses in the field.

The need for ecosystem-level testing is clearer when assessing effects on processes performed by ecosystem components that are not included in conventional organism- and population-level tests. The most obvious examples are the sediment, litter, and soil microcommunities. Decomposers and chemoautotrophs are valued for the processes that they perform, not as organisms or populations, so there is no reason to perform routine testing of these communities below the ecosystem level. However, the problem of extrapolating among ecosystems still applies. For example, reduction of pH may increase or decrease the rate of leaf decomposition in aquatic systems, depending on the tree species, and, even when limited to birch leaves, tests of pH effects may show an increase or decrease in decomposition depending on the microcosm design and field test site (Perry and Troelstrup 1988).

Even if the relationship between the test system and systems of interest is known, the interpretation of results from tests of microcommunity processes is not straightforward. For example, it is not at all clear whether promotion or inhibition of decomposition is an adverse effect. Leaf litter in a forest is a nutrient pool that, depending on condition, may either be slowly releasing nutrients to the soil, thereby minimizing loss and increasing tree growth, or may be immobilizing nutrients thereby reducing tree growth. Litter may enhance or inhibit seed sprouting, reduce erosion, promote fire, moderate soil climate, and in other ways interact with the rest of the ecosystem. Most reports of soil and litter tests implicitly assume that a decrease in decomposition is adverse, but the EPA (1978) suggests that a decrease in litter biomass is a serious adverse effect on forest ecosystems. It is tempting to assume that a change in either direction of any magnitude is adverse. However, that assumption implies a vision of perfectly coadapted and precariously balanced ecosystems that is not supported by ecological science, ignores the considerable resistance and resilience of ecosystems, and is irrelevant to managed ecosystems. Interpretation of effects on ecosystem processes is a complex and ecosystem-specific problem, but that recognition does not help a regulatory decisionmaker to decide how to interpret decomposition test results or other tests of ecosystem processes. Clearly it is not sufficient to assume that whatever is measured in an ecosystem-level test is useful for making a decision concerning the acceptability of the tested chemical. More thought concerning the use of results in assessment needs to precede routine ecosystem-level testing.

These problems are significant but not insurmountable. Similar problems have been overcome and some still must be overcome in the development and application of organism-level toxicity tests. The principal need is for greater recognition that microcosm and mesocosm research and development cannot be limited to developing and operating the physical test systems. Problems of analysis, interpretation, and extrapolation must also be addressed.

Extrapolation from Organism-Level to Ecosystem-Level Test Endpoints

Although there is no reason why statistical models might not predict ecosystem-level effects from lower-level test data, this assessment approach has received little attention. An exception is the study of Sloof et al. (1986). Those authors regressed no observed effect concentrations (NOEC) from ecosystem-level tests ($NOEC_e$) against the lowest reported single-species acute LC_{50}s and EC_{50}s and against the lowest reported single species chronic NOEC ($NOEC_s$) for the same chemical, which resulted in the following equations:

$$\log NOEC_e = -0.55 + 0.81 \log LC_{50} \tag{9.1}$$

$$\log NOEC_e = 0.63 + 0.85 \log NOEC_s. \tag{9.2}$$

The units are $\mu g/L$, the correlation coefficients are 0.77 and 0.85, respectively, and the arithmetic-scale prediction intervals are ± 85.7 and 33.5, respectively. The authors conclude that ecosystem-level responses in chronic tests are more sensitive than single species responses in acute tests, but less sensitive than single species responses in chronic tests. These models are discussed at length not because they are particularly useful, but because the fact that they work at all suggests that ecosystem responses are not all that mysterious and unpredictable. Effective concentrations for ecosystem tests can be grossly estimated from a single organism-level test endpoint using a simple statistical model.

Statistical models of distributions of species sensitivity have been discussed earlier in this chapter as estimators of the response of individuals of a most-sensitive species or of any particular untested species. Another interpretation allows them to be used as models of the response of ecosystems to pollutants, if ecosystems are treated as an aggregation of species with some distribution of sensitivity. Clearly, if a criterion concentration is set at a concentration so low that there is a high probability that all species will be protected, then no ecosystem-level effects will occur. This is the premise of Kooijman's (1987) analysis which calculated concentrations that would protect all species with 95% confidence. However, such conservatism results in extremely low criteria. Where species sensitivity distributions have been used for regulation, some risk of affecting some species has been allowed. For U.S. National Water Quality Criteria the best estimate (fiftieth percentile) of the concentration protecting 95% of species is used (Stephan et al. 1985). The Dutch require 95% confidence of protecting 95% of species (Van Leeuwen 1990).

Statistical models like these could be used in ecosystem-level risk assessments. For example, if an LC_{50} value or single species NOEC value is available for a chemical, equations like 9.1 and 9.2 and their associated variances could be used to estimate the probability that effects (expressed as NOEC values) would be seen in an ecosystem-level test. If the probabilities are near 1 or 0, then the test need not be required; otherwise, the test may be desirable to resolve the uncertainty. However, as Sloof et al. (1986) acknowledge, it is difficult to interpret the output of their model because of the use of lowest test endpoints from various numbers of tests and because of the various test systems and endpoints for the $NOEC_e$ values.

Ecosystem Models

The purpose of this section is to highlight several ecosystem models that might be used to estimate ecological risks. Ecosystem models attempt to represent mathematically much of the same ecological structure and processes represented physically in microcosms, mesocosms, and field tests. Unlike toxicity tests, mathematical models are not constrained to the physical dimensions, time scales, and kinds of organisms that can be conveniently isolated for study in these artificial ecosystems. That is, ecosystem models can be developed to estimate risks for spatial-temporal scales appropriate to the particular environmental concern. Additionally, mathematical models can be developed to examine ecosystem perturbations that cannot be addressed experimentally, due either to the severity of the expected impacts or to the scale of the necessary experiment. Finally, models offer the potential for assisting in the completion of large numbers of assessments (e.g., premanufacturing notifications under TSCA) that cannot be reasonably performed using ecosystem-level tests.

The following collection of models results from an initial attempt to (1) identify existing models designed expressly for simulating or forecasting adverse ecological effects in ecosystems, and (2) select basic research models that may be useful with a reasonable effort expended in their modification. If model selection is limited to those designed expressly to simulate the environmental fate or ecological effects of toxic chemicals and other anthropogenic disturbances relevant to society's concern (e.g., pesticides, water quality, acid deposition, climate change), the resulting collection of models is relatively small. Alternatively, many ecosystem models constructed originally for basic research models offer some potential for modification and use in estimating risk. The set identified from this perspective could include tens to hundreds of models. For example, consider the wealth of models directed at plankton production dynamics (Patten 1968, Swartzman and Bentley 1979). Thus, the number of potential models for ecosystem risk analysis certainly depends on the criteria for model selection, particularly the effort necessary to modify basic research models.

Chemical transport and fate models (Chapter 5) have not been included in this discussion, even though some have been developed in the spirit of risk analysis (e.g., Olsen and Wise 1979, Onishi et al. 1982, Parkhurst et al. 1981). Risk, as estimated in these models, refers mainly to the probability of exceeding some water quality criterion or toxicity benchmark. Biological components are typically absent from these fate models and adverse ecological effects are not considered.

In most cases, even the ecosystem models that simulate the effects of toxic chemicals or other disturbances require some minor modifications to translate calculated effects into estimates of risk. Depending upon information supplied by the authors, the general nature of each model (i.e., applicable environments, spatial-temporal scales), its major ecological state variables, necessary input data, and model output that might be used for risk analysis are summarized below. Reported efforts in model evaluation and sensitivity analyses are also mentioned.

The key distinguishing criteria used to classify models as ecosystem models were the explicit representation of abiotic system components (e.g., light, temperature, nutrients), in addition to representation of ecological populations from one or several trophic levels (e.g., primary producers, consumers, decomposers).

The following text briefly identifies and outlines some of the ecosystem models that might be used to address problems in risk analysis. The models have been classified into energy processing, material cycling, and naturalistic categories identified in the preceding discussion of ecosystem perspectives. Some of the models might be classified in more than one category, and the resulting assignment reflects the author's interpretation of best fit. The models are summarized in Table 9.2.

Energy Flow Models

White (1984a,b,c) developed a model of a dense multispecies grass community (sward) and multispecies consumer carbon flux through a generalized temperate grassland ecosystem. State variables for the primary producer submodel (White 1984b) include free water, soil water, and biomass of root, shoot, flower, litter, and standing dead for a variety of perennial plants. Changes in these state variables are determined by difference equations using one-day time steps. Input data consist of cloud cover, relative humidity, rainfall, snowfall, snow depth, air temperature, and wind velocities. The consumer submodel (White 1984c) permits definition of aboveground and below-ground herbivores, grazers, and sap-feeders, both homeotherms and poikilotherms. Time-varying overlapping life-stage models simulate the mean population size within a designated area for the user-defined consumer assemblage. The life-stage models also use difference equations and a one-day time step. Model testing consisted of comparing model predictions with measured values of liveweight, tussock, and intertussock over a factorial design that varied sheep grazing intensity, stocking rates, and grazing management alternatives for a New Zealand montane trial site (White 1984a).

Table 9.2. Summary of Ecosystem Models for Use in Ecological Risk Analysis

Reference	Environment	Disturbance Type
Terrestrial ecosystem models		
Goodman 1982	orchards	pesticides
Heasley et al. 1981	grasslands	sulfur dioxide
White 1984a,b,c	temperate grasslands	none
Coughenour 1981	grasslands	sulfur deposition
Aquatic ecosystem models		
Benz 1986	general food webs	toxic chemicals
O'Neill et al. 1982	pelagic food webs	toxic chemicals
Bartell, Gardner and O'Neill, 1992	general food webs	naphthalene
Park et al. in press	lotic food webs	toxic chemicals
Rose 1985	defined microcosm	copper
Horwitz 1981	estuarine food web	heavy metals
Harris et al. 1984	macrotidal estuary	aromatics, Cd
Svirezhev et al. 1984	fish ponds	none
Wolfe et al. 1986	ponds	none
Sciandra 1986	700-L seawater tanks	none
Herrick et al. 1982	woodland streams	mercury
McIntire and Colby 1978	small streams	none
Park et al. 1980	general aquatic systems	pesticides, toxic chemicals
Thomann 1981, 1984	general food webs	PCBs, plutonium, cesium
Brown et al. 1982	general food webs	PCBs
Hopkinson and Day 1977	Barataria Bay, LA	none
Kelly and Spoford 1977	Delaware Estuary	wasteloads, BOD
Nixon and Kremer 1977	Narragansett Bay	none

Benz (1986) modeled the changes in primary producers, consumers, and decomposers in relation to biotic and abiotic stress. The model also considers dead organic matter and inorganic nutrients. Individuals in the model populations occupy three states (reproductive, nonreproductive, and dead) determined by Markovian transition matrices. Effects of toxic chemicals are modeled as transition probabilities using dose-response functions. The model uses arbitrary time units and simulates the changes in population sizes for a hypothetical aquatic system. No model testing or sensitivity analyses were reported.

Park et al. (in press) constructed a one-dimensional stream reach model (AQUATOX) that simulates the effects of toxic chemicals in aquatic food webs. Model state variables include populations of phytoplankton, zooplankton/zoobenthos, two functionally defined fish populations, detritus, ammonia, nitrate, phosphate, and dissolved oxygen. The model consists of coupled differential equations using a basic time step of one day. Toxic effects were modeled as changes in mortality rate in relation to body burden of toxic chemicals. The model was designed for general application to canonical lotic systems (streams and reservoirs). Necessary input data include physical-chemical parameters for degradation of the toxicant (modeled after PEST, Park et al. 1980) and default values for the biological production equations, which are modified from CLEAN (Park et al. 1974) and its subsequent versions. Site-specific information can be incorporated if available. Initial biomass values and model parameters for site-specific applications are also required; default values are used for general model application. The model will simulate the changes in biomass and toxicant concentration for each population.

O'Neill et al. (1982) simplified CLEAN (Park et al. 1974) with the express purpose of extrapolating the results of acute toxicity assays to probabilistic estimates of observing specific toxic effects in aquatic systems. The result was the Standard Water Column Model (SWACOM). The aquatic food web consists of ten functionally defined populations of phytoplankton, five zooplankton populations, three planktivorous fish populations, and a single population of piscivores. The methodology, Ecosystem Uncertainty Analysis, requires toxicity data for species representative of the food web populations (e.g., *Daphnia*, algae, fishes) and an estimated exposure concentration (assumed time invariant) for the specific toxic chemical. The model uses difference equations to simulate daily changes in biomass of the food web populations. Input requirements include daily values of light, temperature, and nutrients (default values are used in the absence of site-specific data). The model reproduces the qualitative pattern of seasonal production dynamics commonly measured for northern dimictic lakes, although the model was neither designed nor calibrated for a specific system. The sensitivity of forecasted risks has been examined in relation to model assumptions and the availability of toxicity data for several heavy metals and organic pollutants (O'Neill et al. 1983). The water column model itself has been subjected to detailed numerical sensitivity analyses (Bartell et al. 1988a). Predicted toxic effects on selected model components have been compared to effects measured for phenolic compounds in experimental ponds (Bartell 1986). Additional littoral and benthic food web components have been included in an expansion of SWACOM to produce a Comprehensive Aquatic Simulation Model (CASM) for use in ecological risk analysis (DeAngelis et al. 1989).

Bartell, Gardner and O'Neill (1988) integrated an approach for simulating toxic effects (O'Neill et al. 1982) with a dynamic chemical fate model (FOAM, Bartell et al. 1981) to produce an integrated fate and effects model (IFEM) for an aquatic ecosystem. The model is currently constrained to polycyclic aromatic hydrocarbons and was implemented to examine the fate and toxic effects of naphthalene in a hypothetical aquatic system. The food web consists of single populations of algae, periphyton, macrophytes, bacteria, zooplankton, benthic insects, larger benthic invertebrates, detritivorous fish, and omnivorous fish. The model simulates the dynamics of suspended particulate matter, detritus, and sediments. The model predicts the time varying concentration of toxicant in each model component and the time varying change in population size that results from the sublethal effects on basic growth processes.

The IFEM model is unique in that population-specific toxic effects are calculated as a function of body burden for each population. Body burden changes in relation to the population-specific uptake and depuration rates for the toxicant. The IFEM requires toxicity assay data and kinetic information describing the net accumulation of toxicant by modeled populations. The model consists of coupled difference equations using a daily time step. The model integrates production and toxic chemical fate and effects under one square meter of a user-defined water column depth. The model operates within a Monte Carlo framework that permits examination of parameter uncertainties, estimation of risk, and sensitivity analysis. Model predictions have not been compared to data.

Rose (1985) derived a set of ordinary differential equations used to simulate the toxic effects of copper on aquatic populations in Taub's standard aquatic microcosm (Swartzman and Rose 1984). State variables include eight functional groups of phytoplankton (*Nitzschia*, blue greens, *Chlorella*, *Chlamydomonas*, *Selenastrum*, *Ankistrodesmus*, *Scenedesmus*, and filamentous greens), seven groups of zooplankton (small, medium, and large *Daphnia*, ostracods, amphipods, rotifers, and hypotrichs) and several forms of detrital and dissolved organic and inorganic phosphorus and nitrogen. The model was calibrated in a stepwise, iterative fashion using the results of laboratory experiments and statistical goodness-of-fit tests and graphical comparisons (Rose et al. in press). Toxic effects on phytoplankton were modeled as a reduction in growth rate (photosynthesis — respiration); effects on zooplankton were implemented as a direct mortality term added to the natural mortality rate (Swartzman and Rose 1984). The toxic effects model was calibrated to results of experiments performed with copper in laboratory microcosms (Taub and Read 1982). Initial values for phytoplankton and zooplankton were obtained from the experiments. Constant light and temperature conditions were also defined by the experiments. The model employs a daily time scale and has a spatial scale defined by the physical dimensions of the microcosms.

Svirezhev et al. (1984) describe a model of a fish pond ecosystem. Ordinary differential equations determine changes in the concentration of phosphorus and nitrogen and changes in the biomass of the state variables that include forage fish, silver carp, bighead, phytoplankton, zooplankton, detritus, and bacteria. The process equations lend themselves to modification to include direct and indirect effects. The model has 112 parameters. Daily values of light, temperature, phosphorus, nitrogen, aeration rate, and fish feeding constitute the necessary input data. The model was designed for shallow ponds less than 1 meter deep and less than 1 hectare in surface area. The model time step is one day. No detailed analysis or comparisons of model results to data were provided.

Another pond model was constructed by Wolfe et al. (1986). This model considers changes in fish, algae, suspended organic particulate matter, ammonia, nitrate, dissolved oxygen, carbon dioxide, alkalinity, and five bacterial populations. Input data include the schedule for feeding the fish, the pond aeration rate, light intensity, and water exchange rate. The ponds were 2300 L in volume. A daily time scale was used. No sensitivity analyses were reported. Comparisons between model predictions and pond data were used to test the model.

Sciandra (1986) constructed a model to simulate the dynamics of phytoplankton, zooplankton, detritus, bacteria, and zooplankton fecal material in 700-L seawater tanks. Daily values of light, temperature, and nitrate loads drive the model, which has been calibrated to data measured in the tanks. No sensitivity analyses were documented.

Herrick et al. (1982) report a model of mercury contamination of a small woodland stream. Model components include sediments, detritus, invertebrates, and fish. Model inputs are concentration of dissolved monomethyl mercury and dissolved inorganic mercury in the water. The model predicts the concentration of mercury in system components. Applicable environments are woodland streams dominated by detritus-based food webs, and ranging from pH 4.5 to 8.3. The model operates on a daily time step. Model evaluation was qualitative only, and no sensitivity analyses were reported.

McIntire and Colby (1978) describe an ecological process model that simulates the dynamics of small streams characteristic of the northwestern United States. Processes included in the model are periphyton dynamics, grazing, shredding, collecting, invertebrate predation, vertebrate predation, and detrital conditioning. The model has 14 functionally defined state variables whose dynamics are determined by the model processes. The model was programmed as a discrete time representation of the stream ecosystem with a basic time step of one day. The model has been used to examine hypotheses concerning the role of various consumer processes (e.g., grazing, shredding, predation) in determining energy flux through streams (McIntire 1983). The model has been run extensively under a wide range of scenarios and stream types, although no formal sensitivity analysis was reported.

Other stream ecosystem models that might be adapted for ecological risk analysis, but not detailed here, include Boling et al. (1975), Webster (1983), and Elwood et al. (1983).

This survey does not include the large number of plankton production models (see Patten 1968), eutrophication models (DiToro et al. 1975), or large lake simulation models (e.g., Patten et al. 1975, Park et al. 1974). In some instances (e.g., O'Neill et al. 1982), the directly applicable models represent structural simplifications of these larger system models. A large number of models have been constructed to examine the flow of energy and/or the processing of nutrients in aquatic ecosystems (primarily lakes). The eutrophication literature offers many of these models and there is a great deal of similarity among them. Some of these models have been modified to simulate the transport and fate of toxic chemicals. Treatment of this literature is beyond the scope of this chapter.

The following models might be modified to address ecological risk in coastal wetlands: Hopkinson and Day (1977), Kelly and Spoford (1977), and Nixon and Kremer (1977). The spatial scale of the last two models suggests their applicability for regional risk analysis.

Hopkinson and Day (1977) describe the development of a large-scale comprehensive model of carbon and nitrogen dynamics in the Barataria coastal ecosystem in Louisiana. Ecological state variables were aggregated as (1) primary producers, (2) dead standing marsh grass, (3) a detritus-microbial complex, and (4) macrofauna. Coupled differential equations determine the flow of C and N through this system. The spatial scale of this model is 1 square meter; the time scale is 1 day. In addition to initial conditions and model parameter values, the model requires field data for insolation, sea level, and water temperature. A sine curve simulates the riverine inputs of nitrogen into the system. Model predictions were compared to measured monthly values for several nitrogen and carbon pools; model results were generally within the ranges of the measured values. The sensitivity of marsh grass standing crop to variations in the light, temperature, and nitrogen-forcing functions was reported, with temperature being identified as the key variable for producing model agreement with data. The Barataria model is an example of a model that might be classified as an energy flow or a material cycling model.

Kelly and Spoford (1977) constructed a reach model of the Delaware Estuary for application within a water quality management scheme. The model divides the estuary into approximately 22 spatial segments and simulates the steady-state behavior within each segment using differential equations. The ecological state variables include one compartment each for primary producers, herbivores, omnivorous fish, decomposers, nitrogen, phosphorus, organic matter (BOD), and oxygen. Turbidity, toxic chemicals, and temperature are considered as exogenous variables that potentially modify water quality. Nonlinear approximations of physiological (i.e., photosynthesis, respiration, feeding, excretion, death) and population processes (i.e., predation, upstream inputs, downstream advection) structure the differential equations that determine the daily changes in the model state variables. Units within each reach are mg/L. Model predictions of oxygen concentration, BOD, and phosphorus concentration have been compared with measured values along the length of the estuary. Model values lie generally within the range of the reported data. Predictions of algae, zooplankton, bacteria, and fish were not validated. No sensitivity analyses were reported. The estuary model serves as a component in a set of linear and nonlinear optimization models used to determine wastewater discharge levels that minimize regional costs of waste treatment while meeting required water quality standards.

Nixon and Kremer (1977) describe a model of Narragansett Bay. This model comprises three interrelated submodels: an economic activity/effluent load model, a hydrodynamics model, and a model of the chemistry and biology within the Bay. To model the complicated geometry of the Bay, a 19×48 element grid was superimposed on the system. Each element is 0.93 km (0.5 nautical miles) on a side. Differential equations using a time step of 4 minutes (or less) represent the flow velocities through the grid. Driving variables include the tide at the mouth of the estuary, discharge from the two largest rivers, and tide at the mouth of St. Hope Bay. The ecological submodel consists of winter and summer phytoplankton, copepods (eggs, juveniles, adults), benthos, ctenophores, fish, and nutrients (phosphate, silica, nitrogen). Biological growth processes are modeled on a daily time scale. External driving forces are light, water temperature, and nutrients. Predicted spatial-temporal patterns of phytoplankton production were compared with values measured at different points in the Bay. Model values typically were within the range of reported measurements, except for autumn data. The need to consider the implications of possible ranges of parameter values on model results was discussed, but no formal sensitivity analyses were reported. The modeling approach might be amenable to other large-scale estuarine systems, although the data requirements and site-specific modifications may restrict other applications.

Material Cycling Models

Heasley et al. (1981) used difference equations to simulate the flow of carbon, nitrogen, and sulfur through a grassland ecosystem. The SAGE model components include abiotic, primary producer, soil, and ruminant subsystems. In addition to model parameters and initial conditions for the compartments, necessary input data include sulfur dioxide deposition, solar radiation, air temperature, precipitation, relative humidity, and wind speed. The SAGE model simulates the response of a "typical" square meter of grassland at a daily time scale. Data from 1975 were used to calibrate the model parameters; 1976 and 1977 data were used for subsequent model testing of predicted aboveground biomass, soil water content, and sulfur content in leaves. SAGE predictions for 1976–1977 were nearly always within the standard errors of the measured values.

Coughenour (1981) constructed a model that simulates the effects of sulfur dioxide dry deposition on the sulfur cycle for a grassland ecosystem. Modeled state variables include live shoot and root biomass, dead leaves, litter, nematodes, bacteria, fungi, and soil sulfate. Wind speed profiles, leaf resistances, deposition rates, and soil moisture data drive this model. Several model processes employ an hourly time scale, while the dominant time scale is one day. The model simulates the dynamics associated with a single square meter of grassland. Model testing consisted of comparing predicted and measured time-varying concentrations (May – September) of sulfur in shoots of the dominant grass, *Agropyron smithii*.

Additional simulation models of decomposition (Hunt 1977), nitrogen cycling (Reuss and Innis 1977, Risser and Parton 1982), and phosphorus cycling (Cole et al. 1977) have also been constructed for grasslands. Given their focus, these models have not been detailed here. If changes in rates of decomposition and nutrient cycling become generally accepted endpoints for ecological risk analyses, these models may be applicable.

Horwitz (1981) presents a model of the direct effects of power plant entrainment on planktonic populations with subsequent indirect effects on populations of predators in an estuarine ecosystem. Coupled differential equations determine the population dynamics of phytoplankton; zooplankton; adult, juvenile, and larval planktivorous fishes; larvae, juveniles, and adult omnivorous fishes; benthos; refractory organic matter; labile organic matter and associated bacteria; and dissolved nutrients (nitrate, nitrite, ammonia, urea). The model requires initial conditions for the above components, parameters that determine their dynamics (too numerous to cite individually), and external inputs of light, temperature, and day length. The model employs a daily time scale, and simulations of up to 10 years are routinely used. Output variables consist of the magnitude (mg/1000 m^3) of the state variables outlined above. The model was designed for application to large estuaries (e.g., the Chesapeake Bay). Evaluation of the model was qualitative; patterns of production were discussed in light of current system understanding; however, no direct model to data comparisons were provided.

Naturalistic Models

Naturalists have historically emphasized the autecology of particular taxa of interest, with minimal consideration of the ecosystem context that largely determines the persistence and abundance of these organisms. In a sense, models are unnatural because they necessarily simplify nature and models might therefore lack in appeal or apparent utility from this perspective. A method of ecosystem analysis of potential use in estimating ecosystem risk that provides a link to the concerns of naturalists is food web assembly models.

Food web assembly models have been developed to explore alternative hypotheses of factors controlling the fundamental structure of food webs and ecological communities (e.g., Yodzis 1981, 1982). These models address food web metrics including species richness, number of trophic links, food chain lengths, proportion of herbivores, ratios of prey to predator species, and connectance. While the controversy concerning what regulates the structure of food webs in nature remains largely unresolved (DeAngelis et al. 1983), the various assembly models appear ripe for exploring possible implications of disturbance on species interactions from a more naturalistic viewpoint than traditional biomass or energy flow models.

Selecting and Developing Models

This section briefly outlines important considerations in selecting or developing a model for ecosystem risk estimation. As previously concluded, the number of models available for ecosystem risk analysis varies with the criteria of the prospective user. In selecting or developing a model, the risk analyst should consider:

1. What risks? What are the specific endpoints of interest in the overall analysis? What are the relevant space-time dimensions for the selected endpoints?

2. What information is available to use in the analysis? This includes ecological as well as toxicological data, and the sources and magnitudes of uncertainties associated with those data.

Obviously, selecting a model will depend on how closely the model outputs address the risk endpoints and on whether there is sufficient information to use the model in the particular application. It might often be the case that selecting a basic research model for use in risk estimation will depend on how easily the model can be modified. This will, of course, depend on the talents and resources available to the risk analyst. Any model changes must be thoroughly examined and, if possible, tested. As a minimum, the model should be subject to detailed sensitivity analysis prior to implementation for the production simulations that will provide the basis for risk estimation.

In developing a model, many of the same considerations apply as in selecting one. The principal advantage is that the user may design the model specifically for the problem at hand. Again, the key is to precisely specify the risk endpoint(s) and determine the subsequent information requirements. It makes little sense to design a model for assessment that cannot be used for lack of information. In constructing a model, one must recognize that incorporating ecological complexity for its own sake may inhibit model performance to the same degree as oversimplification. Correspondingly, a model that becomes overdetermined may be able to match any set of data with sufficient calibration efforts. Similarly, agreement between observations and the model might be produced by more than one set of parameter values (i.e., nonuniqueness). Overly simple models might produce the general pattern of the data, but fail to agree with any specific observations. A prudent procedure is to explore different model structures and add complexity only as it moves the model results closer to the observations for reasons consistent with process level understanding of the ecological/toxicological phenomena represented in the model. The important issue is not simple versus complex models, but models of necessary and sufficient detail to meet the risk estimation objectives.

Summary of Ecosystem Modeling

The results of this survey suggest that there are a few models of immediate use in risk analysis. Additionally, there is a considerably larger set of potentially useful models for ecological risk estimation. The size of this set depends primarily on the degree of willingness to modify existing codes for simulation of toxic or physical effects.

Considering the increasing dependence of risk analysis on the use of mathematical models, we offer several suggestions concerning the development, modification, use, and evaluation of these models.

The development of ecosystem models (and models in general) for risk assessment would benefit from precise definition of the criteria for decisions and statement of how the models will be used in the assessments. The power of systems analysis can be best realized when the modeling objectives are clearly stated as quantitative questions (Berlinski 1976). The modeling objectives (i.e., assessment endpoints) will largely determine the necessary model structure, and perhaps alleviate some problems associated with ponderous, data-intensive models, which offer little promise of implementation, testing and evaluation, or routine application.

The degree of testing required before a model can be judged as useful in the context of risk estimation and assessment must be defined. Methods of model testing certainly exist and have been applied to various models (e.g., Balci and Sargent 1981, Burns 1986, Games 1983, Mankin et al. 1975, Schaeffer 1980). Nonetheless, without clearly stated requirements for accuracy and precision, the evaluation question cannot be objectively addressed by model builders. How good is good enough? The use of ecosystem models in risk assessment will likely benefit from the development of a set of models scaled appropriately to endpoints of interest (e.g., Kercher et al. 1981). No single model will likely provide accurate estimates across a wide range of relevant spatial and temporal scales associated with different hazards, endpoints, and mandates for risk estimation and regulation.

Finally, more data and more relevant data are needed. Routine toxicity tests might be modified to collect data that are more amenable to process-level modeling and extrapolation. For example, measuring rates of photosynthesis, feeding, or respiration in relation to chemical exposure, instead of merely counting the dead, would generate process-level data evaluating the stress syndrome used in some ecosystem risk models (O'Neill et al. 1982, Bartell et al. 1988). Performing similar tests in a nested spatial design using increasingly larger containers (e.g., Perez 1991) can provide some initial data for deriving extrapolation rules. If protocols, standardization, and consensus are demanded, then the resulting assays should provide meaningful data for estimating risk.

CONCLUSIONS

A principal reason why both ecosystem-level testing and ecosystem models have not fulfilled their promise as tools for assessment and regulation is the lack of clear goals. At the organism level, the goals are to prevent chemical-induced mortality, slowed growth, or infertility, and both tests and models are available to estimate those responses (Chapter 7). Similarly, at the population level, the goals are to prevent reductions in abundance and production of valued populations, and models are available to estimate those responses from organism-level test data (Chapter 8). However, at the ecosystem level many endpoints have been used but none have the endorsement of either the regulatory agencies or the scientific community. As a result, it is not possible to endorse any particular ecosystem-level test or model because it is not possible to say that one estimates risks to ecosystem endpoints better than another. In other words, without clear goals comparisons of microcosms, mesocosms, and ecosystem models tend to degenerate into arguments about means with no ends.

In this situation, ecological risk assessors must provide their own clarity about endpoints (Chapter 2). If ecosystem-level endpoints are needed, then the assessor must judge which tests and models are feasible and provide suitable estimates of the endpoints. If lower-level endpoints are chosen, the role of ecosystem processes in the imposition of effects must be considered, and appropriate tests or models selected for estimating those ecosystem-mediated effects. In any case, the limitations on ecosystem-level risk assessments is more a lack of vision than a lack of tests or models.

PART IV

UNCONVENTIONAL ECOLOGICAL RISK ASSESSMENT

The conventional paradigm for risk assessment is predictive and deals with the localized effects of a particular action that may result in toxic effects. This part deals with ecological risk assessments that do not fit that paradigm. The first chapter (Chapter 10) addresses retrospective assessments of events that occurred in the past or are ongoing. These assessments do not follow any existing paradigm because, unlike the human health risk assessment paradigm which has been the model for most ecological risk assessments (Chapter 1) and the hazard assessment paradigm which has been the basis for most assessments of ecotoxicological effects, a paradigm for retrospective ecological risk assessments must incorporate the use data from biological studies contaminated sites to estimate risks. Risk assessments of ecological effects occurring at regional scales (Chapter 11) are unconventional primarily in terms of the need to spatially and temporally integrate toxic effects with other factors operating in the region. Chapter 12 discusses the need to develop a risk-based approach to analysis of the environmental trends revealed by surveillance monitoring. Finally, risk assessments of exotic organisms differ from conventional risk assessments in that the organisms reproduce, actively move, and evolve (Chapter 13). Because none of these problems are currently addressed in terms of a standard formal risk-based paradigm, these discussions are brief and conceptual relative to the presentation of the conventional predictive paradigm in Chapters 3–9. Each topic will be worthy of its own text when methods are sufficiently developed.

10

Retrospective Risk Assessment

Glenn Suter

INTRODUCTION

The emphasis on predictive risk assessments in previous chapters of this text reflects their predominance in regulatory agencies and the fact that nearly all projects to develop methods for ecological risk assessment have addressed predictive assessments. However, emphasis has been shifting to assessments of the effects of pollution that began in the past and may have ongoing consequences such as waste sites, acid rain, and existing pesticides. These assessments, which are termed retrospective, are distinguished by two facts: the scope of the problem has been defined by events, and there is a polluted environment to be examined and measured. These facts make retrospective assessments more diverse than predictive assessments in both their methods and logic (Table 10.1).

All predictive assessments begin with a proposed source (a new chemical, effluent, etc.) and proceed to estimate risks of effects (Chapter 3). In contrast, the impetus for a retrospective assessment may be a source, observed effects, or evidence of exposure (Table 10.2).

Source-driven retrospective assessments result from observed pollution that requires elucidation of possible effects. They include assessments of past pollution effects such as spills and accidental releases of untreated effluents (i.e., damage assessments, Chapter 1) and of ongoing pollution such as hazardous waste sites, existing effluents, persistent residues from spills, and chemicals that are currently in use.

Effects-driven retrospective assessments result from the observation of apparent effects in the field that require explanation (Chapter 1). Examples include fish or bird kills, declining populations of a species, and changes in the state of ecosystems such as acidification of lakes.

Exposure-driven retrospective assessments are prompted by evidence of exposure without prior evidence of a source or effects. An example from human-health assessment is the scare over mercury in swordfish. Exposure-

311

Table 10.1. Comparison of Predictive and Retrospective Assessments

Predictive Assessments	Retrospective Assessments
Motivations for Assessment	
Source-driven	Source-driven, Effects-driven, or Exposure-driven
Potentially Available Information	
Environmental description	
Properties of the proposed receiving environment	Properties of the exposed environment, or Observations of the influence of environmental properties on chemical fate and effects
Source term	
Estimated sources	Estimated sources, or Measured sources
Exposure assessment	
Physical/chemical properties	Physical/chemical properties, or Measured ambient concentrations and distributions
Distribution and behavior of predicted receptors	Distribution and behavior of predicted receptors, or Body burdens and biomarkers in actual receptors
Effects assessment	
Toxicity tests	Toxicity tests (including ambient tests) Measured responses in field

driven ecological assessments are rare, but as environmental surveillance programs that monitor exposure become more common (Chapter 12), the discovery of high levels of exposure should prompt more exposure-driven retrospective assessments.

Because the sources of information for predictive assessments are limited (Table 10.1), the logical structure and procedures for predictive assessments tend to be relatively uniform. In contrast, all of the tools of predictive assessment (toxicity testing, measurement of chemical properties, and mathematical models of fate and effects) are available in retrospective assessment, plus all of the methods for measuring (1) the distribution and behavior of pollutants in

Table 10.2. Types of Retrospective Risk Assessment and the Associated Flow of Inference About Sources, Exposure, and Effects

Source Driven Assessments:
$$KnownSource \rightarrow UnknownExposure \rightarrow UnknownEffects$$

Effects Driven Assessments:
$$ObservedEffects \rightarrow UnknownExposure \rightarrow UnknownSource$$

Exposure Driven Assessments:
$$UnknownSource \leftarrow ObservedExposure \rightarrow UnknownEffects$$

the receiving environment, (2) the responses of the biota to pollutants, and (3) the properties of the environment that control fate and effects.

The logic of retrospective assessments is somewhat different from that of predictive risk assessments (Chapter 3). Predictive risk assessments begin with a possible future source and perform independent exposure and effects assessments that are combined to estimate the risks posed (Figure 3.1). In retrospective assessments, source, exposure, and effects assessments are logically parallel in that release from sources, exposure of organisms, and effects on organisms have all already occurred and any one may be inferred from measurements or estimates of the others (Figure 10.1).* That is, one may measure a source such as waste-contaminated soil and make inferences about exposure and effects, measure an indicator of exposure such as body burden and make inferences about sources and effects, measure sources and effects and infer exposure, etc.

Because of the diversity in logical structure and sources of information available to assessors who work with various retrospective assessment problems, the common logic portrayed in Figure 10.1 is not generally recognized. Therefore, the relationship of the assessment paradigm discussed below to specific retrospective assessments may not be obvious to the reader. Consequently, following the discussion of retrospective assessment data and methods, specific examples of retrospective ecological risk assessments are presented in terms of this paradigm.

Hazard Definition

The hazard definition component of retrospective risk assessments includes defining the impetus and goals for the assessment (motive definition), description of the environment to be assessed, and choice of endpoints (Figure 10.1). As was explained above, the motivation for the assessment determines the appropriate strategy for the assessment. If the assessment is source-driven, then the source must be defined as part of the hazard definition. If it is effects-driven, then the affected species or ecosystems and the nature of the effects must be defined. If it is exposure-driven, then the evidence for exposure must be defined.

Whereas predictive risk assessments often must devise reference environments that are generic or hypothetical (Chapter 3), retrospective risk assessments have a real polluted environment to assess. Therefore, the principal task for defining the assessment environment is to appropriately delimit the portion of the environment that is considered to be at risk. For effects-driven assessments, the environment to be assessed is defined by the distribution of the effects. For example, if a population is declining, the range of the assessment is the range of the population. For source-driven assessments, the environment

---

*Although the logical structure for predictive risk assessment presented in Chapter 3 is similar to standard paradigms, the logical structure for retrospective risk assessment is original to this volume.

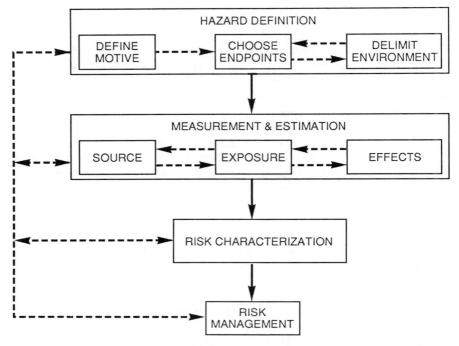

Figure 10.1. A diagrammatic representation of the process of retrospective ecological risk assessment. The dashed arrows indicate constraints (e.g., the choice of measures of source and effects limits the choice of indicators or exposure, and the desires of the risk manager constrain all other choices). The heavy vertical arrows indicate the procedural sequence.

to be assessed is the area exposed to potentially significant levels of the pollutant, which may be established in different ways.

1. If the pollutant has not spread beyond a limited area and is readily measured in environmental media, sampling and chemical analysis of the contaminated media can establish the bounds of the polluted environment.
2. If the pollutant is used in identifiable areas, such as a pesticide that has been used on certain crops, then the areas or regions of use are assessed.
3. In some cases, regulations may provide bounds on the area assessed. For example, assessment of the effects of aqueous effluents begins beyond a zone of initial dilution (OWRS 1985).
4. Toxicity testing can be used to establish the extent of the toxicity. For example, tests of soil toxicity can be used to map the extent of contamination at waste sites. At the Rocky Mountain Arsenal, soil samples were taken from arrays of points in potentially contaminated areas and subjected to a set of toxicity tests (Thomas et al. 1986). The proportional responses obtained from the tests were then mapped back onto the sites and these were used to create contour maps of soil toxicity (Figure 10.2). For purposes of screening or mapping it is not necessary that the tests be used to predict effects, but they

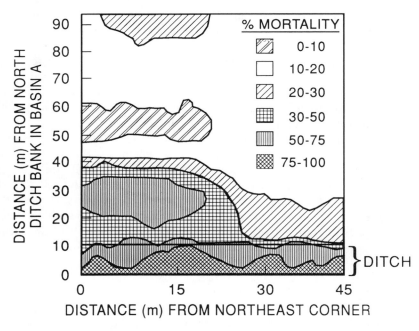

Figure 10.2. A map, derived by kriging of percent lettuce seed mortality in surface soil, from the Rocky Mountain Arsenal (Thomas et al. 1986).

must at least be correlated with effects of interest (i.e., assessment endpoints), if the results are to be useful.

5. In some cases, the environment to be assessed cannot be delimited by measurement or by any a priori concept of appropriate bounds on the pollutant exposure. In those cases, it will be necessary to use some sort of conceptual or mathematical model of the fate of the pollutant to establish the area to be assessed. For example, a fate model could be used to estimate how far a spill traveled before it was diluted to negligible concentrations.

The selection of endpoints for retrospective assessments requires consideration of the same criteria as predictive assessments (Chapter 3). The assessment endpoints for effects-driven assessments are formal, quantifiable statements of the environmental values that have been observed to be affected. For example, the endpoint for an assessment of the secondary pesticide poisoning of raptors might be the reduction of the number of nesting pairs by as much as X percent in the area where the pesticide is used, or simply, the occurrence of more than Y mortalities per year. The endpoints for source-driven assessments are not as obvious. They are limited to the environmental values that occurred in the exposed area but otherwise may be constrained only by the practical criteria that apply to all assessment endpoints (Chapter 2). However, a preliminary ecological survey of the site may indicate which populations or processes appear to be affected and those would be candidate assessment endpoints (Baker 1989).

In many cases the best way to define the hazard is to perform a screening risk assessment. For example, most ecological risk assessments of waste sites are source-driven but the emissions from the source are unknown. A screening assessment would gather existing information concerning chemicals in the site, the potential transport pathways, ambient contamination, and effects. Conservative assumptions and simple assessment models (e.g., the quotient method—Chapter 7) could be used to eliminate chemicals that clearly could not cause significant effects on any endpoint, endpoints that could not be significantly exposed to any chemical, and pathways that could not serve as significant routes of exposure. The results of a screening assessment serve as the hazard definition phase of the assessment and guide the planning of measurement and estimation activities for definitive risk assessments of the chemicals, routes, and endpoints that were retained by the screen.

MEASUREMENT AND ESTIMATION

Retrospective ecological risk assessments are primarily concerned with establishing that a relationship exists between a pollutant source and an ecological effect that is caused by exposure of organisms to the pollutant (Figure 10.1). The methods used to elucidate the relationship depend on the amount and relative reliability of the information available concerning sources, exposure, and effects. Because the concepts of source, exposure, and effects are somewhat different in retrospective assessments, it is important to define the types of information that are potentially available.

Whereas the source in a predictive risk assessment can be defined simply as that release of a chemical or effluent to the environment that someone wishes to be permitted, the source in a retrospective risk assessment is often less clear. It is broadly defined as that aspect of a pollution that is subject to management and to which the assessor attempts to relate exposure and effects. This vague definition is necessary because the conceptualization of the source varies among assessments and can even vary over the course of time within an assessment. For example, assessment of the effect, lake acidification in the northeastern U.S., began with atmospheric deposition of inorganic acids and natural formation of acids in the watersheds as alternate possible sources. Determination that atmospheric deposition is the primary source has led to the assessment of possible categories of sources such as local pulp mills, smelters, regional automotive traffic, and large coal-burning facilities in other regions. Ultimately, when individual regulated sources are repermitted, they must be assessed as sources of acidifying gases.

Although the concept of source cannot be clearly defined for retrospective assessments in general, the source can be clearly defined for most individual assessments. For source-driven assessments, the source is the environmental pollution that prompted the assessment. Examples are the amount and rate of spillage in a tanker accident, the spatial distribution of the concentration of waste chemicals in the soil of waste disposal sites, and the flow rate and concentrations of chemicals in an effluent. For effects-driven assessments the sources are the hypothesized anthropogenic causes of the observed effect. Like the causes of a physical phenomenon in pure science, their elucidation is the object of the investigation. Potential sources are hypothesized at the beginning of the assessment, and as work progresses sources are added, refined, and eliminated until only one source (or a set of jointly acting sources) remains.

Exposure, as for predictive assessments, is the process that links sources with effects. Also, as with predictive assessments, the actual process by which the organism becomes exposed to and takes up a chemical is often a subject for vague, unstated assumptions. If the source is identified and is temporally or spatially remote from the target, exposure assessment may be dominated by the same sort of transport and fate models discussed in Chapter 5. However, if the source is not known, it may be possible to develop receptor models, models that use the chemical composition, isotopic composition, or physical characteristics of the pollutant mixture occurring at the point of exposure to identify the source or apportion responsibility among multiple sources (Gordon 1988). If contaminated media are available to be sampled and measured and if the ultimate source is not in question, the primary issue in exposure assessment is whether and to what extent organisms are exposed to the various media in such a way as to take up the contaminants. In that case, exposure can be assessed in terms of the relationship among the concentrations of various contaminants in the media, the behavior and physiology of the organisms, and the indicators of exposure measured in the organisms. The indicators of exposure include whole body or tissue concentrations of contaminants (body burdens) and biomarkers of exposure. Biomarkers, which will be discussed below, are "measurements of body fluids, cells, or tissues that indicate in biochemical or cellular terms the presence and magnitude of toxicants or of host response" (Committee on Biological Markers 1987).

Effects, as with predictive assessments, are the changes in the ecological values specified by the assessment endpoint that are potentially attributable to a pollutant. If measurement of effects in the affected environment is practically impossible, the effects assessment is based on toxicity testing. In those cases, it is necessary to extrapolate between the test endpoint and the assessment endpoint (Chapters 7–9). If effects can be measured in the environment, the effects assessment is based on those measurements. If the measurement endpoints do not correspond to an assessment endpoint, effects must be extrapolated (e.g., from frequencies of histopathology to reductions in population production).

Measurable Source, Exposure, and Effects

Measuring Sources

The ideal situation for a retrospective assessment is one in which the source of pollution and the affected biotic community are available to be measured and the resources are available to perform the measurements. In many cases, source measurement is a relatively straightforward combination of sampling and analytical chemistry. Point-source effluents can be sampled and analyzed periodically to determine the variability of their emission rates and composition. Nonpoint-sources such as storm runoff require considerably more ingenuity to sample. Sampling of contaminated media such as soil at waste sites, contaminated sediments, or a plume of spill-contaminated water requires that the investigator define the relevant media, forms, horizontal extent, and vertical extent of the contamination. An example of the types of error that can be committed when sampling contaminated media is use of the extent of the surface layer of spilled oil to define the extent of solution-phase or sediment sampling. Recommended sampling designs and standard methods for collecting, handling, and analysis of environmental media and effluents are available from the Environmental Protection Agency (EPA 1983b, EPA 1986, Ford and Turina 1985, Ford et al. 1984, etc.), Nuclear Regulatory Commission (Eberhardt and Thomas 1986), ASTM (1991), and others (e.g., APHA 1985, Keith 1991).

Measuring Exposure

Sometimes exposure can be determined by observation. Examples include observation of oil on aquatic birds, animals drinking from sumps, or pesticide granules in the crops of birds. However, the principal measurable indicators of exposure are body burdens of chemicals and biomarkers of exposure. Guidance is available for sampling, preparation, and analysis of fish and shellfish (EPA 1981, 1982e; Tetra Tech 1986). Methods developed for monitoring human workplace exposure (e.g., Lauwerys 1983) are not well suited to nonhuman organisms because they use fluids that a human is willing to part with but which are relatively difficult to obtain from a nonhuman organism (saliva, urine, expired air, milk, and semen). Nonhuman organisms, however, can be killed so all of their tissues are available, and the volume of the sample is limited only by the size of organs.

The choice of species in which to measure body burdens depends on the assessment problem and on practical considerations. If the assessment is prompted by concern for particular species, body burdens for those species and of their food species should be measured. If the assessment is source-driven, species should be selected from the exposed community on the basis of their relationship to the assessment endpoint and the following practical considerations adapted from Phillips (1978).

1. The indicator species should exhibit a relatively constant correlation between pollutant concentrations and body burdens across sites.
2. The species should persist at the maximum concentrations encountered in the environment.
3. The organisms should be relatively sedentary in order to relate body burdens to specific sites.
4. The species should be abundant throughout the study area, including reference sites.
5. The species should be long-lived so as to provide chronically exposed individuals.
6. The species should be large enough to provide adequate samples for analysis.
7. The species should be easy to collect.
8. The species should survive well in the laboratory to allow evacuation of gut contents before analysis or to allow studies of uptake and depuration under controlled conditions.

The tissues chosen for analysis should include tissues that are known to accumulate the chemical or target tissues for the toxic effects of the chemical. In many cases, whole-organism analysis is appropriate because the species is an intermediate food-chain species that is consumed whole, or because toxicity has been measured as a function of whole body concentrations. For lipophilic organic chemicals, it is necessary to also determine the fat content of the analyzed organisms or tissues to derive a lipid normalized concentration (Schnoor 1982, Walker 1990). If endangered species are involved, collection of tissues is severely regulated (in the U.S. by the Fish and Wildlife Service) and may be limited to tissues that can be collected without injury such as fur, feathers, excreta, and possibly blood.

Finally, the individual organisms selected for analysis should be matched or stratified to reduce extraneous variance in body burdens. Factors to be considered include age, sex, reproductive (or phenetic) state, condition, diet, substrate soil or sediment composition, and exposure to chemicals other than the chemicals of interest that may interfere with uptake, metabolism, or excretion of the chemicals of interest (Hellawell 1986).

Biomarkers of exposure are biochemical or physiological changes which indicate that an organism has received an internal dose of a chemical (Committee on Biological Markers 1987).* One example is the hepatic mixed function oxidases which are induced by a variety of xenobiotic chemicals (Payne et al. 1987, Jimenez et al. 1988, Rattner et al. 1989). Many of these inducible enzymes and electron transport molecules have been identified and associated with various classes of inducer chemicals. Another example is DNA or protein

---

*Biomarkers of exposure and those of effects are distinguished by the way they are used, not by an inherent dichotomy. Levels of all biomarkers are biological effects of exposure, so they are all potentially useful as indicators of either exposure or effects. However, many biomarkers are more useful in one role because they have clearer relationships to one process (either exposure or induction of ultimate effects) than to the other.

adducts, electrophilic chemicals or chemicals metabolically activated to an electrophilic state that have become bound to a macromolecule (Shugart et al. 1987). Potential examples of biomarkers for plants include nitrate reductase as a marker of exposure to nitrogen oxides (Norby 1989) and free radical scavengers as indicators of exposure to photochemical oxidants or other pollutants that induce the production of free radicals (Richardson et al. 1989).

The advantages of biomarkers of exposure are that they indicate that exposure has occurred in such a way as to cause a response, and they may persist much longer than the chemical itself. The disadvantages are that the relationship of the measured level of a biomarker and either the externally-delivered dose or the level of toxic effect are poorly defined, and biomarkers are sensitive to extraneous variables such as age, sex, and diet (Jimenez and Burtis 1989).

Because biomarkers are a new component of environmental monitoring, guidance is not available for sampling. However, most of the criteria for selecting species and individuals for monitoring body burdens (discussed above) should also apply to monitoring biomarkers. Biomarkers are an active area of research and are increasingly used in environmental studies (DiGuilio 1989, Vouk et al. 1987, Shugart et al. 1989).

Measuring Effects

Indicators of toxic effects include changes in population, community, or ecosystem characteristics, overt toxic injury in individual organisms, and biomarkers of effects. The obvious way to measure effects is to measure the population and community characteristics that constitute or are indicative of the assessment endpoints. That is, if you want to know whether an effluent reduces the number of species in the receiving community, you might inventory the species, and if you were concerned about the yield of a game fish population you might measure the abundance and size distribution of the population. Methods for measuring the properties of populations, communities, and ecosystems are numerous and abundant in the literature of ecology, natural resources management, and environmental assessment. Methods deemed to be useful specifically for measuring effects of toxicants are described by Adams (1990), Kaputska et al. (1989), ASTM (1991), Hellawell (1986), APHA (1985), and EPA (1973).

Because assessment endpoints are often vaguely defined or undefined, it has not always been clear what should be measured in retrospective risk assessments. In addition, the endpoint species or characteristics may be difficult to measure. In such cases surrogates, indicator species, or indicator parameters are measured. These may be chosen because they are thought to be sensitive to pollution or indicative of a high-quality ecosystem, or because they are easily sampled and characterized. Much effort has been devoted to developing generic indicators of pollutant effects, but they have generally been found to be applicable to a narrow range of communities and pollutants (Sheehan et al.

1984, Dayton 1986). Therefore, an effort should be made to develop clear assessment endpoints and measure those endpoints or parameters that are indicative of the endpoints, rather than resorting to generic pollution indicators.

Because the variation in population and community characteristics is difficult to attribute to particular causes, it is desirable to measure toxic responses in individual organisms. The most obvious example is mortality; fish and bird kills are classic indicators of pollutant effects. Other examples are erratic behavior, lesions, gross deformities, and chlorosis of plant leaves. These responses are clearly negative and more likely to be attributable to pollution than are changes in populations and communities. The nature and pattern of visible injury of plant leaves and flowers may be characteristic of specific air pollutants (Jacobson and Hill 1970, Heck et al. 1979, Malhotra and Blauel 1980, Skelly et al. 1990), and behavioral abnormalities are characteristic of certain classes of pesticides (Hudson et al. 1984). However, all have potential causes other than pollution and must be interpreted in light of supporting evidence.

Suborganismal indicators of effects include biomarkers of effects (biochemical or physiological indicators of a toxicological response), and histopathologies. These responses are more diagnostic of particular classes of pollutants than organismal responses but more distantly related to endpoint effects. Potential biomarkers of effects include the frequency of DNA breaks (measured by the alkaline unwinding assay — Shugart 1988), delta-ALAD reduction (a mechanism of lead toxicity — Dieter and Finley 1979), and acetyl cholinesterase reduction (the mechanism of organophosphate and carbamate pesticide toxicity — Coopage et al. 1975). Histopathological techniques are well developed and the results are often interpretable in terms of the known effects of specific classes of pollutants (Meyers and Hendricks 1986). However, histopathology is more laborious and costly than analysis of biochemical biomarkers. Suborganismal indicators of effects were reviewed by DiGuilio (1989).

Biomarkers of effects are not nearly as well developed for plants as for animals. The proposed biomarkers of effects on plants are general indicators of stress or abnormal physiology that are not reliable markers of pollutant effects (NRC 1989). The best-studied example is the production of ethylene, which is a general indicator of plant stress, and may be a mediator of stress effects (Taylor et al. 1988), but is not characteristic of pollution.

The relative utility of these measurements of effects are predictable from the general characteristics of the levels of biological organization addressed (Chapter 2). Population- and ecosystem-level measures are relevant to assessment endpoints but, because of compensatory and adaptive mechanisms, are often resistant to effects and are not diagnostic of pollutants. Suborganism-level measures are potentially much more diagnostic and sensitive to pollutants but the relevance of a biochemical or histological response to population and community level assessment endpoints is poorly defined. Organism-level measures are intermediate in relevance, sensitivity, and diagnostic utility.

Contaminated Media Testing

Tests of contaminated ambient media (i.e., water, sediment, soil, and air) can be performed in situ or in the laboratory. In situ toxicity tests involve planting, caging, or otherwise placing organisms in ambient media that are contaminated to various degrees. Plants are easily used for this purpose. Potted plants can be placed in areas subject to air pollution and seeds or seedlings can be planted in polluted soils (Manning and Feder 1980, Heck et al. 1979). Their condition can be compared to plants placed in clean reference sites. Where air pollution is widespread, plants receiving filtered or unfiltered ambient air or clean or acidified rain can be compared (Heagle et al. 1979, Miller et al. 1977, Laurence et al. 1982) (Figure 10.3). In addition, air pollutants can be added to ambient air to create an exposure series using chambers or arrays of tubes (Lee and Lewis 1978, Miller et al. 1977, Ormrod et al. 1988). Fish and invertebrates can be placed in cages to test the toxicity of waters (Newbry and Lee 1984, Schofield and Driscoll 1987, Munawar et al. 1989) or sediments (Sasson-Brickson and Burton 1991), and terrestrial animals can be caged on waste sites or in fields receiving pesticides (Fink 1979). Standard substrates for aquatic invertebrates can be placed at polluted and reference sites or can be colonized at clean sites and then placed in polluted and reference sites (Cairns

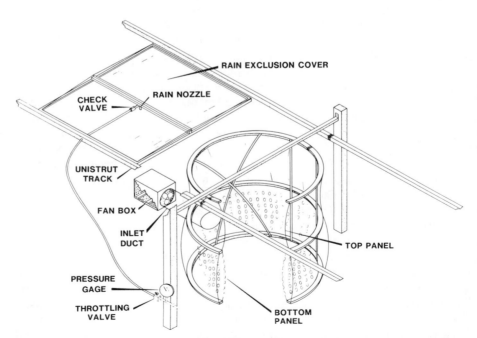

Figure 10.3. A device for controlling the exposure of plants to air pollutants by (a) either filtering the air in the fan box or adding gaseous pollutants and (b) covering the chamber during natural rain and substituting formulated rain.

1982b, Pontasch et al. 1989b). Colonial invertebrates can be easily introduced at terrestrial sites and individuals can be sampled from the colonies after they have been exposed during foraging. Honey bees and harvester ants have been used for this purpose (Bromenshenk et al. 1985, Gano et al. 1985).

Laboratory tests of media can be conducted in four ways, depending on the treatment of the media:

1. If a natural gradient of contamination can be identified, samples of media can be collected from a series of sites and used as treatment levels in the test. For example, if a point source effluent to a stream is being assessed, water or sediment can be collected at a series of sites upstream and downstream of the source.
2. If there is no natural gradient because the contaminant is either present or absent (e.g., where a liquid has been spilled on soil) or if the contamination is spotty (as at many hazardous waste sites) samples of the contaminated media and clean media can be tested. If the contaminated medium is toxic in that initial test, a test to establish exposure-response relationships can be conducted by mixing the contaminated medium with samples of clean medium in a dilution series (Greene et al. 1988). A mixture may not exactly simulate a less contaminated medium, but it will serve to demonstrate that effects are a function of exposure and can provide a better estimate of the degree of toxicity than a test of one contamination level.
3. If an effluent or other pollution source can be sampled, the samples can be taken to the laboratory and mixed with the receiving medium to create a dilution series of treatment levels. This approach has become a popular means of assessing the toxicity of liquid effluents (Bergman et al. 1986) and is used in the U.S. to regulate surface discharges under the National Pollutant Discharge Elimination System (NPDES) permit program (EPA 1985b).
4. If the appropriate contaminated medium is not available to be sampled, it may be simulated. The most common example is the testing of contaminated water created as a leachate or elutriate of contaminated soil (Porcella 1983, Greene et al. 1988, Parkhurst et al. 1989) or as pore water extracts or elutriates of sediments (Shuba et al. 1977, Sasson-Brickson and Burton 1991). Application of this approach to soil is most appropriate when the assessor is concerned that leachate from contaminated soil will reach aquatic ecosystems but cannot identify where the leachate is entering surface water, or cannot obtain samples because the contamination is episodic (i.e., occurs following storm events or snow melt). It has been used less appropriately to create a liquid medium so that standard aquatic tests can be used to test the toxicity of soil. Testing of water derived from sediments is controversial because the ability of various pore water extracts and sediment elutriates to represent the toxicity of either in place or resuspended sediments is not well established.

Contaminated media may be tested using the same toxicity tests that have been developed for testing chemicals (Part III). However, contaminated media create additional procedural problems such as changes in the composition of the contaminants and media during collection, transportation, and storage, and the effects on the contaminants of the vigorous aeration required to test waters and aqueous effluents with high BOD. These problems plus the tendency of surface water and effluent characteristics to vary widely over time, and concerns about costs of testing have led to the development of short-term toxicity tests for effluents and ambient waters (Peltier and Weber 1985, Weber et al. 1988, Weber et al. 1989) and soil (Greene et al. 1988). However, if the goal of the assessment is to determine the chronic toxic effects of a contaminated medium, then variability is realistic and true chronic tests should be conducted. For example, Poels et al. (1980) exposed trout to Rhine River water for 18 months.

Estimated Exposure and Effects

Retrospective assessments may not be based on measurements if it is not feasible to measure the pollution and the target organisms or if there simply are no resources available to perform measurements. Measurements may be infeasible because the events are physically remote, the contaminants are quickly degraded or otherwise lost from the affected system, the organisms are difficult to sample, or the association between the pollutant and exposed organisms is transient and cannot be defined after the fact. The best example is a pelagic marine spill of a degradable organic chemical; it is difficult to reach the site, sample the contaminated water, or identify and sample fish or plankton that were exposed to the spill. In such cases, it is often necessary to estimate the magnitude of the source, model the fate of the chemical to assess exposure, assess the exposure-response relationship for the assessment endpoints, and then estimate the probabilities of the endpoint effects. In other words, the assessor treats the retrospective assessment as if it were a predictive assessment. Such inferences are termed retrodictive. This process has been formalized for marine chemical spills in the Type A Natural Resource Damage Assessment Procedure (DOI 1987a, Economic Analysis, Inc. 1987).

RISK CHARACTERIZATION

The integration of information concerning sources, exposure, and effects to estimate risks can be performed in four basic ways: (1) If effects are measured, an epidemiological approach should be used. (2) If effects are not measured, but media can be tested, a toxicological approach should be used. (3) If sources, exposure, and effects are not measured, a retrodictive approach is needed. (4) If major uncertainties inhibit decisionmaking, the existing pollution can be experimentally regulated to determine the effects of treatment on exposure and effects.

Ecological Epidemiology

Epidemiology is the study of the causes, frequency, and distribution of human morbidity and mortality. Because retrospective ecological risk assessments that attempt to determine the causes, magnitude, and extent of effects

Table 10.3. Epidemiological Study Designs

Ecological Designs—those studies that attempt to relate the level or frequency of an effect in a human population to occurrence of an environmental factor. An example would be a study comparing cancer incidence in counties that contain petrochemical plants to those that do not. An analogous ecological use of an ecological study design is the attempt to relate the frequency of declining trees in forest stands to certain air pollution parameters (Innes 1988, Schulze 1989). This is the least reliable type of design because the investigator has no control of the study but simply compares what is known about the state of a community with its current exposure to one or more suspected agents. It is applicable to effects-driven assessments.

Proportional Designs—those studies that compare the proportional distributions of total cases among populations with different levels of exposure (e.g., proportion of deaths attributable to a particular disease). An ecological use would be determining the proportion of injured or recently dead trees on plots at varying distances from a pollution source. The ability to choose the levels of exposure gives the investigator more control than in the ecological design (e.g., to distribute the sampled sites evenly along a gradient of exposure).

Cohort Designs—those studies in which the level of exposure of the subjects is known and their health status is monitored during a follow-up period. An ecological use would be monitoring forest stands that are maximally exposed to a new pollution source along with stands that are minimally exposed, to determine the death rate or rate of decline of the trees. The advantages of this design are that the onset of symptoms can be monitored so that diagnostic symptoms can be noted, and the time and order of responses can be monitored so that indirect or contributory causes can be elucidated. The major limitation is that the study must begin at the initiation of the pollution, otherwise the community will have adapted to the pollution and little or no further change will be noted. Long-lived and slowly-responding species such as trees are potential exceptions.

Cross-sectional Designs—those studies in which individuals are sampled from a target population and the potentially causative factors and disease state are determined simultaneously. An example would be selecting a sample of people and determining whether they were intravenous drug users, prostitutes, etc., while taking a blood sample to analyze for HIV antibodies. An ecological use would be randomly sampling forest stands and determining their state of growth or decline while determining pollution loading. These designs are useful if an effect has been observed but its extent is unknown or its cause is undetermined.

Case-Control Designs—those studies in which a group of cases (individuals with an identified disease) is compared to one or more groups of controls (individuals without the disease) to determine how they differ. An ecological use would be a comparison of the environmental conditions of sugar maples that are declining to those of equal-aged trees that are still vigorous.

measured in the field are analogous to epidemiology, they may be termed ecological epidemiology. The statistical methods of epidemiology are not useful in general for ecological assessments because they deal with frequencies of dichotomous responses (e.g., frequency of individuals with disease *a* or dying of cause *b*), whereas ecological effects measured in the field are usually count or continuous variables (e.g., number of organisms or rate of production). However, the standard epidemiological designs are applicable. In some cases, ecological risk assessors have consciously used epidemiological designs (e.g., Medeiros et al. 1984). A review of epidemiological designs serves to clarify the alternate ways that data can be collected and the constraints that data collection places on assessments (Table 10.3).

All assessments in ecological epidemiology address observed and measurable reductions in an environmental value that are believed to potentially be caused by a human action. For such studies, risk characterization is not simply

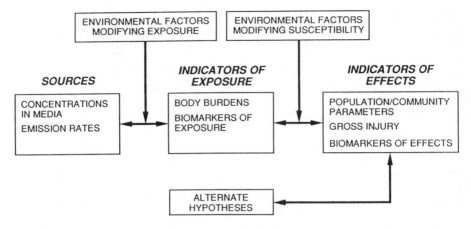

Figure 10.4. A diagrammatic representation of the relationship among the components of the risk characterization stage of retrospective assessments based on ecological epidemiology.

a matter of relating exposure to effects (Chapter 3), but rather must consider the relationship of putative sources to indicators of exposure and effects, the role of environmental factors, and the occurrence of alternate causal factors (Figure 10.4). The risk characterization attempts to answer one or more of the following four questions about the putative effect: (1) what is the likelihood that the apparent effect is real rather than simply a result of natural variation in the environment, (2) what is its likely cause, (3) what is its magnitude, and (4) what are its ultimate consequences?

Is There a Real Effect?

In human health epidemiology, the reality of an effect is established by comparing the frequency of a disease or cause of death or injury in a subject population with the expected frequency. The expected frequency is based on health records for other human populations and is corrected for differences in age distribution, sex, and other potentially confounding variables. The deaths, cancers, and other pathologies are clearly undesirable, so there is assumed to be a real effect if it is demonstrated that the difference in frequency is likely to be nonrandom. Although there are analogous situations in ecological epidemiology such as studies of tree mortality, the endpoints are often changes in the relative abundance of organisms, in decomposition rates, or similar measures that are not inherently pathological. As a result, it is not possible to say with certainty whether a change or difference between sites represents a pathology or simply a different normal state of the community type. Even in the case of tree mortality surveys, the assessor is unlikely to know the cause of death of a tree or whether the observed death rate of trees in a stand is normal. In other words, in ecological epidemiology it is often not clear that there is an effect that needs a causal explanation. The apparent anthropogenic effect may be the result of normal spatial and temporal variation. One can only say that, given the natural variability in the endpoint parameter across space and time, the likelihood that the observed mean and variance of the endpoint parameter at the exposed site belongs to the same distribution as the unexposed sites is some value p.

If, as is often the case in ecological epidemiology, an exposed community is simply compared to an unexposed or less exposed reference community, it is not possible to say that either or both are pathological or normal. However, if the attributes of a number of reference communities are measured, it is possible to identify normality for a particular community type in a particular region. An example is the use of regional reference sites with the index of biotic integrity which is an aggregate of fish community measures that distinguishes disturbed from undisturbed midwestern streams (Karr et al. 1986). Streams are scored with respect to twelve metrics based on their similarity to nominally undisturbed streams of similar size in the same region. For example, species richness is scored relative to the maximum species richness of streams of the same order in the region. These scores are then combined into a total score which is converted to one of six integrity classes (excellent, good, etc.).

A more informative approach to establishing departure from normality is to eschew indices and use multivariate statistics to determine whether the exposed community differs from unexposed communities (Green 1980, 1987). Perhaps the most useful general approach is to establish the n-dimensional state-space of the reference communities, where n is the number of measured variables believed to be indicative to effects of the pollutant of interest and a state-space is a mathematical volume occupied by a collection of entities, each defined by the values assumed by the n variables. The probability that the measured attributes of a particular community fall outside the reference state-space indicates the risk that it is an abnormal community (Johnson et al. 1988). Other approaches include similarity measures, principal components analysis, and discriminant analysis (Boyle et al. 1984, Gauch 1982, Smith et al. 1989).

The advantages of multivariate statistical models over indices are (1) information about which attributes account for the difference is retained, (2) the choice of attributes is not constrained as in indices, so the attributes that are most likely to respond to the pollutant of interest can be used and the assessment can still be performed if some attributes have not been measured, cannot be measured, or are irrelevant to the communities being assessed, and (3) they lend themselves to standard statistical tests and probability estimates. The disadvantages are that they require relatively large sample sizes, and the statistical assumptions may be violated in many cases.

For regional baselines, it is necessary to carefully define the reference communities to ensure comparability. The EPA has been developing standard assessment regions for the United States and identifying reference sites within some of those regions (Hughes et al. 1986).

Another potential means of determining the normal state of a community is habitat evaluation. If a habitat is within the range of a species and is suitable for the species, absence of the species can be taken as evidence of an effect. The U. S. Fish and Wildlife Service has devoted considerable effort to the development of procedures for habitat evaluation and of habitat profiles for individual species (FWS 1980, Bovee and Zuboy 1988). In particular, instream flow models have been developed for estimating presence or abundance of fish following flow modification (Bovee 1982, Gore and Nestler 1988, Orth 1987, Sale 1985), and have been recommended for evaluation of habitats receiving aqueous effluents (Dickson and Rogers 1986). The River Invertebrate Prediction and Classification System (RIVPACS) provides estimates of the probability of occurrence of invertebrate species in British streams, given 11 physical and chemical parameters (Wright et al. 1989). Some of these habitat models can estimate the abundance of a species in a habitat. However, these estimates are generally not sufficiently accurate to permit the assumption that an observed abundance lower than predicted by the habitat model (short of local extirpation) indicates a real effect.

Finally, if the exposed site has been studied for some considerable period of time prior to the apparent effect, it may be possible to establish the abnormality of the apparent effect as a component of the time series. That is, one can analyze the time series to remove trends and cycles and then determine the probability that the event in question significantly departs from the remaining random variance.

The alternative to defining the normal state of a community is to choose as endpoints states of the community that are clearly abnormal such as fish kills, bird kills, fishless streams, etc. They may not be caused by pollution or any other human action, but it is likely that they have some cause rather than simply resulting from random variation in mortality rates or the distribution of fish populations. In practice, use of this approach to proving that an effect has occurred is an environmental application of Descartes' method of doubt. That is, if one skeptically examines the evidence and still cannot doubt that an effect has occurred, then one must accept that it has. The reliability of this method depends on the expertise and experience of the individual involved.

What Is the Likely Cause?

Probably the most common flaw in both human-health and ecological epidemiology is what epidemiologists refer to as the "ecological fallacy," the assumption that measured differences between the environments inhabited by subject populations are the cause of differences in the frequency of disease (Morgenstern 1982). The fallacy arises from the possibility that the differences in health are caused by some unconsidered factor, and that the association of the putative cause with the disease is either coincidental or results from an association between the putative cause and the true cause. In other words, correlation does not prove causation. A classic example is the once common belief that breathing swamp air caused malaria.

In ecological assessments, the "ecological fallacy" occurs when populations and communities found in association with pollution are compared with populations and communities at less polluted sites, and any biological differences are attributed to the pollution. However, the differences could be caused by anthropogenic factors associated with the pollution, to natural gradients, or simply to random variation in population and community characteristics. For example, biological differences between communities upstream and downstream of an effluent discharge may be due to toxic chemicals in the effluent, to differences in stream flow resulting from augmentation by the effluent, to physical disturbance of the substrate during construction, to coating of the substrate by particles in the effluent, to toxic effects not attributable to the effluent (e.g., spills, or storm drainage), to the natural gradient in stream communities, to differences in stream morphology, or to random differences in the biotic communities of different stream reaches.

The "ecological fallacy" is compounded by the inappropriate use of hypothesis-testing statistics in ecological epidemiology. ANOVA, t tests, etc. were designed for experimental studies in which treatments are replicated and subjects are randomly assigned to treatments. In such studies, rejection of the null hypothesis that the populations of subjects are the same in all treatments with respect to some measured attribute, indicates that the populations are different (with probability $1-\alpha$), and that difference is attributable to the treatment. However, in ecological epidemiology there is no random assignment of populations or communities to treatments, and treatments are almost never replicated so we cannot use statistics to test the hypothesis that populations or communities treated with a pollutant are different from those that are not.

The nonrandom assignment of treatments is self-evident. The investigator cannot assign some stream reaches to receive an effluent and others to be untreated. Nonrandom assignment means that differences cannot be attributed to treatments because they may be an artifact of the assignment. For example, untreated sites are nearly always upstream of treated sites and any effects of constructing and operating a facility other than the effluents, such as siltation, will affect the effluent-treated downstream site more than the untreated upstream site.

The lack of replication is less obvious. For example, investigators studying the effects of effluents on streams commonly take multiple samples of benthic invertebrates above and below the outfall, treat the samples at each site as replicates, and use them in hypothesis tests to determine whether attributes of the communities are significantly affected by the effluent. However, the downstream samples are drawn from a single community affected by a single effluent and the samples are pseudoreplicates from that one treatment (Eberhardt 1976, Hurlbert 1984) (Box 10.1). One can draw conclusions about whether the populations of samples from the upstream and downstream sites are different, but not about the treatment. Clearly, the object of the study is the treatment, not some academic comparison of two stream segments.

Box 10.1. A Pseudocure for Pseudoreplication

Stewart-Oaten et al. (1986) responded to Hurlbert's (1984) criticism of hypothesis testing with pseudoreplication and nonrandom assignment of treatments by proposing that exposed sites and "control" sites be paired, and that their properties be measured before and after the treatment in a before-after-control-impact (BACI) design. Smith et al. (1990) have shown that not only does the BACI design not eliminate the violation of assumptions, but it can fail to produce reasonable results in real applications.

A study of the effects of aqueous effluents from a new power plant employed the BACI design, and found statistically significant change in the abundance of various fish species at the downstream site following plant start-up (Smith et al. 1990). However, if the last few years before start-up of the plant were used as the "after" case, there were nearly as many statistically significant changes from the period before construction as there were following start-up. The changes were attributed to confounding variables and to the influence of a large number of zero-abundance observations for some species. Clearly, the hypothesis tests were not actually addressing effluent effects.

Trends were detected in the before data for some species and this constitutes another violation of the assumptions underlying BACI (Smith et al. 1990). Trends are nearly inevitable in natural communities, but few studies start long enough before the treatment to detect them. Therefore, this assumption would be routinely, but unknowingly, violated in practice.

The dangers of hypothesis testing with pseudoreplicates is illustrated by a study of the survival of lake trout eggs in two acidified lakes and a circumneutral lake (Gunn 1989). Individual in situ incubators were treated as replicates and statistically significant differences were found among the lakes ($p < 0.001$). However, the mortality was highest in the circumneutral lake. The author concluded that the differences were due to extraneous factors. Although the hypothesis being tested was not formally stated, the topic of the paper was effects of acidified water, so one must suspect that the author would have attributed the mortality to acidified water if the results had been reversed. The use of ANOVA to test the hypothesis that acidification reduces survival of lake trout eggs was inappropriate without true replication.

Another potential flaw in the use of observational studies to establish causation is "data dredging" (Feinstein 1988). If a large number of potential associations between causes and effects are statistically examined, some will appear to be "significant" by chance alone. In comparisons of communities upstream and downstream of an effluent, investigators commonly determine the number of individuals of various taxa and age classes, the diversity and perhaps evenness of all taxa and of each taxon, perhaps the numbers of individuals and species of functional guilds, etc. At the same time, other investigators may be accumulating numerous response parameters for fish, algae, and ecosystem processes, so it is not unusual for a large monitoring study to generate more than 50 measurement endpoints. In such studies, apparent differences are likely to be found even if the sampled communities are identical (Type I error). Similarly, in studies to determine the causes of observed effects, measurement of a large number of potential causes can increase the chance that one will appear to be significantly associated with an effect by chance alone. Type I error for the study as a whole can be reduced by applying Bonferroni's adjustment, but that greatly increases the probability of not detecting a real difference in an individual response parameter or a real causal association (Type II error).

The solutions to these problems are clearer definition of the problem and rigorous standards of proof. As we have emphasized repeatedly in this volume, good risk assessments require clearly defined assessment endpoints and explicit linkage of measurement endpoints to assessment endpoints. If the assessor has defined in advance (1) what ecological values are to be protected, (2) what ecological measurements best represent those values, and (3) what environmental characteristics are potentially causative, based on knowledge of mechanisms of action, routes of exposure, and ecological linkages, then data dredging can be avoided and the costs of monitoring can be reduced.

In addition, clear problem definition encourages the use of more appropriate statistical methods. The problem of demonstrating that an observed effect is associated with a cause, is logically distinct from testing for the occurrence of an experimental effect. The standard proof of toxic effect is demonstration of an increase in effect with an increase in exposure. Therefore, rather than testing the null hypothesis that exposed communities are the same as unexposed communities, the assessor should hypothesize and statistically test a model describing the relationship between the measurement endpoints and the hypothesized causes of variation in the measurement endpoints (i.e., pollutant exposure and environmental variables that control response) (Stevens et al. 1989).

These descriptive statistical models may range from simple correlations of a measurement endpoint with distance from a source to complex multivariate models. Hakanson (1984) presents an approach in which effects that are measurable in the field are modeled as a function of three classes of variables: pollutant load, the inherent toxicity of the pollutant (which may be treated as a constant), and characteristics of the receiving environment. He showed that methyl mercury concentration in predatory fish in several lakes was a function of mercury load, lake water pH, and lake productivity. Similarly, several investigators have attempted to assess lake acidification by empirically regressing lake pH on data for atmospheric, watershed, and lake characteristics (e.g., Hunsaker et al. 1986, Schnoor et al. 1986). Jones et al. (1979) surveyed visible injury to plants in the vicinity of Tennessee Valley Authority power plants, and concluded that the threshold for visible injury was 0.32 ppm (13.3 μmol/m^3) SO$_2$ for a one-hour averaging period. McLaughlin used a similar survey to establish the concentration response models shown in Figure 7.5. A final example is regression of pelican eggshell thickness against DDE residues (Blus et al. 1972, Blus 1982). A good discussion of the development of such models for human-health effects of pollutants is presented in Lave and Seskin (1979). Kranz (1990) provides discussions of statistical models in plant epidemiology. Although that book is concerned with pathogens rather than pollution, its treatment of population processes and spatial issues is relevant.

Clear problem definition and proper statistical analysis are prerequisites but they do not eliminate the "ecological fallacy." That is, a model associating a clearly defined endpoint with one or more environmental factors is important evidence of causation, but by itself it does not prove causation. What is needed is a clear rigorous line of reasoning employing sufficient evidence to establish causation. As some plant toxicologists have pointed out (Adams 1963, Woodman and Cowling 1987), Koch's postulates for proving that a particular pathogen causes a disease are also applicable to environmental toxicology. Analogs of Koch's postulates for ecological epidemiology of pollution effects are:

1. The injury, dysfunction, or other putative effects of the toxicant must be regularly associated with exposure to the toxicant and any contributory causal factors.
2. Indicators of exposure to the toxicant must be found in the affected organisms.
3. The toxic effects must be seen when normal organisms or communities are exposed to the toxicant under controlled conditions, and any contributory factors should contribute in the same way during the controlled exposures.
4. The same indicators of exposure and effects must be identified in the controlled exposures as in the field.*

Rule 1 amounts to a requirement of well-designed field studies. As far as possible they should have true replication of treatments (e.g., multiple lakes receiving mercury or multiple forests near coal-fired power plants) to provide generality and reduce the chance of random correlations. In any case, they should use appropriate statistical designs in sampling and analysis of data as discussed above. To satisfy the requirement that exposure to the toxicant be regularly associated with some apparent effect, the effect should be well-defined. This is relatively easy in the case of human disease where a particular consistent syndrome can be identified and diagnosed, but biotic communities differ in their species composition and structure so that symptoms vary due to variation in the subjects. However, it should be possible in most cases to identify characteristically sensitive species, taxa, or processes such as

---

*There is no clear simple formulation of Koch's postulates in his writings. Some formulations do not include the fourth postulate (e.g., Woodman and Cowling 1987), but it can be an important component of an argument for causation.

the sensitivity of blue-green algae and nitrogen fixation to copper. The ability to identify characteristic responses is increased by measuring biomarkers of effects and other suborganismal effects. The diagnostic syndromes being developed to identify mode of action of chemicals (Drummond et al. 1986) might be adapted for field diagnosis. Suborganismal responses are more likely to be characteristic than the population and community responses that are used as indicators of the assessment endpoints. This is because changes in population or community parameters have many possible causes, most of which are natural, and few of which change the parameters in distinctive ways.

Given a well-defined effect and a hypothesized cause, regular association can be defined as consisting of Hume's (1748) criteria for causation: (1) spatial and temporal contiguity, (2) temporal succession, and (3) consistent conjunction. This means that the cause must be associated with the effect in space and time, the effect must follow rather than preceding the cause, and the cause and effect must always occur together. As Holland (1986) has pointed out, consistent conjunction may be difficult to demonstrate because measurement error or variation in the way that individual units respond to exposure may obscure a true conjunction. This difficulty can be reduced by relaxing the requirement (i.e., a high frequency of conjunction rather than consistent conjunction), and by restricting the population of interest to those units that are most likely to respond similarly (i.e., susceptible units). Note that Hume's criteria do not actually prove causation in our sense; in Hume's empirical philosophy causation is no more than regular association. However, they constitute a complete set of criteria for satisfying Koch's first postulate. In some cases, particularly when the time of onset of the effects is unknown, regular association must be accepted without satisfying all of Hume's criteria.

If a characteristic effect or set of effects analogous to a disease syndrome can be identified, satisfaction of Hume's criteria is conceptually straightforward. For example, the thinning of peregrine falcon egg shells occurred in regions where DDT had been used, followed the introduction of DDT in those regions, and occurred in all regions where peregrines were significantly exposed to DDT and its metabolites (Hickey and Anderson 1968, Cade and Fyfe 1970). Inference is relatively simple in such cases, because the fact that the assessment begins with a characteristic effect implies that the units are responding consistently, so Holland's (1986) concerns about consistent conjunction are alleviated.

If, rather than identifying a characteristic effect with which a toxicant can be associated, one begins with a generic effect such as death or decline in a forest condition index that can occur by natural causes in the absence of the pollutant, consistent conjunction must be established in the opposite direction. That is, every time a particular exposure level occurs, the effect must also occur. This association is more difficult to establish because it depends on the presence of susceptible organisms (in the other case, susceptibility is a given because the characteristic effect has occurred). That requires the presence of susceptible species, genotypes, and life stages. In addition, if the exposure falls in the interval between exposures that are always safe and exposures that will always injure all organisms of the susceptible species and life stages, consistent conjunction also requires occurrence of population and habitat conditions that induce susceptibility (Box 10.2). Examples include drought, crowding, low or high nutrient levels, high background levels of toxic chemicals, low dissolved oxygen, and infections by pathogens or parasites (Chapter 7). In addition, if the effect is indirectly caused by the pollutant, the environmental components responsible for the indirect mechanism must be present and susceptible before the endpoint can be considered susceptible. A good example of environmental control of susceptibility is the ecological effects of acid deposition. Aquatic communities receiving approximately the same loading of acids or having the same aqueous pH and containing the same potentially susceptible species do not all experience effects. The occurrence of pH effects in fish depends on aluminum and calcium concentrations in the water; and surface water pH, Al, and Ca depend in turn on characteristics of the watershed, as well as on acid deposition rates.

If a variety of factors potentially contribute to the occurrence of effects, a synoptic approach may be employed (Wallace 1978, Schulze 1989). In synoptic studies, all factors that are believed to contribute to the probability of occurrence of effects (i.e., component causes) are measured at each study site and multivariate statistics are used to determine which factors and interaction terms contribute to estimating the level of effect. For example, a decline in tree growth may be a function of air pollutants, climate, nutrient availability, and crowding. Statistical models can help to sort out complex causation and establish the regularity of association, as well as describing the relationship of exposure to effects, as discussed above. Synoptic studies must not be focused on pollutants as causes. It is necessary to consider that the susceptibility factors may themselves be sufficient to cause the observed damage and the pollutant may simply be correlated with those causes. An additional problem is the possibility that a parameter that was not even measured may be the cause, and the cause identified by the synoptic study is simply correlated with the true but unmeasured cause.

A major difficulty in synoptic studies is the need for very large data sets because of the multiple hypothesized causal variables and interaction terms. Hence, synoptic studies typically require long time series of measurements, a large number of spatially replicated measurements, or both. These are most likely to be available for resource species such as fish, game mammals, and trees. For example, time series analysis of fish landings was used to demonstrate that production of various fish populations in estuaries of the northeastern United States was related to indicators of pollution loading as well as hydrologic factors (Rose et al. 1986, Summers and Rose 1987a, 1987b).

Box 10.2. Multiple Causal Factors

The discussion of causation in this section is framed in terms of individual causes and variance in susceptibility to those causes, as is common in epidemiology and toxicology, because the assessor is usually interested in managing only one type of cause (e.g., a pollutant, pathogen, or drug). However, it is more accurate to speak in terms of multiple causal factors. Diseases or damaged states of ecosystems require multiple events, conditions, or characteristics (component causes) which together constitute a sufficient cause for the observed condition. If these component causes are not accounted for in the analysis, they are known as "confounding factors," and if they are assumed to simply modify the sensitivity of the biota to a particular cause that is of interest, they are termed "susceptibility factors," but an unbiased analysis of causation will treat all component causes as logically equivalent.

For example, sufficient causes for a local die-off of mussels might be presence of a sensitive mussel species, high temperatures, low dissolved oxygen, and high dissolved metals from a plating facility. A comparison of the site of the local extinction with similar sites where mussels survived might show chromium to be the cause, based on survival of mussels in the presence of similar temperatures and DO but no chromium. However, similar levels of temperature, DO, and metals might induce the same effect as a result of naturally high levels of metals and temperature along with high BOD effluents. In that case, a field survey would identify low DO as the cause. In both cases, all of the component causes are necessary, but the "strongest" cause, as indicated by the epidemiological analysis, is simply the rarest one, because it will be most strongly correlated with effects. Fortunately, pollutants are often less frequent and more localized than natural causes.

Difficult epidemiological cases arise when anthropogenic causes and natural causes have approximately the same spatial scale and frequency. For example, sufficient causes of death of local populations (stands) of trees might be any of the following: (a) species of tree, age, drought, and fungal pathogens; (b) species, age, drought, and ozone; or (c) species, age, fungal pathogens, and ozone. Tree species, age of stands, drought, pathogen occurrence, and ozone levels tend to have regional distributions, so no single cause may stand out as highly positively correlated with effects, and no plausible cause may be eliminated as negatively correlated with effects.

In the worst case, component causes that are extraneous to the assessment may be stronger than the causes of interest. In the NCLAN assessment of ozone effects on crops, epidemiological approaches were eliminated from the study as too weak (Heck et al. 1988). This conclusion was reasonable because of the strong causal effect on crop yield of weather which varies at approximately the same scale as ozone concentrations and of agronomic factors such as soil type, fertilizer, pesticides, and tillage which have high spatial variability relative to ozone exposure. This example illustrates another aspect of estimating strength of causation: imprecisely specified contributing causes will appear to be weak relative to precisely measured causes. In this case, the weather and agronomic factors can be well-specified for individual fields relative to ozone concentrations which are measured at few points in the rural U.S.

In general, the number of component causes necessary for a pollutant to cause an effect is determined by the magnitude of exposure to the pollutant. At a high enough concentration any pollutant can kill all organisms, even under optimal conditions (i.e., no other component causes are acting). As pollutant exposures decline, other causes must contribute and the magnitude of those causes must increase (the exposed species must be more sensitive, the physical conditions more stressful, the pathogens more virulent, etc.).

Rule 2 is the equivalent of Koch's requirement that the pathogen be isolated from the diseased subjects. The obvious analog is finding elevated concentrations of the toxicant or characteristic metabolites in organisms in the affected community. However, biomarkers that are diagnostic of exposure to a particular toxicant or class of toxicants serve the same purpose. In the case of the decline of peregrine falcons, DDT metabolites were shown to be elevated in the parents and in the eggs (Hickey and Anderson 1968, Cade et al. 1971).

In some cases, body burdens require careful interpretation. Walker (1990) presents an example involving a large kill of marine birds (mostly guillemots). The whole-body concentrations of PCBs in the dead birds was approximately equal to that of apparently healthy birds that had been shot for analysis. Therefore the official report concluded that PCBs were unlikely to have been a major factor in the deaths. However, Walker points out that (1) the dead birds, unlike the shot birds, had virtually no storage fat and therefore comparison of whole body concentrations are uninformative, (2) the PCB concentrations in liver were much higher in the dead birds than the shot birds and were comparable to those in cormorants that had died in toxicity tests, and (3) metabolism of fat by anorexic birds would mobilize stored PCBs, leading to both toxic effects and excretion.

Rule 3 is the equivalent of Koch's requirement that inoculation of a susceptible host with the pathogen should induce the disease. The analog is toxicity testing, and the requirement of a susceptible host implies not only appropriate species and life stages, but also equivalent conditions to those that are thought to result in effects in the field. In the peregrine falcon case, because standard avian test species are more resistant to DDT than the Falconiformes, the attribution of effects to DDT was not established until tests performed on falcons (kestrels) demonstrated effects at realistic exposure levels (Wiemeyer and Porter 1970, Lincer 1975). This requirement may be met by existing toxicity data or may require new tests including tests of contaminated media. It may be met by studies that control exposure by reducing the exposure of organisms in the polluted environment (e.g., exposure of plants to charcoal-filtered air), as well as by conventional toxicity tests. Ideally, the magnitudes of test responses would correlate with the magnitudes of observed effects in the field (Figure 10.5).

Rule 4 is the equivalent of Koch's requirement that pathogens isolated from the experimentally infected host match those isolated from the original diseased population. The analogy is comparison of the relative sensitivity of species, overt symptomology, body burdens, biomarkers, and other measurable responses in the laboratory and field. In our example, the effects on kestrels were the same as on peregrines (eggshell thinning) and it occurred at similar concentrations of DDT metabolites (Lincer 1975).

Studies satisfying all four of these rules are the gold standard of retrospective ecological risk assessment. Examples, besides DDT effects on raptorial and piscivorous birds, include the damage to Eastern white pine by ozone (Woodman and Cowling 1987), reproductive effects of selenium on birds at the Kesterson National Wildlife Refuge, California (Ohlendorf et al. 1986a, 1986b, Presser and Ohlendorf 1987, Heinz et al. 1987), mass mortality of water birds due to aqueous emissions of alkyl lead (Bull et al. 1983, Osborn et al. 1983, Wilson and Jones 1986), avian mortality due to ingestion of lead shot

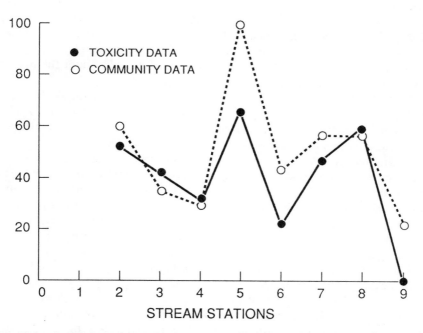

Figure 10.5. A comparison of percent toxicity in a test and percent reduction in the number of taxa in the stream (Norberg-King and Mount 1986).

(Eisler 1988), and two examples discussed below, avian mortality from granular carbofuran, and the depletion of local fish populations by acid deposition.

The requirements of the postulates may seem severe, but they are similar to some existing guidance. Requirements very similar to this version of Koch's postulates are imposed on measured biological responses that are proposed to be indicative of injury from oil or other hazardous materials in natural resource damage assessments (43 CFR 11.62). Injury is indicated if "the biological response is often the result of exposure," if "in free-ranging organisms" there is "correlation of the degree of the biological response to the observed exposure concentration," and if "exposure . . . is known to cause this biological response in controlled experiments." Similarly, the U.S. Fish and Wildlife Service's manual for investigating fish kills (Meyer and Barklay 1990) effectively requires that these postulates be considered when establishing toxic chemicals as the cause of a kill incident. That is, the chemical should be identified in the water, the spatial distribution of the chemical should be associated with the distribution of the kill, the chemical or biomarkers should be analyzed in the fish, the concentrations in the water should be associated with concentrations that kill fish in laboratory tests, and clinical signs of toxicosis should be identified in the fish.

Although Koch's postulates were developed for identification of the causes of infectious diseases (i.e., for effects-driven assessments), they are also applicable to proving that a pollutant has effects (i.e., source-driven assessments) if the effects can be measured in the field. That is, if a toxicant is consistently associated with an effect in the field, if indications of exposure can be identified in the affected organisms, if the same effects are induced in controlled tests, and if the indicators of exposure and effects are the same in the tests as in the uncontrolled field exposure, there can be little doubt about the nature of the effect and its causation.

Assessments seldom meet these criteria. In most cases, the desirability of multiple lines of evidence is not recognized. For example, the EPA's scheme for licensing existing aqueous effluents treats biological surveys and toxicity testing as alternate approaches rather than as complements for establishing the cause and nature of effects (EPA 1984b, 1990). In other cases, it is not practical or even possible to satisfy all of these criteria. Limitations may be technical, such as the inability to perform toxicity tests on certain organisms or modes of exposure, but more often they are due to insufficient time or resources for a complete assessment. In such cases, the assessor should remember that the goal of assessments is not to establish scientific truth but to establish a sufficient body of evidence to allow a decision, or to establish which conclusion is supported by the preponderance of evidence. Therefore, the assessor should carefully examine the existing evidence, determine what additional evidence is obtainable, and consider what degree of proof is attainable with those data. It is helpful to consider risk characterization in ecological epidemiology in terms of establishing the relationships among sources, exposure, effects, and other environmental factors (Figure 10.4). The stronger and

Table 10.4. Hill's (1965) Factors for Evaluating the Likelihood of Causal Association in Epidemiological Assessments

Strength—A stronger response to a hypothesized cause is more likely to indicate true causation. This means either a severe effect or a large proportion of organisms responding in the exposed areas relative to reference areas, and a large increase in response per unit increase in exposure. In other words, a steep exposure-response curve situated low on the exposure scale.

Consistency—A more consistent association of an effect with a hypothesized cause is more likely to indicate true causation. This is a weak form of Hume's third criterion, consistent conjunction. Hill's discussion implies that the case for causation is stronger if the number of instances of consistency is greater, if the systems in which consistency is observed are diverse, and if the methods of measurement are diverse.

Specificity—The more specific the effect, the more likely it is to have a single consistent cause. This is equivalent to our suggestion that regular association is more readily established if a characteristic effect is identified. Also the more specific the cause, the easier it is to associate it with an effect. For example, it is easier to demonstrate that localized pollution caused an effect than that a regional pollutant caused an effect.

Temporality—A cause must always precede its effects. This is Hume's second criterion.

Biological gradient—The effect should increase with increasing exposure. This is the classic requirement of toxicology that effects must be shown to increase with dose.

Plausibility—Given what is known about the biology, physics, and chemistry of the hypothesized cause, the receiving environment, and the affected organisms, is it plausible that the effect resulted from the cause?

Coherence—Is the hypothesized relationship between the cause and effect consistent with all of the available evidence?

Experiment—Changes in effects following changes in the hypothesized cause are strong evidence of causation. Because Hill was concerned with effects on humans, he emphasized "natural experiments" rather than the controlled exposures required by Koch's third postulate. An example would be observations of recovery of a receiving community following abatement of an effluent.

Analogy—Is the hypothesized relationship between cause and effect similar to any well-established cases?

---

more extensive the web of evidence concerning the relationships represented by the arrows in that figure, the stronger the circumstantial case for causation.

Hill's (1965) factors for interpreting evidence of environmentally induced diseases (Table 10.4) are an aid to evaluating the strength of evidence of causation in epidemiological studies. Although some of these factors correspond to considerations that have been previously discussed, others such as "plausibility" and "strength" are new. They are not necessary to the proof of causation (e.g., some true causes are implausible a priori), but they increase confidence in causation.

Paleoecology and historical ecology provide alternative evidence of causation that is seldom available to human health epidemiologists. If a change in a population or community can be shown to be commensurate with a cycle or trend that precedes the putative toxicological cause, then that cause is excluded. However, if the change is unprecedented, and begins at the time that the pollution began, then the attribution of the change to the pollutant is supported. Logically, this is another way of establishing the regularity of association, but in practice the techniques are quite different from those of conventional toxicological field studies.

Lake sediments are particularly useful for paleoecology because they preserve a stratigraphy of (1) aquatic subfossils including the shells of diatoms and other algae, the mandibles, head capsules and other resistant parts of arthropod exoskeletons, and seeds and tissues of macrophytes, (2) terrestrial subfossils including pollen, seeds, and charcoal, (3) chemical indicators of water quality such as nutrient elements and oxidized or reduced metals indicating oxidized or anoxic waters, (4) radioisotopes, and (5) putative pollutant chemicals (Cowgill 1988, Brugam 1988, Crisman 1988). For example, Davis et al. (1988) reconstructed 200 year pH histories of three lakes in the Adirondack Mountains, NY. They used ^{210}Pb, Ca, Mg, K, pollen, and charcoal, to distinguish effects of fire, logging, forest development, and acid deposition on inferred lake pH.

Another important source of historical data is annual growth rings of trees. Tree rings provide a record of the annual growth of a tree and of its exposure to metals, so that responses to a change in pollutant exposure can be studied after the fact (Baes and McLaughlin 1984, Bondietti et al. 1989, Noble et al. 1989). For example, McLaughlin et al. (1987) used radial cores from 1000 red spruce at 48 sites in the eastern U.S. to show that a substantial reduction in growth had occurred in the previous 20 to 24 years, the decline coincided with increases in pollution loading, and was not attributable to climate, stand dynamics, or local factors.

Historical correlations, like other correlations, should not be naively accepted. For example, the decline of the peregrine falcon in North America was believed to have begun following the extermination of the passenger pigeon and appeared to be largely limited to the former range of the passenger pigeon, so the decline appeared to some to be attributable to the loss of the principal prey species rather than to DDT (Beebe 1969).

Another technique for assessing causation is the process of elimination. If the evidence for toxic effects is inconclusive, the assessor may examine alternate possible causes of the observed effects. Elimination of all alternate hypotheses has been described as the best method of scientific inference (Platt 1964, Popper 1968), and Carpenter (1990) has recently presented a Baysian approach for establishing ecological causation by eliminating alternate models of causation. However, the utility of this approach is limited in ecological epidemiology because it is usually not possible to identify a complete finite set of possible causes for an observed change in an ecosystem. In addition, investigating alternate possible causes can consume large amounts of time and resources. However, elimination of any alternate cause is de facto support for toxic effects, and evidence supporting an alternate cause weakens the case for toxic effects. The study of red spruce decline discussed above is an example of strengthening a case for pollution effects by eliminating proposed alternate causes.

What Is the Magnitude of the Effect?

Given that a real effect has been detected (e.g., bird kills) and it has been attributed to a cause (e.g., a particular pesticide use), the next question must be what is the risk that these kills are sufficiently large and frequent to exceed some endpoint (e.g., greater than x% reduction in local populations of any native bird species). The process of answering this question resembles conventional risk assessment.

Establishing the magnitude of effects would seem to simply be a matter of comparing the magnitudes of the measures of effects at exposed and unexposed sites (e.g., the density of fish upstream and downstream of an effluent). However, for the same reason that one cannot simply infer that differences between sites are caused by a pollutant, one cannot simply infer that the magnitude of the differences is the magnitude of the pollutant effect. The effect may be either larger or smaller because of confounding differences in the sites. As with proving causation, establishing the magnitude of effects using a sufficiently high quality and quantity of evidence includes (1) using toxicity test data to support biological survey data, (2) establishing that the reference sites are "normal" for the community type, and (3) accounting for difference in site characteristics other than presence of the pollutant.

Statistical models of the results of either observational studies or toxicity tests can serve to estimate the magnitude of effects.* Ideally, the two would be combined by calibrating the model of the test results to the results of observational studies. Christensen et al. (1988) fit multivariate logistic models to test data for responses of brook trout to various combinations of pH, Al, and Ca. They used these models to calculate the relative reproductive potential corresponding to the water chemistry in each of 67 Adirondack lakes. They then calculated a cumulative probability of occurrence of brook trout (based on a survey of the lakes) as a logistic function of relative reproductive potential (Figure 10.6). In this way, relative reproductive potential is calibrated as an endpoint for tests of acidification effects against the assessment endpoint, loss of brook trout from high-elevation lakes, and the risk to other lakes can be estimated.

Although this approach is promising, it is not well established. More often the assessor must judge which type of study provides the best model. For example, exposure-response models of test results were used to estimate losses in the National Crop Loss Assessment Network (NCLAN) assessment of ozone effects on crops, because relatively realistic field tests could be performed, and the many confounding variables (e.g., soils, weather, agronomic practices) were judged to be too strong for comparison of crop yields among regions with different ozone exposures (Heck et al. 1988). However, when assessing the effects of an aqueous effluent on an otherwise undisturbed or

---

*At this point, the advantage of descriptive statistics over hypothesis-testing statistics in risk assessment is particularly apparent. Hypothesis tests do not provide estimates of the magnitudes of effects.

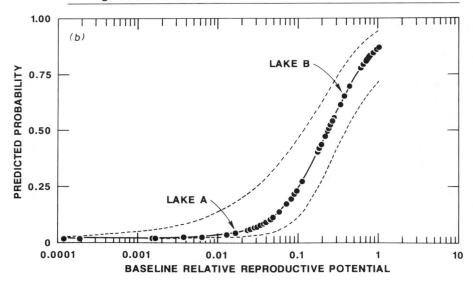

Figure 10.6. Probability of occurrence of brook trout in Adirondack lakes as a function of relative reproductive potential (RRP). RRP was estimated for each lake by substituting the lake's water chemistry into a model of RRP as a function of test water chemistry in laboratory tests of pH and aluminum effects on brook trout. A logistic function (solid line) was then fit to the cumulative proportion of lakes with trout. The dashed lines are 95% confidence bounds (Christensen et al. 1988).

uniformly disturbed stream, a model of the observed contaminant fate and effects in the stream is likely to be more reliable than a model based on laboratory chemistry and test results.

Estimating the magnitude of effects involves not only estimating the magnitude of effects at a point receiving a particular exposure, but also integrating effects over space. Because they can establish a relationship between exposure and effects, statistical models can be used to estimate the magnitude of effects at any point for which exposure can be estimated. In situations where there is an essentially linear gradient of exposure, as in a stream below an outfall or along a line away from an air pollution source in level terrain, it is sufficient to create an exposure-response or distance-response model based on measured responses of the receiving community or in situ toxicity tests. These models can be used to interpolate between points at which effects are measured or to estimate the point at which effects fall below some prescribed magnitude. However, for situations where the pattern of response is more complex, mapping techniques such as kriging may be used to display the areal extent of effects (Clark 1979, Mason 1983, Fig. 10.2). For example, one might estimate the areas of sediment where biomass of invertebrates was reduced more than 75%, 50%, and 25% relative to reference sites.

Although biological responses in the field are typically modeled as responses to concentrations of contaminants or to distance from sources, exposures are often complicated by the temporal variation in exposure, by exposure to multiple contaminants, and by other factors that may need to be included to make models reasonably descriptive. The conventional predictive approaches to incorporating temporal variability (Chapter 3) and chemical interactions (Chapter 7) can also be used in descriptive epidemiological models. However, the fact that epidemiological studies deal with particular relatively well-defined situations encourages the development of ad hoc approaches to quantifying exposure. For example, in a field study of exposure of breeding bird populations to pesticide spraying in a mixed agricultural setting, the exposure was episodic and irregular and included multiple pesticides (Patnode and White 1991). In order to place all of the monitored nests on a common exposure axis, the authors developed a factor F incorporating the normalized toxicity of each pesticide T, the number of exposures (i.e., the number of applications in the field adjoining the nest) N, and the number of days that a nest was active D:

$$F = \Sigma TN/D$$

Production of young mockingbirds, cardinals, and brown thrashers was found to be a function of this factor. Although this next step was not taken by the authors, it would be possible to estimate the reduction in production of young birds on a farm or in a region given a pattern of pesticide application.

What Are the Ultimate Consequences?

Once the reality, cause, and magnitude of effects have been established, it is often desirable to estimate what effects will occur if the source continues to operate or if mitigation or control measures are taken. In effect, this is predictive assessment, but it differs from the predictive assessment discussed in earlier chapters in that it can be based on a temporal extrapolation of observed and measured processes.

To the extent that they capture the relationship of pollutant to effects, statistical models that were developed to describe the existing situation can be used to extrapolate into the future. For example, Hakanson (1984) used his model of mercury body burdens of fish in lakes to predict the reduction in fish contamination that would result from reduced mercury loading. This approach depends on the assumption that recovery is simply the reversal of degradation. That is, following recovery, the level of effects at a site where exposure has been reduced from 2x to x will be the same as that at sites where the exposure level was x prior to the remediation. However, the resilience of ecosystems is limited, and severe effects may not be reversed. Rather, a new ecosystem structure may result from the recovery, or the pollution adapted community may persist (Holling 1978).

Although models of the production of effects can be used in many cases to estimate the ultimate degree of recovery, they do not address the rate of recovery. If persistent pollutants have accumulated in the ecosystem, if species have become locally extinct, or if pollution has modified the physical habitat, recovery may be slow. There are no general methods for predicting the rate of recovery, but some guidance is provided by the existing body of experience (Box 3.2).

Both prediction of recovery and estimation of the extent of recovery following damage are complicated by the fact that the state of ecosystems is variable even in the absence of human disturbance. Ecosystem properties have fluctuations, cycles, and trends that are poorly predictable. Therefore, the base state to which the recovering systems return in Figure 3.11 should be depicted as a confidence band that diverges with time rather than as a constant baseline. One can only say that there is some probability that a particular predicted or measured state of a post-contamination ecosystem corresponds to the state that would have occurred in the absence of contamination (Bartell, 1992). This problem of determining when a system is recovered is similar to the problem of determining whether a system is damaged and can be solved similarly by determining that a system is recovered when its state falls within the state space of reference sites.

If a pollution problem occurs repeatedly and its effects are controlled by a common set of factors in all instances, empirical assessment models can also be used to predict the effects of new instances. For example, Jones et al. (1979) applied their threshold model of SO_2 damage of plants to a new power plant in the same region and estimated that the proportional frequency of occurrence of SO_2 exposures sufficient to cause injury was less than 2%.

Summary of Inference in Ecological Epidemiology

Ecological epidemiology is distinguished from other types of ecological assessment by the use of observational studies in the polluted environment to elucidate effects. The ability to observe and measure effects is potentially a major advantage because none of the complexity of pollutant transport, fate, exposure, proximate effects, and ultimate effects needs to be approximated. The real world is speaking directly to the assessor. However, that real-world complexity can mask effects, amplify effects, or generate pseudo effects.

Toxicity tests can prove causality, but they are physical models of the real world, so the proof is applicable to the field only to the extent that the test incorporates the relevant characteristics of the field. Inevitably, toxicity tests are incomplete and imperfect models. For example, survival and growth of fathead minnows in an effluent toxicity test does not prove that the contamination is not responsible for observed degradation of the fish community in the receiving stream. The fish community might be affected (1) when physical conditions differ, (2) when the composition of the effluent differs, (3) when fish are stressed, (4) when other life stages of fish are exposed, (5) if the fish species in the receiving community are more sensitive, or (6) if food-chain species are affected. Similarly, the test may be too sensitive for a variety of reasons. Tests of organisms confined on the site are likely to be more realistic than laboratory tests, but are still open to criticisms about the effects of caging, feeding, outplanting, etc. Toxicity data extrapolation models (Chapters 7–9) indicate the likely bias and error in test results, but they cannot prove that the same cause is operating in the test as in the polluted environment.

Causality is established by demonstrating concordance between the real but uncontrolled observational studies and the controlled but somewhat unreal toxicity tests. The primary links are indicators of exposure and diagnostic effects. However, any evidence that the same mechanisms are at work in the tests and in the observations (e.g., similar rankings of the sensitivity of species) is supportive of common causation.

Estimation of the magnitude and distribution of effects requires some sort of formal model. The relative utility of models derived from observational studies and toxicity tests is a subject of controversy. Some have argued that models derived from test data should be used because observational studies cannot prove causation and their variables are inevitably confounded (Anderson et al. 1985, Rawlings et al. 1988). However, proving causation is a separate problem from estimating effects. Once causation is established, models based on testing, observation, or a combination of the two can be chosen to estimate the magnitude and extent of effects.

The relevance of the concept of risks as probabilities is not as clear for ecological epidemiology as for the conventional risk paradigm because of the emphasis on using congruence of independent lines of evidence to prove causation. However, formal quantitative models can enter into several components of the process, and models should incorporate uncertainty and estimate outcomes probabilistically. In addition, expressing results as probabilities can contribute to the logical, nonquantitative aspects of epidemiological inference in a manner analogous to the probabilities of disease calculated by epidemiologists. For example, statistical models can be used to estimate the risk that an exposed ecosystem is affected by estimating the probability that it belongs to the same set as the set of reference ecosystems. Similarly, one can quantify the probability that an apparently consistent conjunction of cause and effect is due to chance.

Test Interpretation in Retrospective Assessment

When it is not practical to measure effects in the environment, toxicity tests can be used in place of an epidemiological approach to estimate the magnitude of effects. In such cases, causation is not an issue because no effects have been observed. Rather, the problem in interpretation is to define the nature and distribution of ecological effects implied by the test results. In general, the interpretation and use of toxicity test data for retrospective assessments requires that the same problems be addressed as in predictive assessments. These include relating the concentration and duration dynamics of the exposures and the severity and frequency of effects in the toxicity test to the exposure dynamics and effects of concern in the field (Chapter 3). In addition,

Table 10.5. Uncertainty Factors for the NPDES Assessments (EPA 1985b)

Uncertainty Source	Uncertainty Factor	Rationale
Effluent Variability	10	Range of effluent toxicity not observed to exceed one order of magnitude
	100	Heavy metals observed to vary two orders of magnitude in POTW effluents
Species Sensitivity	10	Acute or chronic tests using one vertebrate and one invertebrate
Acute/Chronic Effluent Ratio	10	Value observed from onsite whole toxicity testing

extrapolations between the test organisms or model ecosystems used in the toxicity tests and the species, life stages, populations, and ecosystems in the field must be addressed (Chapters 7–9). If data from toxicity tests of individual chemicals rather than whole effluents or contaminated media are used, then the possibility of combined toxic effects must be considered (Chapter 7). A difference in the interpretation of test results in retrospective assessments is that the concentration and dynamics of the pollution are often measured rather than estimated.

As with predictive assessments (Chapter 7), most retrospective assessments that are based on toxicity testing use the quotient method with factors to account for extrapolations and uncertainties. An important example is the EPA's use of toxicity testing in the regulation of existing aqueous effluents (EPA 1985b). The procedure derives toxic units (TU) by calculating the quotient of the lowest LC_{50} or NOEC (in percent effluent) divided by the instream waste concentration (the percent effluent at the end of the mixing zone). This procedure is the quotient method (Chapter 7), but with units of effluent concentration rather than chemical concentration. The factors shown in Table 10.5 are applied to the quotient to determine whether the effluent is acceptable, unacceptable, or needs more testing. As with other assessments based on the hazard assessment paradigm (Chapter 1), this scheme sorts pollutants on a common scale, but does not attempt to estimate effects in the field.

Effluent and ambient media tests have the advantage of estimating the toxicity of complex toxicants without requiring assumptions about the nature of the interactions between components (Chapter 7). However, if multiple effluents are added to a body of water, it is necessary to make some assumption about the nature of their interactions if effects on the receiving community are to be estimated from effluent toxicity data. The EPA (1985b) recommends that the toxic units be assumed to be additive. A recent study by DiToro et al. (1991b) showed that in the Naugatuck River the toxicity of combined effluents was considerably less than additive. However, currently there is no basis for alternate assumptions about combining effluent or media test results. The alternatives to assuming additivity are to identify the component chemicals and model their availabilities and interactions, or test the ambient mixtures of mixtures as in DiToro et al. (1991b).

Another difficulty in performing assessments on the basis of effluent and ambient media toxicity tests results from the lack of information about the cause of toxicity. Determination that a contaminated medium is toxic may be sufficient cause for action, but identification of the responsible sources and of appropriate options for treatment, disposal, or remediation requires knowledge of which components of the medium cause the observed toxicity. Methods for identification of the toxic components of mixtures are discussed in Chapter 7. These methods are being adapted for assessment of contaminated ambient media (Ankley et al. 1992).

Most of the uses of toxicity test data in risk assessment depend on establishing a concentration-response function or other exposure-response relationship. In some cases, such as the EPA's procedures for regulation of existing aqueous effluents, the exposure gradient used in the test is established by diluting an effluent or contaminated medium with clean medium or site-specific diluent (e.g., upstream water). However, in many other cases the assessor has only results of tests of a contaminated medium and a nominally clean reference medium, or a set of test results from sites that are not related to a pollution gradient or to concentrations of a particular chemical of concern. It may be possible to establish a concentration-response relationship by correlating toxic effects with chemical concentrations in the tested media if media from a number of sites are tested and if concurrent chemical measurements are obtained (Long and Buchman 1990). However, the concentrations of individual pollutant chemicals in the ambient environment are often correlated with each other so it is hard to associate effects with particular chemicals or classes of chemicals. The attempt to establish correlations can serve to indicate whether pollutants are likely to account for differences in test results among sites or whether they are due to other factors. For example, mortality of the amphipod *Rhepoxynius abronius* in toxicity tests of San Francisco Bay area sediments was more highly correlated with sediment texture than with any of the chemical measurements (Long and Buchman 1990). However, the assessor should be aware that background environmental variables may be correlated with toxicant concentrations in these studies, as in purely observational studies, so inference is tenuous.

If no exposure-response relationships are established, the assessor must find other ways to make inferences about risk. If one or more species are tested at a contaminated site, factors or other extrapolation models (Chapter 7) cannot be used to conservatively estimate a safe level of exposure if there is no exposure gradient. For example, if data are presented to the assessor indicating that water from a site is not lethal to *Daphnia magna*, then one might ask, what is the risk that there is some more sensitive species that would be killed. The maximum risk can be estimated by conservatively assuming that the *D. magna* were just short of dying. The risk that some proportion of the community is significantly more sensitive can then be estimated from the positions of *D.*

magna LC$_{50}$s in the distributions of sensitivity used to establish the U.S. national water quality criteria. If we assume hypothetically that *D. magna* are, on average, the first percentile species in sensitivity distributions, then, if there are 500 species in the community, we would expect four of them to be more sensitive than *D. magna*. Therefore, on average four species in a 500 species community might experience toxic effects even when *D. magna* did not. Risks could be estimated similarly for sets of test species or for risks that more sensitive effects occur, given data on lethality. Such models are analogous to the models developed to derive advisory concentrations for protection of aquatic life (OWRS & ORD 1987), and to develop aquatic life pesticide concentrations (Table 7.2).

To the extent that the effects measured in the tests are relevant to population or ecosystem level effects, their implications at those levels can be estimated. For example, if a test of polluted water yields percentage mortalities of fish larvae, a plot of percent reduction in population size against percent larval mortality could be generated from a fish population model to demonstrate the implications of $x\%$ larval mortality for populations with the modeled life history.

In some cases, indicators of exposure are measured in the field, but effects are not. In such cases the levels of body burdens or other exposure indicators in the field can be related to effects associated with those levels in ad hoc toxicity tests or in tests reported in the toxicological literature (e.g., McKim et al. 1976, Mount and Oehme 1981). For example, Bender and Huggett (1984) related body burdens of kepone in blue crabs from the James River to body burdens in blue crabs experiencing toxic effects in the laboratory (Figure 10.7). A slightly more complex approach is required if body burdens are not provided with the toxicity test results. If the body burden of a chemical in fish is x and the bioaccumulation factor for fish exposed to the chemical in water is y, then (assuming equilibrium has been achieved) the time-weighted average concentration to which the fish has been exposed is approximately x/y. That estimated aqueous concentration can be related to the concentration-response relationship observed in toxicity tests. This use of measurements of internal exposure to estimate external exposure or the magnitude of sources is referred to as reconstructive exposure assessment (EPA 1988).

As in ecological epidemiology, the use of exposure indicators provides a logical link between sources and effects. In addition, they are superior to ambient chemical concentrations for estimating field effects from laboratory toxicity data because they bypass all of the spatial variability, temporal dynamics, and biological processes that determine the level of exposure. However, the use of internal exposure-response relationships is rare in ecological risk assessment, and therefore is not well developed. In particular, we assume that tissue concentration-response relationships are more similar among species and life stages than ambient concentration-response relationships, but the generality of that assumption needs to be verified and quantified.

In sum, although retrospective assessments increasingly rely on toxicity test-

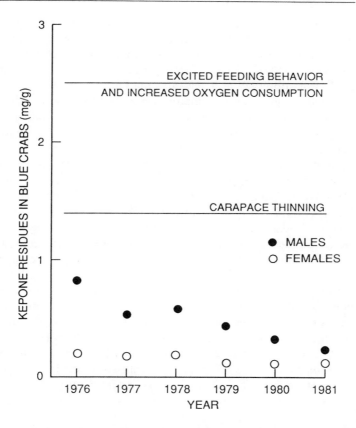

Figure 10.7.   Kepone body burdens (residues) in blue crabs from the James River (points) and in crabs experiencing toxic effects in the laboratory (lines). Redrawn from Bender and Huggett (1984).

ing, risk procedures and models to utilize that data have not been developed to nearly the extent that they have in predictive risk assessments. Although the problems of estimating effects on sensitive species, life stages, and processes and of extrapolating between levels of ecological organization are the same, the test data and assessment problems are sufficiently different that new approaches and models must be developed.

Model Interpretation in Retrospective Assessments

Statistical and mathematical models are potentially as important in retrospective risk assessments as in the predictive assessments discussed in Chapters 3–9. As noted above, a variety of extrapolations across space, time, and conditions are required in retrospective assessments. It may be that adverse ecological effects have not yet occurred but may occur in the future if existing conditions continue. For example, a contaminated groundwater plume may not yet have reached a surface discharge point but could do so at some future date. Because the generation times of important biota such as fish, birds, and trees may be long, stress must often continue for a number of years before detectable population or ecosystem change occurs. Extrapolation in space may also be required. Effects observed at a few sites may have to be projected to the scale of a much larger region. Finally, when environmental change has occurred, evaluation of management alternatives may require that the effects of the stress in question be separated from the effects of other influences such as natural environmental fluctuations or exploitation by man.

Application of models for the above purposes differs little from the predictive assessments described in previous chapters. The most significant difference between predictive and retrospective use of models are that (1) the predictions are extrapolations from existing conditions at a contaminated site so many parameters and functional forms can be derived from measurements, and (2) whereas models used in predictive assessments are typically generic, models used in retrospective assessment are often site-specific or region-specific.

The use of models in assessments of effects of power plants on fish populations provides a valuable example for retrospective assessments. Thermal power plants often kill large numbers of young fish by entraining them through cooling systems or impinging them on intake screens. The numbers of fish killed in this way at operating power plants can be directly measured. The importance of this mortality is often a significant issue in the licensing of these plants, and therefore quantitative assessments are frequently performed. The reductions in short-term abundance or year-class recruitment can often be estimated with a fair degree of precision. However, the long-term consequences of the impacts, which are the objects of concern for regulators, can usually not be observed within the time available for making decisions. When valuable resources are being affected or the potential for reduction is believed to be high, population models are often used to project observed mortality rates to reductions in abundance or yield to fisheries (Barnthouse et al. 1986; Lawler 1988, Northeast Utilities 1990).

Given an observed change, models can sometimes be used to estimate the relative contributions of the various known causes to the change. Standard life table models permit the contributions to changes in population-level parameters (see Chapter 5) to known causes of mortality or reduced reproduction. This approach has been used, for example, by Barnthouse and Van Winkle (1988) to identify the contributions of six different Hudson River power plants to impingement impacts on white perch. Floit and Barnthouse (1991) used the same approach to rank sources of mortality to the San Joaquin kit fox population on the Elk Hills Petroleum Reserve. Models have also been used widely in fish and game management to evaluate the relative importance of exploitation and natural mortality in determining the abundance and yield of exploited populations. Given appropriate models, the same could also be done for populations or ecosystems affected by contaminated sites, habitat degradation, or other forms of stress.

The above discussion has dealt with use of models for prediction or extrapolation. At least in principle, models can be used in ecological epidemiology to determine whether the magnitude or type of observed change is consistent with a hypothesized cause. The procedure (Figure 10.8) requires (1) an observed change in a population or ecosystem attribute, (2) independent information concerning sources such as loading rates or ambient contaminant concentrations, and (3) a model that expresses a hypothesized causal relationship between the source and the change. Using the model, one can calculate the magnitude of source required to produce the observed change and then compare this to the measured source. If the required model inputs are consistent with the magnitude of the observed source, then one obtains partial confirmation that the source could have caused the observed change. If the model is inconsistent with the observed source, then the particular mechanisms included in the model are refuted. Either the source caused the change via some other process not included in the model, or the change results from some other cause. This approach has not actually been applied in ecological assessments, although clear analogies exist both in water quality assessment and in human epidemiology.

A related use of models is estimation of remedial action goals. If a risk assessment indicates that contamination of a medium is causing unacceptable effects on an endpoint population or ecosystem, then the assessor must estimate a level of contamination which should not be exceeded following remediation (the remedial action goal). This is done by, in effect, running the risk assessment in reverse. From the effects model, an exposure level is derived that corresponds to the threshold for unacceptable effects. This exposure level is then entered into the exposure model to estimate the contamination level that

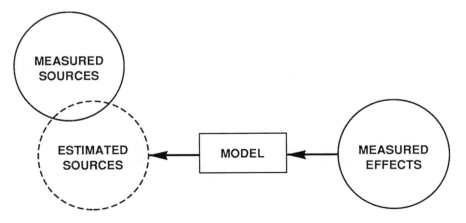

Figure 10.8.   Use of mathematical models in ecological epidemiology to assist inferences about causation.

would induce that level of effect. Fordham and Reagan (1991) used this approach to estimate concentrations of dieldrin in sediments and water that would not pose a hazard to bald eagles. Macintosh et al. (1992) used a probabilistic food web model to estimate the probability density of mercury and PCB concentrations in a reservoir that would minimally reduce fecundity in mink and great blue herons.

This approach has been used on a large scale to estimate the levels of emissions of air pollutants that can be allowed without exceeding critical loading rates for acid, sulfur, and nitrogen. These are rates that are thought to approximate the capacities of the ecosystems in those locations to assimilate the pollutants. The first is to estimate critical loads for individual grid cells based on the soils, geology, precipitation, and land use (Hettelingh et al. 1991). These are then compared to current loading rates. For areas in which critical loads are exceeded by current loads, regional pollutant transport models can be used to identify sources of the deposition.

Adaptive Regulation and Assessment

The concept of adaptive management (Chapter 1) provides a supplemental means of inference in retrospective assessments. If there are significant uncertainties in the assessment, a decisionmaker might perform a regulatory experiment rather than making a final decision about regulation, treatment, or remediation of existing pollution. The design of the experiment would depend on the nature of the action being considered. For example, the relative importance of different sources of acid-forming gases to acid deposition in acid-sensitive regions of Europe and North America is uncertain and controversial. Before requiring the installation and operation of billions of dollars worth of scrubbers on coal-fired power plants in source regions or entire nations, it might be prudent to experimentally burn low-sulfur coal for a few years in each region in rotation, beginning with the most likely major source, to see to what extent deposition is reduced in sensitive receptor areas (Walters 1986). Alternately, scrubbers could be installed in sequence, beginning with plants that are thought to be major sources. Another example would be experimentally restricting in certain areas the use of a pesticide that is suspected of causing aquatic toxic effects. Streams and ponds could be monitored in that area and in areas where the pesticide is still used, both before and after the restriction, to determine whether the restriction has beneficial effects on aquatic communities.

If they are to be useful assessment tools, these regulatory experiments must be designed to illuminate major uncertainties concerning the appropriate course of action. For example, the models of transport of acidifying gases are much smaller sources of uncertainty in estimating the effects of acid deposition than the models of effects of the deposition (Morgan et al. 1985). It would be difficult to experimentally estimate the contributions of sources to effects because of the delays and complex mechanisms involved in acidification effects induction. However, given a determination that acid deposition levels are unacceptable in certain regions, the uncertainty concerning transport becomes critical to deciding which sources should be regulated and to what extent. Experimental regulation of sources could effectively eliminate that uncertainty.

The idea of assessment through regulatory experiments has not been adopted by either regulators or environmental scientists. According to Walters (1986), the idea has been opposed by environmental scientists who do not care to admit that their science is inadequate to provide a clear basis for all decisions. Also, decisionmakers are usually reluctant to appear indecisive or to make decisions that may appear unfair because all sources are not regulated equally. However, regulatory experiments provide the only definitive answer to the question: how do you know it will work until you have tried it?

Similarly, retrospective risk assessment can be used to substitute entirely or in part for predictive risk assessment. If the risks of a new chemical or effluent are uncertain, rather than doing more testing and measurement to reduce the uncertainty, a regulator can approve the release and wait to see what happens. There is a certain element of this strategy in any regulatory program because there are always unresolved uncertainties that can lead to unforeseen effects, and if these effects are detected they may force a reassessment. Substitution of retrospective assessment for predictive assessment has been recommended as a conscious strategy by some authors (Herricks et al. 1981), and it is more common in European countries than in the U.S. For example, the U.S. EPA requires field testing for effects of many pesticides on wildlife, but European countries do not. European governments rely on the public to detect and report pesticide-induced mortalities after the pesticide is in use (Somerville and Walker 1990). The advantage of this strategy is that the observed effects of chemical use are more realistic than any program of testing and predictive assessment. The disadvantages are (1) effects may be quite severe before they are detected by informal public monitoring of the environment, (2) a formal monitoring program to reliably detect effects before they become severe is likely to be expensive, and (3) removing a chemical from the market or retrofitting improved treatment facilities is more expensive than restricting a chemical before it is manufactured and marketed or installing treatment when a facility is constructed. The costs of monitoring effects and retrofitting treatment would be more easily justified if the retrospective assessment was part of a strategy of adaptive management in which the results of the retrospective assessment were used to test predictive assessment models so as to improve future predictive assessments.

EXAMPLES OF RETROSPECTIVE ASSESSMENT

The following three examples were chosen to represent a diversity of retrospective ecological risk assessment in terms of the magnitude of effort, the nature of the sources and endpoints, and the nature of their mandates. For each example, we have discussed the hazard definition phase of the assessments, the estimation and measurement of sources, exposures, and effects, and the risk characterization. Since all of these examples were, at least potentially, problems in ecological epidemiology, we discuss which of Koch's postulates were satisfied by the assessments. Finally, we describe how the extent and magnitude of effects were estimated and conveyed to the risk managers.

Acid Deposition and Fisheries

Acid deposition has been the subject of numerous research and assessment efforts, but the most ambitious and comprehensive is the U.S. National Acid Precipitation Assessment (NAPAP). NAPAP is a 10-year research and assessment program authorized by the U.S. Congress in 1980 and recently reauthorized. It was charged to identify the ultimate sources of acidifying gases (power plants, smelters, automobiles, etc.), assess effects on a wide range of potentially affected values (agricultural crops, forests, aquatic systems, materials, human health, and visibility), model the intermediate processes of transport and transformation, and assess economic costs and benefits of amelioration. To keep this discussion to a manageable length, it is limited to a single assessment endpoint, aquatic systems, and to the proximate source, acidified surface water. Unless otherwise identified, the reference for this example is the aquatic effects chapter of the NAPAP interim assessment (Malanchuk and Turner 1987).

Hazard Definition

The hazard definition consists of defining the motive of the assessment and the assessment endpoints and delimiting the spatial extent of the assessment (Figure 10.1). Although "aquatic systems" is the nominal assessment endpoint, the operational assessment endpoint is loss or severe decline (i.e., functional extinction) of fisheries. This choice is based on the inherent importance of fish and on the fact that fish integrate effects on entire aquatic ecosystems. In addition, loss of fisheries was the first effect credibly attributed to acid deposition, and much of the early research supported effects-driven assessments attempting to determine the cause and extent of loss of fisheries in Scandinavia and the northeastern U.S. However, NAPAP is intended to assess the need for regulation of sources of acidifying gases so it is logically structured as a source-driven assessment.

The spatial extent of NAPAP is the entire U.S., but an early assessment task was to identify susceptible areas so that the assessment could be limited to areas where effects have occurred or are likely to occur. Susceptible areas are those that have experienced acidification or appear susceptible to acidification because of high deposition and low buffering capacity of soils and waters.

Source/Exposure/Effects Measurements

For this case study, the source is acidified water. Feedback from early assessments established that the effects of acidification were due to acid-mobilized aluminum as well as direct effects of H^+, so the source is defined in terms of the pH and concentration of Al species. For NAPAP's assessment of current effects, source characterization is provided by the National Surface Water Survey which measured water chemistry in a stratified random sample of lakes and streams. Because fish are immersed in and respire water, exposure is reasonably assumed to be characterized by surface water chemistry. However, two issues have led to measurement of exposure parameters. First, the extent to which reduced pH has resulted in increased exposure to available forms of metal has been addressed by examining the metal content of fish tissues. These measurements have served to confirm the importance of aluminum, but tissue concentrations have not been used to estimate effects. Second, the differences in exposure of various life-stages of fish have been studied by monitoring water chemistry in spawning and nursery areas during spawning, incubation, hatching, and larval development. The results are used to determine to what lake and stream chemistry each life stage is exposed. Measurements of effects have included surveys of the status of fish populations in lakes and streams, laboratory toxicity tests, ambient toxicity tests (i.e., caged fish and eggs), and whole-ecosystem tests, including acidification or liming of lakes and streams. Figure 10.9 shows the relationships of these measurements and their interpretation.

Risk Characterization

Assessment of acidification effects on fish began in the late 1970s with the observation that fish populations were declining in areas with ostensibly clean water (southern Scandinavia, southeastern Canada, and the Adirondack Mountains of New York), and the association of this observation with the recently recognized phenomenon of acid deposition. Most of the early attempts at assessment tried to associate pH with the status of fish populations or communities measured in field surveys. That is, they attempted to demonstrate a regular association between acidification and fisheries decline by establishing a binary model: lakes or streams with pH $< x$ have degraded fisheries as indicated by measurement endpoint y. A more sophisticated approach to demonstrating regular association and developing a model of acidification effects is use of survey data to establish a continuous model of degradation as a function of exposure. For example, Baker and Harvey (1984) fit a probit model to the probability of occurrence of brook trout in Adirondack lakes as a function of average pH (Fig. 10.10). They were able to derive a reasonably good model by (1) tightly defining the measurement endpoint (absence of brook trout from a lake that had them in the past, rather than absence of trout from a lake), (2) eliminating lakes for which other causes were plausible (cessa-

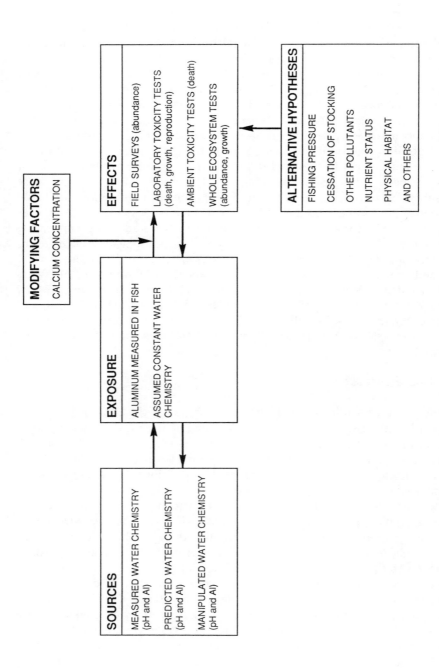

Figure 10.9. Integration of the indicators of source, exposure, and effects for characterization of risks to fisheries in the NAPAP interim assessment.

Predicted probability

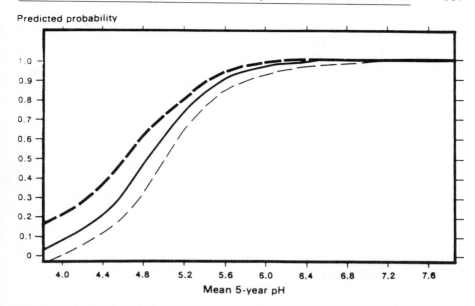

Figure 10.10. Probit model of the probability of occurrence of brook trout in Adirondack lakes as a function of pH (Malanchuck and Turner 1987).

tion of stocking, angler pressure, marginal physical habitat, etc.), and (3) confining the study to a relatively homogeneous region where pH is correlated with other parameters of water chemistry. These empirical approaches could not prove causation because of the large number of factors that influence fisheries, and the surveys gave inconsistent results among regions because of differences in water quality other than acidity.

Laboratory toxicity tests have established that low pH is toxic to fish, that aluminum concentrations found in acidified waters (particularly inorganic monomeric aluminum) are toxic, and that calcium is ameliorative. In these studies, unlike the field surveys, there is no question as to causation, and interactions among the parameters can be identified. Ambient toxicity tests demonstrate that the variation in water quality observed in the field are associated with mortality and other effects in fish, but they do not indicate which water quality characteristics cause the effect or how they interact. Field experiments like the acidification of Lake 223 (Schindler et al. 1985) allow the course of decline of fish populations to be observed along with other responses of the lake ecosystem. However, they do not include the leaching of metals from the watershed, and the absence of replication makes inference about causation and the general applicability of results somewhat uncertain (Carpenter 1989, Carpenter et al. 1989).

Having shown an association between acidification and decline in fisheries, and having shown that acidification can cause biologically significant effects in toxicity tests, only the second and third of Koch's postulates remain to be satisfied. That is, a common set of effects at common levels of exposure must be shown in acidified lakes and streams, and in the tests. That need to link laboratory and field results was satisfied principally by the reasonably good correspondence between the water chemistries that cause effects in the laboratory and those that cause effects in the field. However, this correspondence was based largely on pH levels and it was not clear that the interactions of pH, Al, and Ca were the same. Also, laboratory fish appeared to be a little less sensitive, which was attributed to greater effects of the varying exposures in the field than the constant exposures in the laboratory. The similar relative sensitivity of species in laboratory tests and in the field strengthened the logical link between the two types of studies. The evidence that the mode of action is the same in both types of study was the observation that acid-stressed and acid + aluminum-stressed fish have low blood and tissue ion concentrations both in laboratory tests and in the field, and in both situations calcium is ameliorative. In addition, aluminum and other metals have been shown to occur at elevated concentrations in fish tissues from both the laboratory and field.

By satisfying all four of Koch's postulates, the NAPAP interim assessment demonstrates that acidification of surface waters has caused decline and loss of fisheries in certain regions. The ongoing efforts in the NAPAP assessment of aquatic effects are concerned with quantifying the scope and magnitude of current effects and predicting and monitoring the effects of regulatory options. These goals require defining the current and projected water chemistry in susceptible regions, defining the resources at risk in those waters, creating models relating water chemistry to the condition of fish populations, and integrating the water chemistry and resource data with the effects models. The results are available to the readers of this book in the NAPAP Final Assessment and the State of Science and Technology Reports.

Ecological Effects of a Waste Site

Waste site assessments constitute a large proportion of environmental risk assessments performed in the U.S. Although billions of dollars in remediation costs are at stake, the resources devoted to individual assessments or to development of appropriate assessment methods are small. As a result, risk assessments of waste sites tend to be relatively crude, and, although this example compares poorly with the other two examples, it was a good waste site assessment. The example is a case study prepared by the EPA's Office of Solid Waste and Emergency Response (the source for this section, including all quotations is Bascietto 1988). The case study is an assessment of an unnamed inactive landfill of about 300 acres containing several million cubic yards of municipal waste. The ecological assessment is part of the Remedial Investigation / Feasibility Study, the primary objectives of which are "evaluating the site's potential impacts on public health, welfare, and the environment, and developing a cost-effective remedial action plan."

Hazard Definition

This is a source-driven assessment, motivated by the existence of the contaminated site without prior evidence of exposure or effects. The spatial extent of the assessment is the waste site itself (including freshwater lakes and wetlands), and an adjoining estuarine river and wetlands. This extent was established by observation of leachate flowing from the landfill and by detection of contaminant chemicals in onsite lake water, groundwater, and soils.

The assessment endpoints are not clearly defined. The environmental values are "representative habitats" (the river, onsite and offsite wetlands, an offsite lagoon, and onsite lakes were selected as representative of freshwater, brackish, and saltwater environments), and "representative species" (herons for wetlands and unspecified mammals for uplands). The "specific biological endpoints evaluated can include acute toxicity, chronic and reproductive toxicities, bioaccumulation, and on occasion, mutagenicity." Operationally, the endpoints in this assessment were water quality criteria and acute and chronic test endpoints.

Measurement and Estimation

The source was characterized by analyzing soil, water, sediment, and air for EPA hazardous substance list chemicals and ammonia. The chemical composition of oysters and fish tissues was measured as sources of exposure for humans and piscivorous wildlife. Acute toxicity tests of water samples were conducted using fresh and saltwater fish, invertebrates, and algae. In addition, existing toxicity data and water quality criteria for waste-associated chemicals were used as indicators of effects. No onsite measurements of effects were obtained. These measurements and their integration are portrayed in Figure 10.11.

Risk Characterization

Ecological epidemiology was not used; effects were not measured onsite, and the measurements of concentrations in oysters and fish were not used as indicators of exposure of aquatic animals. Instead, the assessments were based on toxicity tests. The acute aqueous toxicity tests indicated that the onsite lake was highly toxic and tidal wetlands samples caused 5%-15% lethality in three of the ten species tested. Other water samples were not acutely lethal, but this result does not preclude chronic lethality or sublethal effects. Because there was no accepted method for estimating chronic or sublethal effects from these tests, the assessors resorted to comparing concentrations of individual chemicals to their water quality criteria using the quotient method (Chapter 7). Several chemicals occurred at concentrations exceeding acute or chronic water quality criteria.

Effects on herons and mammals were also estimated using the quotient

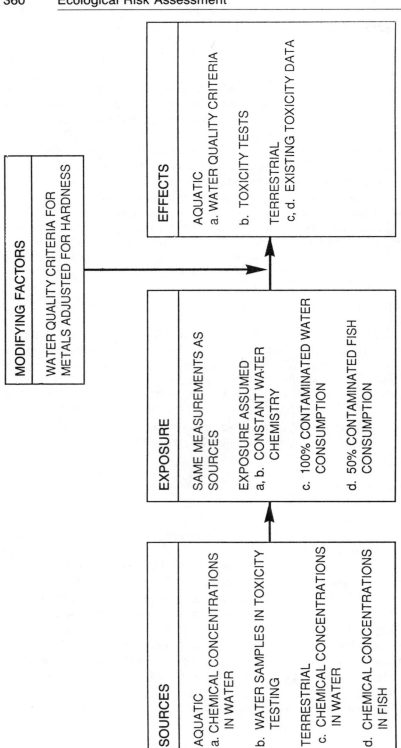

Figure 10.11. Integration of the indicators of source, exposure, and effects in risk characterization for a hazardous waste site.

method and existing toxicity data for individual chemicals. Herons were assumed to get 35% of their diet from contaminated fish and the expression of toxicity was the highest NOEL from a chronic test of any bird species, or, if only acute data was available for a chemical, 1/5 of an acute LC_{50}.* Hence the equation for the heron quotients was [0.35(concentration in fish)] / (NOEC or LC_{50}/5). The heron quotients were in the range 10^{-1} to 10^{-3} for the six chemicals assessed. Mammals were assumed to drink onsite lake water equal to 1% of their body weight daily, and the toxicity data were LD_{50}s. Hence the equation for the mammalian quotients was [0.01(concentration in water/weight)] / [(LD_{50}/5)weight]. The mammalian quotients were in the range 10^{-5} to 10^{-6} for the four chemicals that were assessed.

Risk Characterization

The rapid death of all fish in water from the onsite lakes suggests fairly clearly that the lakes would not support fish and probably would support a severely depauperate aquatic community. The observations of up to 15% mortality in some tests of some offsite water and of chemical concentrations that exceeded local or national water quality criteria suggest that toxic effects occur in offsite aquatic communities. However, because of (1) the absence of information concerning the biotic communities in the vicinity of the site, (2) the fact that the assessors did not use.the concentrations in fish to interpret the magnitude and significance of exposure, and (3) the rather meager analysis and interpretation of the toxicity data, the assessor could not make conclusions concerning the extent, magnitude, or nature of the effects.

The quotients for herons and mammals were judged to be low enough to conclude that the wastes did not pose a risk to these animals. This conclusion is suspect because the safety factors (the inverse of the quotient) were only 1–3 orders-of-magnitude for effects of the individual chemicals on birds. The uncertainties due to differences among bird species and life stages, the small fraction of the chemicals in the waste that were assessed, the possibility of combined toxicity of the waste chemicals, possible routes of exposure other than fish in the diet, and the poorly defined levels of exposure could overwhelm an order-of-magnitude safety factor. The assessor acknowledged these and other uncertainties, but, because of "a lack of an appropriate risk assessment methodology," he had to make a professional judgment.

Effects of Granular Carbofuran on Birds

Pesticides are subject to retrospective assessment when they are reregistered and when, as in this case, they are the subject of a "special review" because of an apparent hazard that was not apparent in the initial registration. The special review of granular carbofuran (henceforth simply carbofuran) was

---

*The factor of 1/5 is Urban and Cook's (1986) factor for correction of differences among species, derived from slopes of dose-response curves (Chapter 7), but apparently used here as an acute/chronic correction factor.

prompted by apparent cases of carbofuran-induced bird kills and laboratory data indicating that carbofuran is exceptionally toxic to birds. Carbofuran is a carbamate insecticide and nematocide which acts by inhibiting cholinesterase enzymes. This summary is based on the special review technical support document (OPP 1989). The case for the defense is presented by Rand (1990).

Hazard Definition

The spatial extent of the assessment is the areas in the United States where carbofuran is used. Because the major use is on corn, the assessment is largely concerned with the corn belt. Because this is an effects-driven assessment, the endpoint is constrained to some aspect of the observed effect, avian mortality. Although the endpoint is not explicitly defined, it appears to be reductions in populations of birds, particularly endangered species.

Measurement and Estimation

The primary source is granules on the soil surface, defined in terms of the density of granules resulting from various modes of application as measured in field trails ($334-9006/m^2$). Secondary sources are earthworms and vertebrates, with concentration of carbofuran in worms and dead or moribund vertebrates collected at field tests as the measure of exposure. Exposure was defined in the toxicity tests in terms of oral dose (mg/kg body weight and granules/bird) or concentration in diet (mg/kg food). It was measured in field tests and kill incidents in terms of body burdens, cholinesterase levels, counts of granules or analysis of food in the gastrointestinal tract, and observations of foraging and feeding behavior. Effects were measured in terms of mortality in toxicity tests and carcass counts in field tests and kill incidents.

Risk Characterization

All four of Koch's postulates were satisfied in this study. Dead birds were found in all of the field tests, and a sufficient number of kill incidents have been associated with routine use to establish that bird kills are not peculiar to planned studies (postulate 1). Carbofuran was found at high concentrations in dead and moribund birds from field tests and kill incidents (postulate 2). Toxicity tests established that carbofuran is acutely lethal at small doses including a single granule in passerine birds (postulate 3). Finally, the effective doses (estimated from gut contents and body burdens) and the biochemical and behavioral markers of effects were similar in the laboratory tests and in the field (postulate 4). This analysis established that carbofuran kills birds but does not address the magnitude of effects on the assessment endpoint. The components of risk characterization are presented in Figure 10.12.

To address population-level effects, the EPA used conceptual and mathematical models. First, an estimate of the rate at which birds are killed was

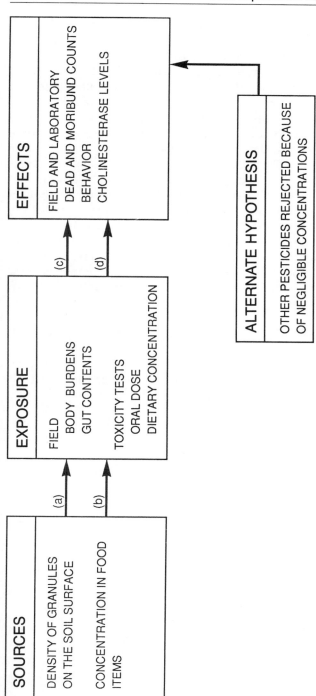

Figure 10.12. Integration of the indicators of source, exposure, and effects for characterization of risks of carbofuran toxicity to birds. (a) Birds seen foraging in treated fields and raptors seen scavenging dead or moribund birds. (b) Measured body burdens are reasonable given the availability of granules. (c) Laboratory and field exposures result in lowered cholinesterase levels, loss of coordination, and rapid death. (d) Death occurs at similar doses or body burdens in the laboratory and the field.

obtained by multiplying the average bird carcass density in field studies (0.3 per acre, with the 799 horned larks killed in one study deleted as atypical) times the acreage treated per annum. The result, 1,667,000 to 2,400,000 deaths per annum, suggests that a large number of individuals are killed but does not address the population-level implications. Second, the following conceptual population model was developed: (1) 13 of the 26 avian species that are declining in the U.S. are exposed to carbofuran (some documented victims of carbofuran are members of these 13 species and all of them frequent agricultural fields in areas of carbofuran use during periods of application); (2) carbofuran may or may not be the principal cause of the observed declines, but any source of mortality hastens the decline of a declining population; so (3) carbofuran is contributing to the decline of avian populations. This model obviously does not establish that carbofuran is contributing significantly to the decline of any species. Third, because the deaths of several bald eagles in eastern Virginia have been attributed to carbofuran, the EPA determined the effects of the observed mortality rate on the lower James River population as represented in Grier's (1980) model. They found that, even assuming maximum baseline survival and reproduction rates, the observed rate of carbofuran-induced mortality (approximately 7%) would reduce the growth rate of the model population to near zero.

The assessment established that carbofuran is killing large numbers of birds and suggested that it is contributing to the decline of avian populations including a local population of an endangered species. Because alternate pesticides are available that appear to pose less risk to avian populations and acceptable risks to other endpoints, the EPA proposed to cancel the registration of granular formulation of carbofuran.

11

Regional Risk Assessment

Glenn Suter

I placed a jar in Tennessee,
And round it was, upon a hill.
It made the slovenly wilderness
Surround that hill.

from *Anecdote of the Jar*, Wallace Stevens

Most risk assessments address the local effects of individual actions, but for several reasons effects at larger scales and of multiple actions need to be assessed. First, large local sources may have regional consequences. Examples include the *Exxon Valdez* oil spill and the Chernobyl reactor accident. Second, the combined effects of multiple acceptable sources within a region may be unacceptable either because of combined toxic effects of the mixtures, or because small effects of each effluent acting independently result in large effects at the regional scale. Third, regional scale processes may affect transport and transformation of pollutants in ways that are not observed at local scales. The obvious examples are the formation of photochemical oxidants from numerous individual sources of hydrocarbons and nitrogen oxides, and the formation of sulfate aerosols from emissions of sulfur dioxide. Fourth, emissions may have effects at regional scales that do not occur at local scales. Examples include depletion of stratospheric ozone by chlorofluorocarbon compounds, and climatic modification by carbon dioxide. Fifth, regions possess characteristics that do not occur at local scales and that must be protected. Finally, assessments of the success of regulatory or resource management programs must be performed at a regional scale in order to integrate the effects of individual decisions.

Regional risk assessment does not constitute a different type of risk assessment. The paradigms of predictive and retrospective risk assessment presented in Chapters 3 and 10 are applicable to regional-scale assessments. However, performing regional-scale assessments requires consideration of a set of issues that do not enter into local assessments. These include scaling, landscape description, and integration of a large number of qualitatively dissimilar stresses. Therefore, this chapter does not present a method for regional risk assessment but rather discusses components and issues of risk assessment in terms of regional-scale issues. A systematic discussion of regional risk assessment was presented by Hunsaker et al. (1990).

Description of the Region

The problem of setting bounds on the region to be assessed is similar to the problem of setting bounds on local assessments (Chapter 3). However, it is complicated by the greater number of factors that operate at larger scales and by the fact that assessment regions are often not as well defined as sites.

Ideally, regions to be assessed should be defined in terms of natural features. Examples include watersheds, airsheds, air basins, physiographic provinces, ecoregions, and the ranges of populations or communities of concern. One advantage of naturally defined regions is that regions so defined are relatively physically and biologically uniform and therefore are easily described and are likely to respond in a predictable way to an imposed stress. A more important advantage is that the processes that control transport, exposure, and effects are encompassed within the scope of the assessment. A good example is the intergovernmental program to assess the degradation of the Chesapeake Bay (Chesapeake Implementation Committee 1988). By including states such as Pennsylvania that do not include any of the bay but which constitute part of the watershed, this program ensures that the bay's hydrology and associated pollutant transport are not artificially truncated. In other cases, particularly those involving air pollution or marine pollution, natural boundaries are not so obvious. It may be necessary to perform a large-scale preliminary assessment to define the region within which the processes that control exposure and effects operate. The region may change depending on the endpoints and sources that are considered. For example, the watershed boundaries are appropriate for assessment of water quality effects on the Chesapeake Bay's aquatic populations, but assessment of the causes of declines in waterfowl abundance in the bay must include areas used during other portions of waterfowl breeding and migratory cycles.

Some ecoregion classifications have been developed specifically for ecological assessments (Omernick 1987, Whittier et al. 1988). These are designed to be relatively uniform with respect to certain ecological properties that are thought to determine response to disturbances (principally vegetation and land form). Before adopting these ecoregions, an assessor should ensure that they are in fact relatively uniform with respect to the properties that control response to the disturbance being assessed, and that they encompass the relevant structures and processes of transport, exposure, and effects.

Although naturally bounded regions are ideal, human actions, like Stevens' jar, inevitably shape regions. The "corn belt" is a human artifact, based on climate and soils but defined by the ingenuity of corn breeders, the market for corn, and other unnatural factors. Nevertheless, the corn belt is more real in the sense of having more influence on the landscape and ecology of the included area than an equivalent natural region, the tall grass prairie. It is clearly the most useful region for assessment of agrochemicals applied to corn.

In addition to setting bounds on regions on the basis of natural or anthropogenic features, the assessor must consider the distribution and magnitude of the sources. Examples of source-defined regions include the southern Appalachian coal basin as a region for assessment of acid mine drainage, and the airshed of the coal-burning plants on the Ohio River for assessment of acid deposition effects. For regional-scale retrospective assessments, the actual distribution of the pollutants may define the assessment region.

Finally, the bounds of a region may be defined by the wholly artificial consideration of political boundaries. For decades, ecologists have been fond of pointing out that pollutants do not recognize political borders, but risk managers are constrained by the fact that the political power to manage pollution ends at the political border. International and interstate treaties and compacts should be encouraged as a means around this problem, but political borders will continue to be a major constraint on risk management. In this situation, assessments should still be done on the region defined by appropriate natural and anthropogenic features, and the political bounds should be applied afterward as an overlay. In this way, the risk manager can be shown the extent to which the risks are being imported or exported across political boundaries.

As the preceding discussion suggests, the definition of an assessment region can be a complex process. In many cases, multiple factors will need to be considered. In general, it is desirable to be inclusive rather than exclusive. Define the assessment region to include the relevant natural and anthropogenic features, including the sources, the transport processes, and the extent of the affected populations and communities.

Endpoints

The number of potential assessment endpoints is larger for regions than for populations or individual ecosystems. The values to be protected in a region may be described in terms of characteristics of its component populations and ecosystems or in terms of characteristics of the region as a whole (Table 11.1). The population-level and community-level endpoints do not differ conceptually from the local level, so the only problem in using them at regional scales is scaling up from local to regional effects (see below). However, the appropriate assessment endpoints for representing the value of a region as a whole are not obvious. The ultimate value of a region, the unquantifiable abstraction contemplated by the risk manager, is the quality of life provided to the region's inhabitants. This is a function of the region's ability to provide food and other biological resources, clean air and water, aesthetic experiences, recreation, and other services without floods, wild fires, and other disservices. An appropriate strategy for selecting regional endpoints is to use measures of these services and disservices. A discussion of regional productivity as an assessment end-

Table 11.1. Types of Potential Assessment Endpoints for Regional Ecological Risk Assessments[a]

Traditional Ecological Endpoints

Population
  Extinction
  Abundance
  Production
  Massive mortality

Community/Ecosystem
  Change in type
  Production

Endpoints Characteristic of Regions

Population/Species
  Range

Productive capability
  Soil loss
  Nutrient loss
  Regional production

Pollution of other regions
  Pollution of outgoing water
  Pollution of outgoing air

Susceptibility
  Pest outbreaks
  Fire
  Flood
  Low flows

Climatic
  Drought
  Glaciation
  Sea level
  UV radiation level

Anthropocentric Endpoints

Population
  Frequent gross morbidity
  Contamination (FDA Action Levels)

Community/Ecosystem
  Market/Sport value
  Recreational quality

Air and water quality standards

Landscape aesthetics

[a]*Source:* Modified from Suter 1990b.

point will serve to illustrate the problems involved. Other endpoints are discussed in Suter (1990b).

Biological productivity is clearly an important regional endpoint but it is not as easily defined as one might suppose. Crop, livestock, and timber production are largely controlled by economic factors and mediated by technology, so realized production is not a good indicator of productive capability or of the influence of pollutants on productivity. In addition, certain increases in productivity such as fertilization of lawns and conversion of old-growth forests to tree plantations are not necessarily desirable. Loss of soil and nutrients might be used as an indicator of loss of regional productive capability, and can be measured by monitoring soil and nutrient loss in rivers draining a region. However, they do not directly indicate loss of productive capability because they are influenced by numerous factors other than soil management such as operation of reservoirs, construction practices, fertilization, and nutrient input in liquid wastes. The proportion of a region devoted to biological production or lost from biological production due to a particular action is relatively easily determined and is also used as a surrogate for regional production. However, land area estimates are insensitive to effects of pollutants or other actions not causing conversions of land use. Better measures of regional productivity need to be developed. One practical approach might be to use production of specific major crops, forest types, or rangeland types normalized to weather, land, water, fertilizer, pesticide, and energy input. Brown's (1987) "ecologically deflated production" goes a step further. It is calculated as realized agricultural production minus production from unsustainable practices such as tillage of highly erodible land.

At the regional scale it is particularly apparent that not all components of the environment can be measured, tested, modeled, or otherwise assessed. In addition, many of the potential endpoints, like regional production, are rather abstract or hard to define in operational terms. Therefore, it is important that risk assessors and risk managers identify the aspects of regions that are most significant at regional scales and are susceptible to the major known hazards, so that the assessment tools can be developed to address them.

Scale

The issues of choosing appropriate spatial and temporal scales for assessments, extrapolating between scales, and combining data collected at different scales are relevant to all assessments (Chapter 2), but they become critical at regional scales (Dale et al. 1989, Frost et al. 1988). As a simple example, consider an assessment to determine whether the range of a species of bird is changing by monitoring its occurrence at particular sites. On one square meter sites the species may occur and then become locally extinct several times a day, while on million square kilometer sites the species is unlikely to become extinct during a period of a thousand years. Clearly, if the changes in the range of the species are to be related to human actions, a scale of resolution somewhere between these extremes must be chosen.

Even within the temporal scale of routine human actions, causation may be scale dependent. For example, agronomic production is determined by agronomic factors on the scale of a field, by microeconomic factors on the scale of a farm, by ecological factors at the scale of a watershed or landscape, and by macroeconomic factors at the regional or national scale (Lowrance et al. 1986). Therefore, to assess the effects of environmental factors such as air pollution or soil loss on production, one should monitor production at the scale of natural landscape units.

The simplest assumption to employ in extrapolating from local to regional scales is that the response of the region is homogeneous. For example, if two birds per hectare are killed in a field test of a pesticide, then the number of birds killed per annum may be estimated as twice the number of hectares treated per annum in the region (Chapter 10). If different ecosystems respond differently or if different effects occur at different seasons, then one might assume that responses are homogeneous within categories and then use the relative abundances of treatment of the categories as weights to calculate the regional effect. For example, Minns et al. (1989) divided the lakes of the shield region of Ontario into categories based on seven factors that were thought to control their response to acid deposition, estimated the yield per hectare for each category at a particular acid deposition rate, and summed across lake categories the products of the estimated yields and the hectares of lakes to obtain the regional yield. Both of these approaches assume that effects are additive and the structure of the region does not matter. However, some exposure and effects-inducing processes clearly are not additive, such as the addition of new sources of oxygen-demanding effluents to a watershed.

If the physical and biological structure of the region modifies the response to a stressor, interactions of responses must be represented in the risk model. The simplest way to incorporate such interactions is to divide the region into cells that can be considered ecologically homogeneous, impose the stressor on each cell, change the state of the cells based on the response of each cell to the stressor, change the state of the cells based on the state of adjacent cells and a model of interaction between ecosystem types, and iterate the process for as many time steps as are required for the state of the region to stabilize. Graham et al. (1991) demonstrated the potential utility of this approach by simulating ozone stress in a landscape corresponding to the Adirondack State Park and in simulated landscapes. Assumptions were made about the increase in the probability and intensity of bark beetle attack in particular conifer species resulting from ozone exposure, the probability of attack in a cell, given the intensity of attack in adjoining cells, and the conifer mortality resulting from beetle attacks. Conifer mortality in turn caused changes in forest cover which affected lake water quality (conifers tend to acidify soils and subsequently water) and wildlife habitat. This heuristic exercise showed that the risks of particular changes in regional land cover, water quality, and landscape pattern in this sort of system depend on the initial landscape pattern.

Landscape Ecology and Ecotoxicological Risks

While ecotoxicological assessments have largely ignored regional-scale dynamics, ecologists concerned with physical disturbance and land use have developed a distinct field of inquiry termed landscape ecology (see Turner 1987a, Turner and Gardner 1991, Forman and Godron 1986, Urban et al.

1987, Zonneveld and Forman 1990, and the journal *Landscape Ecology*). This field is concerned with measuring properties of landscapes, developing landscape descriptors, and elucidating the role of landscape pattern in ecological processes. It has been described as a marriage of ecology and geography (Zonneveld 1990). Its recent popularity is attributable, at least in part, to the availability of powerful new tools: multispectral satellite and aerial imagery, image analysis software, and geographic information analysis software.

To date, landscape ecology has been largely descriptive and has largely been concerned with vegetation and land form. To the extent that it has addressed the assessment of human actions, it has been almost entirely concerned with the effects of physical disturbances such as forest harvesting and highway construction. These actions result in a modified landscape pattern which has aesthetic implications and which affects the frequency and magnitude of fires and floods, disrupts or enhances the spread of animals and seeds, and changes the abundance and probability of extinction of populations. In general, pollutants have not had effects of sufficient magnitude to modify landscapes, and, for this reason, use of landscape ecology as a management paradigm would tend to deemphasize toxicological concerns. However, modification of landscapes affects the transport of pollutants and the susceptibility of populations and communities to exposures and effects, so landscape characteristics should be incorporated into assessments of toxic substances.

An example of interaction of physical transformation of a landscape and toxicological effects is the extinction of the California condor in the wild. Condors are large vultures that require a habitat that generates large carcasses (relative to competing scavengers, they forage over large areas and can open large tough carcasses). These were apparently supplied by large ungulates and sea mammals during pre-Columbian times. Spanish settlers replaced the native ungulate community with cattle and horses, but the slaughtering of cattle for their hides may have actually improved the region as condor habitat. However, "Anglo" settlers converted the rangelands of California's valleys to tilled agriculture. Following intensive sealing, whaling, and shoreline development, the coastal habitat also produced fewer large carcasses. The apparent result was that condor numbers declined, and the remnant population increasingly competed with turkey vultures, coyotes, eagles, and ravens for smaller carcasses. Under these circumstances, carcasses of sheep and of game animals shot but not retrieved by hunters became important resources. However, this led to ingestion of predator control agents and of lead bullets and shot. The discovery of lead and cyanide poisoning in the remnant condor population led to their removal from the wild to a captive breeding program. Although it has been argued that habitat was not limiting (Ogden 1985), it is clear from a broader perspective that regional transformations forced the California condor population into marginal subsistence which exposed it to toxicants and made it highly susceptible to extinction. Attempts to reestablish California condors as a natural population will need to consider the risks of lead and other contaminants to condors, given the relative abundance of clean and contaminated carcasses provided by the region, and the ability of the region as it exists to provide resources for a self-sustaining population of large vultures. Given that the competitive advantage of condors over other scavengers is the ability to forage over very large areas and find large carcasses, it is questionable that the region will support anything but a heavily managed and precarious population.

Consideration of interactions between landscape pattern, population biology, and toxicology could greatly increase the predictive capabilities of ecological risk assessment. For example, in the EPA's review of carbofuran they suggested that it contributed to the decline of those avian species that visited carbofuran-treated fields and that were already declining (Chapter 10). However, they could not substantiate this claim because they did not know the causes of the declines in the avian populations, much less whether carbofuran-induced mortality would, in fact, augment the declines. A serious attempt to predict the effects of a pesticide on a population would require analysis of the population biology of the birds, including limiting factors and compensatory mechanisms (Chapter 8), and of the temporal and spatial distributions of the birds, of the controlling factors, and of pesticide applications. Such assessments would require the combination of landscape ecology, population ecology, and toxicology.

Multiple Stresses

The need to consider the combined effects of multiple and diverse stresses on ecological systems is more compelling at regional than local scales. When assessing the effects of an effluent or waste dump, it is usually possible to ignore the effects of changes in land use, resource management practices, "background" pollution, and other sources of stress on the exposed biota. However, at regional scales the influences of these stresses are apparent, are often greater than the toxicological hazard being assessed, and are often so dynamic that they cannot be treated simply as a background to the toxicological effects. Therefore, the toxic effects on a particular regional-scale endpoint should be assessed as one component of a cumulative effects assessment.

Consideration of the joint effects of multiple actions has become a hot topic in environmental impact assessment under the heading of "cumulative impacts" (Beanlands et al. 1985, Peterson et al. 1987, Bedford and Preston 1988, Gosselink et al. 1990). The following classification of cumulative effects is modified from one developed by a joint Canada—United States workshop (Beanlands et al. 1985).

Nibbling:

The cumulative effects of a number of actions which have similar small incremental effects. The individual actions are typically dismissed with statements of the form: The X hectares of ecosystem type Y that will be damaged represents less than Z percent of the total area of that ecosystem type. This approach can result in either a genuinely minimal effect at a regional scale or the environmental equivalent of the death of a thousand cuts. An ecotoxicological example is allocation of a sacrifice area for each pollutant source such as the allocated impact zones allowed in permits for aqueous effluents in a body of water (EPA 1983a).

Time-Crowded Perturbations:

The cumulative effects that occur when actions are so close in time that the system has not recovered from effects of one before the next one occurs. Examples include the repeated application of a single pesticide or sequential applications of multiple pesticides before prior applications have degraded or before populations of nontarget organisms have recovered.

Space-Crowded Perturbations:

The cumulative effects that occur where actions are so close in space that the areas within which they can induce effects overlap. The obvious ecotoxicological examples are effects of multiple aqueous effluents that mix in a river, and multiple atmospheric effluents that mix in an air basin.

Indirect Effects:

The cumulative effects that occur when the direct effects of actions are not space- or time-crowded, but their indirect effects are.

Assessing interactions of pollution effects on a region with effects of intensive land uses such as agriculture, forestry, and urbanization, requires that some assumptions be made about the form of interactions. As with interactions of individual chemicals (Chapter 7), the simplest assumption is that they are additive, but more complex interactions are possible. The best understood examples of nonadditive effects are thresholds. Once patches of habitat drop below a certain size they will not support populations of a species; once the organic input to a water body rises above a certain level, anoxia results; once mortality rates rise above a certain level in a population, extinction occurs; etc. Identification and quantification of such thresholds is a critical component of cumulative effects assessments. Synergisms and antagonisms are more difficult sorts of nonadditivity in cumulative assessments. As with chemical mixtures, mixtures of diverse disturbances may have more or less than additive effects. In the case of the California condor, habitat degradation and toxic exposures had a joint effect (extinction in the wild) that was greater than would have resulted from simply adding the losses that either would have caused, acting alone.

The assessment of multiple stresses demands that well-defined endpoints be used that are applicable to ways in which all of the stresses act on the target biota. It is not possible to combine toxic effects expressed as multiples of an MATC, fishing effects expressed as tons harvested, and habitat degradation expressed as hectares of salt marsh filled. Instead, an endpoint such as recruit abundance (the abundance of new one-year-olds) must be used so that all effects can be expressed in the same units (Barnthouse et al. 1990). In general, resource managers have developed appropriate endpoints and models for this purpose, individuals engaged in assessment of habitat loss have been working at developing equivalent capabilities (e.g., FWS 1980, Bovee and Zuboy 1988), and ecological toxicologists are only beginning to develop the needed tools.

Even when the same endpoint is used for assessments of multiple stresses, the ways in which the endpoint is estimated may be incompatible. Prediction of the effects of stresses requires that the means of estimating the endpoint be sensitive to the mechanisms by which each stress acts. For example, Turner (1987b) estimated changes in the net primary production of Georgia, USA, by estimating the areal extent of 18 land use categories and assigning a primary production rate to each category. This approach is perfectly adequate if effects of land use changes are the only concern, as in Turner's assessment. However, this means of estimating production cannot be used to assess how land use has combined with air pollution, herbicides, fertilization, etc. to determine changes in production. On the other hand, the National Crop Loss Assessment Network's region-specific estimates of reduction in production of individual crops exposed to air pollutants (Adams et al. 1985b) are readily combined with estimates of the effects of economic, climatic, and management factors to estimate realized production.

Prospects for Regional Risk Assessment

By now the reader of this chapter should have the impression that the prospects for regional ecological risk assessment are rather daunting. The concept of a region is vague and regions are harder to define than sites, potential assessment endpoints are numerous and diverse, very large extrapolations across spatial and temporal scales are necessary, cumulative impacts of diverse human actions must be considered in most cases, and the necessary assessment tools are only beginning to be developed. However, the large effort devoted to the assessment of acid deposition has provided an awareness of regional assessment problems and an appreciation of the need for future regional-scale assessments to use an integrative framework such as risk assessment. In addition, although the bulk of effort in environmental assessment and regulation is still devoted to local problems such as waste disposal sites, public concern for problems with larger scales than their backyards has been stimulated by the issues of acid deposition, ozone depletion, and greenhouse gases. Finally, the EPA and other environmental agencies have come to realize that, to assess how well they are protecting the environment, they must step back and assess trends in environmental quality at the regional scale.

One result of this increased awareness of the importance of regional-scale assessment is the development of the EPA's Environmental Monitoring and Assessment Program (EMAP). EMAP will ecologically characterize a sample of sites including near-coastal marine, inland surface water, wetlands, forests, arid lands, and agroecosystems and monitor status and trends in indicators of exposure, effects, and environmental stressors (OMMSQA 1990). To accomplish its goal of quantifying degradation and improvement in environmental resources and identifying causes of these changes, EMAP will need to perform numerous retrospective regional-scale risk assessments. Because this effort will encompass the full range of ecosystem types, pollutants, and nonpollutant stresses occurring in the USA, it must develop a powerful science of regional ecological risk assessment. The relationship between surveillance monitoring programs such as EMAP and risk assessment is discussed in the next chapter.

12

Environmental Surveillance

Glenn Suter

Environmental biologists worry too much about statistical methodology and too little about what they are monitoring for.

R.H. Green, 1984

Much of environmental science has been devoted to monitoring the environment for purposes of surveillance rather than to support specific risk assessments. Environmental surveillance has at least four general purposes. The first is assessing statutory or regulatory compliance. The second is providing information needed for systems control or management. The third is assessing environmental quality so as to determine whether it is satisfactory or new initiatives are needed. The fourth is detecting unexpected environmental damage.

Monitoring programs have often failed for lack of a well-defined purpose (Vouk et al. 1987, Green 1984, Segar and Stamman 1986, Perry et al. 1987, and many others). Development of a risk-based approach to surveillance monitoring could correct that failing. Surveillance is the repetitive measurement of some environmental characteristic over an extended period of time. It is not another type of risk assessment. Rather, it is a set of activities that bear various relationships to risk assessment and decisionmaking, but have in common concerns with selection of optimum sites and measures, and the extraction of a useful signal from the resulting time series of data. Careful definition of the goals of surveillance monitoring and the assessments needed to achieve those goals ensure the relevance and utility of environmental surveillance programs.

PURPOSES OF SURVEILLANCE MONITORING

Assessing Compliance

Surveillance monitoring can be used to determine whether regulatory criteria or legislative goals are being met. Most commonly, this type of monitoring is applied to ambient air and water quality criteria for specific chemicals which are expressed as concentrations that should not be exceeded when measured with some averaging time and frequency. Such monitoring is conceptually and practically straightforward, but provides an indirect measure of progress toward environmental quality. It is more appropriate to determine whether the environmental goals formulated by environmental legislation and regulations are being achieved. This type of surveillance can be site-specific (i.e., conducted locally to determine whether specific regulatory actions have been effective) or regional (i.e., conducted at state or national scales to determine whether a law or regulatory program is achieving its goals). Given direct measures of the environmental values that are to be protected (e.g., number of species or productivity of resource populations) an agency can determine whether their actions have been effective, or whether standards and other regulatory requirements need to be changed or augmented.

Compliance monitoring occurs after a risk assessment has been conducted to establish criteria, standards, permit requirements, or other actions. If contaminant concentrations are monitored, it must be assumed that the assessment that established the criteria accurately set levels that would be protective of the endpoint values without being excessively overprotective. If the endpoint values or indicators of the endpoints are monitored, then it is possible to evaluate the assessment as well as to evaluate compliance with the regulatory goal.

Surveillance for System Control

Surveillance monitoring may be used as a means of system control. Any monitoring of chemical or biological conditions could be used to determine when an ecosystem's capacity to tolerate pollutants is being reached so that the rate of release could be managed. However, it is clearly preferable to use indicators that respond rapidly, and measurement endpoints that are easily obtained at frequent intervals. Cairns (1976) envisioned a system of telemetered fish distributed in a river whose level of physiological stress would be monitored by a computer. When a sufficient number of fish in a reach displayed stress symptoms, alarms would be activated that would cause effluents at the head of that reach to be reduced. Such a system could detect changes in an emission or mix of emissions that increase toxicity or changes in the environment that reduce the resistance of organisms to the contaminants. Although this goal has not been achieved, analogous systems have been developed to monitor the toxicity of aqueous effluent streams by diluting a side stream of effluent and running it through tanks where the responses of organisms are monitored (Figure 12.1). The most common responses measured in such systems are fish ventilation and fish rheotaxis. Although several techniques had been developed in the '70s for routine biological monitoring of effluents (Cairns et al. 1976), and progress has been made in techniques for rapid automated measurement of biological responses and in telemetry (e.g., Gruber et al. 1988, Heath 1987), biological monitoring of effluents and ambient media as a means of system control has not caught on. Toxicity testing of

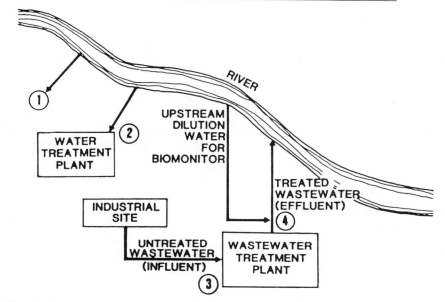

① Surface water biomonitor or upstream biomonitor for a water treatment plant.
② Water treatment plant biomonitor.
③ Influent wastewater biomonitor.
④ Effluent wastewater biomonitor.

Figure 12.1. Diagram of a system to biologically monitor the quality of waste water (from van der Schalie 1987).

effluents and ambient media (Chapter 10) could serve the same control function, but it is generally performed too infrequently to effectively serve that function.

Monitoring for system control is like monitoring for compliance except that the time scale is shorter so the standards or permit conditions need not be so conservative. A risk assessment is performed and the results are used both to design the treatment system and to set standards for the control system (e.g., more than two fish coughing with a frequency greater than x). When the biological stress responses exceed the standard, actions are taken to reduce exposure. The biological standard is like a water quality standard, but it is more biologically relevant and integrates the effects of all components of the effluents.

Assessing Environmental Quality

In addition to determining whether particular criteria or legislative mandates are being achieved, it is desirable to determine whether the existing laws, regulations, and industry practices are resulting in a desirable environmental quality, and to determine whether conditions are improving or not. This goal is often described as "characterizing status and trends." The EPA is currently attempting to achieve this goal in the U.S. by establishing a national Environmental Monitoring and Assessment Program (EMAP)(OMSQA 1990), and the U.S. Geological Survey has proposed an analogous National Water Quality Assessment Program (Hirsch et al. 1988).

This type of surveillance program should be designed to serve as the initial stage of retrospective risk assessments (Chapter 10). That is, detection of apparent temporal or spatial trends should lead to an attempt to determine their reality, causes, extent, and consequences. Therefore, measurements should be chosen for their relationship to clearly defined assessment endpoints, there should be an explicit model of the relationship between the measurement endpoints and the assessment endpoints, associations between sources, exposure, and effects should be analyzed, the uncertainties in the measurements and their relation to the assessments should be defined and, as far as possible, quantified, and the results should be characterized in a way that supports the need of a risk manager to make informed decisions.

Detecting the Unexpected

Unexpected environmental effects inevitably occur, because not all sources of pollution can be identified and assessed; not all environmental pathways, physical-chemical processes, and breakdown products can be predicted; not all species can be tested for responses to all pollutants; and not all pathways of indirect effects can be identified. Examples include the acidification of lakes and streams, the DDT-induced decline of raptors, and more recently, the decline of many populations of frogs and toads. Therefore, it is prudent to broadly monitor the environment to detect significant changes in the environment that may be due to human actions. Like monitoring environmental quality, monitoring to detect the unexpected serves as a stimulus for retrospective risk assessments. However, monitoring for the unexpected is likely to rely on encountered data from academic ecological studies, public environmental activities such as the Christmas bird counts, and incident reports from the public or local agencies, rather than from a consistent, statistically designed program. It is difficult to establish the magnitude, extent or causation of effects from such data.

The remainder of this chapter is primarily concerned with surveillance monitoring for assessment of environmental quality. Programs of this type tend to be much less clearly focused than those for assessing compliance or system control, and would most benefit from a clearer risk assessment context. In contrast, any program to detect unexpected effects would be relatively unfocused, largely undesigned, and would depend on an ability to digest large volumes of heterogeneous data.

SURVEILLANCE AND RISK

The assessments performed with surveillance data are conceptually different from both prospective and retrospective risk assessments. They attempt to detect exposure or effects as they are occurring rather than predict them or explain and describe them after they have occurred. Therefore, surveillance assessments are concerned with trend analysis, pattern recognition, and other techniques for extracting a signal from the noise of background variation. The concept of risk should enter into both the data analyses and the design of surveillance programs. Any surveillance program has limited resources so not everything can be measured, not every location can be measured, and nothing can be measured with perfect accuracy or precision. Therefore, there is always a risk of false positives or false negatives. To estimate these risks it is necessary to completely define the assessment endpoint. That is, one must identify an entity to be assessed (waterfowl), a characteristic of the entity (density of breeding pairs), an amount of change that must be detected (15% change), an area within which change must be detected (20% of the prairie pothole region), a time allowed to detect the change (within 5 years), and statistical confidence (90% certainty of detecting the change at the 0.05 level of significance). With that information, and the results of preliminary measurements to indicate the background variance, one can design a sampling program that reduces the risk of erroneous conclusions to a desired level (Green 1984). When using the results of surveillance monitoring to assess temporal and spatial trends, one must use the variance information to estimate the actual risks that a true trend or pattern was not detected (type two error) or that an apparent trend or pattern was actually due to natural variability (type one error).

Surveillance programs act as instigators of retrospective risk assessments. Once increasing toxicant levels, population declines, local extinctions, or other problems have been identified, a retrospective risk assessment must be performed to determine the cause, magnitude, extent, and implications of the problem (Chapter 10). A well-designed and conducted surveillance program can provide considerable input to risk assessments. Ideally, the surveillance program would provide a sufficient basis for risk assessment. However, because most surveillance programs are not designed to support risk assessment, they focus on only one of the components of risk: sources, exposure, or effects. Therefore, most risk assessments prompted by surveillance results must generate measurements of the other two components of the triad. In addition, the frequency, density, or scale of measurement in a surveillance program are often inappropriate for determining the extent and magnitude of effects. Therefore, it is often necessary to remeasure the measured variables so as to establish the spatial and temporal associations, and develop the conceptual, statistical, or mathematical models that quantify the associations among the components.

ENDPOINTS FOR SURVEILLANCE MONITORING

Much ink has been expended on advocacy of particular measurement endpoints for surveillance monitoring. A review of that literature is beyond the scope of this chapter (see Hunsaker and Carpenter 1990). However, the following general points address the relevance of measurement endpoints for surveillance monitoring in the context of risk assessment.

First, the relationship of source, exposure, and effects is different in surveillance-based assessments. Rather than being components of an assessment that are used together to estimate the risk to an assessment endpoint, they are alternate classes of endpoints. Surveillance-based assessments may estimate trends and patterns in the magnitudes of sources, exposures, or effects. Most environmental surveillance programs have measured proximate sources of exposure (concentrations in water, air, sediment, etc.) or ultimate sources of exposure (concentrations in effluents, runoff, etc.). Source monitoring is simple and directly relevant to air and water quality standards. It has a tenuous relationship to effects but is easily associated with a cause. Indicators of exposure (principally body burdens or tissue concentrations of chemicals but also biomarkers of exposure — Chapter 10) are less commonly measured than sources. They are easily measured but are directly interpretable only in terms of human dietary standards. Although exposure indicators are not directly interpretable in terms of either environmental standards or effects, they have the advantage of being readily associated with sources and potentially serving as predictors of effects. Indicators of effects (biological characteristics thought to be indicative of response to pollutants) have seldom been monitored because the sampling is relatively difficult, the natural variance is high, and the results are difficult to interpret or attribute to a cause (LaPoint et al. 1984). However, only by monitoring effects can a surveillance program directly address environmental goals. Sources, exposure, or effects can be monitored to determine trends in the environment, but determination of the significance of a change in either sources or exposure requires that associated effects be estimated.

There are no magic indicators of environmental quality. No measurement and no summary index of diverse measurements can serve as a surrogate for the various responses of an ecosystem to a disturbance (Green 1984, Dayton 1986). Similarly, any finite array of measurement endpoints only represents a finite array of effects. Even a "holistic" response such as net primary production is simply one of many "holistic" responses and is not representative of all ecosystem-level responses or of responses at other levels of organization. Therefore, in surveillance, as in conventional risk assessments, one must decide what aspects of the ecological response are most important. Then one can decide what the appropriate measurement endpoints are, and how the sampling should be conducted to derive them. For example, rather than trying to decide whether birds can be used as bioindicators of environmental quality (Temple and Wiens 1989), ornithologists should acknowledge that birds are socially valued and therefore worth monitoring in their own right. They would then determine what measures of the state of bird populations and communities best reflect those values, how they should be monitored, and how causes of patterns and trends can be identified.

Although endpoints for surveillance studies must be well defined to properly design the sampling strategy and to estimate the risk of not detecting effects, they have seldom been well defined, in part because it is advantageous to imply that a program will prevent all effects. However, "i) no activity of man is without concomitant effects on man and the environment, ii) some degree of effect on the environment or man is acceptable and may be desirable, and iii) no surveillance program can ever demonstrate that there is ZERO effect on a particular resource due to anthropogenic inputs" (Segar and Stamman 1986). In other words, if you measure anything intensively enough, a statistically significant difference between sites or time intervals can be demonstrated, and the absence of detected differences at any level of effort cannot disprove the existence of differences. Therefore, one must decide in advance what types of effects, what levels of effects, what extent of effects, and what duration of effects must be detected. If many types of effects are deemed to be important, if the levels of effects to be detected are small relative to background variance, and if fine temporal and spatial resolution is desired, then financial resources are likely to become limiting. In extreme cases, the desired assessment endpoints may not be technically achievable.

Measurement endpoints may be appropriate to surveillance programs that are not suitable for risk assessment. Risk assessments are concerned with causation and with estimation of well-defined effects. However, surveillance programs are, at least in part, concerned with reducing large volumes of data to a few numbers or graphs that can be quickly presented to the public and their political representatives. Recently, there has been renewed interest in the old idea of indices of water quality based on the species composition of aquatic communities (Hellawell 1986, Karr et al. 1986). The simplicity of using a single number to express the state of a community is appealing, and, as long as they represent important responses of the community, indices may be useful (Box 12.1). However, once a program passes from presenting trends and patterns to understanding their causes and predicting their consequences, it is necessary to disaggregate the indices.

Box 12.1. Problems with the IBI and Similar Indices

The "index of biotic integrity" (IBI–Karr et al. 1986) illustrates some undesirable properties of environmental indices. It combines dissimilar measures,
which leads to index ambiguity (i.e., you cannot tell why the index is high or
low) and eclipsing (i.e., a serious negative effect on one measure may be
hidden by measures that are not affected or are positively affected) (Ott 1978).
Other problems with indices include sensitivity to the computational scheme
used to aggregate the components (e.g., additive, multiplicative, max/min),
and the high variance on indices due to aggregation of component variances
(Ott 1978).

More importantly, the IBI shares with the more complex diversity indices the
highly undesirable property of not serving as a clear measurement of anything
(Poole 1974, Green 1980, 1987). In other words, it is not clear what, if any,
property of communities is represented by those particular aggregations of
parameters.

The IBI assumes that certain species of fish are sensitive and others are
resistant (Karr et al. 1986). However, the division of fish into these classes
appears to be based largely on tolerance of habitat quality characteristics such
as suspended solids and dissolved oxygen levels that commonly affect fish
communities in midwestern streams. As was discussed in Chapter 7, there are
no consistently sensitive species with respect to toxic chemicals, much less to
all chemicals plus all physical conditions, and all habitat modifications. When
a presumed resistant species is sensitive to a particular pollutant or a presumed
resistant species is sensitive, the IBI values will be misleading. A priori suppositions about sensitivity are appropriate only if specific pollutants or classes of
pollutants are being monitored and the relative sensitivity of the populations in
the monitored community to those pollutants are well-characterized.

Finally, by positing a single scale of "biological integrity" along which
stream communities move as their exposure to anthropogenic stress increases,
the IBI assumes that there is only one type of response of fish communities to
anthropogenic stress. In other words, although the use of 12 parameters
implies that different effects could cause a community to follow different
vectors in a 12-dimensional state space, combining the variables into a single
index requires that there be only one vector. That assumption implies that
there is only one mode of action of physical and chemical agents, that all
modes of action produce the same response at the community level, or that
each mode of action influences one of the components of the index without
influencing the others.

The IBI has been used by some agencies with apparent success (Ohio EPA
1988), but that success is relative to simply measuring chemical concentrations
or to alternate indices with the same or worse problems. Alternatively, it is
possible to summarize effects data while avoiding suppositions, aggregations,

and ambiguity by using the measurements to estimate higher-level system properties that are important in themselves. These might be population characteristics such as mean age, fecundity, and condition factor (Munkittrick and Dixon 1989a, 1989b), stand indices that are aggregations of measures of the health of individual trees (Alexeyev 1989), or unitary, comprehensible, and significant community parameters such as number of species as a proportion of the expected number of species (LaPoint et al. 1984).

Practical technical and economic issues are more important in selecting endpoints for surveillance programs than in taking environmental measurements for risk assessment. In a retrospective risk assessment, there is a particular hazard to be assessed in a relatively short time so relatively difficult and expensive measurement techniques can be employed. However, surveillance does not have the same urgency and needs to be carried out for a long time, ideally at frequent intervals, and potentially over a large area. In addition, long-term routine measurements are not as likely to be carried out by highly qualified and experienced people. Finally, a surveillance program is likely to measure more indicators because it is less focused on particular pollutants or effects. Therefore, measurement endpoints for surveillance monitoring should be obtainable quickly, simply, and at little cost. This is one reason why instrumental analysis of pollutant chemicals has been the most popular monitoring approach. For this reason, automation of biological measurements will be a great boon to surveillance monitoring. Potential examples include telemetry of physiological measurements and computerized analyses of satellite imagery.

Finally, sensitivity of the measurement endpoint is more important in surveillance programs. In most retrospective risk assessments, the hazard has operated for some period of time that is sufficient for effects to have occurred. However, an ideal surveillance program should be concerned with detecting new exposures and effects in their early stages so that severe effects can be prevented. Therefore, it is desirable to select measurement endpoints that are sensitive and respond rapidly to act as sentinels even if they are not significant in themselves. The necessary degree of sensitivity and rapidity depends on the goals and design of the program. For example, surveillance for control of emissions must respond within a matter of minutes if it is to prevent serious effects of emissions from breakdown of a treatment plant or other upset conditions. At the other extreme, for a large-scale program such as EMAP that will monitor at annual or five year intervals, any measurement endpoint that responds within a year is rapid enough. In fact, many responses and source-response associations will be too rapid for EMAP. Sensitivity or rapidity can be obtained by selecting sensitive endpoints for the biological level of interest or by going to lower levels. For example, if one is interested in monitoring trends in fish populations, one could either choose populations with short life cycles that are r-selected, or one could use organismal or sub-organismal level responses such as a condition index or a physiological response.

HYPOTHESES FOR SURVEILLANCE MONITORING

The design of any surveillance program depends on the question to be answered. Most surveillance studies and most of the advice on design of surveillance studies assume that the question is framed in terms of disproving a null hypothesis (Green 1984, Segar and Stamman 1986). As discussed above, this approach is appropriate if a maximum acceptable level of a source, exposure, or effect is identified that needs to be distinguished from background with some probability. In addition to specifying the assessment endpoint in this way, it is necessary to carefully identify what is meant by background (i.e., what is the null case). In practice, the background is typically taken to be a nominally undisturbed reference site. However, the objective is unlikely to be answering the question: does disturbed site x differ by a particular amount from undisturbed site y but rather, does it differ from undisturbed sites in general $y_{1...n}$ (Hughes et al. 1986)? This is equivalent to the problem of determining whether a real effect has occurred in a retrospective risk assessment (Chapter 10).

It is often inappropriate to force a monitoring program into this statistical form. For example, if the question is what is the rate of change in the mean concentration of a pollutant in a reservoir, then the hypothesis that the rate is zero has no particular significance, and if the pollutant is no longer being added, the zero rate is impossible. In general, if the question concerns rates or magnitudes the program should be designed to provide a point estimate with a specified precision. For example, determine the rate of decline (X) with a 95% confidence interval of less than 0.5X to 2X. Once a monitoring-based risk assessment moves from determining whether a real trend or pattern exists to determining its magnitude and biological significance, testing null hypotheses is not particularly helpful.

EXAMPLES

The utility of monitoring the quality of ambient media (proximate sources) is illustrated by the results of a review of water quality data from two surveillance programs conducted by the U.S. Geological Survey, the National Stream Quality Accounting Network, and the National Water Quality Surveillance System (Smith et al. 1987). Some water quality goals were being achieved during the 1974–1981 period; specifically improved sewage treatment lowered fecal coliform levels, and reduction of lead in gasoline reduced lead concentra-

tions. However, the increased regulation and treatment of point-source effluents did not produce the expected improvements in dissolved oxygen, phosphorus, or nitrate because of the importance of nonpoint sources. In addition,
increases in nitrate, arsenic, and cadmium appeared to be associated with
atmospheric deposition, a source of water pollution that had not been of great
concern in the 1970s. The use of a representative sample of streams led to the
conclusion that nonpoint sources dominated national trends. However, if the
goal had been to detect changes in severely impacted areas rather than in the
nation as a whole, it is likely that point sources would have appeared to be
more important. Because exposure and effects data were not collected, it is not
clear which, if any, of the trends were biologically significant. Therefore, a
logical follow-up to these results would have been to perform risk assessments
to determine whether the observed changes in water quality had biological
effects, and, for those pollutants that are increasing, to estimate when significant effects will occur at the current rates of increase.

The National Contaminant Biomonitoring Program (NCBP) serves as an
example of a program devoted to monitoring the degree of exposure of organisms to toxicants. The program, which began in 1969, now monitors concentrations of seven elements and various chlorinated organic chemicals in waterfowl, starlings, and freshwater fish. A summary of elemental analyses of fish
from 112 stations for the period 1978 to 1981 found little evidence of temporal
trends or consistent patterns among regions because differences in species
created considerable extraneous variance (Lowe et al. 1985). Because the goal
of the program was simply to "monitor temporal and geographic trends in
contaminant concentrations" rather than to provide a basis for assessment,
there is no tie to source data or effects. Similar difficulty in interpreting
exposure monitoring data in terms of sources or effects is found in the "mussel
watch" program (Farrington et al. 1987). Risk assessments are concerned with
regulating sources and with preventing effects, but exposure is simply an intermediate step and is not important per se. Therefore, without significant trends
and without correlations with sources or effects, such exposure monitoring
programs become dead ends.

More surveillance data are available for bird populations than for any other
potential class of effects indicators. In the U.S., long-term and large-scale
programs of the U.S. government, state governments, and private organizations (principally the National Audubon Society) monitor the distribution and
abundance of avian species in the U.S. The conclusions of a recent publication
that reviewed the utility of these data illustrate the difficulty of assessing
trends in effects data (Temple and Weins 1989). The authors found that trends
were hard to interpret because often there are delays between environmental
change and avian population response, and because avian populations are
often buffered by density-dependent compensation. They concluded that
proximate responses such as changes in mortality rates or fecundity were more
likely to reveal effects than ultimate population responses such as density,
abundance, or range. However, if the management goal is "to maintain viable

populations over a long period in particular habitats or regions," these proximate responses have little importance. The ability to detect changes and trends is limited by the variance in population parameters and the ability to monitor large numbers of birds. For example, DDT caused approximately a 30% decline in the fecundity of bald eagles. Fecundity is monitored at fewer than 40 bald eagle nests each year, but to detect a 30% decline in fecundity relative to the average rate with 90% confidence at the 0.05 level of significance would require monitoring 120 nests. In contrast, the approximately 1,500 nest records obtained each year for bluebirds would allow a 4% decline in fecundity to be detected at those levels of confidence and significance. Finally, the authors conclude that it is nearly impossible to determine the causes of detected trends because data concerning habitat change and other potential causes are not collected in conjunction with the avian data. This point is illustrated by a review of breeding bird survey data which identified many apparent trends in abundance, but could only attribute causation in cases of mass mortality due to extreme weather events (Robbins et al. 1986). In sum, tremendous effort in collecting and compiling avian population data has been of little use because it has lacked clear purposes and has not been designed to support assessments. One result is that, despite an abundance of monitoring data, it is not clear which North American birds that winter in the tropics are declining and whether tropical deforestation is a causal factor (Terborgh 1989, Wille 1990).

CONCLUSIONS

With the exception of programs to monitor compliance with media quality standards, most environmental surveillance programs have been inconclusive. That is, they have not been able to go beyond describing patterns in the data to determining whether they represent real phenomena and whether anthropogenic causes are responsible. This failing could be eliminated by designing surveillance programs either to serve as the basis for risk assessments, or to identify hazards or damage for subsequent risk assessments. The chief distinction is that to serve as the basis for risk assessment a surveillance program must monitor sources and effects and should monitor exposure, but to serve as a stimulus for risk assessment, only one component of risk need be monitored.

The strategy of surveillance as a stimulus to risk assessment can be pursued only if it is feasible to identify causes and consequences of trends after the fact. That is, if a program monitors only sources, the designers of the program must have a strategy to determine the effects of patterns or trends after they have been detected; if a program monitors only effects, there must be a strategy to determine the causes of patterns or trends after they have occurred; and if a program monitors only exposure, there must be a strategy for determining the sources and effects. Clearly, this is difficult to achieve. Unless a natural historical record such as metal burdens in tree rings or pollutant levels in bedded sediments is available (Chapter 10), or a record is obtained by other programs such as the weather records (Robbins et al. 1986), one cannot use regular spatial association or temporal conjunction to help establish the causal nature of an association.

There are also conceptual difficulties in designing a surveillance program to serve as a basis for risk assessment. In many cases, it is not clear in advance what associations of sources, exposures, and effects need to be detected and it is obviously not feasible to monitor everything. In general, when monitoring organisms to detect effects, it is advisable to also monitor the habitat characteristics that are known to control abundance and production in the absence of pollutant effects. Most variance in the levels of effects indicators are likely to be due to these factors, so the need for risk assessments of anthropogenic effects can be minimized by assessing natural factors first. Another potentially useful strategy when causal associations cannot be hypothesized in advance is specimen banking (Lewis 1987). If biological specimens are archived at ultracold temperatures, they can be used after the appearance of effects or identification of a new pollutant of concern to determine body burdens of contaminants, biomarkers of exposure or effects, or levels of suborganismal injury that occurred before and during the period when the pollutant or effect appeared.

Finally, coordination and integration of surveillance monitoring programs would greatly increase their utility for risk assessment. The return on investment in environmental surveillance is greatly diminished by the way that each program is designed as if no other programs existed. As a result, data from different programs, which should be mutually supportive in terms of identifying and explaining patterns and trends, cannot be used together. Ideally, all environmental surveillance programs within a nation or region should be coordinated (RSC 1988). At minimum, anyone designing a new surveillance program should determine in advance what existing surveillance programs an assessor could draw on to identify potential causal relationships or to define the generality of the patterns and trends detected by the new program. Then the timing, distribution, and methods should be chosen to make integration possible. Even if it is not possible to force coordination of surveillance programs, it would be useful to compile the results of surveillance studies and analyze the compilation for spatial, temporal, and causal patterns. The goal of detecting unexpected effects would be particularly well served by such a metaanalysis of data from all environmental surveillance programs in a region.

13

Exotic Organisms

Glenn Suter

> *Beware the Jabberwock, my son!*
> *The jaws that bite, the claws that catch!*
> *Beware the Jubjub bird, and shun*
> *The frumious Bandersnatch!*
>
> —Lewis Carroll, *Jabberwocky*

Many of the most severe anthropogenic effects on the world's natural biotic communities have resulted from the introduction of exotic organisms (Crosby 1986, Mooney and Drake 1986, Drake et al. 1989). Through competition, predation, and pathogenesis, exotic organisms have extinguished native species or reduced them to inhabiting refugia and have drastically changed the character of the invaded communities. Exotic organisms include domestic species, accidentally introduced species, nonnative game and fish species, biocontrol agents, and, quite recently, species modified by bioengineering. Most introductions of exotic species have been accidental or have been conducted without due regard for possible consequences, but introductions of exotic organisms are now regulated in the U.S. and most other countries. The development of genetic engineering techniques has resulted in concerns about man-made pests and pathogens. These concerns have led to numerous reviews of the bases for concern and methods for responding (Halverson et al. 1985, Gillett 1986a, Fiksel and Covello 1986a and 1988, Office of Technology Assessment 1988, Hodgson and Sugden 1988, Mooney and Bernardi 1990).

In this chapter, introductions of all exotic organisms are discussed together because the effects of an introduced organism depends on its ecological properties and not on whether it is a product of natural selection, breeding, or genetic engineering. It has been claimed that products of genetic engineering are less hazardous than other exotic organisms because their characteristics are well specified (e.g., Davis 1987). However, any organism may display surprising traits.

This chapter attempts to outline a predictive risk assessment methodology for exotic organisms based on the methodology for predictive risk assessment of toxic chemicals (Chapter 3). It draws on previous attempts at guidance by Suter (1985), Gillett (1986b), Fiksel and Covello (1986b), Barnthouse et al.

(1988), and Simonsen and Levin (1988), and on attempts to assess risks of environmental transmission of pathogens (ECAO 1986, 1989). All of these discussions are abstract and conceptual because formal risk assessment methods for exotic organisms have not been developed or applied. Currently, regulatory assessments of exotic organisms are performed on an ad hoc basis and rely more on expert judgment than on analysis.

The methods developed for assessment of chemicals are not directly applicable to organisms because organisms have nonentropic properties including reproduction, evolution, phenotypic plasticity, and adaptation to specific habitats. The implications of these characteristics are developed throughout this chapter but they include the following: (1) Unlike chemicals, which only disperse and degrade, a population of organisms can disperse, reproduce, grow, and disperse again in a cycle of indefinite duration; (2) The phenotypic plasticity of individual organisms and the ability of populations of organisms to evolve increases their ability to persist and grow; (3) Because organisms can evolve, their effects on natural systems can change. Despite these differences, the general paradigm for predictive risk assessment (Figure 3.1) is applicable to exotic organisms.

Source Characterization

Risk assessments of organisms differ from those of chemicals in that the identity of the agent to be released is not easily defined. The members of a biotic species are not as uniform as the atoms of an element or molecules of a chemical compound. Particularly for microbes, the properties of a population are not entirely definable from a species designation. Genetic engineering further complicates the definition of the agent being assessed. The organism must be defined in terms of its source (a wild population, a breeding stock, a type culture, etc.), its phenotypic properties, and its genetic properties, particularly any added genetic material.

Having defined the organism, the properties of the release must be defined. Many deliberately released exotic organisms are agricultural pest control agents or plant symbionts, and their releases can be defined like chemical agents in terms of mass per unit area of spray, seed coating, etc. Accidental releases can be treated like chemical or radiological accidents in terms of probability that containment systems will fail, resulting in a release. For organisms with well-defined containment systems such as genetically engineered microbes in a research laboratory or in an industrial plant, fault tree analysis, event tree analysis, or simulation models can be used (Rasmussen 1981, Lincoln et al. 1983). For other cases such as accidental introduction of exotic pests or escape of exotic pets, examination of the historical record and other informal techniques could be used to estimate the release rate.

Table 13.1. Major Classes of Endpoints for Direct Effects of Exotic Organisms

1. Reduction in populations of nontarget organisms by pathogenicity, parasitism, or predation.
2. Changes in the composition or production of a community due to changes in the physical or chemical characteristics of the habitat (modification of element cycles, pH, soil structure, etc.).
3. Competitive displacement of a valued native species.
4. Proliferation to the extent of becoming a pest.
5. Partial degradation of a chemical by a cleanup organism to form a more hazardous intermediate (i.e., more toxic, more mobile, or more persistent).

Environmental Description

Because of the ability of organisms to reproduce and spread, the methods that serve to delimit the environment significantly contaminated by a chemical will not serve for organisms. The environment to be assessed must be the total area that could serve as habitat for the introduced organism. Hence, environmental description is largely a process of defining the inclusive niche of the organism (its complete set of habitat requirements) and then determining where and in what seasons that requirement is met.

Choosing Endpoints

For chemicals, choosing endpoints is largely a matter of determining what valued environmental components could be significantly exposed to the chemical. For organisms, choosing endpoints is largely a matter of examining the traits of the organism to determine what it might do in the ambient environment and then inferring what valued components of the environment might be affected by those activities. This process of hazard identification has been the principal concern of risk assessments of exotic organisms. That is, because there are no reliable methods for estimating the risk posed by exotic organisms, release of exotic organisms is permitted only in the absence of a credible hazard. If a chemical is toxic to fish, a risk assessment is performed, but if an organism is an inadvertent fish pathogen, it will not be released. Identification of the hazard constitutes a sufficient assessment.

The most easily defined hazards of exotic organisms are the direct effects of the known properties of the organism. Organisms that are deliberately introduced have some purpose, and can cause undesired effects by performing their function inappropriately. For example, biocontrol agents may be pathogenic, parasitic, or predatory to nontarget species. Classes of endpoints for direct effects of organisms are listed in Table 13.1.

The identification of indirect effects of organisms or direct effects of organism functions other than those for which it would be released is much more problematical. A relatively straightforward example is the proposed use of a genetically engineered virus to inoculate raccoons and other wildlife against rabies (Anonymous 1990). The major direct hazard is pathogenicity of the agent, a vaccinia virus with rabies virus genes. An indirect effect that seems more likely than the direct effect is disruption of the population cycles of the raccoons and other predators that are immunized by consuming the vaccine in bait. Raccoons are already pests in many rural and suburban communities and increased populations of raccoons, foxes, etc. could exacerbate that problem, as well as increasing predation on nesting waterfowl and other undesired effects on prey species.

It is the need to identify these indirect or novel hazards that inspires the often-repeated call for "case-by-case" assessment of exotic organisms (Tiedje et al. 1989, Simonsen and Levin 1988). One problem with this case-by-case brainstorming approach to hazard identification is that it is rather haphazard. It provides no means of systematically working through the implications of an organism's properties, or of distinguishing plausible from implausible scenarios. A horror story can be generated for any organism, particularly if gene transfer is hypothesized. For example, it was suggested that the insertion of a lac marker gene in *Pseudomonas aureofaciens* might lead to increased spoilage of lactose-containing food products (Simonsen and Levin 1988). Guidelines need to be developed to make hazard identification efficient and thorough, and risk assessment methods need to be developed to deal with the capacity to hypothesize hazards for exotic organisms.

Hazard Minimization

Genetic engineering potentially provides a means of minimizing hazards that is not available for chemicals. Traits can be engineered into organisms that reduce their ability to (1) perform any function other than the intended function, (2) perform their function in any situation other than the intended one, or (3) transfer genes to other organisms. The extreme example is suicide genes, genes that would cause the death of organisms in prescribed circumstances. Another strategy modifies the organism so that it requires a nutrient that is rare in nature and that is applied with the organism. These strategies lower the risk that an application of an exotic organism will cause effects, but they cannot eliminate the hazard because the selective pressure to evolve out of the limitations will be quite strong.

Exposure Assessment

The process by which ecological systems become exposed to exotic organisms is presented in Figure 13.1. Release is defined by the source term. Dispersal of exotic microorganisms is similar to the transport and fate processes of chemicals (Chapter 5). For example, microbial biocontrol agents sprayed on a crop are dispersed like droplets of pesticide: drifting in the air, depositing on leaves and other surfaces, being inhaled by animals, washing off surfaces, etc. Modeling of microbial dispersal is discussed by Andow (1986) and Barnthouse and Palumbo (1986). Dispersal of organisms must consider two traits that are not characteristic of chemicals. First, organisms are actively mobile or possess

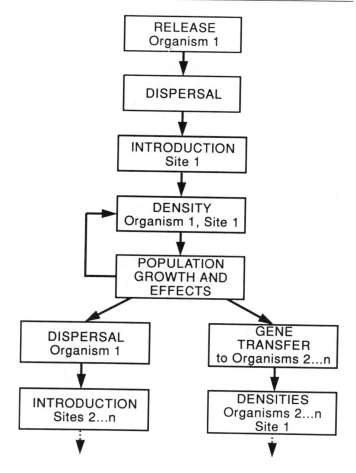

Figure 13.1.   A diagram of the process of exposure of environmental components to an exotic organism.

stages that are adapted to passive dispersal. Second, chemicals are irreparably diluted by transport processes, but organisms are not. A seed, a cluster of invertebrate eggs, or a gravid female vertebrate may be a significant introduction. Much of the literature of biogeography is devoted to documenting the many "sweepstakes routes" that disperse colonist organisms to islands or between continents. Therefore, infrequent routes of dispersal cannot be ignored.

The dispersal processes result in introduction of a particular inoculum density at particular sites. If the site does not constitute a suitable habitat, the organisms will die at some rate, depending on conditions. If the organism has resistant resting stages, the death rate may be quite low. During this period of population decline, the habitat may change to a suitable state, or organisms may be transferred to a suitable habitat. For example, insect pathogens on a leaf may be consumed by a host insect. Suitability of a habitat is not a simple dichotomous variable. A large initial inoculum may colonize a hostile site by modifying the physical environment, resisting or overwhelming predators and competitors, overcoming a host's resistance mechanisms, etc. In addition, a large inoculum is more likely to contain individuals with advantageous traits or rapidly acquire advantageous traits by adaptive mutations.

Upon arrival in a suitable habitat, organisms will use nutrient and energy sources to grow and reproduce. This process, which has no analog in exposure assessment for chemicals, simultaneously induces effects and increases exposure of the environment to the agent of the effects. The appropriate models of this process are population dynamic models for the exotic organism.

The process of replication has two consequences for exposure. First, it results in a greater density of organisms, which results in greater exposure of the habitat to the organism. Second, the greater density of organisms results in a larger source term for dispersal to other sites. This cycle of dispersal, introduction, establishment, and dispersal is repeated until all suitable and accessible sites are colonized, or until some natural or anthropogenic factor causes the extinction of the organism.

Models of the passive physical transport of microorganisms are the most straightforward component of exposure assessment for exotic organisms. Models developed for the transport of chemicals and particulate pollutants can be adapted for transport of organisms by substituting a death rate for degradation rates, adhesion properties for sorption properties, etc. Examples include models of microbial transport in air (Lighthart and Frisch 1976, Peterson and Lighthart 1977) and in groundwater (Harvey and Garabedian 1991).

Active biological and mechanical transport is far more difficult to predict and is likely to be more important than passive transport in many cases. A motile animal can not only transport itself but can also seek out suitable habitats. Microorganisms, invertebrates, and seeds have nearly the same capability. They are carried on the surface or in the guts of animals and, because animals tend to move between similar habitat patches (e.g., from one pond or corn field to another), the passengers tend to be deposited in suitable habitats. Mechanical transport (e.g., ships, trucks, airplanes) is as effective as biological transport in moving organisms long distances (kilometers) and more effective over very long distances (thousands of kilometers). There are no models available to simulate biological or mechanical transport.

The establishment of an organism in a new habitat can be predicted if its niche is known. The niche can be approximated by defining the hyperspace whose dimensions are the physical, chemical, and biological parameters that determine whether the organism can survive, grow, and reproduce. Exposure assessments must include a means of defining that hyperspace. For example, the EPA requires that the growth characteristics of potentially hazardous microbial biocontrol agents be defined with respect to pH, temperature, light, essential nutrients, salinity, antagonists, and potential autotrophic substrates (Betz et al. 1982).

If a site constitutes suitable habitat, then there is some probability that introduced organisms will become established, which is a function of the inoculum density and the quality of the habitat at the time of arrival. Currently, this function cannot be estimated. However, if an organism is repeatedly reintroduced to a suitable habitat (e.g., from an nearby site that is already colonized or from repeated applications), establishment can be treated as inevitable. Once the likelihood of establishment is estimated, population growth and abundance can be estimated using population models derived from laboratory studies. All of this is conceptually straightforward but difficult in practice because of the undeveloped state of microbial population biology.

The processes of establishment and growth of a population and the transfer of genetic material can be physically modeled using microcosms (Liang et al. 1982, Gillett 1986b, Shannon 1989). Because the fate of organisms is highly sensitive to habitat variables, the design of microcosms for this use is of critical importance. Although standard generic designs such as the soil core microcosm (Chapter 9) can be used, they must be designed, within the bounds of realism, to optimize the probability of survival of the test organism. That is, they must include the organism's niche hyperspace. An organism introduced into a microcosm with inappropriate physical, chemical, and biological conditions will quickly become extinct without providing useful information concerning its ability to grow, compete, and proliferate in appropriate habitats. The difficulty of this requirement is illustrated by a study of survival and effects in a mixed flask culture microcosm of *Bacillus thuringiensis* var. *B. israelensis*, a potential mosquito control agent (Shannon 1989). Mosquito pupation, and therefore effects of the microbe, varied among flasks apparently due to differences in the algal species composition.

Another approach to exposure assessment for exotic organisms is analogy to previous introductions. There is a long and often sad history of the introduction and spread of exotic organisms. Study of the movement and fate of introduced pathogens, pests, crops, game species, livestock, etc. can provide general lessons concerning the qualities of organisms that make good invaders and of environments that are easily invaded. Gilpin (1990) has argued that the success of biological invasions cannot be predicted because of the complexity of interactions with biotic and abiotic components of the environment, and argues for statistical analysis of past invasions to estimate probabilities of success of different classes of invasions. Others have argued that mechanistic predictive models can be developed, but they must be based on analysis of past cases (Drake et al. 1989). Certainly, if a specific exotic organism that is proposed for introduction is ecologically similar to previously introduced organisms, the analogy should be used to estimate the likelihood of establishment, rate of spread, and ultimate abundance. The general utility and lessons of these analogies are discussed in several publications including Sharples (1983), Mooney and Drake (1986), Drake (1988), and Drake and Mooney (1989).

The most poorly understood and most controversial aspect of exposure assessment of exotic organisms is the transfer of genetic material among microbes. Genetic transfers between different strains and species occur in nature (Slater 1985), and might potentially move a novel gene from a debilitated strain to one that is not debilitated, or from a species in which it causes benign traits to one in which it causes undesirable traits. Information concerning the potential for transfer of introduced genes is routinely required by regulatory agencies in the U.S. (OSTP 1986). However, there is no consensus on how to interpret the information in terms of the risk of gene transfer in the field.

Effects Assessment

Effects of exotic organisms can be measured in pathogenicity tests, in microcosms, or in field tests. The tests for pathogens and for predators and parasites that are biocontrol agents are relatively well developed. Tests for free-living organisms other than insect predators are not at all well developed.

Effects assessments for pathogens and parasites must determine whether the introduced organism will cause pathology in nontarget organisms and what the nature, severity, frequency, and extent of any pathological effects will be. Similarly for predators and parasitoids, the range of nontarget prey and hosts attacked, the ability to kill the prey or host, and the extent to which prey or host populations may be depressed by the resulting mortality must be determined. The primary tool for this class of effects assessments is laboratory pathogenicity tests. Routine methods have been developed for testing biocontrol agents for insects (Burges 1981, Laird et al. 1989). The EPA requires tests of the pathogenicity of microbial biocontrol agents to birds, mammals, fish, aquatic invertebrates, terrestrial plants, aquatic plants, and nontarget insects (Betz et al. 1982). The mode of exposure in these tests should resemble the mode of exposure expected to occur in the field, and an exposure-response relationship should be derived.

Because only a few potential host or prey species can be tested, it is necessary to extrapolate the results to the range of species and life stages that may be exposed to the exotic pathogen, parasite, or predator. For example, if a biocontrol agent is not pathogenic to an earthworm population in the laboratory, what can we infer about pathogenicity to the same species in the field, to other species of earthworms, or to the thousands of soil insects, arachnids, crustaceans, and myriapods that the earthworm test is said to represent? There are no accepted rules or models for this type of extrapolation.

Microcosms provide a means of testing the effects of free-living nonpredatory organisms and provide a potentially more realistic arena for testing effects of pathogens, parasites, and predators. Effects tests can be performed in the same microcosms as the exposure tests, or in microcosms specifically designed for effects testing (Lightner et al. 1989). In addition to being congenial to the exotic organism, the test must be conducive to the hypothesized effects. In some cases, there is a trade-off between these two criteria. The habitat that provides the best establishment and growth may be the one for which the organism was designed, but the worst effects may occur in suboptimal conditions where the organism utilizes secondary substrates or hosts. Hence, the test system must be designed for the hazard and not just the organism. The problem of extrapolation from these systems to the array of real ecosystems is as unsolved for organism tests as for chemical tests (Chapter 9). However, the required range of extrapolations is narrower. Toxic chemicals can occur and cause effects in nearly any ecosystem, but an organism will only cause effects in systems that constitute habitat, a narrow subset of ecosystems. This generalization is somewhat vitiated by the ability of organisms to expand their habitat range by evolution.

In addition to extrapolating between hosts or ecosystems, the effects assessment must extrapolate from proximate to ultimate effects on the assessment endpoint. The effects of pathogens on host populations are described by epidemiological models, and models of biological pest control can be adapted to estimate effects on populations of susceptible nontarget organisms (e.g., Pinnock and Brand 1981). Ultimate effects of nonpredatory free-living organisms must be estimated from some conceptual or mathematical model of the ecological system in which the effects would occur. For example, if a free-living organism was developed to increase nitrogen fixation in soil, assessors would need to consider not only how much nitrogen it would contribute to various terrestrial ecosystems, but also what the consequences of increased nitrogen would be on the composition of plant communities, their susceptibility to disease, their ability to support wildlife, and water quality.

Risk Characterization

Risk characterization for chemical and physical agents includes estimation of the probability of specific effects by relating the estimated level of exposure from the exposure assessment to the exposure-response relationship from the effects assessment (Chapter 3). However, as is discussed above, exposure to a particular inoculum level of an organism (e.g., initial density in an environmental medium) does not simply cause a predictable level of effects. The level of effects is determined by the level of exposure (population density), but the rate at which effects are imposed (activity) also determines the rate of population growth (increase in exposure). For example, the initial density of an inoculum of a nitrifying bacterium determines the initial rate at which ammonium is nitrified, but the rate at which nitrification occurs, which is a function of that initial density plus environmental factors, determines the rate of proliferation of the bacterial population and the density of bacteria to which the environment is exposed in the next time step. Therefore, exposure (bacterial density) and effects (nitrification rate) are not independent and dependent variables. The level of effects is also not simply a function of the maximum population density. A population that is vigorously growing may have greater effects on its environment than a large stagnant population.

As with chemicals, the form of the exposure-response relationship will be constrained by practical considerations. The expression of exposure must be predictable from existing data and must be measurable in the effects tests. The case for pathogens is closest to that of chemicals. For example, if a microbial pesticide is sprayed on vegetation, the density of cells on the vegetation can be estimated, and, to compare this to an oral inoculation level from a pathogenicity test, the ingestion rate of nontarget herbivores must be estimated. In addition, the temporal aspect of exposure must be considered. As with acute exposures to chemical pesticides, one might assume that one day's ingestion is equivalent to an acute exposure (dose or inoculation).

For other classes of organisms, the correct means of integrating exposure and effects is less clear. The assessor should consider what aspect of the occurrence of the organism in a particular environment is the best expression of exposure. The concept of carrying capacity is useful in this regard. The degree to which an ecosystem is affected by an organism is determined by the amount of support the ecosystem supplies for the organism to exert its hazardous behaviors. Hence, it may be appropriate to express the potential effects of an organism on a system as a function of the level of limiting factors for the organism in the community. For example, if a nitrogen fixing organism is limited by phosphorus, then the degree to which an ecosystem will be exposed to that organism is a function of phosphorus availability.

Another approach to risk estimation is to treat the exposure-response process as occurring in a black box. That is, an assessor can treat a microcosm, mesocosm, or field site in a manner that approximates the commercial application, look for effects, and assume that the same level and type of effects will occur when the organism is in use. This approach simplifies the assessment but it limits the potential for extrapolating to different sites or different uses of the organism.

However the risk is estimated, the assessment must always bear a warning that organisms adapt and evolve unpredictably. The assessment is based on the behavior of the organism as released and not as it may be after a year or a decade in the field. This unpredictable risk that the organism will acquire undesirable traits after release is somewhat balanced by the equally unpredictable ability of ecosystems to adapt and evolve in response to new organisms. Because of this ability of the organism and the receiving community to evolve, there is always uncertainty associated with the results of a risk assessment for an exotic organism. In other words, even if the effects of the organism at the time of release could be perfectly specified, there would still be a risk that must be balanced against the benefits.

Conclusions

Historically, ecosystems have excluded or resisted most introduced organisms (Simberloff 1981). This is also likely to be the case with genetically engineered organisms. In addition, those organisms that become established will often have no significant negative effects. However, if the first 100 or 1,000 releases of genetically engineered organisms occur without negative effects, that must not become an argument for reduced emphasis on risk assessment. One deliberately released chestnut blight, carp, or kudzu could be sufficient to turn the public against biocontrol, bioremediation, and genetic engineering.

As the biotechnology industry develops and as biocontrol becomes a more important agricultural practice, the current ad hoc approach to risk assessment will become unfeasible. It will become necessary to develop standard guidelines, models, and inference procedures. An NSF-sponsored workshop concluded that "development of a generic approach to risk assessment of environmental applications of biotechnology is both feasible and desirable" (Ficksel and Covello 1986b). It would be better to develop those methods before the need becomes desperate.

References

[The numerous reports of United States government agencies, their laboratories, and contractors cited here may be available from the authors listed in the citations but are certainly available from the National Technical Information Service, Springfield, VA 22161, USA.]

Abbott, W. S. 1925. A method of computing the effectiveness of an insecticide. J. Econ. Entomol. 18:265–267.

Adam, D. F. 1963. Recognition of the effects of fluorides on vegetation. J. Air Pollut. Control Assoc. 13:360–362.

Adams, N., K. H. Goulding and A. J. Dobbs. 1985. Toxicity of eight water soluble organic chemicals to *Selenastrum capricornutum*: a study of methods for calculating toxic values using different growth parameters. Arch. Environ. Contam. Toxicol. 14:333–345.

Adams, R. M., T. D. Crocker, and R. W. Katz. 1985b. Yield response data in benefit-cost analyses of pollution-induced vegetation damage. pp. 56–72. In W. E. Winner, H. A. Mooney, and R. A. Goldstein (eds.), Sulfur Dioxide and Vegetation. Stanford University Press, Stanford, California.

Adams, S. M. (ed.). 1990. Biological Indicators of Stress in Fish. American Fisheries Society, Bethesda, Maryland.

Adams, S. M., and D. L. DeAngelis. 1987. Indirect effects of early bass-shad interactions on predator population structure and food web dynamics. pp. 103–117. In W. C. Kerfoot and A. Sih (eds.), Predation in Aquatic Ecosystems, University Press of New Hampshire, Hanover, NH.

Adams, W. J., R. A. Kimerle, B. A. Heidolph, and P. R. Michael. 1983. Field comparison of laboratory-derived acute and chronic toxicity data. pp. 367–385. In W. E. Bishop, R. D. Cardwell, and B. B. Heidolph (eds.), Aquatic Toxicology and Hazard Assessment: Sixth Symposium. American Society for Testing and Materials, Philadelphia.

Adams, W. J., P. S. Ziegenfuss, W. J. Renaudette, and R. G. Mosher. 1985. Comparison of laboratory and field methods for testing the toxicity of chemicals sorbed to sediments. pp. 494–513. In T. M. Posten and Purdy (eds.), Aquatic Toxicology and Environmental Fate: Ninth Volume. American Society for Testing and Materials, Philadelphia.

403

Adriano, D. C. 1986. Trace Elements in the Natural Environment. Springer-Verlag, New York.

Alabaster, J. S. and R. Lloyd. 1982. Water Quality Criteria for Freshwater Fish, Second Edition. Butterworth Scientific, London.

Aldenberg, T., W. Slob, and J. M. Knoop. 1990. Confidence limits for hazardous concentrations based on logistically distributed NOEC toxicity data. (in BKH 1990).

Alexander, M. 1980. Biodegradation of toxic chemicals in water and soil. pp. 179. In R. Haque (ed.), Dynamics, Exposure and Hazard Assessment of Toxic Chemicals, Ann Arbor Science Publishers, Ann Arbor, MI.

Alexander, D. G. and R. Mc.V. Clarke. 1978. The selection and limitations of phenol as a reference toxicant to detect differences in sensitivity among groups of rainbow trout (*Salmo gairdneri*). Water Research 12:1085–1090.

Alexeyev, V. 1989. Forest health diagnosis and its application in air pollution impact studies. pp. 135–141. In R. D. Noble, J. L. Martin, and K. F. Jensen. (eds.), Air Pollution Effects on Vegetation Including Forest Ecosystems. Northeastern Forest Experiment Station, Broomall, Pensylvania.

Allen, T. F. H. and T. B. Starr. 1982. Hierarchy: Perspectives for Ecological Complexity. Univ. of Chicago Press, Chicago.

Altshuler, B. 1981. Modeling of dose-response relationships. Environ. Health Persp. 42:23–27.

AMS (American Management Systems, Inc.). 1987. Review of the literature on ecological end points. Office of Policy, Planning and Evaluation, U.S. Environmental Protection Agency, File Report.

Andersen, K.P., and E. Ursin. 1977. A multispecies extension to the Berton and Holt theory of fishing, with accounts of phosphorus circulation and primary production. Medd. fra Dan. Fisk. Havunders. 7:313–345.

Anderson, F. R. and R. M. Daniels. 1973. NEPA in the Courts: A Legal Analysis of the National Environmental Policy Act. Resources for the Future, Baltimore.

Anderson, J., S. Kieser, R. Bean, R. Riley, and B. Thomas. 1981. Toxicity of chemically dispersed oil to shrimp exposed to constant and decreasing concentrations in a flowing system. pp. 69–75. In Proceedings of the 1981 Oil Spill Conference, American Petroleum Institute, Washington, DC.

Anderson, T. J. and G. W. Barrett. 1982. Effects of dried sewage sludge on meadow vole (*Microtus pennsylvanicus*) populations in two grassland communities. J. Appl. Ecol. 19:759–772.

Anderson, V., T. Bailey, N. R. Draper, and J. O. Rawlings. 1985. Crops research review report. Am. Statistician 39:244–250.

Andow, D. A. 1986. Dispersal of microorganisms with emphasis on bacteria. pp. 470–487. In J.W. Gillett (ed.), Potential Impacts of Environmental Release of Biotechnology Products: Assessment, Regulation, and Research Needs. Environmental Management 10(4).

Andrew, R. W., D. A. Benoit, J. G. Eaton, J. M. McKim, and C. E. Stephan. 1977. Evaluation of an application factor hypothesis. Unpublished manuscript. U.S. Environmental Protection Agency, Duluth, Minnesota.

Andrewartha, H. G. and L. C. Birch. 1984. The Ecological Web. University of Chicago Press, Chicago.

Angus, R. A. 1983. Phenol tolerance in populations of mosquitofish from polluted and unpolluted waters. Trans. Am. Fish. Soc. 112:794–799.

Ankley, G. T., M. K. Schubauer-Berigan, R. A. Hoke. 1992. Use of toxicity identification techniques to identify dredged material disposal options: a proposed approach. Environ. Manage. 16:1–6.

Anonymous. 1990. First genetically engineered vaccine for rabies in wildlife. The Gene Exchange 1:1–2.

APHA (American Public Health Association). 1985. Standard Methods for the Examination of Water and Waste Water. 16th ed. Washington, DC.

Apostolakis, G., 1985. The broadening of failure rate distributions in risk analysis: how good are the experts? *Risk Analysis* 5:89–92.

Asher, S.C., Lloyd, K.M., Mackay, D., Paterson, S., Roberts, J.R. 1985. A critical examination of environmental modeling — modeling the fate of chlorobenzenes using the persistence and fugacity models, Rep. No. NRCC 23990. National Research Council of Canada.

Ashton, W. D. 1972. The Logit Transformation with Special Reference to Its Uses in Biology. Hafner Publishing Co., New York.

ASTM (American Society for Testing and Materials). 1984. Standard practice for evaluating environmental fate models of chemicals. E 978–84. ASTM, Philadelphia, PA.

ASTM (American Society for Testing and Materials). 1991. Annual Book of ASTM Standards, Sec. 11 Water and Environmental Technology. Philadelphia, Pennsylvania.

Atkinson, R. 1986. Kinetics and mechanisms of gas phase reactions of hydroxyl radical with organic compounds. Chem. Rev. 86:69–201.

Atkinson, R. 1987a. A structure-activity relationship for the estimation of rate constants for the gas-phase reactions of OH radicals with organic compounds. Int. J. Chem. Kinetics, 19:799–828.

Atkinson, R. 1987b. Estimation of OH radical reaction rate constants and atmospheric lifetimes for polychlorobiphenyls, dibenzo-p-dioxines and dibenzofurans. Environ. Sci. Technol. 21:305–307.

Aulerich, R. J., R. K. Ringer, and J. Safronoff. 1986. Assessment of primary vs. secondary toxicity of Aroclor 1254 to mink. Arch. Environ. Contam. Toxicol. 15:393–399.

Babich, H., J. A. Puerner, and E. Borenfreund. 1986. In vitro cytotoxicity of metals to bluegill (BF-2) cells. Arch. Environ. Contam. Toxicol. 15:31–37.

Babich, H., C. Shopsis, and E. Borenfreund. 1986. In vitro cytotoxicity testing of aquatic pollutants (cadmium, copper, zinc, nickel) using established fish cell lines. Ecotoxicol. Environ. Safety 11:91–99.

Baccini, P., J. Ruchti, O. Warner, and E. Grieder. 1979. MELIMEX, an experimental heavy metal pollution study: regulation of trace metal concentrations in limnocorrals. Schweiz. Z. Hydrol. 41:202–272.

Baes, C. F., III. 1982. Prediction of radionuclide K_d values from soil-plant concentration ratios. Trans. Amer. Nucl. Soc. 41:53–54.

Baes, C. F., III and S. B. McLaughlin. 1984. Trace elements in tree rings: Evidence of recent and historical air pollution. Science 224:494–497.

Bahner, L. H. and J. L. Oglesby. 1982. Models for predicting bioaccumulation and ecosystem effects of Kepone and other materials. pp. 461–473. In R. A. Conway (ed.), Environmental Risk Analysis for Chemicals. Van Nostrand Reinhold, New York.

Bailey, H. C., D. H. W. Liu, and H. A. Javitz. 1984. Time/Toxicity Relationships in Short-Term Static, Dynamic, and Plug-Flow Bioassays. pp. 193–212. In R. C. Bahner and D. J. Hansen (eds.), Aquatic Toxicology and Hazard Assessment, ASTM STP 891. American Society for Testing and Materials, Philadelphia.

Baker, D.J. 1990. Planet Earth: the view from space. Harvard University Press, Cambridge.

Baker, J. P. and T. B. Harvey. 1984. Critique of acid lakes and fish population status in the Adirondack Region of New York State. NAPAP Project E3-25. Draft Final Report, U. S. Environmental Protection Agency, Corvallis, Oregon.

Baker, J. P. 1989. Assessment strategies and approaches. pp. 3-1-3-15. In W. Warren-Hicks, B. R. Parkhurst, and S. S. Baker, Jr. (eds.), Ecological Assessment of

Hazardous Waste Sites: A Field and Laboratory Reference Document. EPA 600/3-89/013. Corvallis Environmental Research Laboratory, Oregon.

Baker, J. P., D. P. Bernard, S. W. Christensen, M. J. Sale, J. Freda, K. Heltcher, P. Scanlon, P. Stokes, G. Suter, and W. Warren-Hicks. 1990. Biological effects of changes in surface water acid-base chemistry. State-of-Science/Technology Report 13. National Acid Precipitation Program, Washington, DC.

Balci, O. and R.G. Sargent. 1981. A methodology for cost-risk analysis in the statistical validation of simulation models. Communications of the Association for Computing Machinery 24: 190–197.

Balcomb, R., C. A. Bowen III, D. Wright, and M. Law. 1984. Effects on wildlife of at planting corn applications of granular carbofuran. J. Wildl. Manag. 48:1353–1359.

Balcomb, R. 1986. Songbird carcasses disappear rapidly from agricultural fields. Auk 103:817–820.

Baldocchi, D. D. 1988. A multi-layer model for estimating sulfur dioxide deposition in a deciduous oak forest canopy. Atmos. Environ. 22:869–884.

Barber, M. C., L. A. Suarez, and R. R. Lassiter. 1988. Modeling bioconcentration of nonpolar organic pollutants by fish. Environ. Toxicol. Chem. 7:545–558.

Barbour, M. T., C. G. Graves, and W. L. McCulloch. 1989. Evaluation of the intrinsic rate of increase as an endpoint for *Ceriodaphnia* chronic tests. pp. 273–288. In G. W. Suter II and M. A. Lewis, (eds.), Aquatic Toxicology and Environmental Fate, 11th Symposium, ASTM STP 1007, American Society for Testing and Materials, Philadelphia, PA.

Barrett, G. W. 1968. The effect of an acute insecticide stress on a semi-enclosed grassland ecosystem. Ecology 49:1019–1035.

Barnthouse, L. W., D. L. DeAngelis, R. H. Gardner, R. V. O'Neill, G. W. Suter II, and D. S. Vaughan. 1982. Methodology for Environmental Risk Analysis. ORNL/TM-8167. Oak Ridge National Laboratory, Oak Ridge, Tennessee.

Barnthouse, L. W., J. Boreman, S. W. Christensen, C. P. Goodyear, D. S. Vaughan, and W. Van Winkle. 1984. Population biology in the courtroom: the Hudson River controversy. BioScience 34:14–19.

Barnthouse, L. W. and A. V. Palumbo. 1986. Assessing the transport and fate of bioengineered microorganisms in the environment. pp. 109–128. In J. Ficksel, and V. T. Covello (eds.), Biotechnology Risk Assessment: Issues and Methods for Environmental Introductions. Pergamon Press, New York.

Barnthouse, L. W., and G. W. Suter II (eds.). 1986. User's manual for ecological risk assessment. ORNL-6251. Oak Ridge National Laboratory, Oak Ridge, TN.

Barnthouse, L. W., R. V. O'Neill, S. M. Bartell, and G. W. Suter II. 1986. Population and ecosystem theory in ecological risk assessment. pp. 82–96. In T. M. Poston and R. Purdy (eds.), Aquatic Ecology and Hazard Assessment, 9th Symposium. American Society for Testing and Materials, Philadelphia, PA.

Barnthouse, L. W., G. W. Suter II, A. E. Rosen, and J. J. Beauchamp. 1987. Estimating responses of fish populations to toxic contaminants. Environ. Toxicol. Chem. 6:811–824.

Barnthouse, L. W. and W. Van Winkle. 1988. Analysis of impingement impacts on Hudson River fish populations. American Fisheries Society Monograph 4:182–190.

Barnthouse, L. W., G. W. Suter II, and A. E. Rosen. 1988a. Inferring population-level significance from individual-level effects: An extrapolation from fisheries science to ecotoxicology. pp. 289–300. In G. W. Suter II and M. Lewis (eds.), Aquatic Toxicology and Hazard Assessment, 11th Symposium. American Society for Testing and Materials, Philadelphia, PA.

Barnthouse, L. W., G. S. Saylor, and G. W. Suter II. 1988b. A biological approach to assessing environmental risks of engineered microorganisms. pp. 89–98. In J. Ficksel and V. T. Covello (eds.), Safety Assurance for Environmental Introductions of Genetically-Engineered Organisms. Springer-Verlag, Berlin.

Barnthouse, L. W., G. W. Suter II, C. F. Baes III, S. M. Bartell, M. G. Cavendish, R. H. Gardner, R. V. O'Neill, and A. E. Rosen. 1985. Environmental risk analysis for indirect coal liquefaction. ORNL/TM-9120. Oak Ridge National Laboratory, Oak Ridge, Tennessee.

Barnthouse, L.W., G. W. Suter II, and A. E. Rosen. 1989. Inferring population-level significance from individual-level effects: an extrapolation from fisheries science to ecotoxicology.pp. 289–300 in G. W. Suter II and M. A. Lewis (eds.) Aquatic Toxicology and Environmental Fate: 11th Volume, ASTM STP 1007, American Society for Testing and Materials, Philadelphia, PA.

Barnthouse, L. W., G. W. Suter II, and A. E. Rosen. 1990. Risks of toxic contaminants to exploited fish populations: Influence of life history, data uncertainty, and exploitation intensity. Environ. Toxicol. Chem. 9:297–311.

Barron, M. G. 1990. Bioconcentration. Environ. Sci. Technol. 24:1612–1618.

Barron, M. G., M. A. Mayes, P.G. Murphy, and R. J. Nolan. 1990. Pharmacokinetics and metabolism of triclopyr butoxyethyl ester in coho salmon. Aquatic Toxicol. 16:9–32.

Bartell, S.M. 1978. Size-selective planktivory and phosphorus cycling in pelagic ecosystems. Ph.D. Dissertation, University of Wisconsin-Madison. 195 pp.

Bartell, S. M. 1984. Ecosystem uncertainty analysis: potential effects of chloroparaffins on aquatic systems. Report to the Office of Toxic Substances, U. S. Environmental Protection Agency, Washington, D.C.

Bartell, S. M., R. H. Gardner, R. V. O'Neill and J. M. Giddings. 1983. Error analysis of predicted fate of anthracene in a simulated pond. Environ. Toxicol. Chem. 2:19–28.

Bartell, S.M. 1986. Comparison of predicted and measured effects of phenolic compounds in experimental ponds, pp. 324–343. In Environmental Modelling for Priority Setting among Existing Chemicals, Proceedings, Ecomed, Munich.

Bartell, S.M., P.F. Landrum, J.P. Giesy, and G.J. Leversee. 1981. Simulated transport of polycyclic aromatic hydrocarbons in artificial streams, pp. 133–144. In W.J. Mitsch, R.W. Bosserman, and J.M. Klopatek (eds.), Energy and Ecological Modelling. Elsevier, New York.

Bartell, S. M., R. H. Gardner, and R. V. O'Neill. 1988. An integrated fate and effects model for estimation of risk in aquatic systems. pp. 261–274. In W. J. Adams, G. A. Chapman, and W. G. Landis (eds.), Aquatic Toxicology and Hazard Assessment: 10th Volume. American Society for Testing and Materials, Philadelphia, Pennsylvania.

Bartell, S.M., A.L. Brenkert, R.V. O'Neill and R.H. Gardner. 1988a. Temporal variation in regulation of production in a pelagic food web model. In S. R. Carpenter (ed.), Complex Interactions in Lake Communities. Springer-Verlag, New York.

Bartell, S.M., R.H. Gardner and R.V. O'Neill. 1988b. An integrated fate and effects model for estimation of risk in aquatic systems. Aquatic Toxicology and Hazard Assessment: Tenth Volume, ASTM STP Materials, Philadelphia.

Bartell, S. M., R. H. Gardner, and R. V. O'Neill. 1992. Ecological Risk Estimation. Lewis Publishers, Chelsea, Michigan.

Bascietto, J. J. 1988. Ecological evaluation of a municipal landfill in the superfund program. Unpublished file document, U.S. Environmental Protection Agency, Washington, D.C.

Baskin, L. B. and J. W. Falco. 1989. Assessment of human exposure to gaseous pollutants. Risk Analysis 9:365–375.

Baughman, G. L., and R. R. Lassiter. 1978. Prediction of environmental pollution concentration. pp. 35–44. In J. Cairns, K. L. Dickson, and A. W. Mari, (eds.). Estimating the Hazard of Chemical Substances to Aquatic Life. American Society for Testing and Materials, Philadelphia.

Bayne, K. L., K. R. Clark, and M. N. Moore. 1981. Some practical considerations in the measurement of pollution effects on bivalve molluscs, and some possible ecological consequences. Aquatic Toxicol. 1:159–174.

Beanlands, G. E., and P. N. Duinker. 1983. An ecological framework for environmental impact assessment in Canada. Institute for Resources and Environmental Studies, Dalhousie University, Halifax, Nova Scotia, Canada.

Beanlands, G. E., W. J. Erkmann, G. H. Orians, J. O'Riordan, D. Policansky, M. H. Sadar, and B. Sadler (eds.). 1985. Cumulative Environmental Effects: A Binational

Perspective. No. EN 106-2/1985. Minister of Supply and Services Canada, Ottawa.

Beck, L. W., A. W. Maki, N. R. Artman, and E. R. Wilson. 1981. Outline and criteria for evaluating the safety of new chemicals. Regul. Toxicol. Pharmacol. 1:19–58.

Bedford, B. L. and E. M. Preston. (eds.). 1988. Cumulative Impacts to Wetlands. Environ. Manage. 12(5).

Beebe, F. L. 1969. Passenger pigeons and peregrine ecology. pp. 399–402. In J. J. Hickey (ed.), Peregrine Falcon Populations: Their Biology and Decline. U. Wisconsin Press. Madison.

Bell, P. F., B. R. James, and R. C. Chaney. 1991. Heavy metal extractability in longterm sewage sludge and metal salt-amended soils. J. Environ. Qual. 20:481–486.

Bender, M. E. and R. J. Huggett. 1984. Fate and effects of Kepone in the James River. Reviews in Environmental Toxicology 1:5–50.

Bennett, B. G. 1982. The exposure commitment method for pollutant exposure evaluation. Ecotoxicol. Environ. Safety 6:363–368.

Bennett, R. S. 1989. Role of dietary choices on the ability of bobwhite to discriminate between insecticide-treated and untreated food. Environ. Toxicol. Chem. 8:731–738.

Benz, J. 1986. A modelling attempt for estimating toxicity, pp. 354–370. In Environmental Modelling for Priority Setting among Existing Chemicals, Proceedings, Ecomed, Munich.

Bergman, H. L., R. A. Kimerle, and A. W. Maki (eds.). 1986. Environmental Hazard Assessment of Effluents. Pergamon Press, New York.

Berlinski, D. 1976. On systems analysis: an essay concerning the limitations of some mathematical methods in the social, political, and biological sciences. The MIT Press, Cambridge, Massachusetts.

Bernstein, L., L. S. Gold, B. N. Ames, M. C. Pike, and D. G. Hoel. 1985. Some tautologous aspects of the comparison of carcinogenic potency in rats and mice. Fundam. Appl. Toxicol. 5:79–86.

Betz, F. S., W. R. Beusch, E. B. Brittin, R. Carsel, S. Z. Cohen, R. W. Holst, A. Keller, In Mauer, W. Roessler, D. Urban, A. Vaughan, and W. Woodrow. 1982. Pesticide assessment guidelines: Subdivision M: Biorational pesticides. EPA-540/9-82-082. U. S. Environmental Protection Agency, Office of Pesticide Programs, Washington, DC.

Beverton, R. J. H., and S. J. Holt. 1957. On the dynamics of exploited fish populations. Fishery Investigations, Series II, Marine Fisheries, Vol. 19. Great Britain Ministry of Agriculture, Fisheries and Food, London.

Biesinger, K. E., G. M. Christenson, and J. T. Faindt. 1986. Effects of metal salt mixtures on *Daphnia magna* reproduction. Ecotoxicol. Environ. Safety 11:9–14.

Biesinger, K. E., L. R. Williams, W. H. van der Schalie. 1987. Procedures for conducting *Daphnia magna* toxicity bioassays. EPA/600/8-87/011. U.S. Environmental Protection Agency, Las Vegas, Nevada.

Birge, W. J., J. A. Black, and B. A. Ramey. 1981. The reproductive toxicology of aquatic contaminants. pp. 59–115. In J. Saxena and F. Fisher (eds.), Hazard Assessment of Chemicals: Current Developments, Vol. 1. Academic Press, New York.

Birge, W. J., J. A. Black, and B. A. Ramey. 1986. Evaluation of effluent biomonitoring systems. pp. 66–80. In H. L. Bergman, R. A. Kimerle, and A. W. Maki (eds.), Environmental Hazard Assessment of Effluents. Pergamon Press, New York.

Bischoff, K. B. 1980. Current applications of physiological pharmacokinetics. Fed. Proc. 39:2456–2459.

BKH Consulting Engineers. 1990. Ecotoxicological effects assessment: extrapolation from single species toxicity data. 6631H/V3. The Hague.

Blanck, H., G. Wallin, and S.-A. Wangberg. 1984. Species dependent variation in algal sensitivity to chemical compounds. Ecotox. Environ. Safety. 8:339–351.

Blus, L. J., C. D. Gish, A. A. Belisle, and R. M. Prouty. 1972. Further analysis of the logarithmic relationship of DDE residues to eggshell thinning. Nature 240:164–166.

Blus, L. J. 1982. Further interpretation of the relationship of organochlorine residues in brown pelican eggs to reproductive success. Environ. Pollut. (Ser. A) 28:15–33.

Bodek, I., W. J. Lyman, W. F. Reehl and D. J. Rosenblatt, (eds.). 1988. Environmental Inorganic Chemistry. Pergamon Press, New York, Sect. 2.9.

Boersma, L., F. T. Lindstrom, C. McFarlane, and E. L. McCoy. 1988. Model of coupled transport of water and solutes in plants. Special Report 818, Agricultural Experiment Station, Oregon State University, Corvallis.

Boersma, L., T. Lindstrom, C. McFarlane, and E. L. McCoy. 1988. Uptake of organic chemicals by plants: A theoretical model. Soil Sci. 146:403–417.

Boersma, L., C. McFarlane, and T. Lindstrom. 1991. Mathematical model of plant uptake and translocation of organic chemicals: application to experiments. J. Environ. Qual. 20:137–146.

Boese, B. L., H. Lee II, D. T. Sprecht, R. C. Randall, and M. H. Winsor. 1990. Comparison of aqueous and solid phase uptake for hexachlorobenzene in the tellinid clam *Macoma nasuta* (Conrad): A mass balance approach. Environ. Toxicol. Chem. 9:221–231.

Boethling, R. S. 1986. Application of molecular topology to quantitative structure-biodegradability relationships. Environ. Toxicol. Chem. 5: 797–806.

Boethling, R. S., B. Gregg, R. Frederick, N. W. Gabel, S. E. Campbell and A. Sabljic. 1989. Expert systems survey on biodegradation of xenobiotic chemicals. Ecotoxicol. Environ. Safety 18: 252–267.

Bogen, K. T. and R. C. Spear. 1987. Integrating uncertainty and interindividual variability in environmental risk assessment. Risk Analysis 7:427–436.

Boling, R.H. Jr., E.D. Goodman, J.A. Van Sickle, J.O. Zimmer, K.W. Cummins, S.R. Reice, and R.C. Petersen. 1975. Toward a model of detritus processing in a woodland stream. Ecology 56: 141–151.

Bollag, J.-M. 1983. Coupling of humic materials with xenobiotic substances. pp. 127–139. In Aquatic and Terrestrial Humic Materials. R. F. Christman and E. T. Gessing, Eds. Ann Arbor Science, Ann Arbor, MI.

Bonano, E. J., S. C. Hora, R. L. Keeney, and D. von Winterfeldt. 1990. Elicitation and use of expert judgement in performance assessment for high-level radioactive waste repositories. NUREG/CR-5411. Scandia National Laboratory, Albuquerque, New Mexico.

Bonazountas, M., Wagner, J.M. 1984. SESOIL—A seasonal soil compartment model. Arthur D. Little Co., Cambridge, MA.

Bonazountas, M., Brecker, A., Vranka, R.G. 1988. Mathematical environmental fate modeling. Ch. 5 in I. Bodek, W.J. Lyman, W.F. Reehl and D.H. Rosenblatt (eds.). Environmental Inorganic Chemistry, Pergamon Press, New York.

Bondietti, E. A., C. F. Baes III, and S. B. McLaughlin. 1989. Radial trends in cation ratios in tree rings as indicators of the impact of atmospheric deposition on forests. Can. J. For. Res. 19:586–594.

Boreman, J., 1978. Life history and population dynamics of Cayuga Inlet rainbow trout (Salmo gairdneri Richardson). Ph. D. Dissertation, Cornell University, Ithaca, New York. 162 pp.

Botkin, D. B. 1990. Discordant Harmonies. Oxford University Press, New York.

Botkin, D.B., J.F. Janak, and J.R. Wallis. 1972. Some ecological consequences of a computer model of forest growth. Journal of Ecology 60: 849–872.

Bovee, K. D. 1982. A guide to stream habitat analysis using the instream flow incremental methodology. FWS/OBS-82/26. U. S. Fish and Wildlife Service, Fort Collins, Colorado.

Bovee, K. D. and J. R. Zuboy (eds.). 1988. Proceedings of a workshop on the development and evaluation of habitat suitability criteria. Biological Report 88(11). U.S. Fish and Wildlife Service, Fort Collins, Colorado.

Boyle, T. P., J. Sebaugh, and E. Robinson-Wilson. 1984. A hierarchical approach to the measurement of changes in community induced by environmental stress. J. Environ. Test. Eval. 12:241–245.

Boyle, T. P., S. E. Finger, R. L. Paulson, and C. F. Rabeni. 1985. Comparison of laboratory and field assessment of fluorene — Part II: Effects on the ecological structure and function of experimental pond ecosystems. pp. 134–151. In T. P. Boyle (ed.), Validation and Predictability of Laboratory Methods for Assessing the Fate and Effects of Contaminants in Aquatic Ecosystems. American Society for Testing and Materials, Philadelphia.

Brand, R. J., D. E. Pinnock, and K. L. Jackson. 1973. Large sample confidence bands for the logistic response curve and its inverse. American Statistician 27(4):157–160.

Brattsten, L. B., C. W. Holyoke, Jr., J. R. Leeper, and K. F. Raffa. 1986. Insecticide resistance: Challenge to pest management and basic research. Science 231:1255–1260.

Breck, J. E. 1988. Relationships among models for acute toxic effects: applications to fluctuating concentrations. Environ. Toxicol. Chem. 7:775–778.

Breck, J. E., D. L. DeAngelis, and W. Van Winkle. 1986. Simulating fish exposure to toxicants in a heterogeneous body of water. pp. 451–455. In R. Crosbie and P. Luker (eds.), Proceedings of the 1986 Summer Computer Simulation Conference. The Society for Computer Simulation, San Diego, California.

Breck, J. E., D. L. DeAngelis, W. Van Winkle, and S. W. Christensen. 1988. Potential importance of spatial and temporal heterogeneity in pH, Al, and Ca in allowing survival of a fish population: A model demonstration. Ecological Modelling 41:1–16.

Briggs, G. G., R. H. Bromilow, and A. A. Evans. 1982. Relationship between lipophilicity and root uptake and translocation of nonionized chemicals by barley. Pesticide Sci. 13:495–504.

Bringmann, G. and R. Kuhn. 1980. Comparison of the toxicity thresholds of water pollutants to bacteria, algae, and protozoa in the cell multiplication inhibition test. Water Res. 14: 231–241.

Broderious, S. and M. Kahl. 1985. Acute toxicity of organic chemical mixtures to the fathead minnow. Aquatic Toxicology 6: 307–322.

Bromenshenk, J. J., S. R. Carlson, J. C. Simpson, J. M. Thomas. 1985. Pollution monitoring in Puget Sound with honey bees. Science 227:632–634.

Brooks, A. S. and G. L. Seegert. 1977. The effect of intermittent chlorination on rainbow trout and yellow perch. Trans. Am. Fish. Soc. 106:278–286.

Brown, C. C. 1978. The statistical analysis of dose-effect relationships. pp. 115–148. In G. C. Butler (ed.), Principles of Ecotoxicology, SCOPE 12. J. Wiley & Sons, Chichester.

Brown, M.P., J.J.A. McLaughlin, J.M. O'Connor and K. Wyman. 1982. A mathematical model of PCB bioaccumulation in plankton. Ecological Modelling 15: 29–47.

Brown, S.M., Silver, A. 1986. Chemical Spill Exposure Assessment, Risk Analysis 6:291–298.

Brown, D.S., Allison, J.D. 1987. MINTEQAI Equilibrium metal speciation model: A user's manual. U.S. EPA, Athens, Georgia.

Brown, L. (ed.) 1987. State of the World 1987. W. W. Norton, New York.

Brugam, R. B. 1988. Long-term history of eutrophication in Washington lakes. pp. 63–70. In W. J. Adams, G. A. Chapman, and W. G. Landis (eds.), Aquatic Toxicology and Hazard Assessment: 10th Volume. American Society for Testing and Materials, Philadelphia, Pennsylvania.

Bruggeman, W. A., L. B. J. M. Martron, D. Kooijman, and O. Hutzinger. 1981. Accumulation and elimination kinetics of di-, tri-, and tetra-chlorophenols by goldfish after dietary and aqueous exposure. Chemosphere 10:811–832.

Brunner, H. S., E. Horning, H. Santl, E. Wolff and O. G. Piringer. 1990. Henry's law constants for polychlorinated biphenyls. Environ. Sci. Technol. 24:1751–1754.

Buckley, J. J. 1983. Decision making under risk: a comparison of Bayesian and fuzzy set methods. Risk Analysis 3:157–168.

Buikema, A. L. Jr., J. G. Geiger, and D. R. Lee. 1980. *Daphnia* toxicity tests. pp. 48–59. In A. L. Buikema and J. Cairns Jr. (eds.), Aquatic Invertebrate Bioassays. American Society for Testing and Materials, Philadelphia, PA.

Bull, K. R., W. J. Avery, P. Freestone, J. R. Hall, D. Osborn, A. S. Cooke, and T. Stowe. 1983. Alkyl lead pollution and bird mortalities in the Mersey estuary, UK, 1979–1981. Environ. Pollut. 31A:239–254.

Buongiorno, J. and J. K. Gilles. 1987. Forest Management. Macmillan Publishing Co., New York.

Burges, H. D. 1981. Safety, safety testing, and quality control of microbial pesticides. pp. 737–767. In H. D. Burges (ed.), Microbial Control of Pests and Plant Diseases 1970–1980. Academic Press, London.

Burns, L.A. 1986. Validation and verification of aquatic fate models, pp. 148–172. In Environmental Modelling for Priority Setting among Existing Chemicals, Proceedings, Ecomed, Munich.

Burns, L.A., D.M. Cline and R.R. Lassiter. 1982. Exposure analysis modeling system (EXAMS): User Manual and System Documentation. EPA/600/3-82/023, U.S. Environmental Protection Agency, Environmental Research Laboratory, Athens, Georgia.

Burns, T. P. 1989. Lindeman's contradiction and the trophic structure of ecosystems. Ecology 70:1355–1362.

Bussiere, J. L., R. J. Kendall, T. E. Lacher, Jr., and R. S. Bennett. 1989. Effect of methyl parathion on food discrimination in northern bobwhite (Colinus virginianus). Environ. Toxicol. Chem. 8:1125–1131.

Butler, G. C. 1978. Estimation of doses and integrated doses. pp. 91–112. In G. C. Butler (ed.), Principles of Ecotoxicology, SCOPE 12. John Wiley and Sons, New York.

Byrne, S.V., M.M. Wehrle, M.A. Keller and J.F. Reynolds. 1987. Impact of gypsy moth infestation on forest succession in the North Carolina piedmont: a simulation study. Ecological Modelling 35: 63–84.

Cada, G. F. and C. T. Hunsaker. 1990. Cumulative impacts of hydropower development: reaching a watershed in impact assessment. Environ. Profess. 12:2–8.

Cade, T. J. and R. Fyfe. 1970. The North American peregrine survey, 1970. Can. Fld. Nat. 84:231–245.

Cade, T. J., J. L. Lincer, C. M. White, D. G. Roseneau, and L. G. Swartz. 1971. DDE residues and eggshell changes in Alaskan falcons and hawks. Science 172:955–957.

Cairns, J., Jr. 1976. Summary. pp. 235–242. In J. Cairns, Jr., K. L. Dickson, and G. F. Westlake (eds.), Biological Monitoring of Water and Effluent Quality. American Society for Testing and Materials, Philadelphia, Pennsylvania.

Cairns, J. Jr. (ed.). 1980. The Recovery Process in Damaged Ecosystems. Ann Arbor Science, Ann Arbor, Michigan.

Cairns, J. Jr. 1982a. Restoration of damaged ecosystems. pp. 220–239. In W. T. Mason and S. Iker, (eds.), Research on Fish and Wildlife Habitat. EPA-600/8-82-022. U.S. Environmental Protection Agency, Washington, DC.

Cairns, J., Jr. (ed.). 1982b. Artificial Substrates. Ann Arbor Science Pub., Ann Arbor, Michigan.

Cairns, J., Jr. 1986. On the relationship between structural and functional analysis of ecosystems. Environ. Toxicol. Chem. 5:785–786.

Cairns, J., Jr. 1990. Lack of theoretical basis for predicting rate and pathways of recovery. Environ. Manage. 14:517–526.

Cairns, J., Jr., A. G. Heath, and B. C. Parker. 1975a. The effects of temperature upon the toxicity of chemicals to aquatic organisms. Hydrobiologia 47:135–171.

Cairns, J., Jr., A. G. Heath, and B. C. Parker. 1975b. Temperature influence on chemical toxicity to aquatic organisms. J. Water Pollut. Control Fed. 47:267–280.

Cairns, J., Jr., K. L. Dickson, and G. F. Westlake (eds.). 1976. Biological Monitoring of Water and Effluent Quality. American Society for Testing and Materials, Philadelphia, Pennsylvania.

Cairns, J., Jr., K. L. Dickson, and E. E. Herricks (eds.). 1977. Recovery and Restoration of Damaged Ecosystems. University Press of Virginia, Charlottesville.

Cairns, J., Jr., K. L. Dickson, and A. W. Maki (eds.). 1978. Estimating the Hazard of Chemical Substances to Aquatic Life, STP 657. American Society for Testing and Materials, Philadelphia.

Cairns, J., Jr., K. L. Dickson, and A. W. Maki. 1979. Estimating the hazard of chemical substances to aquatic life. Hydrobiologia 64:157–166.

Calabrese, E. J. 1984. Principles of Animal Extrapolation. John Wiley and Sons, New York.

Calabrese, E. J. 1991. Multiple Chemical Interactions. Lewis Pub., Chelsea, Michigan.

Calamari, D., G. Chiaudani, and M. Vighi. 1985. Methods for measuring effects of chemicals on aquatic plants. pp. 549–571. In V. B. Vouk, G. C. Butler, D. G. Hoel, and D. P. Peakall. Methods for Estimating Risk of Chemical Injury: Human and Non-human Biota and Ecosystems, SCOPE 26. John Wiley and Sons, New York.

Calder, W. A. III and E. J. Braun. 1983. Scaling of osmotic regulation in mammals and birds. Regulatory Integrative Comp. Physiol. 13:R601-R606.

Call, D. J., L. T. Brook, M. L. Knuth, S. H. Poirier, and M. D. Hoglund. 1985. Fish subchronic toxicity prediction model for industrial organic chemicals that produce narcosis. Environ. Toxicol. Chem. 4:335–342.

Calvert, J. G., and J. N. Pitts. 1963. Photochemistry. John Wiley, New York.

Campbell, C. 1982. Evaluating propogated and total error in chemical property estimates. Appendix C. In W. J. Lyman, W. F. Reehl, and D. H. Rosenblatt (eds.), Handbook of Chemical Property Estimation Methods. McGraw-Hill Book Co., New York.

Cardwell, R. D., D. G. Foreman, T. R. Payne, and D. J. Wilbur. 1977. Acute and chronic toxicity of chlordane to fish and invertebrates, EPA-600/3–77–019. U.S. Environmental Protection Agency, Duluth, Minnesota.

Carlson, A. R., H. Nelson, and D. Hammermeister. 1986. Development and validation of site-specific water quality criteria for copper. Environ. Toxicol. Chem. 5:977–1012.

Carpenter, S. R. 1989. Replication and treatment strength in whole-lake experiments. Ecology 70:453–463.

Carpenter, S. R., T. M. Frost, D. Heisey, and T. K. Kratz. 1989. Randomized intervention analysis and the interpretation of whole-ecosystem experiments. Ecology 70:1142–1152.

Carpenter, S. R. 1990. Large-scale perturbations: opportunities for innovation. Ecology 71:2038–2043.

Carsel, R.F., Smith, C.N., Mulkey, L.A., Dean, J.D., Jowise, P., 1984. User's Manual for the Pesticide Root Zone Model (PRZM). EPA-600/3-84-109. U.S. EPA, Athens, Ga.

Caswell, H. 1989. Matrix Population Models. Sinauer Associates, Inc., Sunderland, MA.

Caswell, H. and A. M. John. 1992. From the individual to the population in demographic models. pp. 36-61. In D. L. DeAngelis and L. Gross, (eds.), Individual-Based Approaches in Ecology: Concepts and Models. Chapman and Hall, New York.

Caughley, G. 1977. Analysis of Vertebrate Populations. John Wiley & Sons, New York.

CEQ (Council on Environmental Quality). 1986. Regulations for implementing the procedural provisions of the National Environmental Policy Act. 40 CFR Parts 1500-1508, U.S. Government Printing Office, Washington, DC.

Chapman, G. A. 1983. Do organisms in laboratory toxicity tests respond like organisms in the field? pp. 315-327. In W. E. Bishop, R. D. Cardwell, and B. B. Heidolph (eds.), Aquatic Toxicology and Hazard Assessment: Sixth Symposium. American Society for Testing and Materials, Philadelphia.

Chapman, G. A. 1985. Acclimation as a factor influencing metal criteria. pp. 119-136. In R. C. Bahner, and D. J. Hansen (eds.), Aquatic Toxicology and Hazard Assessment: Eighth Symposium. American Society for Testing and Materials, Philadelphia.

Chapman, N. B. and J. Shorter. 1972. Advances in Linear Free Energy Relations. Plenum Press, London.

Chappell, W. R. 1989. Interspecific scaling of toxicity data: A question of interpretation. Risk Analysis 9:13-14.

Chen, C. W. and R. E. Selleck. 1969. A kinetic model of fish toxicity threshold. J. Water Pollut. Control Fed. 41:R294-R308.

Chesapeake Implementation Committee. 1988. The Chesapeake Bay Program: a commitment to renewal. Chesapeake Bay Program, Annapolis, Maryland.

Chiou, G.T., B.H. Freed, D.W. Schmedding, and K.L. Kohnert. 1977. Partition coefficients and bioaccumulation of selected organic chemicals. Environ. Sci. Technol. 11:475.

Christensen, E. K. 1984. Dose-response functions in aquatic toxicology testing and the Weibull model. Water Res. 18:213-221.

Christensen, E. K. and N. Nyholm. 1984. Ecotoxicological assays with algae: Weibull dose-response curves. Environ. Sci. Technol. 18:713-718.

Christensen, E. K. and C.-Y. Chen. 1985. A general noninteractive multiple toxicity model including probit, logit and Weibull transformations. Biometrics 41:711–725.

Christensen, S. W., D. L. DeAngelis, and A. G. Clark. 1977. Development of a stock progeny model for assessing power plant effects on fish populations. pp. 196–226. In W. Van Winkle (ed.), Assessing the Effects of Power-Plant-Induced Mortality on Fish Populations, Pergamon, New York.

Christensen, S. W., J. E. Breck, and W. Van Winkle. 1988. Predicting acidification effects on fish populations, using laboratory data and field information. Environ. Toxicol. Chem. 7:735–747.

Christensen, S.W., W. Van Winkle, and J. S. Mattice. 1976. Defining and determining the significance of impacts: concepts and methods. pp. 191219. In Sharma, R. K., J. D. Buffington, and J. T. McFadden (eds.) The biological significance of environmental impacts. NR-CONF-002, U.S. Nuclear Regulatory Commission, Washington, D.C.

Christian, J. J. 1983. Love Canal's unhealthy voles. Natural History 10:8–16.

Clark, I. 1979. Practical Geostatistics. Applied Science, London.

Clark, J. R., L. R. Goodman, P. W. Borthwick, J. M. Patrick, Jr., J. C. Moore, and E. M. Lores. 1986. Field and laboratory toxicity tests with shrimp, mysids, and sheepshead minnows exposed to fenthion. pp. 161–176. In T. M. Posten and Purdy (eds.), Aquatic Toxicology and Environmental Fate: Ninth Volume, ASTM STP 921. American Society for Testing and Materials, Philadelphia.

Clark, K.E., F.A.P.C. Gobas, D. Mackay. 1990. Model of organic chemical uptake and clearance by fish from food and water. Environ. Sci. Technol. 24:1203–1213.

Clayson, D. B., D. Krewski, and I. Munro, (eds.) 1985. Toxicological Risk Assessment. CRC Press, Boca Raton, FL.

Cleland, J. G. and G. L. Kingsbury. 1977. Multimedia environmental goals for environmental assessment, Vols. I-II. EPA-600/7–77–136. U.S. Environmental Protection Agency, Research Triangle Park, NC.

Clements, R. G. (ed.) 1988. Estimating toxicity of industrial chemicals to aquatic organisms using structure activity relationships. EPA-560-6-88-001. U. S. Environmental Protection Agency, Washington, D.C.

Clements, W. H., D. S. Cherry and J. Cairns, Jr. 1988. Structural alterations in aquatic insect communities exposed to copper in laboratory streams. Environ. Toxicol. Chem. 7:715–722.

Clench, M. H. and R. C. Leberman. 1978. Weights of 151 species of Pennsylvania birds analyzed by month, age, and sex. Bull. Carnegie Museum Natur. Hist. 5:1–85.

Cohen, J. E. 1976. Ergodicity of age structure in populations with Markovian vital rates I: countable states. J. Am. Stat. Assoc. 71:335–339.

Cohen, J. E. 1979. Comparative statics and stochastic dynamics of age-structured populations. Theoret. Pop. Biol.16:159-171

Cohen, J. E., S. W. Christensen, and C. P. Goodyear. 1983. A stochastic age-structured population model of striped bass (Morone saxatilis) in the Potomac River. Can. J. Fish. Aquat. Sci. 40:2170-2183.

Cohen, Y., Tsai, W., Chetty, S.L., Mayer, G.J. 1990. Dynamic Partitioning of Organic Chemicals in Regional Environments: A Multimedia Screening-Level Modeling Approach, Environ. Sci. Technol. 24:1549-1558.

Cohen, Y. 1986. Pollutants in a Multimedia Environment. Plenum Press, New York and London.

Cohen, Y., Ryan, P.A. 1985. Multimedia Modeling of Environmental Transport: Trichloroethylene Test Case. Environ., Sci. Technol. 19:412-417.

Cohen, Y. 1989. The Spatial Multimedia Compartments Model (SMCM), Users Manual Version 3.0, NCITR, UCLA, CA.

Cohrssen, J. J. and V. T. Covello. 1989. Risk Analysis: A Guide to Principles and Methods for Analyzing Health and Environmental Risks. Council on Environmental Quality, Washington, D.C.

Colborn, T. 1991. Global implications of Great Lakes wildlife research. Internat. Environ. Affairs 3:3-25.

Cole, C.V., G.S. Innis, and J.W.B. Stewart. 1977. Simulation of phosphorus cycling in semiarid grasslands. Ecology 58: 1-15.

Cole, L. C. 1954. The population consequences of life history phenomena. Quart. Rev. Biol. 19:103-137.

Colinvaux, P. 1986. Ecology. John Wiley & Sons, New York.

Committee on Biological Markers of the National Research Council. 1987. Biological markers in environmental health research. Environ. Health Persp. 74:3-9.

Connell, D. W. 1989. Bioaccumulation of Xenobiotic Compounds. CRC Press, Boca Raton, Florida.

Connolly, J. P. 1985. Predicting single-species toxicity in natural water systems. Environ. Toxicol. Chem. 4: 573-582.

Conway, R. A. (ed.). 1982. Environmental Risk Analysis for Chemicals. Van Nostrand Reinhold, New York.

Coomer, E. C., Jr., 1976. Population dynamics of black bass in Center Hill Reservoir, Tennessee. TWRA Technical Report No. 76-54. Tennessee Technological University, Cookeville, Tennessee.

Coopage, D. L., E. Mathews, G. H. Cook, and J. Knight. 1975. Brain acetylcholinesterase inhibition in fish as a diagnosis of environmental poisoning by malathion,

O,O-dimethyl S-(1,2-dicarbethoxyethyl) phosphorodithioate. Pesticide Biochem. Physiol. 5:536–542.

Cordes, E. H. 1964. Mechanisms and catalysis of the hydrolysis of acetals, ketals and orthoesters. Prog. Phys. Org. Chem. 4:1–23.

Cordes, E. H. 1967. Hydrolysis of acetals and ketals. Prog. Phys. Org. Chem. 4:1–23.

Cordes, E. H. and W. P. Jencks. 1964. The mechanisms of hydrolysis of Schiff bases derived from aliphatic amines. J. Amer. Chem. Soc. 85:2843–2848.

Costanza, R. 1991. Ecological Economics: The Science and Management of Sustainability. Columbia University Press. New York.

Costerton, J. W., and G. G. Geesey. 1979. Natural Aquatic Bacteria: Activity and Ecology. ASTM, Philadelphia.

Cothern, C. R., W. A. Coniglio, and W. L. Marcus. 1986. Development of quantitative estimates of uncertainty in environmental risk assessments when the scientific data base is inadequate. Environ. Internat. 12:643–647.

Coughenour, M.B. 1981. Relationship of SO_2 dry deposition to a grassland sulfur cycle. Ecological Modelling 13: 1–16.

Cowgill, U. M. 1988. Paleoecology and environmental analysis. pp. 53–62. In W. J. Adams, G. A. Chapmen, and W. G. Landis (eds.), Aquatic Toxicology and Hazard Assessment: 10th Volume. American Society for Testing and Materials, Philadelphia, Pennsylvania.

Cowgill, U. M., D. L. Hopkins, S. L. Applegath, I. T. Takahashi, S. D. Brooks, and D. P. Milazzo. 1985. Brood size and neonate weight of *Daphnia magna* produced by nine diets. pp. 233–244. In R. C. Bahner and D. J. Hansen (eds.), Aquatic Toxicology and Hazard Assessment, Eighth Symposium. American Society for Testing and Materials, Philadelphia.

Cox, D. C. and P. Baybutt. 1981. Methods of uncertainty analysis: a comparative survey. Risk Analysis 1:251–258.

Craig, P. N. and K. Enslein. 1981. Structure-activity in hazard assessment. pp. 389–420. In J. Saxena (ed.), Hazard Assessment of Chemicals: Current Developments. Academic Press, New York.

Crisman, T. L. 1988. The use of subfossil benthic invertebrates in aquatic resource management. pp. 71–88. In W. J. Adams, G. A. Chapman, and W. G. Landis (eds.), Aquatic Toxicology and Hazard Assessment: 10th Volume. American Society for Testing and Materials, Philadelphia, Pennsylvania.

Croft, B. A. and J. G. Morse. 1979. Research advances in pesticide resistance in natural enemies. Entomophaga 24:3–11.

Crosby, A. W. 1986. Ecological Imperialism: The Biological Expansion of Europe, 900–1900. Cambridge University Press, Cambridge.

Crossland, N. O. 1982. Aquatic toxicology of cypermethrin. II. Fate and biological effects in pond experiments. Aquat. Toxicol. 2:205–222.

Crossland, N. O., S. W. Shires, and D. Bennett. 1982. Aquatic toxicology of cypermethrin. III. Fate and biological effects of spray drift deposits in fresh water adjacent to agricultural land. Aquat. Toxicol. 2:205–222.

Crouch, E. A. C. 1983. Uncertainties in interspecies extrapolations of carcinogenicity. Environ. Health Persp. 50:321–327.

Cummins, K. W. 1974. Structure and function of stream ecosystems. Bioscience 24:631–641.

Cushing, D. H. 1979. The monitoring of biological effects: the separation of natural changes from those induced by pollution. Phil. Trans. R. Soc. Lond. 286:597–609.

Cushing, D. H., and J. G. K. Harris. 1973. Stock and recruitment and the problem of density-dependence. Rapports et Proces-Verbaux des Reunions, Conseil International pour l'Exploration de la Mer 164:142–155.

Dale, V. H., T.W. Doyle and H.H. Shugart. 1985. A comparison of tree growth models. Ecological Modelling 29: 145–169.

Dale, V. H., and R. H. Gardner. 1987. Assessing regional impacts of growth declines using a forest succession model. J. Environ. Manage. 24:83–93

Dale, V. H., R. H. Gardner, and M. G. Turner (eds.). 1989. Predicting Across Scales: Theory, Development, and Testing. Landscape Ecology 3:3/4.

Daniels, R. E., and J. D. Allan. 1981. Life table evaluation of chronic exposure to a pesticide. Can. J. Fish. Aquat. Sci. 38:485–494.

Danil'chenko, O. P. 1977. The sensitivity of fish embryos to the effect of toxicants. J. Ichthyo. 17:455–463.

Dave, G. 1981. Influence of diet and starvation on toxicity of endrin to fathead minnows (*Pimephales promelas*). PB 81–244 436. National Technical Information Service, Springfield, VA.

Davis, B. D. 1987. Bacterial domestication: underlying assumptions. Science 235:1329–1335.

Davis, R. B., D. S. Anderson, D. F. Charles, and J. N. Galloway. 1988. Two-hundred-year pH history of Woods, Sagamore, and Panther Lakes in the Adirondack Mountains, New York State. pp. 89–111. In W. J. Adams, G. A. Chapman, and W. G. Landis (eds.), Aquatic Toxicology and Hazard Assessment: 10th Volume. American Society for Testing and Materials, Philadelphia, Pennsylvania.

Dawe, C. J. 1990. Implications of aquatic animal health for human health. Environ. Health Persp. 86:245–255.

Dawson, C. L. and R. A. Hellenthall. 1986. A computerized system for the evaluation of aquatic habitats based on environmental requirements and pollution tolerance associations of resident organisms. EPA-600/3-86-019. U.S. Environmental Protection Agency, Corvallis, OR.

Dawson, W. R., J. D. Ligon, J. R. Murphy, J. P. Myers, D. Simberloff, and J. Verner. 1987. Report of the Advisory Panel on the Spotted Owl. The Condor 89:205–229.

Dayton, P. K. 1986. Cumulative impacts in the marine realm. pp. 79–84. In Proceedings of the Workshop on Cumulative Environmental Effects: Setting the Stage. Minister of Supply and Services Canada Catalog No. En 106-2/1985. Ottawa.

DeAngelis, D. L., R. A. Goldstein, and R. V. O'Neill. 1975. A model for trophic interaction. Ecology 56:881–892.

DeAngelis, D. L., S. W. Christensen, and A. G. Clark. 1977. Responses of a fish population model to young-of-the-year mortality. J. Fish. Res. Board Can. 34:2124–2132.

DeAngelis, D.L., W.M. Post, and G. Sugihara. 1983. Current trends in food web theory. Report on a workshop. ORNL-5983, Oak Ridge National Laboratory, Oak Ridge, Tennessee.

DeAngelis, D.L., L. Godbout, and B. J. Shuter. 1991. An individual-based approach to predicting density-dependent compensation in smallmouth bass populations. Ecol. Model. 57:91–115.

DeAngelis, D. L. and K. A. Rose. 1992. Which individual-based approach is most appropriate for a given problem? pp. 67–87. In D. L. DeAngelis and L. Gross (eds.), Individual Based Approaches in Ecology: Concepts and Models. Chapman and Hall, New York.

DeAngelis, D. L., S. M. Bartell, and A. L. Brenkert. 1989. Effects of nutrient recycling and food-chain length on resilience. Am. Nat. 134:778–805.

DeAngelis, D. L., L. W. Barnthouse, W. Van Winkle, and R. G. Otto. 1990. A critical appraisal of population approaches in assessing fish community health. J. Great Lakes Res. 16:576–590

DeAngelis, D. L. and L. J. Gross (eds.). 1992. Individual-Based Models and Approaches in Ecology. Chapman and Hall, New York.

Dearden, J. C. and R. M. Nicholson. 1986. The prediction of the biodegradability by use of structure activity relationships: correlation of biological oxygen demand with atomic charge difference. Pestic. Sci. 17: 3305–310.

Dedrick, R. L. and K. B. Bischoff. 1980. Species similarities in pharmacokinetics. Federation Proc. 39:54–59.

deMayo, P. 1980. Rearrangements in Ground and Excited States. Academic Press, New York. Vol. 1–3.

Denoyelles, F., Jr., and W. D. Kettle. 1985. Experimental ponds for evaluating bioassay predictions. pp. 91–103. In T. P. Boyle (ed.), Validation and Predictability of Laboratory Methods for Assessing the Fate and Effects of Contaminants in Aquatic Ecosystems. American Society for Testing and Materials, Philadelphia, PA.

De Roos, A.M., O. Diekmann, and J.A.J. Metz. 1992. Studying the dynamics of structured population models: a versatile technique and its application to *Daphnia*. Amer. Natur. 139:123–146.

Desvousges, W. H. and V. A. Skahen. 1987. Techniques to measure damages to natural resources, final report. PB88-100136. National Technical Information Service. Springfield, Virginia.

Detenbeck, N. E., P. W. DeVore, G. J. Niemi, and A. Lima. 1992. Recovery of temperate-stream fish communities from disturbance: a review of case studies and synthesis of theory. Environ. Manage. 16:33–53.

Dew, C. B., 1981. Impact perspective based on reproductive value. pp. 251–256. In L. D. Jensen (ed.), Issues Associated with Impact Assessment. EA Communications, Sparks, Maryland.

DiCarlo, F. J., P. Rickart, and C. M. Auer. 1985. Role of the Structure-Activity Team (SAT) in the premanufacture notification (PMN) process. pp. 433–449. In M. Tichy (ed.), QSAR in Toxicology and Xenochemistry. Elsevier, Amsterdam.

Dickson, K. L., J. Cairns, Jr., and J. C. Arnold. 1971. An evaluation of the use of basket-type artificial substrates for sampling macroinvertebrate organisms. Trans. Amer. Fish. Soc. 100:553–559.

Dickson, K. and A. W. Maki (eds.), Estimating the Hazard of Chemical Substances to Aquatic Life. ASTM STP 657. American Society for Testing and Materials, Philadelphia, Pennsylvania.

Dickson, K. L., A. W. Maki, and J. Cairns, Jr. (eds.). 1979. Analyzing the Hazard Evaluation Process. American Fisheries Society, Washington, D.C.

Dickson, K.L., A. W. Maki, J. Cairns. (eds.). 1982. Modeling the Fate of Chemicals in the Environment. Ann Arbor Science Publishers, Ann Arbor, Michigan.

Dickson, K. L. and J. H. Rodgers, Jr. 1986. Assessing the hazard of effluents in the aquatic environment. pp. 209–215. In H. L. Bergman, R. A. Kimerle, and A. W. Maki (eds.), Environmental Hazard Assessment of Effluents. Pergamon Press, New York.

Dickson, K. L., A. W. Maki, and W. A. Brungs (eds.). 1987. Fate and Effects of Sediment-Bound Chemicals in Aquatic Systems. Pergamon Press, New York.

Dieter, M. P. and M. T. Finley. 1979. Delta-aminolevulinic acid dehydratase enzyme activity in blood, brain, and liver of lead-dosed ducks. Environ. Res. 19:127–135.

DiGulio, R. 1989. Biomarkers. pp. 7-1-7-34. In W. Warren-Hicks, B. R. Parkhurst, and S. S. Baker, Jr. (eds.), Ecological Assessment of Hazardous Waste Sites: A Field and Laboratory Reference Document. EPA 600/3-89/013. Corvallis Environmental Research Laboratory, Oregon.

Dilling, W. L., G. U. Boggs, and C. G. Mendoza. 1986. Relative rate measurements for solution reactions of hydroxyl radical. Abstr. Div. Environ. Chem. (ACS) 26:21-24.

DiToro, D.M., D.J. O'Connor and R.V. Thomann. 1975. Phytoplankton-zooplankton-nutrient interaction model for western Lake Ontario, pp. 424-474. In B.C. Patten (ed.), Systems analysis and simulation in ecology. Volume II. Academic Press, Inc. New York.

DiToro, D. M. 1985. A particle interaction model of reversible organic chemical sorption. Chemosphere 14:1503-1508.

DiToro, D. M., J. D. Mahony, D. J. Hansen, K. J. Scott, M. B. Hicks, S. M. Mayr, and M. S. Redmond. 1990. Toxicity of cadmium in sediments: The role of acid volatile sulfide. Environ. Toxicol. Chem. 9:1487-1502.

DiToro, D. M., C. S. Zarba, D. H. Hansen, W. J. Berry, R. C. Swartz, C. E. Cowan, S. P. Pavlou, H. E. Allen, N. A. Thomas, and P. R. Paquin. 1991a. Technical basis for establishing sediment quality criteria for nonionic organic chemicals using equilibrium partitioning. Environ. Toxicol. Chem. 10:1541-1583.

DiToro, D. M., J. A. Halden, and J. L. Plafkin. 1991b. Modeling *Ceriodaphnia* toxicity in the Naugatuck River II. copper, hardness, and effluent interactions. Environ. Toxicol. Chem. 10:261-274.

DES (Division of Ecological Services). 1980. Habitat evaluation procedure (HEP). ESM 102. U.S. Fish and Wildlife Service, Washington, D.C.

Dixon, R. L. 1976. Problems in extrapolating toxicity data for laboratory animals to man. Environ. Health. Persp. 13:43-50.

Doane, T. R., J. Cairns, Jr., and A. L. Buikema, Jr. 1984. Comparison of biomonitoring techniques for evaluating effects of jet fuel on bluegill sunfish (*Lepomis macrochirus*). pp. 103-112. In D. Pascoe and R. W. Edwards (eds.), Freshwater Biological Monitoring. Pergamon Press, Oxford.

Dobson, S. 1985. Methods for measuring effects of chemicals on terrestrial animals as indicators of ecological hazard. pp. 537-548. In V. B. Vouk, G. C. Butler, D. G. Hoel, and D. B. Peakall (eds.), Methods for Estimating Risk of Chemical Injury: Human and Non-human Biota and Ecosystems, SCOPE 26. John Wiley and Sons, NY.

Doherty, F. G. 1983. Interspecies correlations of acute aquatic median lethal concentration for four standard test species. Environ. Sci. Technol. 17:661-665.

DOI (U.S. Department of the Interior). 1986. Natural resource damage assessments; final rule. Fed. Regist. 51:27674-27753.

DOI (U.S. Department of the Interior). 1987a. Natural resource damage assessments; final rule. Fed. Regist. 52:9042–9100.

DOI (U.S. Department of the Interior). 1987b. Type B technical information documents, PB88-100128 – PB88-100169. Washington, DC.

Doi, J. and D. Grothe. 1988. Use of fractionation/chemical analysis schemes for plant effluent toxicity evaluation. pp. 123–138. In G. W. Suter II and M. Lewis (eds.), Aquatic Toxiciology and Hazard Assessment, Eleventh Symposium. American Society for Testing and Materials, Philadelphia, PA.

Dolan, D.M., Bierman, V.J., Jr. 1982. Mass balance modeling of heavy metals in Saginaw Bay, Lake Huron. J. Great Lakes Res. 8:676–694.

Doty, C. B. and C. C. Travis. 1990. Is EPA's National Priorities List correct? Environ. Sci. Technol. 24:1778–1780.

Dourson, M. L. 1986. New approaches in the derivation of acceptable daily intake (ADI). Comments Toxicology 1:35–48.

Downing, D. J., R. H. Gardner, and F. O. Hoffman. 1985. An examination of response-surface methodologies for uncertainty analysis in assessment models. Technometrics, 27:151–163.

Drake, J. A. (ed.). 1988. Biological Invasions: A Global Perspective. Wiley-Interscience, New York.

Drake, J. A. and H. A. Mooney (eds.). 1989. The Ecology of Biological Invasions. SCOPE 37. John Wiley & Sons Ltd. West Sussex, UK.

Draper, N. R. and H. Smith. 1966. Applied regression analysis. John Wiley and Sons, Inc., New York.

Drossman, J., W. R. Mabey, and T. Mill. 1991. Metal-catalyzed hydrolysis of organic compounds in natural waters. MS in preparation.

Drossman, H., H. Johnson, and T. Mill. 1988. Structure activity relationships for environmental processes. 1: Hydrolysis of esters and carbamates. Chemosphere 17:1509–1530.

Drostan, T. D., R. Maudlin, Z. Jinxian, A. Hipps, E. Wehry and G. Mamontov. Adsorption and photodegradation of pyrene on coal stack ash. Environ. Sci. Technol. 23:303–308.

Drummond, R. A., C. L. Russom, D. L. Geiger, and D. L. DeFoe. 1986. Behavioral and morphological changes in fathead minnow (*Pimephales promelas*) as diagnostic endpoints for screening chemicals according to mode of action. pp. 415–435. In T. M. Poston and R. Purdy (eds.), Aquatic Toxicology and Environmental Fate: Ninth Volume, ASTM, Philadelphia, Pennsylvania.

Dulin, D., and T. Mill. 1982. Development and evaluation of sunlight actinometers. Environ. Sci. Technol. 16:815–820.

Dutka, B. J., N. Nyholm, and J. Petersen. 1983. Comparison of several microbial screening tests. Water Res. 17:1363–1368.

Eadie, J. McA., L. Broekhoven, and P. Colgan. 1987. Size ratios and artifacts: Hutchinson's rule revisited. Amer. Naturalist 129:1–17.

Eaton, J. G., J. M. McKim, and G. W. Holcomb. 1978. Metal toxicity to embryos and larvae of seven freshwater fish species—I. Cadmium. Bull. Environ. Contam. Toxicol. 19:95–103.

Eberhardt, L. L. 1976. Quantitative ecology and impact assessment. J. Environ. Manage. 4:27–70.

Eberhardt, L. L. and J. M. Thomas. 1986. Survey of statistical and sampling needs for environmental monitoring of commercial low-level radioactive waste disposal facilities. NUREG/CR4162. U.S. Nuclear Regulatory Commission, Washington, D.C.

ECAO (Environmental Criteria and Assessment Office). 1986. Qualitative pathogen risk assessment for ocean disposal of municipal sludge. PB89–126593. U.S. Environmental Protection Agency, Cincinnati, Ohio.

ECAO (Environmental Criteria and Assessment Office). 1987. Recommendations for and documentation of biological values for use in risk assessment. EPA/600/6-87-008. U.S. Environmental Protection Agency, Cincinnati, Ohio.

ECAO (Environmental Criteria and Assessment Office). 1989. Pathogen risk assessment for land application of municipal sludge. Volume 1. Methodology and computer model. PB90–171901. U.S. Environmental Protection Agency, Cincinnati, Ohio.

Economic Analysis, Inc. 1987. Measuring damages to coastal and marine natural resources: concepts and data relevant to CERCLA type A damage assessments. DOI/SW/DK-87-002. U. S. Department of Interior, Washington, D. C.

Eisenreich, S.J., Capel, P.D., Robbins, J.A., Bourbonniere R. 1989. Accumulation and diogenesis of chlorinated hydrocarbons in lacustrine sediments. Environ. Sci. Technol. 23:1116–1126.

Eisler, R. 1970. Factors affecting pesticide-induced toxicity in an estuarine fish. Tech. Paper Fish Wildl. Serv. 17. National Technical Information Service, Springfield, Virginia.

Eisler, R. 1985. Selenium hazards to fish, wildlife, and invertebrates: A synoptic review. Biological Report 85(1.5). U. S. Fish and Wildlife Service, Laurel, Maryland.

Eisler, R. 1988. Lead hazards to fish, wildlife, and invertebrates: A synoptic review. Biological Report 85(1.14). U. S. Fish and Wildlife Service, Laurel, Maryland.

El-Amamy, M. M., and T. Mill. 1984. Hydrolysis kinetics of organic chemicals on montmorillonite and kaolinite surfaces as related to moisture content. Clays and Clay Minerals 32:67–73.

Elson, P. E. 1967. Effects on wild young salmon of spraying DDT over New Brunswick forests. J. Fish. Res. Bd. Canada 24:731–767.

Elwood, J.W., J.D. Newbold, R.V. O'Neill, and W. Van Winkle. 1983. Resource spiralling: an operational paradigm for analyzing lotic ecosystems, pp. 3–27. In T.D. Fontaine III and S.M. Bartell (eds.), Dynamics of lotic ecosystems. Ann Arbor Science, Ann Arbor, Michigan.

Emanuel, W.R., G.G. Killough, W.M. Post and H.H. Shugart. 1984. Modeling terrestrial ecosystems in the global carbon cycle with shifts in carbon storage capacity by land-use change. Ecology 65: 970–983.

Emlen, J. M. 1989. Terrestrial population models for ecological risk assessment: A state-of-the-art review. Environ. Toxicol. Chem. 8:831–842.

Emlen, J. M., and E. K. Pikitch. 1989. Animal population dynamics: identification of critical components. Ecol. Model. 44:253–274

Enfield, G.C., Carsel, R.F., Cohen, S.Z., Phon, T., Walters, D.M. 1982. Approximating pollutant transport to groundwater, Ground Water. 20:711–727.

Engel, D. W. and B. A. Fowler. 1979. Factors influencing cadmium accumulation and its toxicity to marine organisms. 28:81–88.

Enslein, K., T. Tuzzeo, H. Borgstedt, B. Blake, and J. Hart. 1988. Prediction of *Daphnia magna* LC_{50} values from rat oral LD_{50} and structure. pp. 397–409. In G. W. Suter II and M. Lewis (eds.), Aquatic Toxicity and Hazard Assessment: Eleventh Symposium. American Society for Testing and Materials, Philadelphia.

Environmental Effects Branch. 1984. Estimating "concern levels" for concentrations of chemical substances in the environment. U. S. Environmental Protection Agency, Washington, D.C.

EPA (U.S. Environmental Protection Agency). 1973. Biological field and laboratory methods for measuring the quality of surface waters and effluents. EPA/670/4–73/001. National Environmental Research Center, Cincinnati, Ohio.

EPA (U.S. Environmental Protection Agency). 1978a. Proposed guidelines for registration of pesticides in the United States. Fed. Regist. 43:29697–29741.

EPA (U.S. Environmental Protection Agency). 1978b. Air quality criteria for ozone and other photochemical oxidants. EPA-600/8–78–004. Environmental Criteria and Assessments Office, Research Triangle Park, NC.

EPA (U.S. Environmental Protection Agency). 1979. Water quality criteria. Part V: Request for comments. Fed. Regist. 44:15926–15981.

EPA (U.S. Environmental Protection Agency). 1980. Water quality criteria documents; availability. Fed. Regist. 45:79318–79379.

EPA (U.S. Environmental Protection Agency). 1981. Interim methods for the sampling and analysis of priority pollutants in sediments and fish tissue. EPA 600/4-81-055. Environmental Monitoring and Support Laboratory, Cincinnati, Ohio.

EPA (U.S. Environmental Protection Agency). 1982a. Appendix A—Uncontrolled hazardous waste site scoring system: A users manual, Fed. Reg. 37(137):31219–31243.

EPA (U.S. Environmental Protection Agency). 1982b. Guidelines and support documents for environmental effects testing. Washington, D.C.

EPA (U.S. Environmental Protection Agency). 1982c. Air Quality Criteria for particulate matter and sulfur oxides, EPA-600/8-84-029f. Research Triangle Park, North Carolina.

EPA (U.S. Environmental Protection Agency). 1982d. Air Quality Criteria for oxides of nitrogen, EPA-600/8-84-026f. Research Triangle Park, North Carolina.

EPA (U.S. Environmental Protection Agency). 1982e. Sampling protocols for collecting surface water, bed sediment, bivalves and fish for priority pollutant analysis; Final report. Office of Water Regulation and Standards, Washington, D.C.

EPA (U.S. Environmental Protection Agency). 1983a. Water Quality Standards Handbook, Office of Water, Washington, D.C.

EPA (U.S. Environmental Protection Agency). 1983b. Methods for chemical analysis of water and wastes. EPA-600/4-79-020. Environmental Monitoring and Support Laboratory, Cincinnati, Ohio.

EPA (U.S. Environmental Protection Agency). 1984a. Air Quality Criteria for ozone and other photochemical oxidants, EPA-600/8-84-020b. Research Triangle Park, North Carolina.

EPA (U.S. Environmental Protection Agency). 1984b. Development of water quality-based permit limitations for toxic pollutants. Fed. Regist. 49:9016-.

EPA (U.S. Environmental Protection Agency). 1984c. Risk assessment and management: framework for decision making. EPA 600/9-85-002. Washington, D.C.

EPA (U.S. Environmental Protection Agency). 1985a. Water quality criteria; availability of documents. Fed. Regist. 50:30784–30796.

EPA (U.S. Environmental Protection Agency). 1985b. Technical support document for water quality-based toxics control. EN-336. Office of Water, Washington, D.C.

EPA (U.S. Environmental Protection Agency). 1985c. Toxic Substance Control Act Test Guidelines, Final Rules. 40 CFR Parts 796, 797, and 798. Fed. Reg. Sept. 27 pp. 39252–39323.

EPA (U.S. Environmental Protection Agency). 1986a. Guidelines for estimating exposures. Fed. Reg. 51:34042–34054.

EPA (U.S. Environmental Protection Agency). 1986b. Test methods for evaluating solid waste. SW-846. Office of Solid Waste, Washington, D.C.

EPA (U.S. Environmental Protection Agency). 1987. Toxic Substances Control Act test guidelines; Proposed rules. Fed. Reg. 52:36334–36371.

EPA (U.S. Environmental Protection Agency). 1988. Toxic Substance Control Act Test Guidelines, Amendments 40 CFR Part 798. Fed. Reg. Sept. 7. pp. 34522–34530.

EPA (U.S. Environmental Protection Agency). 1990a. Hazard ranking system; final rule. Fed. Reg. 55:51532–51667.

EPA (U.S. Environmental Protection Agency). 1990b. National oil and hazardous substances pollution contingency plan; final rule. Fed. Reg. 55:8666–8873.

EPA (U.S. Environmental Protection Agency). 1990c. Biological criteria: national program guidance for surface waters. EPA-440/5-90-004. Office of Water, Washington, D.C.

Erickson, R.J. and C. E. Stephan. 1985. Calculation of the final acute value for water quality criteria for aquatic organisms. National Technical Information Service, Springfield, Virginia.

Eschenroeder, A., E. Irvine, A. Lloyd, C. Tashima and K. Tran. 1980. Computer simulation models for assessment of toxic substances, pp. 323–368. In R. Haque (ed.), Dynamics, Exposure and Hazard Assessment of Toxic Chemicals. Ann Arbor Science Inc. Ann Arbor, Michigan.

Exner, O. 1972. The Hammett equation—the present position. pp. 1–70. In Advances in Linear Free Energy Relationships. N. B. Chapman and J. Shorter, (eds.). Plenum Press, New York.

Exner, O. 1978. A critical compilation of substituent constants. In Correlation Analysis in Chemistry, N. B. Chapman and J. Shorter, (eds.). Plenum Press, New York.

Extrapolation Models Subcommittee. 1987. Review of research in support of extrapolation models by EPA's Office of Research and Development. SAB-EC-87-030. Science Advisory Board, U.S. Environmental Protection Agency, Washington, D.C.

Fairley, W. 1975. Criteria for evaluating the "small" probability of a catastrophic accident from the marine transportation of liquefied natural gas, pp. 135–151. In D. E. Okrent (ed.), Risk-benefit methodology and application. UCLA-ENG-7598. University of California, Los Angeles.

Farrington, J. W., A. C. Davis, B. W. Tripp, D. K. Phelps, and W. B. Galloway. 1987. "Mussel Watch"—Measurement of chemical pollutants in bivalves as one indicator of coastal environmental quality. pp. 125–139. In T. P. Boyle (ed.), New Approaches to Monitoring Aquatic Ecosystems. American Society for Testing and Materials, Philadelphia.

Fava, J. A., W. J. Adams, R. J. Larson, G. W. Dickson, K. L. Dickson, W. E. Bishop (eds.). 1987. Research priorities in environmental risk assessment. Society for Environmental Toxicology and Chemistry, Pensacola, FL.

Fedra, K. 1984. Interactive water quality simulation in a regional framework: a management oriented approach to lake and watershed modeling. Ecological Modeling 21: 209–232.

Feinstein, A. R. 1988. Scientific standards in epidemiological studies of the menace of daily life. Science 242:1257–1263.

Ferry, B. W. 1982. Lichens. pp. 291–319. In R. G. Burns and J. H. Slater (eds.), Experimental Microbial Ecology. Blackwell Scientific Publications, Oxford.

Ferson, S., F. J. Rohlf, L. Ginzburg, and G. Jacquez. 1987. RAMAS/a user manual. Exeter Publishing, Ltd, 100 North Country Road, Building B, Setauket, New York.

Ficksel, J. and V. T. Covello (eds.). 1986a. Biotechnology Risk Assessment: Issues and Methods for Environmental Introductions. Pergamon Press, New York.

Ficksel, J. and V. T. Covello. 1986b. The suitability and applicability of risk assessment methods for environmental applications of biotechnology. pp. 1–34. In J. Ficksel and V. T. Covello (eds.), Biotechnology Risk Assessment: Issues and Methods for Environmental Introductions. Pergamon Press, New York.

Ficksel, J. and V. T. Covello (eds.). 1988. Safety Assurance for Environmental Introductions of Genetically-Engineered Organisms. Springer-Verlag, Berlin.

Filov, V. A., A. A. Golubev, E. I. Liublina, and N. A. Tolokontsev. 1979. Quantitative Toxicology. John Wiley & Sons, New York.

Fink, R. J. 1979. Simulated field studies—Acute hazard assessment. pp. 45–51. In E. E. Kenaga (ed.), Avian and Mammalian Wildlife Toxicology. American Society for Testing and Materials, Philadelphia, Pennsylvania.

Finkel, A. 1990. Confronting uncertainty in risk management. Resources for the Future. Washington, D.C.

Finlayson-Pitts, B. J., and J. N. Pitts. 1986. Atmospheric Chemistry: Fundamentals and Experimental Techniques. John Wiley, New York.

Finn, J.T. 1976. Measures of ecosystem structure and function derived from analysis of flows. Journal of Theoretical Biology 56:363–380.

Finney, D. J. 1964. Statistical Methods in Biological Assay, 2nd Edition. Griffin Press, London.

Finney, D. J. 1971. Probit Analysis, 3rd Edition. Cambridge University Press, Cambridge.

Fischoff, B., S. Lichtenstein, P. Slovic, S. L. Derby, and R. L. Keeney. 1981. Acceptable Risk. Cambridge University Press, Cambridge.

Fisher, R. A. 1930. The Genetical Theory of Natural Selection. Clarendon Press, Oxford, UK (Reprinted in 1958 by Dover Publications, New York.)

Fite, E. C., L. W. Turner, N. J. Cook, and C. Stunkard. 1988. Guidance document for conducting terrestrial field studies. U. S. Environmental Protection Agency, Washington, D.C.

Fletcher, J. S., M. J. Muhitch, D. R. Vann, J. C. Mcfarlane, and F. E. Benenati. 1985. PHYTOTOX data base evaluation of surrogate plant species recommended by the U. S. Environmental Protection Agency and the Organization for Economic Cooperation and Development. Environ. Toxicol. Chem. 4:523–532.

Fletcher, J. S., F. L. Johnson, and J. C. McFarlane. 1990. Influence of greenhouse versus field testing and taxonomic differences on plant sensitivity to chemical treatment. Environ. Toxicol. Chem. 9:769–776.

Floit, S. B. and L. W. Barnthouse. 1991. Demographic analysis of a San Joaquin kit fox population. ORNL/TM-11679. Oak Ridge National Laboratory, Oak Ridge, Tennessee.

Fogels, A. and J. B. Sprague. 1977. Comparative short-term tolerence of zebrafish, flagfish, and rainbow trout to five poisons including potential reference toxicants. Water Research 11: 811–817.

Forbes, S.A. 1887. The lake as a microcosm. Bulletin Science Association of Peoria, Illinois, 77–87.

Ford, P. J., P. J. Turina, and D. E. Seeley. 1984. Characterization of hazardous waste sites — A method manual, Volume II: Available sampling methods. EPA/600/4-84-076. Office of Advanced Monitoring Systems Division, Las Vegas, Nevada.

Ford, P. J. and P. J. Turina. 1985. Characterization of hazardous waste sites — A method manual, Volume I: Site investigations. EPA/600/4-84-075. Office of Advanced Monitoring Systems Division, Las Vegas, Nevada.

Fordham, C. L. and D. G. Reagan. 1991. Pathways analysis method for estimating water and sediment criteria at hazardous waste sites. Environ. Toxicol. Chem. 10:949–960.

Forman, R. T. T. and M. Godron. 1986. Landscape Ecology. John Wiley and Sons, New York.

Fox, M.-A. and S. Olive. 1979. Photooxidation of anthracene on atmospheric particulate matter. Science 205:582–583.

Frost, T. M., D. L. DeAngelis, S. M. Bartell, D. J. Hall, and S. H. Hurlbert. 1988. Scale in the design and interpretation of aquatic community research. pp. 229–258. In S. R. Carpenter (ed.), Complex Interactions in Lake Communities. Springer-Verlag, New York.

FWS (U.S. Fish and Wildlife Service). 1980. Habitat evaluation procedures (HEP). Division of Ecological Services, Washington, D.C.

Gamble, J. C., J. M. Davis, and J. H. Steele. 1977. Loch Ewe bag experiment, 1974. Bull. Mar. Sci. 27:146–175.

Games, L.M. 1983. Practical applications and comparisons of environmental exposure assessment models. pp. 282–299. In Bishop, W.E., R.D. Cardwell, and B.B. Heidolph. 1983. Aquatic Toxicology and Hazard Assessment: Sixth Symposium. ASTM STP 802, Philadelphia, Pennsylvania.

Gano, K. A., D. W. Carlile, and L. E. Rogers. 1985. A harvester ant bioassay for assessing hazardous chemical waste sites. PNL-5434. Pacific Northwest Laboratory, Richland, Washington.

Gardner, R. H., R. V. O'Neill, J. B. Mankin, and J. H. Carney. 1981. A comparison of sensitivity and error analysis based on a stream ecosystem model. Ecological Modelling 12:173–190.

Gardner, R. H., R. V. O'Neill, J. B. Mankin, and K. D. Kumar. 1980. Comparative error analysis of six predator-prey models. Ecology 61:323–332.

Gardner, R. H., R. V. O'Neill, J. B. Mankin and J. H. Carney. 1981. A comparison of sensitivity analysis and error analysis based on a stream ecosystem model. Ecological Modelling, 12:177–194.

Gardner, R. H., B. Rvjder, and U. Bergstrvm. 1983. PRISM: A systematic method for determining the effects of parameter uncertainties on model predictions. Studsvik Energiteknik AB Report/NW-83/555. Nykvping, Sweden.

Garten, C. T., Jr. 1985. Examination of a proposed test for effects of chemicals on a multispecies association: The *Rhizobium*-legume symbiosis. pp. 1-1 — 1-73. In C. T. Garten, Jr., G. W. Suter II, and B. G. Blaylock (eds.), Development and Evaluation of Multispecies Test Protocols for Assessing Chemical Toxicity. ORNL/TM-9225. Environmental Sciences Division, Oak Ridge National Laboratory, Oak Ridge, Tennessee.

Garten, C. T., Jr. and J. R. Trabalka. 1983. Evaluation of models for predicting terrestrial food chain behavior of xenobiotics. Environ. Sci. Technol. 17:590–595.

Gauch, H. G. 1982. Multivariate Analysis in Community Ecology. Cambridge University Press, Cambridge.

Gearing, J. N. 1989. The role of aquatic microcosms in ecotoxicological research as illustrated by large marine systems. pp. 411–470. In S. A. Levin, M. A. Harwell, J. R. Kelly, and K. D. Kimball (eds.), Ecotoxicology: Problems and Approaches. Springer-Verlag, New York.

Geckler, J. R., W. B. Horning, T. M. Neiheisel, Q. H. Pickering, E. L. Robinson and C. E. Stephan. 1976. Validity of laboratory tests for predicting copper toxicity in streams. EPA-600/3-76-116. U. S. Environmental Protection Agency, Duluth, Minnesota.

Gelber, R. D., P. T. Lavin, C. R. Mehta, and D. A. Schoenfeld. 1985. Statistical analysis. pp. 110–123. In G. M. Rand and S. R. Petrocelli (eds.), Fundamentals of Aquatic Toxicology. Hemisphere Publishing Co., Washington, D.C.

Gentile, J. H., S. M. Gentile, and G. Hoffman. 1983. The effects of a chronic mercury exposure on survival, reproduction and population dynamics of *Mysidopsis bahia*. Environ. Toxicol. and Chem. 2:61–68.

Gersich, F. M., F. A. Blanchard, S. L. Applegath, and C. N. Park. 1986. The precision of daphnid (*Daphnia magna* Straus, 1820) static acute toxicity tests. Arch. Environ. Contam. Toxicol. 15:741–749.

Gibaldi, M. and D. Perrier. 1982. Pharmacokinetics, 2nd ed. Dekker, New York.

Giddings, J. M. 1981. Laboratory tests for chemical effects on aquatic population interactions and ecosystem properties. pp. 23–91. In A. S. Hammons (ed.), Methods for Ecological Toxicology. Ann Arbor Science, Ann Arbor, Michigan.

Giddings, J. M., P. J. Franco, R. M. Cushman, L. A. Hook, G. R. Southworth, and A. J. Stewart. 1984. Effects of chronic exposure to coal-derived oil on freshwater ecosystems: II. Experimental ponds. Environ. Toxicol. Chem. 3:465–488.

Giddings, J. M. 1986. A microcosm procedure for determining safe levels of chemical exposure in shallow-water communities. pp. 121–134. In J. Cairns, Jr. (ed.), Community Toxicity Testing. American Society for Testing and Materials. Philadelphia, Pennsylvania.

Giddings, J. M., A. J. Stewart, R. V. O'Neill, R. H. Gardner. 1983. An efficient algal bioassay based on short-term photosynthetic response. pp. 445–459. In W. E. Bishop, R. D. Cardwell, and B. B. Heidolph (eds.), Aquatic Toxicology and Hazard Assessment: Sixth Symposium, ASTM STP 802, American Society for Testing and Materials, Philadelphia, 1983.

Giesey, J. P. (ed.). 1980. Symposium on Microcosms in Ecological Research. CONF-781101. National Technical Information Service, Springfield, Virginia.

Giesey, J. P. 1985. Multispecies tests: Research needs to assess the effects of chemicals on aquatic life. pp. 67–77. In R. C. Bahner and D. J. Hansen (eds.), Aquatic Toxicology and Hazard Assessment. American Society for Testing and Materials, Philadelphia, Pennsylvania.

Giesy, J. P. and R. L. Graney. 1989. Recent developments in the intercomparisons of acute and chronic bioassays and bioindicators. Hydrobiologia 188/189:21–60.

Gile, J. D. and J. W. Gillett. 1979. Fate of ^{14}C-dieldrin in a simulated terrestrial ecosystem. Arch. Environ. Contam. Toxicol. 8:107–124.

Gile, J. D. and S. M. Meyers. 1986. Effect of adult mallard age on avian reproductive tests. Arch. Environ. Contam. Toxicol. 15:751–756.

Gillett, J. W. and J. M. Witt (eds.). 1979. Terrestrial microcosms. NSF/RA 79-0034. National Academy Press, Washington, D.C.

Gillett, J. W. (ed.). 1986a. Potential Impacts of Environmental Release of Biotechnology Products: Assessment, Regulation, and Research Needs. Environmental Management 10(4).

Gillett, J. W.. 1986b. Risk assessment methodologies for biotechnology impact assessment. Environ. Management 10:515–532.

Gillett, J. W. 1989. The role of terrestrial microcosms and mesocosms in ecotoxocological research. pp. 367–410. In S. A. Levin, M. A. Harwell, J. R. Kelly, and K. D. Kimball (eds.), Ecotoxicology: Problems and Approaches. Springer-Verlag, New York.

Gilpin, M. E. 1987. Spatial structure and population vulnerability. pp. 125–140. In M. E. Soule (ed.), Viable Populations for Conservation, Cambridge University Press.

Gilpin, M. E. 1990. Ecological invasions (book review). Science 248:88–89.

Gledhill, W. E. 1982. Microbial toxicity testing of synthetic organic chemicals. pp. 11–14. In B. T. Johnson (ed.), Impact of Xenobiotic Chemicals on Microbial Ecosystems. U.S. Fish and Wildlife Service Washington, D.C.

Goats, G. and C. A. Edwards. 1982. Testing the toxicity of industrial chemicals to earthworms. Rothamsted Exp. Station Report, 1982. pp. 104–105.

Gohre, K., and G. C. Miller. 1986. Singlet oxygen reactions on irradiated soil surfaces. Environ. Sci. Technol. 20:934–938.

Goldstein, R. A. and P. F. Ricci. 1981. Ecological risk uncertainty analysis. pp. 227–232. In J. W. Mitsch, R. W. Bosserman, and J. M. Klopatek (eds.), Energy and Ecological Modeling. Elsevier, NY.

Goodman, D. 1976. Ecological expertise. pp. 317–360. In H. A. Feiveson, F. W. Sinden, and R. H. Socolow (eds.), Boundaries of Analysis, An Enquiry into the Tocks Island Dam Controversy. Ballinger, Cambridge, Massachusetts.

Goodman, D. 1987. The demography of chance extinction. pp. 11–34. In M. E. Soule, ed., Viable Populations for Conservation, Cambridge University Press.

Goodman, E. D. 1982. Modeling the effects of pesticides on populations of soil/litter invertebrates in an orchard ecosystem. Environ. Toxicol. Chem. 1:45–60.

Goodyear, C. P., 1972. A simple technique for detecting effects of toxicants or other stresses on a predator-prey interaction. Trans. Am. Fish. Soc. 101:367–370.

Goodyear, C. P., 1977. Assessing the impact of power plant mortality on the compensatory reserve of fish populations. pp. 186–195. In Van Winkle, W. (ed.), Assessing the Effects of Power-Plant-Induced Mortality on Fish Populations. Pergamon, New York.

Goodyear, C. P. 1983. Measuring effects of contaminant stress on fish populations. pp. 414–424. In Bishop, W.E., R. D. Cardwell, and B. B. Heidolph (eds.) Aquatic

Toxicology and Hazard Assessment: Sixth Symposium, ASTM-STP 802, American Society for Testing and Materials, Philadelphia, Pennsylvania.

Goodyear, C. P., and S. W. Christensen. 1984. Bias-elimination in fish population models with stochastic variation in survival of the young. Trans. A. Fish. Soc. 113:627–632.

Gordon, G. E. 1988. Receptor models. Environ. Sci. Technol. 22:1132–1142.

Gore, J. A. and J. M. Nestler. 1988. Instream flow studies in perspective. Regulated Rivers Research and Management 2:93–102.

Gosselink, J. G., L. C. Lee, and T. Muir. 1990. Ecological Processes and Cumulative Impacts: Illustrated by Bottomland Hardwood Ecosystems. Lewis Publishers, Chelsea, Michigan.

Gough, L. P., H. T. Shacklette, and A. A. Case. 1979. Element concentrations toxic to plants, animals, and man. Geological Survey Bulletin 1466. U.S. Government Printing Office, Washington, DC.

Graham, R. L., C. T. Hunsaker, R. V. O'Neill, and B. L. Jackson. 1991. Ecological risk assessment at the regional scale. Ecological Applications 1:196–206.

Grant, W. D., and P. E. Long. 1981. Environmental Microbiology, Halsted Press, New York.

Grant, W.E., S.O. Fraser and K.G. Isakson. 1984. Effect of vertebrate pesticides on non-target wildlife populations: evaluations through modelling. Ecological Modelling 21: 85–108.

Gray, R. H. 1990. Fish behavior and environmental assessment. Environ. Toxicol. Chem. 9:53–67.

Green, P. E. 1978. Analyzing Multivariate Data. Dryden Press, Ninsdale, New York.

Green, R. H. 1965. Estimation of tolerance over an indefinite time period. Ecology 46:887.

Green, R. H. 1976. Some methods for hypothesis testing and analysis with biological monitoring data. pp. 200–211. In J. Cairns, Jr., K. L. Dickson, and G. F. Westlake (eds.), Biological Monitoring of Water and Effluent Quality. American Society for Testing and Materials, Philadelphia, Pennsylvania.

Green, R. H. 1979. Sampling Design and Statistical Methods for Environmental Biologists. John Wiley & Sons, New York.

Green, R. H. 1980. Multivariate approaches in ecology: the assessment of ecologic similarity. Annu. Rev. Ecol. Syst. 11:1–14.

Green, R. H. 1984. Some guidelines for the design of biological monitoring programs in the marine environment. pp. 647–655. In H. H. White (ed.), Concepts in Marine Pollution Measurements. Maryland Sea Grant College, College Park, Maryland.

Green, R. H. 1987. Statistical and mathematical aspects: distinction between natural and induced variation. pp. 335–354. In V. B. Vouk, G. C. Butler, A. C. Upton, D. V. Parke, and S. C. Asher (eds.), Methods for Assessing the Effects of Mixtures of Chemicals. SCOPE 30. John Wiley & Sons, Chichester, UK.

Greenberg, H. and J. J. Cramer (eds.). 1991. Risk Assessment and Risk Management for the Chemical Process Industry. Van Nostrand Reinhold, New York.

Greene, J. C., C. L. Bartels, W. J. Warren-Hicks, B. R. Parkhurst, G. L. Linder, S. A. Peterson, and W. E. Miller. 1988. Protocols for short-term toxicity screening of hazardous waste sites. U.S. Environmental Protection Agency, Corvallis, Oregon.

Grice, G. D. and M. R. Reeve, eds. 1982. Marine Mesocosms: Biological and Chemical Research in Experimental Ecosystems. Springer Verlag, New York.

Grier, J. 1980. Modeling approaches to bald eagle population dynamics. Wildlife Soc. Bull. 8:316–322.

Gruber, D., J. Diamond, and D. Johnson. 1988. Performance and validation of an on-line fish ventilatory early warning system. pp. 215–230. In G. W. Suter II and M. A. Lewis (eds.), Aquatic Toxicology and Environmental Fate: Eleventh Volume. American Society for Testing and Materials, Philadelphia.

Gulland, J. A. 1977. Fish Population Dynamics. John Wiley & Sons, London.

Gunn, J. M. 1989. Survival of lake char (*Salvelinus namaycush*) embryos under pulse exposure to acidic runoff water. pp. 23–45. In J. O. Nriagu and J. S. S. Lakshminarayana (eds.), Aquatic Toxicology and Water Quality Management. John Wiley & Sons, New York.

Haag W. R., P. Penwell, H. Johnson, and T. Mill. 1989. Structure activity relationships for alkyl halides and epoxides. Preprints Sympos. Structure Activity in Chem. Toxicol. 1989 Internal. Chem. Congress Pacific Basin Soc. Honolulu, Hawaii, 17–22 December.

Haag, W. R., and T. Mill. 1988a. Effect of a subsurface sediment on hydrolysis of haloalkanes and epoxides. Environ. Sci. Technol. 22:658–663.

Haag, W. R., and T. Mill. 1988b. Some reactions of naturally occurring nucleophiles with haloalkanes in water. Environ. Toxicol. Chem. 7:17–924.

Haag, W. R., D. Dulin, and T. Mill. 1987. Direct and indirect photooxidation of azodyes: kinetic measurements and structure activity relationships. Environ. Toxicol. Chem. 6:359–369.

Haag, W. R., and J. Hoigne. 1986. Singlet oxygen in surface waters – Part III: Steady state concentrations in various types of waters. Environ. Sci. Technol. 20:341–348.

Hadden, K.F. 1984. Multimedia fate and transport models. J. Toxicol.-Clin. Toxicol. 21:65–95.

Hakanson, L. 1984. Aquatic contamination and ecological risk: an attempt to a conceptual framework. Water Res. 18:1107–1118.

Halfon, E. 1985. The bootstrap and jackknife in ecotoxicology or parametric estimates of standard error. Chemosphere 14:1433–1440.

Hall, C. A. S., and J. W. Day, Jr. 1977. Ecosystem Modeling in Theory and Practice. John Wiley & Sons, New York.

Hall, L. W., Jr., D. T. Burton, S. L. Margrey, and W. C. Graves. 1983. The effect of acclimation temperature on the interactions of chlorine, delta T and exposure duration to eggs, protolarvae, and larvae of striped bass, *Morone saxatilis*. Water Res. 17:309–317.

Hall, L. W., Jr., L. O. Horseman and S. Zeger. 1984. Effects of organic and inorganic contaminants on fertilization, hatching success, and prolarval survival of striped bass. Arch. Environ. Contam. Toxicol. 13:723–729.

Hall, L. W., Jr., A. E. Pinkney, and L. O. Horseman. 1985. Mortality of striped bass larvae in relation to contaminants and water quality in a Chesapeake Bay estuary. Trans. Amer. Fish. Soc. 114:861–868.

Hallam, T. G., R. R. Lassiter, J. Li, and W. McKinney. 1990. Toxicant-induced mortality in models of *Daphnia* populations. Environ. Toxicol. Chem. 9:597–621.

Hallenbeck, W. H. and K. M. Cunningham. 1986. Quantitative Risk Assessment for Environmental and Occupational Health. Lewis Publishers, Chelsea, Michigan.

Halverson, H.O., D. Pramer, and M. Rogul (eds.). 1985. Engineered Organisms in the Environment: Scientific Issues. American Society for Microbiology. Washington, D.C.

Hamaker, J. W., and C. A. I. Goring. 1976. Turnover of pesticide residues in soil. p. 219. In Bound and Conjugated Residues, ACS Symp. Ser. No. 29, D. D., Kaufman, G. G. Still, G. D. Paulson, and S. K. Bandal, Eds, American Chemical Society, Washington, D.C.

Hammond, M. W. and M. Alexander. 1972. Effect of chemical structure on microbial degradation of aliphatic acids. Environ. Sci. Technol. 6: 732–735.

Hansen, D. J., S. Schimmel, and F. Mathews. 1974. Avoidance of Arochlor 1254 by shrimp and fishes. Bull. Environ. Contam. Toxicol. 12:253–256.

Hansen, S. R. and R. G. Garton. 1982. Ability of standard toxicity tests to predict the effects of the insecticide diflubenzuron on laboratory stream communities. Can. J. Fish. Aquat. Sci. 39:1273–1288.

Hanson, P. J. and G. E. Taylor, Jr. in press. Modeling pollutant gas uptake by leaves: An approach based on physicochemical properties. In Forest Growth: Process Modeling of Responses to Environmental Stress. Timber Press/Auburn University, Auburn, Alabama.

Hara, T. J., S. B. Brown, and R. E. Evans. 1983. Pollutants and chemoreception in aquatic organisms. pp. 247–306. In J. O. Nriagu (ed.), Aquatic Toxicology. John Wiley and Sons, New York.

Harras, M. C. and F. B. Taub. 1985. Comparisons of laboratory microcosms and field responses to copper. pp. 57–74. In T. P. Boyle, ed. Validation and Predictability of Laboratory Methods for Assessing the Fate and Effects of Contaminants in Aquatic Ecosystems. American Society for Testing and Materials, Philadelphia, Pennsylvania.

Harras, M. C., A. C. Kindig, and F. B. Taub. 1985. Response of blue-green and green algae to streptomycin in unialgal and paired culture. Aquatic Toxicology 6: 1–11.

Harris, J.R.W., A.J. Bale, B.J. Bayne, R.F.C. Mantoura, A.W. Morris, L.A. Nelson, P.J. Radford, R.J. Uncles, S.A. Weston and J. Widows. 1984. A preliminary model of the dispersal and biological effect of toxins in the Tamar estuary, England. Ecological Modelling 22: 253–284.

Harrison, H.L., O. L. Loucks, J. W. Mitchell, D. F. Parkhurst, C.R. Tracy, D. Watts, and V. J. Yannacone, Jr. 1970. Systems studies of DDT transport. Science 170:503–508.

Hart, W. B., P. Douderoff, and J. Greenback. 1945. The evaluation of toxicity of industrial wastes, chemicals and other substances to fresh-water fishes. Atlantic Refining Co., Philadelphia, Pennsylvania.

Hartung, R. and P. R. Durkin. 1986. Ranking the severity of toxic effects: potential applications to risk assessment. Comments Toxicology 1:49–63.

Harvey, R. W. and S. P. Garabedian. 1991. Use of colloid filtration theory in modeling movement of bacteria through a contaminated sandy aquifer. Environ. Sci. Technol. 25:178–185.

Havas, M. and T. C. Hutchinson. 1982. Aquatic invertebrates from the Smoking Hills, N.W.T.: effect of pH and metals on mortality. Can J. Fish. Aquat. Sci. 39:890–903.

HCN (Health Council of the Netherlands). 1989. Assessing the risk of toxic chemicals for ecosystems. Report No. 1988/28E. The Hague, The Netherlands.

Heagle, A. S., R. B. Philbeck, H. H. Rogers, and M. B. Letchworth. 1979. Dispensing and monitoring ozone in open-top field chambers for plant-effects studies. Phytopathology 69:15–20.

Heasley, J.E., W.K. Lauenroth, and J.L. Dodd. 1981. Systems analysis of potential air pollution impacts on grassland ecosystems, pp. 347–359. In Mitsch, W.J., R.W. Bosserman, and J.M. Klopatek (eds.), Energy and Ecological Modelling. Elsevier, New York.

Heath, A. G. 1977. Toxicity of intermittent chlorination to freshwater fish: Influence of temperature and chemical form. Hydrobiologia 56:39–47.

Heath, A. G. 1987. Water Pollution and Fish Physiology. CRC Press, Boca Raton, Florida.

Heck, W. W., S. V. Krupa and S. N. Linzon (eds.). 1979. Methodology for the Assessment of air pollution effects on vegetation. Air Pollution Control Association, Pittsburgh, Pennsylvania.

Heck, W. W., O. C. Taylor, and D. T. Tingey, (eds.). 1988. Assessment of Crop Loss from Air Pollutants. Elsevier Publishers, New York.

Hedtke, S. F. and J. W. Arthur. 1985. Evaluation of a site-specific water quality criterion for pentachlorophenol using outdoor experimental streams. pp. 551–563. In R. D. Cardwell, R. Purdy, and R. C. Bahner (eds.), Aquatic Toxicology and Hazard Assessmant: Seventh Symposium. American Society for Testing and Materials, Philadelphia, Pennsylvania.

Hegdahl, P. L. and B. A. Colvin. 1988. Potential hazard to eastern screech-owls and other raptors of brodifacoum bait used for vole control in orchards. Environ. Toxicol. Chem. 7:245–260.

Heinz, G. H., D. J. Hoffman, A. J. Krynitsky, and D. M. G. Weller. 1987. Reproduction in mallards fed selenium. Environ. Toxicol. Chem. 6:423–433.

Hellawell, J. M. 1986. Biological Indicators of Freshwater Pollution and Environmental Management. Elsevier Applied Science Publishers, London.

Henderson, C. 1957. Application factors to be applied to bioassays for the safe disposal of toxic wastes. pp. 31–37. In C. M. Tarzwell (ed.), Biological Problems of Water Pollution. U.S. Department of Health Education and Welfare, Cincinnati, Ohio.

Henderson, C., Q. H. Pickering, and J. M. Cohen. 1959. The toxicity of synthetic detergents and soaps to fish. Sewage Ind. Wastes 31:295–306.

Hendry, D. G., T. Mill, L. Piszkiewicz, J. A. Howard, and H. K. Eigenmann. 1974. A critical review of H-atom transfer in the liquid phase. J. Phys. Chem. Ref. Data 3:937–978.

Herbes, S. E., W. H. Greist and G. R. Southworth. 1978. Field site evaluation of aquatic transport of polycyclic aromatic hydrocarbons. pp. 221–230. In Proceedings of the Symposium on Potential Health and Environmental Effects of Synthetic Fossil Fuel Technologies, CONF-78093. Oak Ridge National Laboratory, Oak Ridge, Tennessee.

Hermanutz, R. O., J. G. Eaton, and L. H. Mueller. 1985. Toxicity of endrin and malathion mixtures to flagfish (*Jordanella floridae*). Arch. Environ. Contam. Toxicol. 14:307–314.

Hermens, J., H. Canton, P. Janssen, and R. De Jong. 1984a. Quantitative structure-activity relationships and toxicity studies of mixtures of chemicals with anesthetic potency: Acute lethal and sublethal toxicity to *Daphnia magna*. Aquatic Toxicology 5: 143–154.

Hermens, J., H. Canton, N. Steyger, and R. Wegman. 1984b. Joint effects of a mixture of 14 chemicals on mortality and inhibition of reproduction of *Daphnia magna*. Aquatic Toxicology 5: 315–322.

Herrick, C.J., E.D. Goodman, C.A. Guthrie, R.H. Blythe, G.A. Hendrix, R.L. Smith, and J.E. Galloway. 1982. A model of mercury contamination in a woodland stream. Ecological Modelling 15: 1–28.

Herricks, E. E., M. J. Sale, and E. D. Smith. 1981. Environmental impact analysis of aquatic ecosystems using rational threshold value methodologies. pp. 14–31. In J. M. Bates and C. I. Weber (eds.), Ecological Assessments of Effluent Impacts on Communities of Indigenous Aquatic Organisms. American Society for Testing and Materials, Philadelphia, Pennsylvania.

Herting, G. E. and A. Witt, Jr. 1967. The role of physical fitness of forage fishes in relation to their vulnerability to predation by the bowfin (*Amia calva*). Trans. Am. Fish. Soc. 96:427–430.

Hess, K. W., M. P. Sissenwine, and S. B. Saila. 1975. Simulating the impact of the entrainment of winter flounder larvae. pp. 1–29. In Saila, S. B., ed., Fisheries and Energy Production, Lexington Books, D.C. Heath & Co., Lexington, Massachusetts.

Hetrick, D.M., McDowell-Boyer, L.M. 1983. User's Manual for TOXSCREEN: A multimedia screening-level program for assessing the potential fate of chemicals released to the environment. ORNL/TM-8570. Oak Ridge National Laboratory, Oak Ridge, Tennessee.

Hettelingh, J.-P., R. J. Downing, and P. A. M. de Smet. 1991. Mapping critical loads for Europe. CCE Tech. Rpt. No. 1. Coordination Center for Effects, National Institute of Public Health and Environment Protection, Bilthoven, The Netherlands.

Hewlett, P. S. and R. L. Plackett. 1952. Similar joint action of insecticides. Nature (London) 169:198–199.

Hewlett, P. S. and R. L. Plackett. 1979. The Interpretation of Quantal Responses in Biology. Edward Arnold, London.

Heyde, C. C., and J. E. Cohen. 1985. Confidence intervals for demographic projections based on products of random matrices. Theoret. Pop. Biol. 27:120–153.

Hickey, J. J. 1969. Peregrine Falcon Populations: Their Biology and Decline. University of Wisconsin Press, Madison.

Hickey, J. J. and D. W. Anderson. 1968. Chlorinated hydrocarbons and eggshell changes in raptorial and fish-eating birds. Science 162:271–273.

Hill, A. B. 1965. The environment and disease: Association or causation. Proceed. Royal Soc. Medicine 58:295–300.

Hill, A. C., S. Hill, C. Lamb, and T. W. Barrett. 1974. Sensitivity of native desert vegetation to SO_2 and SO_2 and NO_2 combined. J. Air Pollut. Control Assoc. 24:153–157.

Hill, E. F., and M. B. Camardese. 1986. Lethal dietary toxicities of environmental contaminants and pesticides to *Coturnix*, Fish and Wildlife Technical Report 2. U. S. Fish and Wildlife Service, Laurel, Maryland.

Hill, E. F., R. G. Heath, J. W. Spann, and J. D. Williams. 1975. Lethal dietary toxicities of environmental pollutants, Spec. Sci. Rept.–Wildl. 191. U.S. Fish and Wildlife Service, Washington, D.C.

Hill, J. and R. G. Weigert. 1980. Microcosms in ecological modelling. pp. 138–163. In J. P. Giesy (ed.), Microcosms in Ecological Research. ERDA CONF-781101. U.S. Department of Energy.

Hirsch, R. M., W. M. Alley, and W. G. Wilber. 1988. Concepts of a national water-quality assessment program. U.S. Geological Survey Circular 1021. Denver, Colorado.

Hites, R.A., Eisenreich, S.J. (eds.). 1987. Sources and Fates of Aquatic Pollutants, Advances in Chemistry Series 216, Amer. Chem. Soc., Washington, D.C.

Hivey, L. M., L. W. Barnthouse, D. C. Kocher, N. B. Munro, E. D. Smith, and C. C. Travis. 1985. Assessment of Risk Methodologies for DOE Hazardous Chemical Waste Sites, ORNL/TM-9953. Oak Ridge National Laboratory, Oak Ridge, Tennessee.

Hodgson, E., A. P. Kulknari, D. L. Fabacher, and K. M. Robacker. 1980. Induction of hepatic drug metabolizing enzymes in mammals by pesticides: A review. J. Environ. Sci. Health. B15:723–754.

Hodgson, J. and A. M. Sugden (eds.). 1988. Planned Release of Genetically Engineered Organisms. Trends in Biotechnology/Trends in Ecology Special Publication. Elsevier Publications, Cambridge.

Hodson, P. V., D. G. Dixon, D. J. Spry, D. M. Whittle, and J. B. Sprague. 1982. Effect of growth rate and size of fish on rate of intoxication by waterborne lead. Can. J. Fish. Aquat. Sci. 39:1243–1251.

Hoffer-Bosse, T. and R. Kroker. 1985. Acute toxicity of chemicals: SAR studies for toxicological estimation. pp. 65–72. In M. Tichy (ed.), QSAR in Toxicology and Xenochemistry. Elsevier, Amsterdam.

Hoffman, F. O., D. L. Schaeffer, C. W. Miller, and C. T. Garten, Jr. (eds.). 1978. Proceedings of a Workshop on the Evaluation of Models Used for the Environmental Assessment of Radionuclide Releases. CONF-770901. Oak Ridge National Laboratory, Oak Ridge, Tennessee.

Hoffman, F. O., and R. H. Gardner. 1983. Evaluation of uncertainties in environmental radiological assessment models. pp. 11-1-11-55. In Radiological Assessment: A Textbook on Environmental Dose Assessment, J. E. Till and H. R. Meyer (eds.), U.

S. Nuclear Regulatory Commission, Washington, D.C. NUREG/CR-3332, ORNL-5968.

Hoffman, M. R. 1980. Trace metal catalysis in aquatic environment. Environ. Sci. Technol. 14:1061–1066.

Hokanson, K. E. F. and L. L. Smith, Jr. 1971. Some factors influencing toxicity of linear alkylate sulfonate (LAS) to the bluegill. Trans. Am. Fish. Soc. 100:1–12.

Holcombe, G. W., G. L. Phipps, and G. D. Veith. 1988. Use of aquatic lethality tests to estimate safe chronic concentrations of chemicals in initial ecological risk assessments. pp. 442–467. In G. W. Suter II and M. Lewis (eds.), Aquatic Toxicity and Hazard Assessment: Eleventh Symposium. American Society for Testing and Materials, Philadelphia, Pennsylvania.

Holland, P. W. 1986. Statistics and causal inference. JASA 81:945–960.

Holling, C. S. (ed.). 1978. Adaptive Environmental Assessment and Management. John Wiley and Sons, Chichester.

Hopkinson, C.S., Jr. and J.W. Day, Jr. 1977. A model of the Barataria Bay salt march ecosystem, pp. 235–265. In C.A.S. Hall and J.W. Day, Jr. (eds.), Ecosystem modeling in theory and practice. Wiley-Interscience, New York.

Horwitz, R.J. 1981. The direct and indirect impacts of entrainment on estuarine communities—transfer of impacts between trophic levels, pp. 185–197. In Mitsch, W.J., R.W. Bosserman, and J.M. Klopatek (eds.), Energy and Ecological Modelling. Elsevier, New York.

Hose, J. E., T. D. King, K. E. Zerba, R. J. Stoffel, J. S. Stephans, and J. A. Dickson. 1981. Does avoidance of chlorinated seawater protect fish against toxicity? Laboratory and field observations. pp. 967–982. In R. L. Jolly, et al. (eds.), Water Chlorination, Environmental Impact and Health Effects, Vol. 4, Book 2, Environment, Health, and Risk, Ann Arbor Science, Ann Arbor, Michigan.

Hosker, R. P. and S. E. Lindberg. 1982. Review: Atmospheric deposition and plant assimilation of gases and particles. Atmos. Environ. 5:889–910.

Host, G. E., R. R. Regal, and C. E. Stephan. in press. Analysis of acute and chronic data for aquatic life. U.S. EPA Report, Duluth, Minnesota.

Hudson, R. H., R. K. Tucker, and M. A. Heagele. 1972. Effect of age on sensitivity: Acute oral toxicity of 14 pesticides to mallard ducks of several ages. Toxicol. Appl. Pharmacol. 22:556–561.

Hudson, R. H., M. A. Heagele, and R. K. Tucker. 1979. Acute oral and percutaneous toxicity of pesticides to mallards: Correlations with mammalian toxicity data. Toxicol. Appl. Pharmacol. 47: 451–460.

Hudson, R. H., R. K. Tucker, and M. A. Heagele. 1984. Handbook of toxicity of pesticides to wildlife. Resource Publication 153. U. S. Fish and Wildlife Service, Washington, D.C.

Hughes, R. M., D. P. Larsen, and J. M. Omernik. 1986. Regional reference sites: a method for assessing stream potentials. Environ. Manage. 10:629–635.

Hume, D. 1748. An Inquiry Concerning Human Understanding.

Hunsaker, C. T., S. W. Christensen, J. J. Beauchamp, R. J. Olson, R. S. Turner, and J. L. Malanchuk. 1986. Empirical relationships between watershed attributes and headwater lake chemistry in the Adirondack region, ORNL/TM-9838. Oak Ridge National Laboratory, Oak Ridge, Tennessee.

Hunsaker, C. T., R. L. Graham, G. W. Suter II, R. V. O'Neill, L. W. Barnthouse, and R. H. Gardner. 1990. Assessing ecological risk on a regional scale. Environ. Manage. 14:325–332.

Hunsaker, C. T. and R. L. Graham. 1991. Regional ecological assessment for air pollution. pp. 312–334. In S. K. Majumdar, E. W. Miller, and J. Cahir (eds.), Air Pollution: Environmental Issues and Health Effects. Pennsylvania Academy of Sciences, Easton, Pennsylvania.

Hunt, H.W. 1977. A simulation model for decomposition in grasslands. Ecology 58: 469–484.

Hurlbert, S. H. 1984. Pseudoreplication and the design of ecological field experiments. Ecological Mono. 54:187–211.

Huston, M. A., and T. M. Smith. 1987. Plant succession: life history and competition. *Amer. Natur.* 130:168–198.

Huston, M., D. L. DeAngelis, and W. Post. 1987. New computer models unify ecological theory. BioScience 38:682–691.

Hutchinson, T. C., J. A. Hellerbust, D. McKay, D. Tam, and P. Kauss. 1979. Relationship of hydrocarbon solubility to toxicity in algae and cellular membrane effects. pp. 383–427. Proceedings of the 1979 Oil Spill Conference. American Petroleum Institute. Washington, D.C.

Hutton, M. 1982. The role of wildlife species in the assessment of biological impact of chronic exposure to persistent chemicals. Ecotox. Environ. Safety 6:471–478.

IAEA (International Atomic Energy Agency). 1982. Generic models and parameters for assessing the environmental transfer of radionuclides from routine releases. Safety Series No. 57. Vienna.

Iman, R. L. and Conover, W. J. 1982. A distribution-free approach to inducing rank correlation among input variables for simulation studies. Communications in Statistics, B11(3).

Iman, R. L., and J. C. Helton. 1988. An investigation of uncertainty and sensitivity analysis techniques for computer models. *Risk Analysis* 8:71–90.

Innes, J. L. 1988. Forest health surveys—A critique. Environ. Pollut. 54:1–15.

Jacobson, J. S. and A. C. Hill (eds.) 1970. Recognition of Air Pollution Injury to Vegetation: A Pictorial Atlas. Air Pollution Control Association, Pittsburgh, Pennsylvania.

Javitz, H. S. 1982. Relationship between response parameter hierarchies, statistical procedures, and biological judgement in the NOEL determination. pp. 17–31. In J. G. Pearson, R. B. Foster, and W. E. Bishop (eds.), Aquatic Toxicology and Hazard Assessment, Fifth Symposium. American Society for Testing and Materials, Philadelphia, Pennsylvania.

Jensen, A. L., and J. S. Marshall. 1983. Toxicant-induced fecundity compensation: a model of population responses. Environmental Management 7:171–175.

Jensen, K. F. and L. S. Dochinger. 1979. Growth responses of woody species to long-short-term fumigation with sulfur dioxide, Forest Service Research Paper NE-442. Northeastern Forest Experiment Station, Broomall, Pennsylvania.

Jimenez, B. D., L. S. Burtis, G. H. Ezell, B. Z. Egan, N. E. Lee, J. J. Beauchamp, and J. F. McCarthy. 1988. The mixed function oxidase system of bluegill sunfish, *Lepomis macrochirus*: Correlation of activities in experimental and wild fish. Environ. Toxicol. Chem. 7:623–634.

Jimenez, B. D. and L. S. Burtis. 1989. Influence of environmental variables on the hepatic mixed-function oxidase system in bluegill sunfish, *Lepomis macrochirus*. Comp. Biochem. Physiol. 39C:11–21.

Johnson, A. R. 1988. Diagnostic variables as predictors of ecological risk. Environ. Management 12:515–523.

Jones, C. J. and J. S. Coleman. 1989. Biochemical indicators of air pollution effects on trees: Unambiguous signals based on secondary metabolites and nitrogen in fast-growing trees. pp. 261–274. In National Research Council. Biological Markers of Air-Pollution Stress and Damage in Forests. National Academy Press, Washington, D.C.

Jones, H. C., F. P. Weatherford, J. C. Noggle, N. T. Lee, and J. R. Cunningham. 1979. Power plant siting: Assessing risks of sulfur dioxide effects on agriculture. Presented at the 72nd Annual Meeting of the Air Pollution Control Association, Cincinnati, Ohio, June 24–29, 1979.

Jop, K. M., J. H. Rodgers, Jr., P. B. Dorn, and K. L. Dickson. 1986. Use of hexavalent chromium as a reference toxicant in aquatic toxicity tests. pp. 390–403. In T. M. Posten and R. Purdy, (eds.). Aquatic Toxicology and Environmental Fate: Ninth Volume. American Society for Testing and Materials, Philadelphia, Pennsylvania.

Jordan III,, W. R., M. E. Gilpin, and J. D. Aber (eds.). 1987. Restoration Ecology. Cambridge U. Press, Cambridge, U.K.

Jorgensen, S.E. (ed.). 1984. Modelling the Fate and Effect of Toxic Substances in the Environment. Elsevier, Amsterdam.

Jorgensen, S. E. 1988. Fundamentals of Ecological Modelling, 2nd ed. Elsevier, Amsterdam.

Jorgensen, S.E., and M.J. Gromiec. (eds.). 1989. Mathematical Submodels in Water Quality Analysis, Elsevier, Amsterdam.

Jury, W.A., W. F. Spencer, and W. J. Farmer. 1983. Behavior assessment model for trace organics in soil, J. Environ. Qual. 12:558–564.

Kabata-Pendias, A. and H. Pendias. 1984. Trace Elements in Soils and Plants. CRC Press, Boca Raton, Florida.

Kandel, A. and E. Avni (eds.). 1988. Engineering Risk and Hazard Assessment. CRC Press, Boca Raton, Florida.

Kaputska, L., T. LaPoint, J. Fairchild, K. McBee, and J. Bromenshenk. 1989. Field assessments. pp. 8-1-8–88. In W. Warren-Hicks, B. R. Parkhurst, and S. S. Baker, Jr. (eds.), Ecological Assessment of Hazardous Waste Sites: A Field and Laboratory Reference Document. EPA 600/3–89/013. Corvallis Environmental Research Laboratory, Oregon.

Kara, A. H. and W. L. Hayton. 1984. Pharmacokinetic model for the uptake and disposition of di-2-ethylhexyl phthalate in sheepshead minnow *Cyprinodon variegatus*. Aquat. Toxicol. 5:181–195.

Karickhoff, W. W., D. S. Brown, and T. A. Scott. 1979. Sorption of hydrophobic pollutants on natural sediments. Water Res. 13:241.

Karr, J. R., K. D. Fausch, P. L. Angermeier, P. R. Yant, and I. J. Schlosser. 1986. Assessing biological integrity in running waters; a method and its rationale. Illinois Natural History Survey Special Pub. 5. Champaign, Illinois.

Kates, R. W. 1978. Risk Assessment of Environmental Hazard, SCOPE 8. John Wiley & Sons, New York.

Keith, L. H. 1991. Environmental Sampling and Analysis: A Practical Guide. Lewis Publishers, Chelsea, Michigan.

Keller, T. 1984. Direct effects of sulphur dioxide on trees. Phil. Trans. R. Soc. Lond. B 305:317–326.

Kelly, R.A. and W.O. Spoford, Jr. 1977. Application of an ecosystem model to water quality management: the Delaware Estuary, pp. 419–443. In C.A.S. Hall and J.W. Day, Jr. (eds.), Ecosystem modeling in theory and practice. Wiley-Interscience, New York.

Kenaga, E. E. 1973. Factors to be considered in the evaluation of the toxicity of pesticides to birds in the environment. Environ. Qual. Saf. 2:166–181.

Kenaga, E. E. 1977. Evaluation of the hazard of pesticide residues in the environment. pp. 51–94. In D. L. Watson and A. W. A. Brown (eds.), Pesticide Management and Insecticide Resistance. Academic Press, New York.

Kenaga, E. E. 1978. Test organisms and methods useful for early assessment of acute toxicity of chemicals. Environ. Sci. Technol. 12: 1322–1329.

Kenaga, E. E. 1979a. Acute and chronic toxicity of 75 pesticides to various animal species. Down Earth 35:25–31.

Kenaga, E. E. 1979b. Aquatic test organisms and methods useful for assessment of chronic toxicity of chemicals. pp. 101–111. In K. L. Dickson, A. W. Maki, and J. Cairns, Jr. (eds.), Analyzing the Hazard Evaluation Process. American Fisheries Society, Washington, D.C.

Kenaga, E. E. 1982. Predictability of chronic toxicity from acute toxicity of chemicals in fish and aquatic invertebrates. Environ. Toxicol. Chem. 1:347–358.

Kenaga, E. E., and C. A. I. Goring. 1980. Relationship between water solubility, soil sorption, octanol-water partitioning, and concentrations of chemicals in biota. pp. 78–115. In J. G. Eaton, P. R. Parrish, and A. C. Hendricks (eds.), Aquatic Toxicology. ASTM STP 707. American Society for Testing and Materials, Philadelphia, Pennsylvania.

Kenaga, E. E., and D. W. Lamb. 1981. Introduction. pp. 1–3. In D. W. Lamb and E. E. Kenaga (eds.), Avian and Mammalian Wildlife Toxicology: Second Conference. American Society for Testing and Materials, Philadelphia, Pennsylvania.

Kenaga, E. E., and R. J. Moolenaar. 1979. Fish and Daphnia toxicity as surrogates for aquatic vascular plants and algae. Environ. Sci. Technol. 13: 1479–1480.

Kenavanga, E. E. 1977. Evaluation of the hazard of pesticide residues in the environment. pp. 51–94. In D. L. Watson and A. W. A. Brown (eds.), Pesticide Management and Insect Resistance. Academic Press, New York.

Kercher, J.R., M.C. Axelrod and G.E. Bingham. 1981. Hierarchical set of models for estimating the effects of air pollution on vegetation, pp. 401–409. In Mitsch, W.J., R.W. Bosserman, and J.M. Klopatek (eds.), Energy and Ecological Modelling. Elsevier, New York.

Kercher, J.R. and M.C. Axelrod. 1984. Analysis of SILVA: a model for forecasting the effects of SO_2 pollution on Western coniferous forests. Ecological Modelling 23: 165–184.

Kerswill, C. J. and H. E. Edwards. 1967. Fish losses after forest spraying with insecticides in New Brunswick, 1952–62, as shown by caged specimens and other observations. J. Fish. Res. Bd. Canada 24:709–729.

Kettle, W. D., F. deNoyles, Jr., B. D. Heacock, and A. M. Kadoum. 1987. Diet and reproductive success of bluegill recovered from experimental ponds treated with atrazine. Bull. Environ. Contam. Toxicol. 38:47–52.

Kimball, K. D. and S. A. Levin. 1985. Limitations of laboratory bioassays:The need for ecosystem-level testing. BioScience 35:165-171.

Kimerle, R. A., A. F. Werner, and W. J. Adams. 1983. Aquatic hazard evaluation principles applied to the development of water quality criteria. pp. 538-547. In R. D. Cardwell, R. Purdy, and R. C. Bahner (eds.), Aquatic Toxicology and Hazard Assessment, Seventh Symposium. American Society for Testing and Materials, Philadelphia, Pennsylvania.

King, K. A. and A.J. Krynitsky. 1986. Population trends, reproductive success, and organochloride chemical contaminants in waterbirds nesting in Galveston Bay, Texas. Arch. Environ. Contam. Toxicol. 15:367-376.

Kirby, A. J. 1972. In C. H. Bamford, C. F. H. Tipper, (eds.). Comprehensive Chemical Kinetics, Elsevier Publishing Company, Amsterdam, Vol. 10.

Kitchell, J.F., D. J. Stewart, and D. Weininger. 1977. Applications of a bioenergetics model to yellow perch (Perca flavescens) and walleye (Stizostedion vitreum vitreum). Journal of the Fisheries Research Board of Canada 34:1922-1935.

Klapow, L. A. and R. H. Lewis. 1979. Analysis of toxicity data for California marine water quality standards. J. Water Pollut. Control Fed. 51:2054-2070.

Klecka, G. M. 1985. Biodegradation. In W. B. Neely and G. E. Blaue, (eds.). Environmental Exposure from Chemicals, Vol. I. CRC Press, Boca Raton, Florida.

Klee, A. J. and M. U. Flanders. 1980. Classification of hazardous wastes. J. Environ. Eng. Div., Proc. Am. Soc. Civil Eng. 106(EEI):163-175.

Klein, A. W., W. Klein, W. Kordel, and M. Weiss. 1988b. Structure-activity relationships for selecting and setting priorities for existing chemicals — A computer-assisted approach. Environ. Toxicol. Chem. 7:455-467.

Klein, W., W. Koerdel, A. W. Klein, D. Kuhnen-Clausen, and M. Weiss. 1988a. Systematic Approach for Environmental Hazard Ranking of New Chemicals. Chemosphere 7:1445-1462.

Klerks, P. L., and J. S. Levinton. 1988. Effects of heavy metals in a polluted aquatic ecosystem. pp. 41-67. In S. A. Levin, M. A. Harwell, J. R. Kelly, and K. D. Kimball (eds.), Ecotoxicology: Problems and Approaches. Springer-Verlag, New York.

Koch, A. L. 1966. The logarithm in biology 1. Mechanisms generating the log-normal distribution exactly. J. Theoret. Biol. 12:276-290.

Kocher, D. C. and F. O. Hoffman. 1991. Regulating environmental carcinogens: where do we draw the line? Environ. Sci. Technol. 25:1986-1989.

Kodell, R. L. and J. G. Pounds. 1985. Characterization of the joint action of two chemicals in an in vitro test system. 1985 Proceedings of the Biopharmaceutical Section, American Statistical Associastion:48-53.

Kodell, R. L. 1987. Modeling the joint action of toxicants: Basic concepts and approaches. pp. 1–8. In E. Landau and D. G. Wellington, (eds.), Current Assessment of Combined Toxic Effects, EPA-230–03–87–027. U.S. Environmental Protection Agency, Washington, D.C.

Koeman, J. H. 1983. Ecotoxicological evaluation: the eco-side of the problem. Ecotoxicol. Environ. Safety 6:358–362.

Konemann, H. 1980. Structure-activity relationships and additivity in fish toxicities of environmental pollutants. Ecotoxicol. Environ. Safety 4: 415–421.

Konemann, H. 1981. Quantitative structure-activity relationships in fish toxicity studies. Part 1: Relationship for 50 industrial pollutants. Toxicology 19:209–221.

Konemann, H. and A. Musch. 1981. Quantitative structure-activity relationships in fish toxicity studies. Part 2: The influence of pH on the SAR of chlorophenols. Toxicology 19:223–228.

Kooijman, S. A. L. M. 1981. Parametric analysis of mortality rates in bioassays. Water Res. 15:107–119.

Kooijman, S. A. L. M. 1983a. Parametric analyses of population growth in bio-assays. Water Res. 17:527–538.

Kooijman, S. A. L. M. 1983b. Statistical aspects of the determination of mortality rates in bioassays. Water Res. 17:749–759.

Kooijman, S. A. L. M. 1987. A safety factor for LC_{50} values allowing for differences in sensitivity among species. Water Res. 21:269–276.

Kooijman, S. A. L. M., A. O. Hanstveit, and H. Oldersma. 1983. Parametric analysis of population growth in bio-assays. Water Res. 17:527–538.

Kooijman, S. A. L. M., and J. A. J. Metz. 1984. On the dynamics of chemically stressed populations: the deduction of population consequences from effects on individuals. Ecotox. Environ. Safety 8:254–274.

Kranz, J. (ed.). 1990. Epidemics of Plant Diseases (2nd Ed.). Springer-Verlag, Berlin.

Krasovskij, G. N. 1976. Extrapolation of experimental data from animals to man. Environ. Health Persp. 13:51–58.

Krebs, C. J. 1985. Ecology: The Experimental Analysis of Distribution and Abundance. Harper & Row, New York.

Kremer, J. N., and S. W. Nixon. 1978. A Coastal Marine Ecosystem: Simulation and Analysis. Springer-Verlag, New York.

Krome, E. C. 1990. Chesapeake Bay ambient toxicity assessment workshop report. CBP/TRS 42/90. U.S. Environmental Protection Agency/Chesapeake Bay Program, Baltimore, Maryland.

Lacroix, G. L. 1985. Survival of eggs and alevins of atlantic salmon (*Salmo salar*) in relation to the chemistry of interstitial water in reeds in some acidic streams of Atlantic Canada. Can. J. Fish. Aquatic Sci. 42:292–299.

Ladd, W. 1986. New hope for the southern sea otter. Endangered Species Tech. Bull. 11:12–14.

Landau, E. and D. G. Wellington, (eds.). Current assessment of combined toxicant effects, EPA-230–03–87–027. U.S. Environmental Protection Agency, Washington, D.C.

Lande, R. 1988. Demographic models of the northern spotted owl (*Stix occidentalis caurina*). Oecologia (Berlin) 75:601–607.

Lande, R. and S. H. Orzack 1988. Extinction dynamics of age-structured populations in a fluctuating environment. Proc. Nat. Acad. Sci. USA 85:7418–7421.

Landrum, P. F. and J. A. Robbins. 1990. Bioavailability of sediment-associated contaminants to benthic invertebrates. pp. 237–263. In R. Baudo, J. Giesy, and H. Muntau (eds.), Sediments: Chemistry and Toxicity of In-Place Pollutants. Lewis Publishers, Chelsea, Michigan.

Laird, M., L. A. Lacey, and E. W. Davidson. 1989. Safety of Microbial Insecticides. CRC Press, Boca Raton, Florida.

La Point, T. P. and J. A. Perry. 1989. Use of experimental ecosystems in regulatory decision making. Environ. Manag. 13:539–544.

La Point, T. W., S. M. Melancon, and M. K. Morris. 1984. Relationships among observed metal concentrations, criteria, and benthic community structural responses in 15 streams. J. WPCF 56:1030–1038.

La Point, T. W. and J. A. Perry. 1989. Use of experimental ecosystems in regulatory decision making. Environ. Manage. 13:539–544.

Larson, G. L., C. E. Warren, F. E. Hutchins, L. P. Lamberti, and D. A. Schlesinger. 1978. Toxicity of residual chlorine compounds to aquatic organisms, EPA-600/3–78–023. U. S. Environmental Protection Agency, Washington, D.C.

Larson, R. J. 1979. Estimation of biodegradation potential of xenobiotic organic chemicals. Appl. Environ. Microbiol., 38:1153.

Laskowski, D. A., C. A. I. Goring, P. J. McCall, and R. L. Swann. 1982. Terrestrial environment. pp. 198–256. In R. Conway (ed.). Environmental Risk Analysis for Chemicals. Van Nostrand Rheinhold, New York.

Lassiter, R. R. 1985. Design criteria for a predictive ecological effects modeling system. pp. 42–54. In T. M. Posten and Purdy (eds.) Aquatic Toxicology and Environmental Fate: Ninth Volume, ASTM STP 921. American Society for Testing and Materials, Philadelphia.

Lassiter, R.R., and T. G. Hallam. 1990. Survival of the fattest: implications for acute effects of lipophilic chemicals on aquatic populations. *Environ. Toxicol. Chem.* 9:585596.

Lauer, G.J., J. R. Young, and J. S. Suffern. 1981. The best way to assess environmental impacts is through the use of generic and site-specific data. pp. 21–34. In L. D. Jensen, (ed.). Issues Associated with Impact Assessment, EA Communications, Sparks, Maryland.

Laurence, J. A., D. C. Maclean, R. H. Mandl, R. E. Schneider, and K. S. Hansen. 1982. Field tests of a linear gradient system for exposure of row crops to SO_2 and HF. Water Air Soil Pollut. 17:399–407.

Lauwerys, R. R. 1983. Industrial Chemical Exposure: Guidelines for Biological Monitoring. Biomedical Publications, Davis, California.

Lave, L. B. 1982. Quantitative Risk Assessment in Regulation, Brookings Institute, Washington, D.C.

Lave, L. B. and E. P. Seskin. 1979. Epidemiology, causality, and public policy. American Scientist 67:178–186.

Law, R. 1983. A model for the dynamics of a plant population containing individuals classified by age and size. Ecology 64:224–230.

Law, R. and M. T. Edley. 1990. Transient dynamics of populations with age- and size-dependent vital rates. Ecology 71:1863–1970.

Lawler, J. 1988. Some considerations in applying stock-recruitment models to multiple-age spawning populations. Am. Fish. Soc. Monogr. 4:204–218.

Layher, W. G. and O. E. Maughan. 1988. Using stream survey data to predict directional changes in fish populations following physicochemical stream perturbations. Environ. Toxicol. Chem. 7:689–699.

LeBlanc, G. A. 1984. Interspecies relationships in acute toxicity of chemicals to aquatic organisms. Environ. Toxicol. Chem. 3: 47–60.

LeBlanc, G. A., B. Hilgenberg, and B. J. Cochrane. 1988. Relationships between the structure of chlorinated phenols, their toxicity, and their ability to induce glutathione S-transferase activity in *Daphnia magna*. Aquatic Toxicol. 12:147–156.

Lebsack, M. E., A. D. Anderson, G. M. DeGraeve, and H. L. Bergman. 1981. Comparison of bacterial luminescence and fish bioassay results for fossil-fuel process waters and phenolic constituents. pp. 348–356. In D. R. Branson and K. L. Dickson (eds.), Aquatic Toxicology and Hazard Assessment: Fourth Conference, STP 737. American Society for Testing and Materials, Philadelphia.

Lee, G. F. and R. A. Jones. 1982. An approach to evaluating the potential significance of chemical contaminants in aquatic habitat assessment. pp. 294–302. In Aquisition and Utilization of Aquatic Habitat Inventory Information. American Fisheries Society, Washington, D.C.

Lee, J. J. and R. A. Lewis. 1978. Zonal air pollution system: Design and performance. pp. 322–344. In E. M. Preston and R. A. Lewis, eds. The Bioenvironmental Impact of a Coal-Fired Power Plant. EPA-600/3-79-044. U. S. Environmental Protection Agency, Corvallis, Oregon.

Lefohn, A. S., S. V. Krupa, V. C. Runeckles, and D. S. Shadwick. 1989. Important considerations for establishing a secondary ozone standard for protection of vegetation. JAPCA 39:1039–1045.

Lehman, J.T., D.B. Botkin and G.E. Likens. 1975. The assumptions and rationales of a computer model of phytoplankton population dynamics. Limnology and Oceanography 20:343–364.

Lemke, A. E. 1981. Interlaboratory comparison acute testing set, EPA 600/3-81-005. U.S. Environmental Protection Agency, Duluth, Minnesota.

Lemke, A. E. 1983. Interlaboratory comparison of continuous flow, early life stage testing with fathead minnows, EPA 600/3-84-005. U.S. Environmental Protection Agency, Duluth, Minnesota.

Lemke, A. E., E. Durhan, and T. Felhaber. 1983. Evaluation of a fathead minnow Pimephales promelas embryo-larval test guideline using acenaphthene and isophorone. EPA 600/3-83-062. U.S. Environmental Protection Agency, Duluth, Minnesota.

Lemly, A. D. 1985. Toxicology of selenium in a freshwater reservoir: implications for environmental hazard evaluation and safety. Ecotoxicology and Environmental Safety 10:314–338.

Lemly, D. A. and R. J. F. Smith. Effects of exposure to acidified water on the behavioral response of fathead minnows, Pimephales promelas, to chemical feeding stimuli. Aquatic Toxicology 6:25–36.

Leo, A., C. Hansch, and D. Elkins. 1971. Partition coefficients and their uses. Chem. Rev. 71:525–615.

Leopold, L. B., F. E. Clarke, B. B. Hanshaw, and J. R. Balsley. 1971. A procedure for evaluating environmental impact. Circular 645. U.S. Geological Survey, Washington, D.C.

Lepp, N. W. 1981. Effect of Heavy Metal Pollution on Plants. Applied Science Publishers, London.

Leslie, P. H. 1945. On the use of matrices in certain population mathematics. Biometrika 33:183–212.

Levin, S. A. 1976. Population dynamic models in heterogeneous environments. Ann. Rev. Ecol. Syst. 7:287–310

Levin, S. A., and C. P. Goodyear. 1980. Analysis of an age-structured fishery model. J. Math. Biol 9:245–274.

Levin, S. A., and K. D. Kimball. 1984. New perspectives in ecotoxicology. Environ. Manage. 8:375–442.

Levin, S. A. and M. A. Harwell. 1986. Potential ecological consequences of genetically engineered organisms. pp. 495–514. In J.W. Gillett (ed.), Potential Impacts of Environmental Release of Biotechnology Products: Assessment, Regulation, and Research Needs. Environmental Management 10(4).

Levin, S. A. 1987. Scale and predictability in ecological modeling. pp. 2–8. In T. L. Vincent, Y. Cohen, W. J. Grantham, G. P. Kirkwood. and J. M. Skowronski (eds.), Modeling and Management of Resources Under Uncertainty. Lecture Notes in Biomathematics 72. Springer-Verlag, Berlin.

Levins, R. 1966. The strategy of model building in population biology. American Scientist 54:421–431.

Levins, R. 1968. Evolution in Changing Environments. Princeton University Press, Princeton, New Jersey.

Lewis, M. A. 1986. Comparison of effects of surfactants on freshwater phytoplankton communities in experimental enclosures and on algal population growth in the laboratory. Environ. Toxicol. Chem. 5: 319–332.

Lewis, M. A. 1990. Are laboratory-derived toxicity data for freshwater algae worth the effort? Environ. Toxicol. Chem. 9:1279–1284.

Lewis, M. A. and B. G. Hamm. 1986. Environmental modification of the photosynthetic response of lake plankton to surfactants and significance to laboratory-field comparison. Wat. Res. 20: 1575–1582.

Lewis, M. A., M. J. Taylor, and R. J. Larson. 1986. Structural and functional response of natural phytoplankton and periphyton communities to a cationic surfactant with considerations on environmental fate. pp. 241–268. In J. Cairns, Jr. (ed.), Community Toxicity Testing, ASTM STP 920. American Society for Testing and Materials, Philadelphia, Pennsylvania.

Lewis, R. A. 1987. Guidelines for environmental specimen banking with special reference to the Federal Republic of Germany: Ecological and managerial aspects. U.S. Department of Interior, Washington, D.C.

Li, W. K. W. 1984. A modified logistic growth equation: Effects of cadmium chloride on the diatom, *Thalassiosira weissflogii*, and the dinoflagelate, *Amphidinium carteri*, in unialgal and bialgal batch cultures. Aquatic Toxicology 5:307–313.

Liang, L. N., J. L. Sinclair, L. N. Mallory, and M. Alexander. 1982. Fate in model exosystem of microbial species of potential use in genetic engineering. Appl. Environ. Microbiol. 44:708–714.

Lichtenstein, E. P., K. R. Schulz, and T. T. Liang. 1977. Fate of fresh and aged soil residues of the insecticide (^{14}C)-N-2596 in a soil-corn-water ecosystem. J. Econ. Entomol 70:169–175.

Liefer, A. 1988. The Kinetics of Environmental Aquatic Photochemistry. American Chemical Society, Washington, D.C.

Lighthart, B. and A. S. Frisch. 1976. Estimation of viable airborne microbes downwind from a point source. Appl. Environ. Microbiol. 31:700–704.

Lightner, D. V., R. B. Thurman, and B. B. Trumper. 1989. An enclosed multispecies test system for testing microbial pest control agents with non-target species. EPA/600/S4–89/027. U.S. Environmental Protection Agency, Gulf Breeze, Florida.

Likens, G.E., F.H. Bormann, R.S. Pierce, J.S. Eaton, and N.M. Johnson. 1977. Biogeochemistry of a forested ecosystem. Springer-Verlag, New York. 146 p.

Limburg, K. E. 1986. PCBs in the Hudson. pp. 83–130 In Limburg, K.E., M. A. Moran, and W. H. McDowell (eds.) The Hudson River Ecosystem. Springer-Verlag, New York.

Lincer, J. L. 1975. DDE-induced eggshell-thinning in the American kestrel: A comparison of the field situation and laboratory results. J. Appl. Ecol. 12:781–793.

Lincoln, D. R., E. S. Friber, D. Lambert, M. A. Chatigney, and M. A. Levin. 1983. Release and containment of microorganisms from applied genetics activities. Carnegie-Mellon University, Pittsburgh, Pennsylvania.

Linder, E. 1987. Statistical Inference in the Linear Errors-in-Variables Model Using the Bootstrap, With Applications in Environmental Risk Analysis. The Pennsylvania State University, University Park.

Lindstrom, F. T., L. Boersma, C. McFarlane, K. P. Suen, and D. Cawlfield. 1988. Uptake and transport of chemicals by plants (version 2.1). Special Report 819, Agricultural Experiment Station, Oregon State University, Corvallis.

Linthurst, R. A., D. H. Landers, J. M. Eilers, D. F. Brakke, W. S. Overton, E. P. Meier, and R. E. Crowe. 1986. Characteristics of lakes in the eastern United States. Vol. I: Population relationships and physicochemical relationships. EPA/600/4-86/007a. U. S. Environmental Protection Agency, Washington, D.C.

Lipnick, R. L. 1985. Validation and extension of fish toxicity QSARs and interspecies comparisons for certain classes of organic chemicals. pp. 39–52. In M. Tichy (ed.), QSAR in Toxicology and Xenochemistry. Elsevier, Amsterdam.

Lipton, J., and J. W. Gillett. 1991. Uncertainty in ocean dumping health risks: influence of bioconcentration, commercial fish landings and seafood consumption. Environ. Toxicol. Chem. 10:967–976.

Liss, P. S., and P. G. Slater. 1974. Flux of gasses across the air-sea interface. Nature 247:181–184.

Lloyd, R. 1979. Toxicity tests with aquatic organisms. pp. 165–178. In Lectures Presented at the Sixth FAO/SIDA Workshop on Aquatic Pollution in Relation to Protecting Living Resources. U.N. Food and Agricultural Organization, Rome.

Lloyd, R. 1986. The toxicity of mixtures of chemicals to fish: An overview of European laboratory and field experience. pp. 42–53. In H. L. Bergman, R. A. Kimerle, and A. W. Maki, (eds.), Environmental Hazard Assessment of Effluents. Pergamon Press, New York.

Logan, D.T. 1986. Use of size-dependent mortality models to estimate reductions in fish populations resulting from toxicant exposure. Environ. Toxicol. Chem. 5:769–775.

Long, E. R. and M. F. Buchman. 1990. A comparative evaluation of selected measures of biological effects of exposure of marine organisms to toxic chemicals. pp. 355–394. In J. F. McCarthy and L. R. Shugart (eds.), Biomarkers of Environmental Contamination. Lewis Publishers, Chelsea, Michigan.

Lotka 1924. Elements of Physical Biology. Williams and Wilkins, Baltimore (Reprinted in 1956 by Dover Publications, New York, as Elements of Mathematical Biology.)

Lowe, T. P., T. W. May, W. G. Brumbaugh, and D. A. Kane. 1985. National Contaminant Biomonitoring Program: Concentrations of seven elements in freshwater fish. Arch. Environ. Contam. Toxicol. 14:363–368.

Lowrance, R., P. F. Hendrix, and E. P. Odum. 1986. A hierarchical approach to sustainable agriculture. Amer. J. Alternative Agricul. 1:169–173.

Lowry, T. H. and K. S. Richardson. 1981. Mechanism and Theory in Organic Chemistry. Harper and Row, New York.

Lucas, P. W., D. A. Cottam, and T. A. Mansfield. 1987. A large-scale fumigation system for investigating interactions between air pollution and cold stress on plants. Environ. Pollut. 43:15–28.

Lucky, T. D. and B. Venugopal. 1977. Metal Toxicity in Mammals — 1: Physiological and Chemical Basis for Metal Toxicity. Plenum Press, New York.

Ludwig, J. A. and J. F. Reynolds. 1988. Statistical Ecology: A Primer on Methods and Computing. John Wiley & Sons, New York.

Luoma, S. N. 1989. Can we determine the biological availability of sediment-bound trace elements? Hydrobiologia 176/177:379–396.

Lyman, W. J., W. F. Reehl, and D. H. Rosenblatt (eds.). 1982. Handbook of Chemical Property Estimation Methods. McGraw-Hill Book Company, New York.

Mabey, W., and T. Mill. 1978. Critical review of hydrolysis of organic compounds in water under environmental conditions. J. Phys. Chem. Ref. Data 7:383–415.

Macalady, D. and N. L. Wolfe. 1984. Abiotic hydrolysis of sorbed pesticides. pp. 221–244. In R. F. Knueger and J. N. Seiber, (eds.). Treatment of Disposal of Pesticide Wastes. ACS Symposium Series 259. Washington, D.C.

Macek, K. J. and W. A. McAllister. 1970. Insecticide susceptibility of some common fish family representatives. Trans. Amer. Fish. Soc. 1970:20–27.

MacIntosh, D. L., G. W. Suter II, and F. O. Hoffman. 1992. Model of PCB and mercury exposure to mink and great blue heron inhabiting the off-site environment downstream from the U. S. Department of Energy Oak Ridge Reservation. ORNL/ER-90. Oak Ridge National Laboratory, Oak Ridge, Tennessee.

Mackay, D. 1979. Finding fugacity feasible. Environ. Sci. Technol. 13:1218–1223.

Mackay, D. 1980. Solubility, partition coefficients, volatility and evaporation rates. pp. 31–45. In The Handbook of Environmental Chemistry. Vol. 2, part A. O. Hutzinger, Ed. Springer-Verlag, Berlin.

Mackay, D. 1991. Multimedia Environmental Models: The Fugacity Approach. Lewis Publisher, Chelsea, MI.

Mackay, D. and W. Y. Shiu. 1981. A critical review of Henry's law constants for chemicals of environmental interest. J. Phys. Chem. Ref. Data 10: 1175–1199.

Mackay, D. 1987. The holistic assessment of toxic organic chemicals in Canadian waters. Canadian Water Res. J. 12:14–20.

Mackay, D. 1989. Modeling the long term behavior of an organic contaminant in a large lake: Application to PCBs in Lake Ontario. J. Great Lakes Res. 15:283–297.

Mackay, D. and A. I. Hughes. 1984. Three-parameter equation describing the uptake of organic compounds by fish. Environ. Sci. Technol. 18:439–444.

Mackay, D., S. Joy, and S. Paterson. 1983. A quantitative water air sediment interaction (QWASI) model for describing the fate of chemicals in lakes. Chemosphere. 12:981–997.

Mackay, D., and S. Paterson. 1982. Fugacity revisited, Environ. Sci. Technol. 16:654A-660A.

Mackay, D., and S. Paterson. 1988. Partitioning models. In C. C. Travis (ed.), Carcinogen Risk Assessment. Plenum Publishing Corporation, New York and London.

Mackay, D., S. Paterson, B. Cheung, and W. B. Neely. 1985. Evaluating the Environmental Behavior of Chemicals with a Level III Fugacity Model. Chemosphere 14:335–374.

MacKenzie, J. J. and M. T. El-Ashry (eds.). 1989. Air Pollution's Toll on Forests and Crops. Yale University Press, New Haven, Connecticut.

Maki, A. W. 1979a. An analysis of decision criteria in environmental hazard evaluation programs. pp. 83–100. In K. L. Dickson, A. W. Maki (eds.), Analyzing the Hazard Evaluation Process. American Fisheries Society, Washington, D.C.

Maki, A. W. 1979b. Correlation between *Daphnia magna* and fathead minnow (*Pimephales promelas*) chronic toxicity values for several classes of test substances. J. Fish. Res. Board Can. 36: 411–421.

Maki, A. W. and W. E. Bishop. 1985. Chemical safety evaluation. pp. 619–635. In G. A. Rand and S. A. Petrocelli (eds.), Fundamentals of Aquatic Toxicology. Hemisphere Publishing Co., Washington, D.C.

Malanchuk, J. L. and H. P. Kollig. 1985. Integrated use of physical and mathematical models to evaluate ecological effects. Water Air Soil Pollut. 24:267–281.

Malanchuk, J. L. and R. S. Turner. 1987. Effects on aquatic systems. pp. 8-1 – 8-81. In NAPAP Interim Assessment, Volume IV: Effects of Acidic Deposition, National Acid Precipitation Assessment Program. Washington, D.C.

Malhotra, S. S. and R. A. Blauel. 1980. Diagnosis of air pollutant and natural stress symptoms on forest vegetation in western Canada. Northern Forest Research Center, Edmonton, Canada.

Maloney, T. E. and C. M. Palmer. 1956. Toxicity of six chemical compounds to thirty cultures of algae. Water & Sewage Works 103: 509–513.

Mancini, J. L. 1983. A method for calculating effects on aquatic organisms of time varying concentrations. Water Res. 17:1355–1362.

Mandel, J. 1984. Fitting straight lines when both variables are subject to error. J. Qual. Technol. 16:1–14.

Mankin, J. B., R. V. O'Neill, H. H. Shugart, and B. W. Rust. 1975. The importance of validation in ecosystem analysis. pp. 63–72. In G. S. Innis (ed.), New Directions in the Analysis of Ecological Systems. Simulation Councils Proc. Ser. 1(1). Simulation Councils, Inc., La Jolla, California.

Manning, W. J. and W. A. Feder. 1980. Biomonitoring Air Pollutants with Plants. Applied Science Publishers, Ltd., London.

Marcot, B.G., and R. Holthausen. 1987. Analyzing population viability of the spotted owl in the Pacific Northwest. Trans. N. Am. Wildl. Nat. Res. Conf. 52:333–347.

Maren, T. H., R. Embry, and L. E. Broder. 1968. The excretion of drugs across the gill of the dogfish, *Squalus acanthias*. Comp. Biochem. Physiol. 26:853–864.

Margalef, R. 1968. Perspectives in ecological theory. University of Chicago Press, Chicago. 111 p.

Martin, L. R. 1984. Kinetics of sulfite oxidation in aqueous solution. pp. 63–100. In J. G. Calvert, Ed. SO_2, NO and NO_2 Oxidation Mechanisms. Butterworth, Boston, Massachusetts.

Martin, M. H. and P. J. Coughtrey. 1982. Biological Monitoring of Heavy Metal Pollution. Applied Science Publishers, London.

Martin, Y. C. 1978. Quantitative Drug Design. Marcel Dekker, New York.

Martz, H. F. 1984. On broadening failure rate distributions in PRA uncertainty analyses. Risk Analysis 4:15-24.

Martz, H. F. 1985. Response to "The broadening of failure rate distributions in risk analysis: how good are the experts?" Risk Analysis 5:93-96

Mason, B. J. 1983. Preparation of soil sampling protocol: Techniques and strategies. EPA-600/4-83-020, Environmental Monitoring Systems Laboratory, Las Vegas, Nevada.

Mastroiacovo, P., A. Spagnolo, E. Marni, L. Meazza, R. Bertollini, and G. Segni. 1988. Birth defects in the Seveso area after TCDD contamination. JAMA 259:1668-1672.

Mayer, F. L., Jr., and M. R. Ellersieck. 1986. Manual of acute toxicity: Interpretation and data base for 410 chemicals and 66 species of freshwater animals. Resource Pub. 160. U.S. Fish and Wildlife Service, Washington, D.C.

Mayer, F. L., K. S. Mayer, and M. R. Ellersieck. 1986. Relationship of survival to other endpoints in chronic toxicity tests with fish. Environ. Toxicol. Chem. 5:737-748.

Mayer, F. L., C. H. Deans, and A. G. Smith. 1987. Inter-taxa correlations for toxicity to aquatic organisms. EPA/600/X-87/332. Environmental Research Laboratory, Gulf Breeze, Florida.

McCann, J. A., W. Teeters, D. J. Urban, and N. Cook. 1981. A short-term dietary toxicity test on small mammals. pp. 132-142. In D. W. Lamb and E. E. Kenaga (eds.), Avian and Mammalian Wildlife Toxicology, Second Conference. American Society for Testing and Materials, Philadelphia, Pennsylvania.

McCarty, L. S. 1986. The relationship between aquatic toxicity QSARs and bioconcentration for some organic chemicals. Environ. Toxicol. Chem. 5: 1071-1079.

McCauley, E., W. W. Murdoch, R. M. Nisbet, and W.S. C. Gurney 1990. The physiological ecology of *Daphnia*: development of a model of growth and reproduction. Ecology 71:703-715.

McDowell-Boyer, L.M. and D.M. Hetrick. 1982. A multimedia screening-level model for assessing the potential fate of chemicals released to the environment. ORNL/TM-8334. Oak Ridge National Laboratory, Oak Ridge, TN.

McEwen, L. C., R. K. Tucker, J. O. Ells, and M. A. Haegele. 1971. Mercury-wildlife studies at the Denver Wildlife Research Center. pp. 146-156. In D. R. Buhler (ed.), Mercury in the Western Environment. Continuing Education Books, Corvallis, Oregon.

McFadden, J. T., 1977. An argument supporting the reality of compensation in fish populations and a plea to let them exercise it. pp. 153-183. In W. Van Winkle (ed.),

Assessing the Effects of Power-Plant-Induced Mortality on Fish Populations. Pergamon, New York.

McFarlane, C. T. Pfleeger, and J. S. Fletcher. 1987. Transpiration effects on the uptake and distribution of bromacil, nitrobenzene, and phenol in soybean plants. J. Environ. Qual. 15:372–376.

McFarlane, G. A., and W. G. Franzin. 1978. Elevated heavy metals: a stress on a population of white suckers, *Catastomus commersoni*, in Hammell Lake, Saskatchewan. J. Fish. Res. Board Can. 35:963–970.

McGrath, E. G., S. L. Basin, R. W. Burton, D. C. Irving, S. C. Jaquette, and W. R. Ketler. 1975. Techniques for efficient Monte Carlo simulation. Vol 1: Selected probability distributions. ARNL/RSIC-38. Oak Ridge National Laboratory, Oak Ridge, Tennessee.

McIntire, C.D. 1983. A conceptual framework for process studies in lotic ecosystems, pp. 43–68. In T.D. Fontaine III and S.M. Bartell (eds.), Dynamics of lotic ecosystems. Ann Arbor Science, Ann Arbor, Michigan.

McIntire, C.D. and J.A. Colby. 1978. A hierarchical model of lotic ecosystems. Ecological Monographs 48: 167–190.

McIntosh, R. P. 1987. Pluralism in ecology. Ann. Rev. Ecol. Syst. 18:321–341.

McIntosh, R. P. 1985. The Background of Ecology: Concept and Theory. Cambridge University Press, Cambridge.

McKay, M. D., Conover, W. J., and Beckman, R. J. 1979. A comparison of three methods for selecting values of input variables in the analysis of output from a computer code. Technometrics, 21:239–245.

McKim, J. M., G. F. Olson, G. H. Holcombe, and E. P. Hunt. 1976. Long-term effects of methylmercuric chloride on three generations of brook trout (*Salvelinus fontinalis*): toxicity, accumulation, distribution, and elimination. J. Fish. Res. Board Can. 33:2726–2739.

McKim, J. M., P. K. Schmieder, and R. J. Erickson. 1986. Toxicokinetic modeling of [^{14}C]pentachlorophenol in the rainbow trout (*Salmo gairdneri*). Aquat. Toxicol. 9:59–80.

McKim, J. M. 1985. Early life stage toxicity tests. pp. 58–95. In G. M. Rand and S. R. Petrocelli (eds.), Fundamentals of Aquatic Toxicology. Hemisphere Publishing Corp., Washington, D.C.

McKone, T.E. and W. E. Kastenberg. 1986. Application of multimedia pollutant transport models to risk analysis. pp. 167–190 In Y. Cohen (ed.), Pollutants in a Multimedia Environment, Plenum Press, New York and London.

McKone, T.E. and D. W. Layton. 1986. Screening the potential risks of toxic substances using a multimedia compartment model: Estimation of human exposure. Regul. Toxicol. Pharmacol. 6:359–380.

McLaughlin, S. B., T. J. Blasing, L. K. Mann, and D. N. Duvick. 1983. Effects of acid rain and gaseous pollutants on forest productivity: A regional scale approach. APCA Journal 33:1042–1049.

McLaughlin, S. B. and O. U. Braker. 1985. Methods for evaluating and predicting forest growth response to air pollution. Experientia 41:310–319.

McLaughlin, S. B., Jr., and G. E. Taylor, Jr. 1981. Relative humidity: Important modifier of pollutant uptake by plants. Science 211: 167–169.

McLaughlin, S. B., Jr., and G. E. Taylor, Jr. 1985. SO_2 effects on dicot crops: Some issues, mechanisms, and indicators. pp. 227–249. In W. E. Winner, H. A. Mooney, and R. A. Goldstein, eds. Sulfur Dioxide and Vegetation. Stanford University Press, Stanford, California.

McLaughlin, S. B, D. J. Downing, T. J. Blasing, E. R. Cook, and H. S. Adams. 1987. An analysis of climate and competition as contributors to decline of red spruce in high elevation Appalachian forests of the Eastern United States. Oecologia 72:487–501.

McPhee, J. A., A. Panaye, and J. E. Dubois. 1978. Steric effects—I. A critical examination of the Taft steric parameter—E_s. Tetrahedron 34:3353–3662.

Medeiros, W. H., P. D. Moscowitz, E. A. Coveny, H. C. Thode Jr., and N. L. Oden. 1984. Oxidant and acid precipitation effects on soy bean yield: Cross-sectional model development. Environment International 10:27–33.

Mehrle, P. M. and F. L. Mayer. 1985. Biochemistry/physiology. pp. 264–282. In G. M. Rand and S. R. Petrocelli (eds.), Fundamentals of Aquatic Toxicology, Hemisphere, Washington, D.C.

Melancorn, S.M., J. E. Pollard, and S. C. Hern. 1986. Evaluation of SESOIL, PRZM and PESTAN in a laboratory leaching experiment. Environ. Toxicol. Chem. 5:865–878.

Menzel, D. B. 1987. Physiological pharmacokinetic modeling. Environ. Sci. Technol. 21:944–950.

Menzel, D. B. and E. D. Smolco. 1984. Interspecies extrapolation. pp. 23–35. In J. R. Rodricks and R. G. Tardiff (eds.), Assessment and Management of Chemical Risks. American Chemical Society, Washington, D.C.

Mercer, E. H. 1981. The Foundations of Biological Theory. J. Wiley & Sons, New York.

Mertz, D. B. 1971, The mathematical demography of the California Condor. Amer. Natur. 105:437–453

Metcalf, R. L. 1971. A model ecosystem for the evaluation of pesticide biodegradability and ecological magnification. Outlook on Agriculture 7:55–69.

Meyer, F. P. and L. A. Barklay. 1990. Field manual for the investigation of fish kills. Resource Pub. 177, U.S. Fish and Wildlife Service, Washington, D.C.

Meyer, J. S., D. A. Sanchez, John A. Brookman, D. B. McWhorter, and H. A. Bergman. 1985. Chemistry and aquatic toxicity of raw oil shale leachates from Piceance Basin, Colorado. Environ. Toxicol. Chem. 4:559–572.

Meyer, J. S., C. G. Ingersoll, L. L. McDonald, and M. S. Boyce. 1986. Estimating uncertainty in population growth rates: Jacknife vs. Bootstrap techniques. Ecology 67:1156–1166

Meyers, S. M. and J. D. Gile. 1986. Mallard reproductive testing in a pond environment: a preliminary study. Arch. Environ. Contam. Toxicol. 15:757–761.

Meyers, T. R. and J. D. Hendricks. 1986. Histopathology. pp. 283–331. In G. M. Rand and S. R. Petrocelli (eds.), Fundamentals of Aquatic Toxicology. Hemisphere Publishing Corp., Washington, D. C.

Meyrahn, J. J. Pauly, W. Schneider, and P. Warneck. 1986. Quantum yields for the photodissociation of acetone in air and an estimate of the lifetime of acetone in the lower troposphere. J. Atmos. Chem. 4:277–291.

Mill, T. 1988. Structure-activity relationships for photooxidation processes in the environment. Environ. Toxicol. Chem. 8:31–43.

Mill, T. 1989. Structure-activity relationships for photooxidation processes in the environment. Environ. Toxicol. Chem. 8:31–43.

Mill, T., D. G. Hendry, and H. Richardson. 1980. Free-radical oxidants in natural waters. Science 207:886–889.

Mill, T., and W. Mabey. 1988. Hydrolysis of organic chemicals. In Handbook of Environmental Chemistry, Vol. 2/D, O. Hutzinger, Ed., Springer-Verlag, Heidelberg.

Mill, T., and W. Mabey. 1985. Photodegradation in water. pp. 175–216. In Environmental Exposure from Chemicals, Vol. I. W. B. Neely and G. E. Blaue, (eds.). CRC Press, Boca Raton, Florida.

Mill, T., J. S. Winterle, A. Fischer, D. S. Tse, W. R. Mabey, H. Drossman, A. Liu and J. E. Davenport. 1984. Toxic substances process data generation and protocol development. EPA Contract 68-03-2981. EPA Report, Office of Research and Development.

Miller, J. E., R. N. Miller, D. G. Sprugel, H. J. Smith, and P. B. Xerikos. 1977. An open air fumigation system for the study of sulfur dioxide effects on crop plants.

pp. 2–6. In Radiological and Environmental Division Annual Report, Part III, January-December, 1977. ANL-77-65. Argonne National Laboratory, Argonne, Illinois.

Miller, P. R. (ed.). 1977. Photochemical oxidant air pollutant effects on a mixed conifer forest ecosystem: A progress report. EPA-600/3-77-058. Environmental Research Laboratory, Corvallis, Oregon.

Miller, W. E., S. A. Peterson, J. C. Greene, and C. A. Callahan. 1985. Comparative toxicity of laboratory organisms for assessing hazardous waste sites. J. Environ. Qual. 14:569–574.

Minns, C. K., J. E. Moore, D. W. Schindler, and M. L. Jones. 1989. Assessing the potential extent of damage to inland lakes of eastern Canada due to acid deposition. III. Predicted impacts on species richness in seven groups of aquatic biota. Can. J. Fish. Aquat. Sci. 47:821–830.

Mitsch, W.J., R.W. Bosserman, and J.M. Klopatek (eds.). 1981. Energy and ecological modelling. Elsevier, Amsterdam. 839 p.

Mittelman, A., J. Settel, K. Plourd, R. S. Fulton III, G. Sun, S. Chaube, and P. Sheehan. 1987. Ecological endpoint selection criteria. Technical Resources, Inc. Rockville, Maryland.

Moghissi, A. A., 1984. Risk management—practice and prospects. Mechanical Engineering 106(11):21–23.

Monroe, B. M. 1985. Singlet oxygen in solution. pp. 177–224. In A. A. Frimer, (ed.)., Singlet Oxygen, CRC Press, Boca Raton, Florida.

Montgomery, R. H. and T. G. Sanders. 1986. Uncertainty in water quality data. pp. 17–29, In A. H. El-Shaarawi and R. E. Kwiatkowski, (eds.), Statistical Aspects of Water Quality Monitoring. Elsevier, Amsterdam.

Mooney, H. A. and G. Bernardi (eds.). 1990. Introduction of Genetically Modified Organisms into the Environment. John Wiley & Sons, New York.

Mooney, H. A. and J. A. Drake (eds.). 1986. Ecology of Biological Invasions of North America and Hawaii. Springer-Verlag, New York.

Moore, R. E., C. F. Baes III, L. M. McDowell-Boyer, A. P. Watson, F. O. Hoffman, J. C. Pleasant, and C. W. Miller. 1979. AIRDOS-EPA: A Computerized Methodology for Estimating Environmental Concentrations and Dose to Man from Airborne Releases of Radionuclides. EPA-520/1-79-009. U.S. Environmental Protection Agency Office of Radiation Programs, Washington, D.C.

Mopper, K. and X. Zhou. 1990. Hydroxyl radical photoproduction in the sea and its potential impact on marine processes. Science 250:661–664.

Morgan, M. G., M. Henrion, and S. C. Morris. 1979. Expert judgements for policy analysis. BNL 51358. Brookhaven National Laboratory, Upton, New York.

Morgan, M. G., M. Henrion, S. C. Morris, and D. A. L. Amaral. 1985. Uncertainty in risk assessment. Environ. Sci. Technol. 19:662–667.

Morgenstern, H. 1982. Uses of ecological analysis in epidemiologic research. Am. J. Public Health 72:1336–1344.

Moriarty, F. 1988. Ecotoxicology: The Study of Pollutants in Ecosystems (Second Edition). Academic Press, New York.

Moulton, M. P. and T. W. Schultz. 1986. Comparison of several structure-toxicity relationships for chlorophenols. Aquat. Toxicol. 8:121–128.

Mount, D. I. 1977. An assessment of application factors in aquatic toxicology. pp. 183–190. In R. A. Tubb, (ed.). Recent Advances in Fish Toxicology. EPA-600/3-77-085. U. S. Environmental Protection Agency, Corvallis, Oregon.

Mount, D. I. 1982. Aquatic surrogates. pp. A6-2-A6-4. In Surrogate Species Workshop Report. TR-507-36B. U.S. Environmental Protection Agency, Washington, D.C.

Mount, D. I. 1985. Scientific problems in using multispecies toxicity tests for regulatory purposes. pp. 13–18. In J. Cairns, Jr., (ed.). Multispecies Toxicity Testing. Pergamon Press, New York.

Mount, D. I. 1989. Methods for aquatic toxicity identification evaluations: Phase III toxicity confirmation procedures. EPA-600/3-88-036. U.S. Environmental Protection Agency, Duluth, Minnesota.

Mount, D. I., L. M. Vigor, and M. L. Schafer. 1966. Endrin: Use of concentration in the blood to diagnose acute toxicity to fish. Science 152:1388–1390.

Mount, D. I., and C. E. Stephan. 1967. A method for establishing acceptable toxicant limits for fish—malathion and butoxyethanol ester of 2,4-D. Trans. Am. Fish. Soc. 96:185–193.

Mount, D. I., N. A. Thomas, T. J. Norberg, M. T. Barbour, T. H. Roush, and W. F. Brandes. 1984. Effluent and ambient toxicity testing and instream community response on the Ottawa River, Lima Ohio. EPA-600/3-84-080. U.S. Environmental Protection Agency, Duluth, Minnesota.

Mount, D. I. and L. Anderson-Carnahan. 1988. Methods for aquatic toxicity identification evaluations: Phase I toxicity characterization procedures. EPA-600/3-88-034. U.S. Environmental Protection Agency, Duluth, Minnesota.

Mount, D. I. and L. Anderson-Carnahan. 1989. Methods for aquatic toxicity identification evaluations: Phase II toxicity identification procedures. EPA-600/3-88-035. U.S. Environmental Protection Agency, Duluth, Minnesota.

Mount, M. E. and F. W. Oehme. 1981. Insecticide levels in tissues associated with toxicity: A literature review. Vet. Hum. Toxicol. 23:34–42.

Mullin, C. A., B. A. Croft, K. Strickler, F. Matsumura, J. R. Miller. 1982. Detoxification enzyme differences between a herbivorous and predatory mite. Science 217:1270–1272.

Munawar, M., I. F. Munawar, and G. G. Leppard. 1989. Early warning assays: an overview of testing with phytoplankton in the North American Great Lakes. Hydrobiologia 188/189:237–246.

Munkittrick, K. R. and D. G. Dixon. 1989a. A holistic approach to ecosystem health assessment using fish population characteristics. Hydrobiologia 188/189:123–135.

Munkittrick, K. R. and D. G. Dixon. 1989b. Use of white sucker (*Catastomus commersoni*) populations to assess the health of aquatic ecosystems exposed to low-level contaminant stress. Can. J. Fish. Aquat. Sci. 46:1455–1462.

Munkittrick, K. R., P. A. Miller, D. R. Barton, and D. G. Dixon. 1991. Altered performance of white sucker populations in the Manitouwadge chain of lakes is associated with changes in benthic macroinvertebrates communities as a result of copper and zinc contamination. Ecotoxicol. Environ. Safety 21:318–326.

Munn, R. E. 1975. Environmental Impact Assessment, SCOPE 5. Scientific Committee on Problems in the Environment, Paris.

Munro, J.K., Jr., R.J. Luxmoore, C.L. Begovich, K.R. Dixon, A.P. Watson, M.R. Patterson, and D.R. Jackson. 1977. Transport model to predict the movement of Pb, Cd, Zn, Cu, and S through a forested watershed, pp. 45–53. In Disposal of Residues on Land. Proceedings of the National Conference on Disposal of Residues on Land, Information Transfer, Inc., Rockville, Maryland.

Murray, T. M. 1987. Monitoring the nation's waters — A new perspective. pp. 1–11. In T. P. Boyle (ed.), New Approaches to Monitoring Aquatic Ecosystems, STP 940. American Society for Testing and Materials, Philadelphia, Pennsylvania.

Nagy, K. A. 1987. Field metabolic rate and food requirement scaling in mammals and birds. Ecological Mono. 57:111–128.

Nash, R. G., M. L. Beall, Jr., and W. G. Harris. 1977. Toxaphene and 1,1,1-trichloro-2,2-bis(p-chlorophenyl)ethane (DDT) losses from cotton in an agroecosystem chamber. J. Agric. Food Chem. 25:336–341.

National Academy of Sciences. 1972. Degradation of Synthetic Organic Molecules in the Biosphere. pp. 56.

Neeley, W. B. 1984. An analysis of aquatic toxicity data: water solubility and acute LC50 fish data. Chemosphere 13:813–819.

Neff, J. M. 1985. Use of biochemical measurements to detect pollutant-mediated damage to fish. pp. 155–183. In R. D. Cardwell, R. Purdy, and R. C. Bahner (eds.), Aquatic Toxicology and Hazard Assessment, Seventh Symposium. ASTM, Philadelphia, Pennsylvania.

Nendza, M., J. Volmer, and W. Klein. 1990. Risk assessment based on QSAR estimates. pp. 213–240. In W. Karcher and J. Devillers (eds.), Practical Applications of Quantitative Structure-Activity Relationships. ECSC, EEC, EAEC. Brussels.

Neuhold, J. M. 1987. The relationship of life history attributes to toxicant tolerance in fishes. Environ. Toxicol. Chem. 6:709–716.

Newbry, B. W. and G. F. Lee. 1984. A simple apparatus for conducting in-stream toxicity tests. J. Testing Evaluat. 12:51–53.

Newcombe, C. P. and D. D. MacDonald. 1991. Effects of suspended sediments on aquatic ecosystems. N. Am. J. Fish. Manage. 11:72–82.

Newman, J. R. 1979. The effect of air pollution on wildlife and their use as biological indicators. pp. 223–232. In Animals as Monitors of Environmental Pollutants. National Academy Press, Washington, D.C.

Newman, J. R. 1980. Effects of air emissions on wildlife resources, FWS/OBS-80/40.1. U. S. Fish and Wildlife Service, Washington, D.C.

Nichols, J. D., G. L. Hensler, and P. W. Sykes, Jr. 1980. Demography of the everglade kite: implications for population management. Ecol. Model. 9:215–232.

Niemi, G. J., P. DeVore, N. Detenbeck, D. Taylor, A. Lima, J. Pastor, J. D. Yount, and R. J. Naiman. 1990. Overview of case studies on recovery of aquatic systems from disturbance. Environ. Management 14:571–587.

Nimmo, D. R., L. H. Bahner, R. A. Rigby, J. M. Sheppard, and A. J. Wilson, Jr. 1977. *Mysidopsis bahia:* An estuarine species suitable for life-cycle toxicity tests to determine the effect of a pollutant. pp. 109–116. In F. L. Mayer and J. L. Hamelink (eds.), Aquatic Toxicology and Hazard Evaluation. ASTM STP 634. American Society for Testing and Materials, Philadelphia, Pennsylvania.

Nixon, S.W. and J.N. Kremer. 1977. Narragansett Bay — the development of a composite simulation model for a New England estuary, pp. 621–673. In C.A.S. Hall and J.W. Day, Jr. (eds.), Ecosystem modeling in theory and practice. Wiley-Interscience, New York.

Noble, R. D., J. L. Martin, and K. F. Jensen (eds.). 1989. Air Pollution Effects on Vegetation, Including Forest Ecosystems. Northeastern Forest Experiment Station. Broomall, Pennsylvania.

Norberg, T. J. and D. I. Mount. 1985. A new fathead minnow (*Pimephales promelas*) subchronic toxicity test. Environ. Toxicol. Chem. 4:711–718.

Norberg-King, T. and D. I. Mount. 1986. Validity of effluent and ambient toxicity tests for predicting biological impact, Skeleton Creek, Enid, Oklahoma. EPA/600/30-85/044. Duluth Environmental Research Laboratory, Minnesota.

Norby, R. J. and T. T. Kozlowski. 1981a. Interactions of SO_2-concentration and post-fumigation temperature on growth of five species of woody plants. Environ. Pollut. Ser. A 25:27–39.

Norby, R. J. and T. T. Kozlowski. 1981b. Relative sensitivity of three species of woody plants to SO_2 at high or low exposure temperature. Oecologia 51:33–36.

Norby, R. J. 1989. Foliar nitrate reductase: A marker for assimilation of atmospheric nitrogen oxides. pp. 245–250. In National Research Council. Biological Markers of Air-Pollution Stress and Damage in Forests. National Academy Press, Washington, D.C.

Norris, R. A. and D. W. Johnston. 1958. Weights and weight variation in summer birds from Georgia and South Carolina. Wilson Bull. 70:114–129.

NRC (National Research Council). 1971. Fluorides. National Academy Press, Washington, D.C.

NRC (National Research Council). 1972a. Degradation of Synthetic Organic Molecules in the Biosphere. National Academy Press, Washington, D.C.

NRC (Research Council). 1972b. Water quality criteria 1972. EPA-R3-73-033. U.S. Environmental Protection Agency, Washington, D.C.

NRC (National Research Council). 1975. Decision making for regulating chemicals in the environment. National Academy Press, Washington, D.C.

NRC (National Research Council). 1981a. Strategies for determining needs and priorities for toxicity testing. National Academy Press, Washington, D.C.

NRC (National Research Council). 1981b. Testing for Effects of Chemicals on Ecosystems. National Academy Press, Washington, D.C.

NRC (National Research Council). 1983. Risk assessment in the federal government: Managing the process. National Academy Press, Washington, D.C.

NRC (National Research Council). 1989a. Improving risk communication. National Academy Press, Washington, D.C.

NRC (National Research Council). 1989b. Biological Markers of Air-Pollution Stress and Damage in Forests. National Academy Press, Washington, D.C.

National Research Council Committee on the Applications of Ecological Theory to Environmental Problems. 1986. Ecological Knowledge and Environmental Problem Solving. National Academy Press, Washington, D.C.

O'Bryan, T. R. and R. H. Ross. 1988. Chemical scoring system for hazard and exposure identification. J. Toxicol. Environ. Health 1:119–134.

Odum, E. P. 1969. The strategy of ecosystem development. Science 164:262–270.

Odum, E.P. 1971. Fundamentals of ecology. W.B. Saunders Company, Philadelphia. 574 p.

Odum, E. P. 1985. Trends expected in stressed ecosystems. Bioscience 35:419–422.

OECD (Organization for Economic Cooperation and Development). 1981. OECD guidelines for testing of chemicals. OECD, Paris.

OECD (Organization for Economic Cooperation and Development). 1984. OECD guidelines for testing of chemicals, updated guidelines. OECD, Paris.

OECD (Organization for Economic Cooperation and Development). 1989. Compendium of Environmental Exposure Assessment Methods for Chemicals. Environment Monographs, No. 27, OECD, Paris.

Office of Technology Assessment. 1988. New developments in biotechnology — Field-testing engineered organisms: Genetic and ecological issues. OTA-BA-350. U.S. Government Printing Office, Washington, D.C.

O'Flaherty, E. J. 1986. Dose dependent toxicity. Comments Toxicology 1:23–34.

Ogden, J. 1985. The California condor. pp. 389–399. In R. L. Di Silvestro (ed.), Audubon Wildlife Report 1985. The National Audubon Society, New York.

Ohio EPA. 1988. User's manual for biological field assessment of Ohio surface waters. Division of Water Quality Monitoring and Assessment. Columbus, Ohio.

Ohlendorf, H. M., D. J. Hoffman, M. K. Saiki, and T. W. Aldrich. 1986a. Embryonic mortality and abnormalities of aquatic birds: Apparent impacts of selenium from irrigation drain waters. Sci. Total Environ. 52:49–53.

Ohlendorf, H. M., R. L. Hothem, C. M. Bunck, T. W. Aldrich, and J. F. Moore. 1986a. Relationships between selenium concentrations and avian reproduction. Trans. 51st N. A. Wildl. & Nat. Res. Conf. 330–342.

Oliver, B. G. and A. J. Niimi. 1983. Bioconcentration of chlorobenzenes from water by rainbow trout: correlations with partition coefficients and environmental residues. Environ. Sci. Technol. 17:287–291.

Oliver, G. R. and D. A. Laskowski. 1986. Development of environmental scenarios for modeling fate of agricultural chemicals in soil. Environ. Toxicol. Chem. 5:225–232.

Olsen, A.R. and S.E. Wise. 1979. Frequency analysis of pesticide concentrations for risk assessment. Prepared for the U.S. Environmental Protection Agency under Contract 23111 03242.

Olszyk, D. M., A. Bytnerowicz, and C. A. Fox. 1987. Sulfur dioxide effects on plants exhibiting crassulacean acid metabolism. Environ. Pollut. 43:47–62.

Omernik, J. M. 1987. Ecoregions of the conterminous Uniter States. Annals Assoc. Amer. Geogr. 77:118–125.

OMMSQA (Office of Modeling, Monitoring Systems and Quality Assurance). 1990. Environmental Monitoring and Assessment Program: Overview. EPA/600/9-90/001. U.S. Environmental Protection Agency, Washington, D.C.

Onishi, Y., S. M. Brown, A. R. Olsen, M. A. Parkhurst, S. E. Wise, and W. H. Walters. 1982. Methodology for overland and instream migration and risk assessment of pesticides, EPA-600/3-82-024. U.S. Environmental Protection Agency, Athens, Georgia.

O'Neill, R. V. 1973. Error analysis of ecological models. pp. 898-908. In D. J. Nelson (ed.), Radionuclides in Ecosystems. CONF-710501. National Technical Information Service, Springfield, Virginia.

O'Neill, R.V. 1976. Ecosystem persistence and heterotrophic regulation. Ecology 57:1244-1253.

O'Neill, R.V. and D.E. Reichle. 1980. Dimensions of ecosystem theory, pp. 11-26. In R.W. Waring (ed.), Forests: fresh perspectives from ecosystem analysis. Oregon State University Press, Corvallis.

O'Neill, R. V. and J. B. Waide. 1981. Ecosystem theory and the unexpected: Implications for environmental toxicology. pp. 43-73. In B. W. Cornaby (ed.), Management of Toxic Substances in Our Ecosystems. Ann Arbor Science, Ann Arbor, Michigan.

O'Neill, R. V., R. H. Gardner, L. W. Barnthouse, G. W. Suter, S. G. Hildebrand, and C. W. Gehrs. 1982. Ecosystem risk analysis: a new methodology. Environ. Toxicol. Chem. 1:167-177.

O'Neill, R. V., S. M. Bartell, and R. H. Gardner. 1983. Patterns of toxicological effects in ecosystems: a modeling study. Environ. Toxicol. Chem. 2:451-461.

O'Neill, R. V., D. L. DeAngelis, J. B. Waide, and T. F. H. Allen. 1986. A Hierarchical Concept of Ecosystems. Monograph in Population Biology 23. Princeton University Press, Princeton, New Jersey.

OPP (Office of Pesticide Programs). 1982. Pesticide assessment guidelines, Subdivision E, Hazard evaluation: Wildlife and Aquatic Organisms. EPA-540/9-82-024. U.S. Environmental Protection Agency, Washington, D.C.

OPP (Office of Pesticide Programs). 1989. Carbofuran: Special review technical support document. NTIS: PB 89168884. Environmental Protection Agency, Washington, D.C.

ORB (Oversight Review Board of the National Acid Precipitation Assessment Program). 1991. The experience and legacy of NAPAP. NAPAP Office of the Director, Washington, D.C.

Orians, G. H. 1990. Ecological concepts of sustainability. Environment 32:10-39.

Ormrod, D. P., B. A. Marie, and O. B. Allen. 1988. Research approaches to pollutant crop loss functions. pp. 27-44. In W. A. Heck, O. C. Taylor, and D. T. Tingey (eds.), Assessment of Crop Loss from Air Pollutants. Elsevier Applied Science, New York.

Orth, D. J. 1987. Ecological considerations in the development and application of instream flow-habitat models. Regulated Rivers Research and Management 1:171–181.

Osborn, D., W. J. Eney, and K. R. Bull. 1983. The toxicity of trialkyl lead compounds to birds. Environ. Pollut. 31A:261–275.

OSTP (Office of Science and Technology Policy). 1986. Coordinated framework for the regulation of biotechnology. Fed. Regist. 51:23302–23393.

Ott, W. R. 1978. Environmental Indices—Theory and Practice. Ann Arbor Science Publishers, Ann Arbor, Michigan.

Overton, W. S. and R. R. Lassiter. 1985. The concept of prognostic model assessment of toxic chemical fate. pp. 191–203. In T. P. Boyle, ed. Validation and Predictability of Laboratory Methods for Assessing the Fate and Effects of Contaminants in Aquatic Ecosystems. American Society for Testing and Materials, Philadelphia, Pennsylvania.

OWRS (Office of Water Regulations and Standards). 1985. Technical support document for water quality-based toxics control. U.S. Environmental Protection Agency, Washington, D.C.

OWRS (Office of Water Regulations and Standards). 1989. Briefing report to the EPA Science Advisory Board on the equilibrium partitioning approach to generating sediment quality criteria. EPA 440/5-89-002. Washington, D.C.

OWRS & ORD (Office of Water Regulations and Standards and Office of Research and Development). 1987. Guidelines for deriving ambient aquatic life advisory concentrations. U.S. Environmental Protection Agency, Washington, D.C.

Pahlsson, A.-M. B. 1989. Toxicity of metals (Zn, Cu, Cd, Pb) to vascular plants. Water Air Soil Pollut. 47:287–319.

Palawski, D., J. B. Hunn, and F. J. Dwyer. 1985. Sensitivity of young striped bass to organic and inorganic contaminants in fresh and saline waters. Trans. Amer. Fisheries Soc. 114:748–753.

Paris, E. F., W. C. Steen, G. L. Baughman, and J. T. Barnett Jr. 1981. Second-order model to predict microbial degradation of organic compounds in natural waters. Appl. Environ. Microbiol., 41, 603.

Paris, E. F., N. L. Wolfe, W. C. Steen and G. L. Baughman. 1983. Effect of phenol molecular structure on bacterial transformation rate constants. Appl. Environ. Microbiol. 45: 1153–1155.

Park, R.A. (and 24 co-authors). 1974. A generalized model for simulating lake ecosystems. Simulation 23: 33–50.

Park, R.A., C.I. Connolly, J.R. Albanese, L.S. Clesceri, G.W. Heitzman, H.H. Herbrandson, B.H. Indyke, J,R, Loehe, S. Ross, D.D. Sharma, and W.W. Shuster. 1980. Modeling transport and behavior of pesticides and other toxic organic materials in aquatic environments. Center for Ecological Modeling, Rensselaer Polytechnic Institute, Troy, New York.

Park, R.A., J.J. Anderson, G.L. Swartzman, R. Morison and J.M. Emlen. (In press). Assessment of risks of toxic pollutants to aquatic organisms and ecosystems using a sequential modeling approach. USA-USSR Symposium: Fate and Effects of Pollutants on Aquatic Organisms and Ecosystems. October 19–21, 1987.

Parkhurst, B. R. 1986. The role of fractionation in hazard assessments of complex materials. pp. 92–106. In H. L. Bergman, R. A. Kimerle, and A. W. Maki (eds.), Environmental Hazard Assessment of Effluents. Pergamon Press, NY.

Parkhurst, B. R., G. Linder, K. McBee, G. Bitton, B. Dutka, and C. Hendricks. 1989. Toxicity tests. pp. 6-1-6–66. In W. Warren-Hicks, B. R. Parkhurst, and S. S. Baker, Jr. (eds.), Ecological Assessment of Hazardous Waste Sites: A Field and Laboratory Reference Document. EPA 600/3-89/013. Corvallis Environmental Research Laboratory, Oregon.

Parkhurst, D. F. 1986. Internal leaf structure: A three-dimensional perspective. pp. 215–249. In T. J. Givnish, (ed.), On the Economy of Plant Form and Function. Cambridge University Press, New York.

Parkhurst, D. F. 1990. Statistical hypothesis tests and statistical power in pure and applied science. pp. 181–201. In G. M. von Furstenberg (ed.), Acting Under Uncertainty: Multidiciplinary Conceptions. Kluwer Academic Publishers, Boston.

Parkhurst, M. A., Y. Onishi, and A. R. Olsen. 1981. A risk assessment of toxicants to aquatic life using environmental exposure estimates and laboratory toxicity data. pp. 59–71. In D. R. Branson and K. L. Dickson (eds.), Aquatic Toxicology and Hazard Assessment. ASTM/STP 737. American Society for Testing and Materials, Philadelphia, Pennsylvania.

Parrish, P.R., E. E. Dyar, J. M. Enos, and W. G. Wilson. 1978. Chronic toxicity of chlordane, trifluralin, and pentachlorophenol to sheepshead minnows (*Cyprinodon variegatus*). Technical Report EPA-600/3-78-010. U.S. Environmental Protection Agency, Gulf Breeze, Florida.

Pascoe, D., S. A. Evans, and J. Woodworth. 1986. Heavy metal toxicity to fish and the influence of water hardness. Arch. Environ. Contam. Toxicol. 15:481–487.

Pascoe, D. and N. A. M. Shazili. 1986. Episodic pollution—a comparison of brief and continuous exposure of rainbow trout to cadmium. Ecotoxicol. Environ. Saf. 12:189–198.

Patin, S. A. 1982. Pollution and the Biological Resources of the Oceans. Butterworth Scientific, London.

Patnode, K. A. and D. H. White. 1991. Effects of pesticides on songbird productivity in conjunction with pecan cultivation in southern Georgia: a multiple-exposure experimental design. Environ. Toxicol. Chem. 10:1479–1486.

Patrick, R., J. Cairns, and A. Scheier. 1968. The relative sensitivity of diatoms, snails, and fish. Progressive Fish-Culturist 1968: 137–140.

Patten, B.C. 1968. Mathematical models of plankton production. Int. Revue ges. Hydrobiol. 53: 357–408.

Patten, B.C., D.A. Egloff, T.H. Richardson, (and 38 co-authors). 1975. Total ecosystem model for a cove in Lake Texoma, pp. 206–421. In B.C. Patten (ed.), Systems analysis and simulation in ecology. Volume II. Academic Press, Inc. New York.

Patterson, M.R., TG. J. Sworski, A. L. Sjoreen, M. G. Browman, C. C. Coutant, D. M. Hetrick, E. D. Murphy, and R. J. Paridon. 1982. A User Manual for UTM-TOX: A Unified Transport Model, draft report prepared for the Office of Toxic Substances, U.S EPA.

Paustenbach, D. J. (ed.). 1990. The Risk Assessment of Environmental and Human Health Hazards: A Textbook of Case Studies. John Wiley & Sons, New York.

Payne, J. F. L. L. Fancey, A. D. Rahimtula, and E. L. Porter. 1987. Review and perspective on the use of mixed-function oxygenase enzymes in biological monitoring. Comp. Biochem. Physiol. 86C:233–245.

Peakall, D. B. and R. K. Tucker. 1985. Extrapolation from single species studies to populations, communities and ecosystems. pp. 611–636. In V. B. Vouk, G. C. Butler, D. G. Hoel, and D. B. Peakall (eds.), Methods for Estimating Risk of Chemical Injury: Human and Non-human Biota and Ecosystems, SCOPE 26. John Wiley and Sons, New York.

Peltier, W. and C. I. Weber (eds.). 1985. Methods for measuring the acute toxicity of effluents to freshwater and marine organisms, Third edition. EPA-600/4-85-013. U.S. Environmental Protection Agency, Cincinnati, Ohio.

Perez, K. T., E. W. Davey, N. F. Lackie, G. M. Morrison, P. G. Murphy, A. E. Soper, D. L. Winslow. 1984. Environmental assessment of a phthalate ester, di(2-ethylhexyl) phthalate (DEHP), derived from a marine microcosm. pp. 180–191. In W. E. Bishop, R. D. Cardwell, and B. B. Heidolph, (eds.), Aquatic Toxicology and Hazard Assessment: Sixth Symposium. American Society for Testing and Materials, Philadelphia, Pennsylvania.

Perez, K. T., G. M. Morrison, E. W. Davey, N. F. Lackie, A. E. Soper, R. J. Blasco, D. L. Winslow, R. L. Johnson, P. G. Murphy, and J. F. Heltshe. 1991. Influence of size on ecological effects of kepone in physical models. Ecol. Applic. 1:237–248.

Perrin, D. D., B. Dempsey, and E. P. Serjeant. 1981. pK_a Prediction for Organic Acids and Bases. Chapman and Hall, London.

Perry, J. A., D. J. Schaeffer, and E. E. Herricks. 1987. Innovative designs for water quality monitoring: Are we asking the questions before the data are collected? In T.

P. Boyle (ed.), New Approaches to Monitoring Aquatic Ecosystems. American Society for Testing and Materials, Philadelphia, Pennsylvania.

Perry, J. A. and N. H. Troelstrop, Jr. 1988. Whole ecosystem manipulation: A productive avenue for test system research? Environ. Toxicol. Chem. 7:941–951.

Petak, W. J. 1982. Natural Hazard Risk Assessment and Public Policy. Springer-Verlag, New York.

Peterman, R. M., and M. J. Bradford. 1987. Statistical power of trends in fish abundance. Canadian Journal of Fisheries and Aquatic Science. 44:1879–1889,

Peters, R. H. 1983. The Ecological Implications of Body Size. Cambridge University Press, Cambridge, U.K.

Peterson, E. W. and B. Lighthart. 1977. Estimation of downwind viable airborne microbes from a wet cooling tower, including settling. Microbial Ecol. 4:67–79.

Peterson, E. B., Y. -H. Chan, N. M. Peterson, G. A. Constable, R. B. Caton, C. S. Davis, R. R. Wallace, and G. A. Yarranton. 1987. Cumulative effects assessment in Canada: An agenda for action and research. Cat. No. EN106-7/1987E. Minister of Supply and Services Canada, Ottawa.

Phillips, D. J. H. 1978. Use of biological indicator organisms to quantitate organochlorine pollutants in aquatic environments: A review. Environ. Pollut. 13:281–317.

Pickering, Q. H. and C. Henderson. 1966. Acute toxicity of some important petrochemicals to fish. Journal WPCF 38:1419–1429.

Pierce, L.L. and R.G. Congalton. 1988. A methodology for mapping forest latent heat flux densities using remote sensing. Remote sensing in the environment 34: 405–418.

Pinnock, D. E. and R. J. Brand. 1981. A quantitative approach to the ecology of the use of pathogens in insect control. pp. 655–665. In H. D. Burges (ed.), Microbial Control of Pests and Plant Diseases. Academic Press, London.

Plackett, R. L. and P. S. Hewlett. 1952. Quantal responses to mixtures of poisons. J. Royal Statistical Soc. B14:141–163.

Plapp, F. W., Jr. 1981. Ways and means of avoiding or ameliorating resistance to insecticides. pp. 244–249. In International Congress of Plant Protection, Proceedings Symposia IX.

Platt, J. R. 1964. Strong inference. Science 146:347–353.

Podoll, R. T., H. M. Jaber and T. Mill. 1986. Tetrachlorodibenzodioxin: rates of volatilization and photolysis in the environment. Environ. Sci. Technol. 20:490–492.

Poels, C. L. M., M. A. van der Gaag, and J. F. J. van der Kerkhoff. 1980. An investigation of the long-term effects of Rhine River water on rainbow trout. Water Research 14:1029–1035.

Pomeroy, L.R. 1970. The strategy of mineral cycling. Annual Review of Ecology and Systematics 1:171–190.

Pontasch, K. W., E. P. Smith, and J. Cairns, Jr. 1989a. Diversity indices, community comparison indices. and canonical discriminant analysis: interpreting the results of multispecies toxicity tests. Wat. Res. 23:1129–1238.

Pontasch, K. W., B. R. Niederlehner, and J. Cairns, Jr. 1989b. Comparisons of single-species, microcosm, and field responses to a complex effluent. Environ. Toxicol. Chem. 8:521–532.

Poole, R. W. 1974. An Introduction to Quantitative Ecology. McGraw Hill Book Company, New York.

Popper, K. R. 1968. The Logic of Scientific Discovery. Harper and Row, New York.

Porcella, D. B. 1983. Protocol for bioassessment of hazardous waste sites. EPA 600/2-83-054. Corvallis Environmental Research Laboratory, Oregon.

Porter, W. P., R. Hinsdill, A. Fairbrother, L. J. Olson, J. Jaeger, T. Yuill, S. Bisgaard, W. G. Hunter, and K. Nolan. 1984. Toxicant-disease-environment interactions associated with suppression of immune system, growth, and reproduction. Science 224:1014–1017.

Pratt, J. R., N. J. Bowers, B. R. Niederlehner, and J. Cairns, Jr. 1988. Effects of chlorine on microbial communities in naturally derived microcosms. Environ. Toxicol. Chem. 7:679–687.

Presser, T. A. and H. A. Ohlendorf. 1987. Biogeochemical cycling of selenium in the San Joaquin Valley, California, USA. Environ. Manage. 11:805–821.

Price, E. E. and M. C. Swift. 1985. Inter- and intra-specific variability in the response of zooplankton to acid stress. Can. J. Fish. Aquat. Sci. 42:1749–1754.

Prigogine, I. 1982. Order out of chaos, pp. 13–32. In Mitsch, W.J, R.K. Ragade, R.W. Bosserman, and J.A. Dillon, Jr. (eds.), Energetics and Systems. Ann Arbor Science, Ann Arbor, Michigan.

Pritchard P. H. and A. W. Borquin. 1985. Microbial toxicity studies. pp. 177–220. In G. M. Rand and S. R. Petrocelli (eds.), Fundamentals of Aquatic Toxicology. Hemisphere, Washington.

Pulliam, H.R. 1988. Sources, sinks, and population regulation. Amer. Natur. 132:652–661.

Rago, P.J., and R. M. Dorazio. 1984. Statistical inference in life-table experiments: the finite rate of increase. Can. J. Fish. Aquat. Sci. 41:1361–1374

Ramsey, J. C. and P. J. Gehring. 1980. Application of pharmacokinetic principles in practice. Federation Proc. 39: 60–65.

Rand, G. M. 1985. Behavior. pp. 221–263. In G. M. Rand and S. R. Petrocelli. Fundamentals of Aquatic Toxicology. Hemisphere Publishing Corp., Washington.

Rand, G. M. and S. R. Petrocelli. 1985. Fundamentals of Aquatic Toxicology. Hemisphere Publishing Corp., Washington, D.C.

Rand, G. M. 1990. An environmental risk assessment of a pesticide. pp. 899–934. In D. J. Paustenbach (ed.), The Risk Assessment of Environmental and Human Health Hazards: A Textbook of Case Studies. John Wiley & Sons, New York.

Randall, A. 1991. The value of biodiversity. AMBIO 20:64–68.

Rao, C. R. 1964. The use and interpretation of principal component analysis in applied research. Sankhya. A26, 329–358.

Rao, D. N., M. Agrawal, and P. K. Nandi. 1988. Air pollutant mixtures and their effects on plants: a review. Perspec. Environ. Botany 2:217–249.

Rapport, D. J. 1989. What constitutes ecosystem health? Perspectives in Biology and Medicine 33:120–132.

Rapport, D.J., H.A. Regier, and T.C. Hutchinson. 1985. Ecosystem behavior under stress. American Naturalist 125:617–640.

Rasmussen, N. C. 1981. The application of probabilistic risk assessment techniques to energy technologies. Ann. Rev. Energy 6:123–138.

Rathbun, R. E., D. W. Stephens, D. J. Schultz, and D. Y. Tai. 1988. Fate of acetone in a model stream. J. Hydrology 104:1181–1199.

Ratsch, H. C., D. J. Johndro, and J. C. McFarlane. 1986. Growth inhibition and morphological effects of several chemicals in *Arabidopsis thaliana* (L.) Heynh. Environ. Toxicol. Chem. 5:55–60.

Rattner, B. A., D. J. Hoffman, and C. M. Marn. 1989. Use of mixed-function oxygenases to monitor contaminant exposure in wildlife. Environ. Toxicol. Chem. 8:1093–1102.

Rawlings, J. O., V. M. Lesser, A. S. Heagele, and W. W. Heck. 1988. Alternative ozone dose metrics to characterize ozone impact on crop yield loss. J. Environ. Qual. 17:285–291.

Rawlings, J. O., V. M. Lesser, and K. A. Dassel. 1988. Statistical approaches to assessing crop losses. pp. 389–416. In W. A. Heck, O. C. Taylor, and D. T. Tingey (eds.), Assessment of Crop Loss from Air Pollutants. Elsevier Applied Science, New York.

Reckhow, K. H. 1983a. A method for the reduction of lake model prediction error. Water Res. 17:911–916.

Reckhow, K. H. 1983b. A Baysian framework for information transferability in risk analysis. pp. 675–678, In W. K. Lauenroth, G. V. Skogerboe, and M. Flug (eds.), Analysis of Ecological Systems: State-of-the-Art in Ecological Modeling, Elsevier Scientific Publishing, New York.

Reckhow, K. H., and S. C. Chapra. 1983. Confirmation of water quality models. Ecological Modelling 20:113–133.

Reed, M., D. P. French, J. Calambodokidis, and J. C. Cubbage. 1989. Simulation of the effects of oil spills on population dynamics of northern fur seals. Ecological Modelling 49:49–72.

Reese, C. H. and S. L. Burks. 1985. Isolation and chemical characterization of petroleum refinery wastewater fractions acutely lethal to *Daphnia magna*. pp. 319–334. In R. D. Cardwell, R. Purdy, and R. C. Bahner (eds.), Aquatic Toxicology and Hazard Assessment: Seventh Symposium. American Society for Testing and Materials, Philadelphia, Pennsylvania.

Reiners, W.A. 1986. Complementary models for ecosystems. American Naturalist 127:59–73.

Rekker, R. F. 1985. The principles of congenericity, a frequently overlooked prerequisite in quantitative structure-activity and structure-toxicity relationship studies. pp. 3–24. In M. Tichy (ed.), QSAR in Toxicology and Xenochemistry. Elsevier, Amsterdam.

Resh, V. H. and J. D. Unzicker. 1975. Water quality monitoring and aquatic organisms: the importance of species identification. J. Water Poll. Control Fed. 47: 9–19.

Reuss, J.O. and G.S. Innis. 1977. A grassland nitrogen flow simulation model. Ecology 58: 379–388.

Ribbons, D. W., and Eaton, R. W. 1982. Chemical transformations of aromatic hydrocarbons that support the growth of microorganisms. p. 59. In A. M. Chakrabarty, (ed.). Biodegration and Detoxification of Environmental Pollutants. CRC Press, Boca Raton, Florida.

Rice, J.A., J. E. Breck, S. M. Bartell, and J. F. Kitchell. 1983. Evaluating the constraints of temperature, activity and consumption on growth of largemouth bass. Environmental Biology of Fishes 9:263–275.

Richardson, C. J., R. T. DiGiulio, and N. J. Tandy. 1989. Free-radical mediated processes as markers of air pollution stress in trees. pp. 251–260. In National Research Council. Biological Markers of Air-Pollution Stress and Damage in Forests. National Academy Press, Washington, D.C.

Richardson, L. B., and D. T. Burton. 1981. Toxicity of ozonated estuarine water to juvenile blue crabs (*Calinectes sapidus*) and juvenile Atlantic menhaden (*Brevoortia tyrannus*). Bull. Environ. Contam. Toxicol. 26:171–178.

Richardson, L. B., D. T. Burton, R. M. Block, and A. M. Stalova. 1983. Lethal and sublethal exposure and recovery effects of ozone-produced oxidants on adult white perch (*Morone americana* Gmelin). Water Res. 17:205–213.

Ricker, W. E. 1954. Stock and recruitment. J. Fish. Res. Board Can. 11:559–623.

Ricker, W. E. 1975. Computation and interpretation of biological statistics of fish populations. Fish. Res. Board Can. Bull. 191:1–302

Riggins, R., E. Herricks, and M. J. Sale. 1982. Quantitative assessment of environmental impacts in the aquatic environment, CERL-TR-N-114. U.S. Army Construction Engineering Research Lab., Champaign, Illinois.

Rish, W. R., S. A. Schaffer, M. Marchlik, and M. Amdurer. 1985. A risk analysis approach to "How clean is clean"? pp. 333–339. In Proceedings of the Conference on Hazardous Wastes and Environmental Emergencies, May 14–16. Cincinnati, Ohio. Hazardous Materials Control Institute.

Risser, P.G. and W.J. Parton. 1982. Ecosystem analysis of the tallgrass prairie: nitrogen cycle. Ecology 63: 1342–1351.

Robbins, C. S., D. Bystrak, and P. H. Geissler. 1986. The breeding bird survey: its first 15 years, 1965–1979. Resource Pub. 157. U. S. Fish and Wildlife Service, Washington, D.C.

Roberts, B. L. and H. W. Dorough. 1985. Hazards of chemicals to earthworms. Environ. Toxicol. Chem. 4:307–323.

Roberts, J.R., M. S. Mitchell, M. J. Boddington, and J. M. Ridgeway. 1981. A screen for the relative persistence of lipophilic organic chemicals in aquatic ecosystems — An analysis of the role of a simple computer model in screening, Part I. National Research Council of Canada, Rep. No. NRCC 18570, Ottawa, Canada.

Roberts, L. 1991. Learning from an acid rain program. Science 251:1302–1305.

Roberts, T. M. 1984. Long-term effects of sulphur dioxide on crops: an analysis dose-response relations. Phil. Trans. R. Soc. Lond. B 305:299–316.

Roch, M. R. N. Nordin, A. Austin, C. J. P. McKean, J. Denisegert, R. D. Kathman. 1985. The effects of heavy metal contamination on the aquatic biota of Buttle Lake and Campbell River drainage (Canada). Arch. Environ. Contam. Toxicol. 14:347–362.

Rolston, H., III. 1988. Environmental Ethics. Temple University Press. Philadelphia, Pennsylvania.

Roose, M. L., A. D. Bradshall, and T. M. Roberts. 1982. Evolution of resistance to gaseous air pollutants. pp. 379–409. In M. H. Unsworth and D. P. Ormrod, eds., Effects of Gaseous Air Pollution on Agriculture and Horticulture. Butterworth Scientific, Boston, Massachusetts.

Root, R. B. 1967. The behavior and reproductive success of the blue-gray gnatcatcher. Ecol. Monogr. 37:317–350.

Root, R. B. 1973. Organization of a plant-arthropod association in simple and diverse habitats: the fauna of collards (*Brasica oleracea*). Ecol. Monogr. 43:95–120.

Roscoe, D. E. and S. W. Nielsen. 1979. Lead poisoning in mallard ducks (*Anas platyrhynchos*). pp. 165–176. In Animals as Monitors of Environmental Pollutants. National Academy of Sciences, Washington, D.C.

Rose, K. A. 1985. Evaluation of nutrient-phytoplankton-zooplankton models and simulation of the ecological effects of toxicants using laboratory microcosms. PhD Dissertation, University of Washington, Seattle.

Rose, K. A. and G. L. Swartzman. 1981. A review of parameter sensitivity methods applicable to ecosystem models. NUREG/CR-2016. U. S. Nuclear Regulatory Commission, Washington, D.C.

Rose, K. A., J. K. Summers, R. A. Cummins, and D. G. Heimbuch. 1986. Analysis of long-term ecological data using categorical time series regression. Can. J. Fish. Aquat. Sci. 43:2418–2426.

Rose, K.A., G. L. Swartzman, A. C. Kindig, and F. B. Taub. 1988. Stepwise iterative calibration of a multi-species phytoplankton-zooplankton simulation model using laboratory data. Ecological Modelling 42:1–32.

Rosen, R. 1975. Complexity as a system property. International Journal of General Systems 3:227–232.

Rowe, W. D. 1977. An Anatomy of Risk. John Wiley & Sons, New York.

Rowley, M. H., J. J. Christian, D. K. Basu, M. A. Pawlitowski, and B. Paigen. 1983. Use of small mammals (voles) to assess a hazardous waste site at Love Canal, Niagara Falls, New York. Arch. Environ. Contam. Toxicol. 12:383–379.

Royama, T. 1977. Population persistence and density-dependence. Ecol. Monogr. 47:135.

RSC (Research Strategies Committee, Science Advisory Board, U.S. EPA). 1988. Future risk: Research strategies for the 1990s. SAB-EC-88-040. U.S. Environmental Protection Agency, Washington, D.C.

Rubin, I. B., M. R. Guerin, S. A. Hardigree, and J. L. Epler. 1976. Fractionation of synthetic crude oils from coal for biological testing. Environ. Res. 12:358–365.

Rubinstein R. Y. 1981. Simulation and Monte Carlo Method. John Wiley and Sons, New York.

Ruckelshaus, W. D., 1983. Science, risk, and public policy. Science 221:1026–1028.

Ruckelshaus, W. D., 1984. Risk in a free society. Risk Anal. 4:157–162.

Ruesink, R. G. and L. L. Smith. 1975. The relationship of the 96-hour LC_{50} to the lethal threshold concentration of hexavalent chromium, phenol, and sodium pentachlorophenate for fathead minnows (*Pimephales promelas* Rafinesque). Trans. Am. Fish. Soc. 1975:567–570.

Russell, B. 1948. Human Knowledge, Its Scope and Limits, Part V. Simon and Schuster, New York.

Saarikowski, J. and M. Viluksela. 1982. Relation between physiochemical properties of phenols and their toxicity and accumulation in fish. Ecotoxicol. Environmen. Safety 6:501–512.

Saaty, T. L., 1980. The Analytic Hierarchy Process. McGraw-Hill, New York.

Sabljic, A. 1983. Quantitative structure-toxicity relationship of chlorinated compounds: a molecular connectivity investigation. Bull. Environ. Contam. Toxicol., 30: 80–83.

Sale, M. J. 1985. Aquatic ecosystem response to flow modification: An overview of the issues. pp. 25–31. In F. W. Olson, R. G. White, and R. H. Hamre (eds.), Small Hydropower and Fisheries. The American Fisheries Society, Bethesda, Maryland.

Saltzman, S., U. Mingelgrin, and B. Yaron. 1976. Role of water in the hydrolysis of parathion and methylparathion on kaolinite. J. Agric. Food Chem. 24:739–743.

Salwasser, H. 1986. Conserving a regional spotted owl population. pp. 227–247. In National Research Council Committee on the Applications of Ecological Theory to Environmental Problems, Ecological Knowledge and Environmental Problem Solving, National Academy Press, Washington, D.C.

Samuels, W. B., and A. Ladino. 1983. Calculation of seabird population recovery from potential oilspills in the mid-Atlantic region of the United States. Ecol. Model. 21:63–84

SAS Institute. 1985. SAS User's Guide: Statistics, Version 5 Edition. Cary, North Carolina, pp. 575–606.

Sasson-Brickson, G. and G. A. Burton. 1991. In situ and laboratory toxicity testing with *Ceriodaphnia dubia*. Environ. Toxicol. Chem. 9:1523–1530.

Scavia, D., and A. Robertson. 1979. Perspectives on Lake Ecosystem Modeling. Ann Arbor Science Publishers, Ann Arbor, Michigan.

Schaeffer, D.L. 1980. A model evaluation methodology applicable to environmental assessment models. Ecological Modelling 8: 275–295.

Schaeffer, D. J., E. E. Herricks, and H. W. Kerster. 1988. Ecosystem health 1: measuring ecosystem health. Environ. Manage. 12:445–455.

Schafer, E. W. Jr. and R. B. Brunton. 1979. Indicator bird species for toxicity determinations: Is the technique useable in test method development? pp. 157–168. In J. R.

Beck (ed.), Vertebrate Pest Control and Management Materials. American Society for Testing and Materials, Philadelphia, Pennsylvania.

Schafer, E. W. Jr., R. B. Brunton, E. C. Schafer, and J. Chevez. 1982. Effects of 77 chemicals on reproduction in male and female *Coturnix* quail. Ecotoxicol. Environm. Safety 6:149–330.

Schafer, E. W. Jr., W. A. Bowles, Jr., and J. Hurlbet. 1983. The acute oral toxicity, repellency, and hazard potential of 998 chemicals to one or more species of wild and domestic birds. Arch. Environm. Contam. Toxicol. 12:355–328.

Schafer, E. W. Jr. and W. A. Bowles, Jr. 1985. Acute oral toxicity and repellency of 933 chemicals to house and deer mice. Arch. Environm. Contam. Toxicol. 14:111–129.

Schimmel, S. C., G. E. Morrison, and M. A. Heber. 1989. Marine complex effluent toxicity program: Test sensitivity, repeatability, and relevance to receiving water toxicity. Environ. Toxicol. Chem. 8:739–746.

Schindler, D. W., K. H. Mills, D. F. Malley, D. L. Findlay, J. A. Shearer, I. J. Davies, M. A. Turner, G. A. Linsey, D. R. Cruikshank. 1985. Long-term ecosystem stress: The effects of years of experimental acidification on a small lake. Science 228:1395–1401.

Schmidt-Bleek, F. and W. Haberland. 1980. The yardstick concept for the hazard evaluation of substances. Ecotoxicol. Environ. Safety 4:455–465.

Schnoor, J. L. 1982. Field validation of water quality criteria for hydrophobic pollutants. pp. 302–315. In J. G. Pearson, R. B. Foster, and W. E. Bishop (eds.), Aquatic Toxicology and Hazard Assessment, Fifth Symposium. ASTM, Philadelphia, Pennsylvania.

Schnoor, J. L., N. P. Nikolaidis, and G. E. Glass. 1986. Lake resources at risk to acidic deposition in the Upper Midwest. J. Water Pollut. Control Fed. 58:139–148.

Schoener, T. W. 1968. Sizes of feeding territories among birds. Ecology 49:123–136.

Schofield, C. L. and C. T. Driscoll. 1987. Fish species distribution in relation to water quality gradients in the North Branch of the Moose River Basin. Biogeochemistry 3:63–85.

Schrader-Frechette, K. S. 1985. Risk Analysis and Scientific Method. D. Reidel Publishing Co., Dordrecht, The Netherlands.

Schulze, E.-D. 1989. Air pollution and forest decline in a spruce (*Picea abies*) forest. Science 244:776–783.

Schuytema, G. S. 1982. A review of aquatic habitat assessment methods. EPA-600/3-82-002. U.S. Environmental Protection Agency, Corvallis, Oregon.

Schwartz, S. 1988. Mass transport limitation in the rate of incloud oxidation of SO_2. Atmos. Environ. 22:2491–2499.

Schwartz, S. 1984. Gas-aqueous reactions of sulfur and nitrogen oxides in liquid water clouds. pp. 173–207. In SO_2, NO and NO_2 Oxidation Mechanisms. J. G. Calvert (ed.), Butterworth, New York.

Sciandra, A. 1986. Study and model of a simple planktonic system reconstituted in an experimental microcosm. Ecological Modelling 34: 61–82.

Seegert, G., J. A. Fava, and P. M. Crumbie. 1985. How representative are the data sets used to derive national water quality criteria? pp. 527–537. In R. D. Cardwell, R. Purdy, and R. C. Bahner. Aquatic Toxicology and Hazard Assessment: Seventh Symposium. American Society for Testing and Materials, Philadelphia, Pennsylvania.

Segar, D. A. and E. Stamman. 1986. A strategy for design of marine pollution monitoring studies. Wat. Sci. Tech. 18:15–26.

Seip, K.L. 1979. A mathematical model for the uptake of heavy metals in benthic algae. Ecological Modelling 6: 183–197.

Seip, K.L. 1980. A mathematical model of competition and colonization in a community of marine benthic algae. Ecological Modelling 10: 77–104.

Sekiya, J., L. G. Wilson, and P. Filner. 1982. Resistance to injury by sulfur dioxide. Plant Physiol. 70: 437–441.

Selye, H. 1950. The Physiology and Pathology of Exposure to Stress. Acta Inc., Montreal, Canada.

Senes Consultants. 1989. Contaminated Soil Cleanup in Canada, Vol. 5, Development of the AERIS Model, Final Report prepared for the Decommissioning Steering Committee.

Severinghaus, W. D. 1981. Guild theory development as a mechanism for assessing environmental impact. Environ. Management 5:187–190.

Shannon, L. J., T. E. Flum, R. L. Anderson, and J. D. Yount. 1989. Adaptation of mixed flask culture microcosms to testing the survival and effects of introduced microorganisms. pp. 224–239. In U. M. Cowgill and L. W. Williams (eds.), Aquatic Toxicology and Hazard Assessment: 12th Volume. ASTM, Philadelphia, Pennsylvania.

Sharples, F. E. 1983. Spread of organisms with novel genotypes: Thoughts from an ecological perspective. DNA Technical Bull. 6:43–56.

Shaw, A. J. (ed.). 1990. Heavy Metal Tolerance in Plants: Evolutionary Aspects. CRC Press, Boca Raton, Florida.

Sheehan, P. J., D. R. Miller, G. C. Butler, and P. Bordeau (eds.). 1984. Effects of Pollutants at the Ecosystem Level. SCOPE 22. John Wiley and Sons, New York.

Shigan, S. S. 1976. Methods for predicting chronic toxicity parameters of substances in the area of water hygiene. Environ. Health Persp. 13:83–89.

Shinn, J. H., B. R. Clegg, M. L. Stuart, and S. E. Thompson. 1976. Exposure of field-grown lettuce to geothermal air pollution — photosynthetic and stomatal responses. J. Environ. Sci. Health All(10&11): 603–612.

Shiu, W. Y. and D. Mackay. 1986. A critical review of aqueous solubilities, vapor pressures, Henry's law constants and octanol-water partition coefficients of the polychlorinated biphenyls. J. Phys. Chem. Ref. Data 15:911–929.

Shu, H., D. Paustenbach, F. J. Murray, L. Marple, B. Brunck, D. Dei Rossi, and P. Teitelbaum. 1988. Bioavailability of soil-bound TCDD: oral bioavailability in the rat. Fundament. Appl. Toxicol. 10:648–654.

Shuba, P. J., H. J. Carroll, and K. L. Wong. 1977. Biological assessment of the soluble fraction of the standard elutriate test. Tech. Rpt. D-77-3. U.S. Army Engineer Waterways Experiment Station, Vicksburg, Mississippi.

Shugart, H. H. 1984. A Theory of Forest Dynamics. Springer-Verlag, New York.

Shugart, H.H. and D.C. West. 1977. Development and application of an Appalachian deciduous forest succession model and its application to assessment of the impact of the chestnut blight. J. Environ. Manage. 5: 161–179.

Shugart, H. H. and D. C. West. 1980. Forest succession models. BioScience. 30:308–313.

Shugart, L. R. 1988. An alkaline unwinding assay for the detection of DNA damage in aquatic organisms. Mar. Environ. Res. 2:321–325.

Shugart, L. R. 1988. Quantitation of chemically induced damage to DNA of aquatic organisms by alkaline unwinding assay. Aquatic Toxicol. 13:43–52.

Shugart, L., J. McCarthy, B. Jimenez, and J. Daniels. 1987. Analysis of adduct formation in the bluegill sunfish (*Lepomis macrochirus*) between benzo[a]pyrene and DNA of the liver and hemoglobin of the erythrocyte. Aquatic Toxicol. 9:319–325.

Shugart, L. R., S. M. Adams, B. D. Jimenez, S. S. Talmage, and J. F. McCarthy. 1989. Biological markers to study exposure in animals and bioavailability of environmental contaminants. pp. 86–97. In R. G. M. Wang, C. A. Franklin, R. C. Honeycutt, and J. C. Reinert (eds.), Biological Monitoring for Pesticide Exposure: Measurement, Estimation, and Risk Reduction. American Chemical Society, Washington, D.C.

Shupe, J. L., H. B. Peterson, and A. E. Olson. 1980. Fluoride toxicosis in wild ungulates of the western United States. pp. 253–266. In Animals as Monitors of Environmental Pollutants. National Academy Press, Washington, D.C.

Sielken, R. J., Jr. 1987. Cancer dose-response extrapolations. Environ. Sci. Technol. 21:1033–1039.

Sigal, L. L. 1988. The relationship of lichen and bryophyte research to regulatory decisions in the United States. pp. 269–287. In T. H. Nash III and V. Wirth (eds.), Lichens, Bryophytes and Air Quality. J. Cramer, Berlin.

Sigal, L. L. and G. W. Suter II. 1987. Evaluation of methods for determining adverse impacts of air pollution on terrestrial ecosystems. Environ. Manage. 11:675–694.

Sigal, L. L. and G. W. Suter II. 1989. Potential effects of chemical agents on terrestrial resources. Environ. Profess. 11:376–384.

Sigmon, C. F., H. F. Kania, and R. J. Beyers. 1977. Reduction in biomass and diversity resulting from exposure to mercury in artificial streams. J. Fish. Res. Bd. Can. 34:493–500.

Simberloff, D. 1981. Community effects of introduced species. pp. 53–81. In M. H. Nitecki (ed.), Biotic Crises in Ecological and Evolutionary Time. Academic Press, New York.

Simonsen, L. and B. R. Levin. 1988. Evaluating the risk of releasing genetically engineered organisms. pp. S27-S30. In J. Hodgson and A. M. Sugden (eds.). Planned Release of Genetically Engineered Organisms (Trends in Biotechnology/Trends in Ecology and Evolution Special Publication). Elsevier Publications, Cambridge.

Sims, J. T. and J. S. Kline. 1991. Chemical fractionation and plant uptake of heavy metals in soils amended with co-composted sewage sludge. J. Environ. Qual. 20:387395.

Singh, H. B. 1977. Preliminary estimation of average tropospheric HO concentrations in the Northern and Southern hemispheres. Geophys. Res. Lett. 4:453–456.

Sinko, J. W., and W. Streifer. 1967. A new model for age-structure of a population. Ecology 48:910–918

Sinko, J. W., and W. Streifer 1969. Applying models incorporating age-size structure of a population of Daphnia. Ecology 50:608–615

Skelly, J. M., D. D. Davis, W. Merrill, E. A. Cameron, H. D. Brown, D. B. Drummond, and L. S. Dochinger (eds.). 1990. Diagnosing Injury to Eastern Forest Trees. Pennsylvania State University, University Park.

Skutsch, M. M. and R. T. N. Flowerdew. 1976. Measurement techniques in environmental impact assessment. Environ. Conservation 3:209–217.

Slater, J. H. 1985. Gene transfer in microbial communities. pp. 89–98. In H. O. Halvorson, D. Pramer, and M. Rogul (eds.), Engineered Organisms in the Environment: Scientific Issues. American Society for Microbiology, Washington, D.C.

Slatkin, M. 1974. Competition and regional coexistence. Ecology 55:128–134.

Slobodkin, L. B. 1961. Growth and Regulation of Animal Populations. Holt, Rinehart, and Winston, New York.

Sloof, W., J. H. Canton, and J. L. M. Hermens. 1983. Comparison of the susceptibility of 22 freshwater species to 15 chemical compounds. I (Sub)acute toxicity tests. Aquatic Toxicol. 4:113–128.

Sloof, W. and J. H. Canton. 1983. Comparison of the susceptibility of 11 freshwater species to 8 chemical compounds. II (Semi)chronic toxicity tests. Aquatic Toxicol. 4:271–282.

Sloof, W., J. A. M. van Oers, and D. de Zwart. 1986. Margins of uncertainty in ecotoxicological hazard assessment. Environ. Toxicol. Chem. 5:841–852.

Smies, M. 1983. On the relevance of microecosystems for risk assessment: Some considerations for environmental toxicology. Ecotoxicol. Environ. Safety 7:355–365.

Smissaert, H. R. and A. A. M. Jansen. 1984. On the variation of toxic effects over species, its cause, and analysis by "structure-selectivity relations." Ecotoxicol. Environ. Safety 8:294–302.

Smith, E. D. and L. W. Barnthouse. 1987. User's manual for the defense priority model. ORNL-6411. Oak Ridge National Laboratory, Oak Ridge, Tennessee.

Smith, E. H. and H. C. Bailey. 1989. Preference/avoidance testing of waste discharges on anadromous fish. Environ. Toxicol. Chem. 9:77–86.

Smith, E. P., K. W. Pontasch, and J. Cairns Jr. 1989. Community similarity and the analysis of multispecies environmental data: a unified statistical approach. Wat. Res. 24:507–514.

Smith, E. P., D. R. Orvos, and J. Cairns, Jr. 1990. Some comments and criticisms of the before-after:control-impact model for impact assessment of a nuclear power plant. p. 36. In Abstracts, Society for Environmental Toxicology and Chemistry, Eleventh Annual Meeting. SETAC, Washington, D.C.

Smith, J. H., D. Mackay, and C. W. K. Ng. 1983. Volatilization of pesticides from water. Residue Reviews 85:73–88.

Smith, J. H., D. C. Bomberger, and D. L. Haynes. 1981. Volatilization of intermediate and low volatility chemicals. Chemosphere 10:281–287.

Smith, J. H., W. R. Mabey, N. Bohonos, B. R. Holt, S. S. Keem, T..-W. Chou, D. C. Bomberger, and T. Mill. 1978. Environmental pathways of selected chemicals in freshwater systems. II. Laboratory studies. Report EPA-600/7-78-074. U.S. Environmental Protection Agency, Athens, Georgia.

Smith, W. H. 1990. Air Pollution and Forests, Second Edition. Springer-Verlag, New York.

Smith, R. A., R. B. Alexander, and M. G. Wolman. 1987. Water-quality trends in the nation's rivers. Science 235:1607–1615.

Snider, J. R., and G. A. Dawson. 1985. Trophopheric light alcohols, carbonyls, and acetonitrile: concentrations in the Southwestern U.S. and Henry's law data. Geophys. Res. 90:3797–3802.

Sniezko, S. F. 1974. The effects of environmental stress on outbreaks of infectious diseases in fishes. J. Fish Biol. 6:197–208.

Somerville, L. and C. H. Walker (eds.). 1990. Pesticide Effects on Terrestrial Wildlife. Taylor & Francis, London.

Sparks, R. E., W. T. Waller, and J. Cairns, Jr. 1972. Effects of shelters on the resistance of dominant and submissive bluegills (*Lepomis macrochirus*) to a lethal concentration of zinc. J. Fisheries Res. Bd. Canada. 29:1356–1358.

Spector, W. S. (ed.).1956. Handbook of Biological Data. W. B. Saunders Company, Philadelphia, Pennsylvania.

Spehar, R. L., E. N. Leonard, and D. L. Defoe. 1978. Chronic effects of cadmium and zinc mixtures on flagfish (*Jordanella floridae*). Trans. Am. Fish. Soc. 107:354–360.

Spehar, R. L. and A. R. Carlson. 1984. Derivation of site-specific water quality criteria for cadmium and the St. Louis River basin, Duluth, Minnesota. EPA-600/3-84-029. U.S. Environmental Protection Agency, Duluth, Minnesota.

Spencer, W. F., W. Farmer and M. M. Cliath. 1973. Pesticide volatization. Residue Rev. 49:1–27.

Sperber, O., J. From, and P. Sparre. 1977. A method to estimate the growth rate of fishes, as a function of temperature and feeding level, applied to rainbow trout. Meddelerser fra Danmarks Fiskeriog Havundersogelser 7:275–317.

Sprague, J. B. 1969. Measurement of pollutant toxicity to fish. I. Bioassay methods for acute toxicity. Water Res. 3:793–821.

Sprague, J. B. 1970. Measurement of pollutant toxicity to fish. II. Utilizing and applying bioassay results. Water Res. 4:3–32.

Sprague, J. B., P. F. Elson and R. L. Saunders. 1965. Sublethal copper-zinc pollution in a salmon river—A field and laboratory study. Int. J. Air Wat. Poll. 9:531–543.

Sprague, J. B. 1985. Factors that modify toxicity. pp. 124–176. In G. M. Rand and S. R. Petrocelli (eds.), Fundamentals of Aquatic Toxicology, Hemisphere Publishing, Washington, D.C.

Starfield, A.M. and A. L. Beloch. 1986. Building Models for Conservation and Wildlife Management. Macmillan Publishing Co., New York.

Staufer, J. R., Jr., R. L. Kaesler, J. Cairns, Jr., and K. L. Dickson. 1980. Selecting groups of fish to optimize acquisition of information on thermal discharges. Water Resources Bulletin 16: 1097–1101.

Stearns, S. C. 1977. The evolution of life history traits: a critique of the theory and a review of the data. Ann. Rev. Ecol. Syst. 8:145–171.

Stebbing, A. R. D. 1982. Hormesis—the stimulation of growth by low levels of inhibitors. Sci. Total Environ. 22:213–234.

Steffans, J. J., E. J. Sampson, I. J. Siewers, and S. J. Bankanc. 1973. Effect of divalent metal ions on intramolecular nucleophilic catalysts of phosphate diester hydrolysis. J. Amer. Chem. Soc. 95:936–938.

Stene, A. and S. Lonning. 1984. Effects of 2-methylnaphthalene on eggs and larvae of six marine fish species. Sarsia 69:199–203.

Stephan, C. E. 1977. Methods for calculating an LC_{50}. pp. 65–84. In F. L. Mayer and J. L. Hamelink (eds.), Aquatic Toxicology and Hazard Assessment, ASTM STP 634. American Society for Testing and Materials, Philadelphia.

Stephan, C. E., D. I. Mount, D. J. Hanson, J. H. Gentile, G. A. Chapman, and W. A. Brungs. 1985. Guidelines for deriving numeric National Water Quality Criteria for the protection of aquatic organisms and their uses, PB85-227049. U.S. Environmental Protection Agency, Duluth, Minnesota.

Stephan, C. E. and J. R. Rogers. 1985. Advantages of using regression analysis to calculate results of chronic toxicity tests. pp. 328–339. In R. C. Bahner and D. J. Hansen (eds.), Aquatic Toxicology and Hazard Assessment: Eighth Symposium. American Society for Testing and Materials, Philadelphia, Pennsylvania.

Stephan, C. E. and R. J. Erickson. n.d. Guidelines for deriving an aquatic life pesticide concentration. U.S. Environmental Protection Agency, Environmental Research Laboratory, Duluth, MN, unpublished document.

Stephenson, R. R., 1982. Aquatic toxicology of cypermethrin. I. Acute toxicity to some freshwater fish and invertebrates in laboratory tests. Aquatic Toxicol. 2:175–185.

Stevens, D., G. Linder, and W. Warren-Hicks. 1989. Data interpretation. pp. 9-1-9-25. In W. Warren-Hicks, B. R. Parkhurst, and S. S. Baker, Jr. (eds.), Ecological Assessment of Harardous Waste Sites: A Field and Laboratory Reference Document. EPA 600/3-89/013. Corvallis Environmental Research Laboratory, Oregon.

Stewart, D.J. 1980. Salmonid predators and their forage base in Lake Michigan: a bioenergetics-modeling synthesis. Ph.D. Dissertation, University of Wisconsin, Madison.

Stewart, D.J., J.F. Kitchell and L.B. Crowder. 1981. Forage fishes and their salmonid predators in Lake Michigan. Transactions of the American Fisheries Society 110: 751–763.

Stewart-Oaten, A., W. W. Murdoch, and K. R. Parker. 1986. Environmental impact assessment: "pseudoreplication" in time? Ecology 67:929–940.

Stockner, J. G. and N. J. Antia. 1976. Phytoplankton adaptation to environmental stresses from toxicants, nutrients, and pollutants—a warning. J. Fish. Res. Board Can. 33:2089–2096.

Stotzky, G. 1965. Microbial respiration. pp. 1550–1572. In C. A. Black (ed.), Methods of Soil Analysis; Part 2, Chemical and Microbiological Properties. American Society of Agronomy, Madison, Wisconsin.

Stout, R. J. and W. E. Cooper. 1983. Effect of p-cresol on leaf decomposition and invertebrate colonization in experimental outdoor streams. Can. J. Fish. Aquat. Sci. 40:1647–1657.

Streitweiser, A. 1962. Solvolytic Displacement Reactions. McGraw-Hill, New York.

Sullivan, J. F., G. J. Atchison, D. J. Kolar, and A. W. MacIntosh. 1978. Changes in the predator-prey behavior of fathead minnows *Pimephales promelas*) and largemouth bass (*Micropterus salmoides*) caused by cadmium. J. Fish. Res. Board Can. 35:446–451.

Summers, J. K. and K. A. Rose. 1987a. The role of interactions among environmental conditions in controlling historical fisheries variability. Estuaries 10:255–266.

Summers, J. K. and K. A. Rose. 1987b. Historical relationships among fisheries abundance and hydrologic and pollution variables in northeastern estuaries. pp. 16621669. In Oceans 87 Proceedings. Marine Technology Society, Washington, D.C.

Sun, M. 1983. Missouri's costly dioxin lesson. Science 219:367–369.

Sunda, W. G., S. A. Huntsman, and G. Harvey. 1983. Photoreduction of manganese oxides in seawater. Nature 301:234–236.

Suter, G. W., II. 1981a. Ecosystem theory and NEPA assessment. Bull. Ecol. Soc. Amer. 63:186–192.

Suter, G. W., II. 1981b. Laboratory tests for effects of chemicals on terrestrial population interactions and ecosystem properties. pp. 93–154. In A. S. Hammons (ed.), Methods for Ecological Toxicology. Ann Arbor Science, Ann Arbor, Michigan.

Suter, G. W. II. 1985. Application of environmental risk analysis to engineered organisms. pp. 211–219. In H. O. Halverson, D. Pramer, and M. Rogul (eds.), Engineered Organisms in the Environment: Scientific Issues. American Society for Microbiology, Washington, D.C.

Suter, G. W. II. 1987. Species interactions. pp. 745–758. In V. B. Vouk, G. C. Butler, A. C. Upton, D. V. Parke, and S. C. Asher (eds.), Methods for Assessing the Effects of Mixtures of Chemicals, SCOPE 30. John Wiley and Sons, Chichester, UK.

Suter, G. W. II. 1989. Ecological endpoints. pp. 2-1-2-28. In W. Warren-Hicks, B. R. Parkhurst, and S. S. Baker, Jr. (eds.), Ecological Assessment of Harardous Waste Sites: A Field and Laboratory Reference Document. EPA 600/3-89/013. Corvallis Environmental Research Laboratory, Oregon.

Suter, G. W. II. 1990a. Uncertainty in environmental risk assessment. Chapter 9, pp. 203–230. In G. M. von Furstenberg (ed.), Acting Under Uncertainty: Multidisciplinary Conceptions. Kluwer Academic Publishers, Boston, Massachusetts

Suter, G. W. II. 1990b. Endpoints for regional ecological risk assessments. Environ. Manage. 14:9–23.

Suter, G. W. II. 1990c. Environmental risk assessment/environmental hazard assessment: similarities and differences. pp. 5–15. In W. G. Landis and W. H. van der Schalie (eds.), Aquatic Toxicology and Risk Assessment: Thirteenth Volume. ASTM, Philadelphia, Pennsylvania.

Suter, G. W. II, D. S. Vaughan, and R. H. Gardner. 1983. Risk assessment by analysis of extrapolation error, a demonstration for effects of pollutants on fish. Environmental Toxicology and Chemistry 2:369–378.

Suter, G. W. II and F. E. Sharples. 1984. Examination of a proposed test for effects of toxicants on soil microbial processes. pp. 327–344. In D. Liu and B. J. Dutka. Toxicity Screening Procedures Using Bacterial Systems. Marcel Dekker, New York.

Suter, G. W. II, L. W. Barnthouse, C. F. Baes III, S. M. Bartell, M. G. Cavendish, R. H. Gardner, R. V. O'Neill, and A. E. Rosen. 1984. Environmental risk analysis for direct coal liquefaction. ORNL/TM-9074. Oak Ridge National Laboratory, Oak Ridge, Tennessee.

Suter, G. W. II, L. W. Barnthouse, J. E. Breck, R. H. Gardner, and R. V. O'Neill. 1985. Extrapolating from the laboratory to the field: How uncertain are you? pp. 400–413. In R. D. Cardwell, R. Purdy, and R. C. Bahner (eds.), Aquatic Toxicology and Hazard Assessment: Seventh Symposium. ASTM STP 854. American Society for Testing and Materials, Philadelphia, Pennsylvania.

Suter, G. W., II, A. E. Rosen, and E. Linder. 1986. Analysis of extrapolation error. pp. 49–81. In L. W. Barnthouse and G. W. Suter II (eds.), User's Manual for Ecological Risk Assessment, ORNL-6251. Oal Ridge National Laboratory, Oak Ridge, Tennessee.

Suter, G. W., II, L. W. Barnthouse, and R. V. O'Neill. 1987a. Treatment of risk in environmental impact assessment. Environ. Manage. 11:295–303.

Suter, G. W., II, A. E. Rosen, E. Linder, and D. F. Parkhurst. 1987b. Endpoints for responses of fish to chronic toxic exposures. Environ. Toxicol. Chem. 6:793–809.

Suter, G. W. II and A. E. Rosen. 1988. Comparative toxicology for risk assessment of marine fishes and crustaceans. Environ. Sci. Technol. 22:548–556.

Svirezhev, Yu.M., V.P. Krysanova and A.A. Voinov. 1984. Mathematical modelling of a fish pond ecosystem. Ecological Modelling 21:315–337.

Swann, R.L. and A. Eschenroeder.1983. Fate of chemicals in the environment. ACS Symposium Series, No. 225, American Chemical Society, Washington, D.C.

Swartzman, G.L. and R. Bentley. 1979. A review and comparison of plankton simulation models. ISEM Journal 1:30–81.

Swartzman, G. and K.A. Rose. 1984. Simulating the biological effects of toxicants in aquatic microsystems. Ecological Modelling 22:123–134.

Swartzman, G. L., and S. P. Kaluzny. 1987. Ecological Simulation Primer. Macmillan Publishing, New York, New York.

Szaro, R. C. and R. P. Balda. 1979. Bird Community Dynamics in a Ponderosa Pine Forest. Studies in Avian Biology No. 3. Cooper Ornithological Society.

Tamm, C. O. 1989. Comparative and experimental approaches to the study of acid deposition effects on soils as substrate for forest growth. AMBIO 18:184–191.

Tanner, J. T. 1967. Effects of population density on growth rates of animal populations. Ecology 47:733–745.

Tate, J., Jr. 1986. The blue list for 1986. Am. Birds 40:227–236.

Taub, F. B. 1969. Gnotobiotic models of freshwater communities. Verh. Internat. Verein. Limnol. 17:485–496.

Taub, F.B. and P.L. Read. 1982. Model Ecosystems: Standardized Aquatic Microcosm Protocol, Vol. II, Final Report, Food and Drug Administration Contract No. 223–80–2352, Washington, D.C.

Taylor, F. 1979. Convergence to the stable age distribution in populations of insects. Amer. Natur. 113:511–530.

Taylor, G. E., Jr. 1978. Generic analysis of ecotypic differentiation within an annual plant species, Geranium carolinianunum L., in response to sulfur dioxide. Botanical Gazette 139:362–368.

Taylor, G. E., Jr., P. J. Hanson, and D. D. Baldocchi. 1988. Pollutant deposition to individual leaves and plant canopies: Sites of regulation and relationship to injury. pp. 227–257. In W. W. Heck, O. C. Taylor, and D. T. Tingey, (eds.), Assessment of Crop Loss from Air Pollutants. Elsevier Publishers, New York.

Tebo, L. B. Jr. 1986. Effluent monitoring: Historical perspective. pp. 13–31. In H. L. Bergman, R. A. Kimerle, and A. W. Maki (eds.), Environmental Hazard Assessment of Effluents. Pergamon Press, New York.

Temple, S. A. and J. A. Weins. 1989. Bird populations and environmental changes: can birds be bio-indicators? Amer. Birds 43:260–270.

Terborgh, J. 1989. Where Have All the Birds Gone? Princeton University Press, Princeton, New Jersey.

Tetra Tech. 1986. Bioaccumulation monitoring guidance: 4. Analytical methods for U.S. EPA priority pollutants and 301(h) pesticides in tissues from estuarine and marine organisms. Final Report. Tetra Tech Inc., Bellevue, Washington.

Thibodeaux, L.J. 1979. Chemodynamics. John Wiley and Sons, New York.

Thomann, R. V. 1972. Systems Analysis and Water Quality Management. McGraw-Hill, New York.

Thomann, R.V. 1981. Equilibrium model of fate of microcontaminants in diverse aquatic food chains. Can. J. Fish. Aquat. Sci. 38: 280–296.

Thomann, R.V. 1984. Physico-chemical and ecological modeling the fate of toxic substances in natural water systems. Ecological Modelling 22: 145–170.

Thomann, R.V. 1989. Bioaccumulation model of organic chemical distribution in aquatic food chains, Environ. Sci. Technol. 23:699–707.

Thomann, R.V., and J. P. Connolly. 1984. Model of PCB in the Lake Michigan lake trout food chain. Environ. Sci. Technol. 18:65–71.

Thomas, J. M., J. R. Skalski, J. F. Cline, M. C. McShane, J. C. Simpson, W. E. Miller, S. A. Peterson, C. A. Callahan, and J. C. Greene. 1986. Characterization of chemical waste site contamination and determination of its extent using bioassays. Environ. Toxicol. Chem. 5:487–501.

Thornton, E. R. 1964. Solvolytic Mechanisms. Ronald Press, New York.

Thron, C. D. 1980. Linear pharmacokinetic systems. Federation Proc. 39: 2443–2449.

Thurman, E. M. 1985. Organic geochemistry of natural waters. Matinus Nijhoff/Dr. Junk Publishers, Boston.

Thurston, R. V., R. C. Russo, and G. R. Phillips. 1983. Acute toxicity of ammonia to fathead minnows. Trans. Am. Fish. Soc. 112:705–711.

Thurston, R. V., T. A. Gilfoil, E. L. Meyn, R. K. Zajdel, T. I. Aoki, and G. D. Veith. 1985. Comparative toxicity of ten organic chemicals to ten common aquatic species. Water Res. 19:1145–1155.

Tiedje, J. M., R. K. Cowell, Y. L. Grossman, R. E. Hodson, R. E. Lenski, R. N. Mack, and P. J. Regal. 1989. The planned release of genetically engineered organisms: Ecological considerations and recommendations. Ecology 70:298–315.

Tipton, A. R. 1980. Mathematical modeling in wildlife management. pp. 211–220. In S. D. Schemnitz and L. Toschik (eds.), Wildlife Management Techniques Manual, Fourth edition, revised. The Wildlife Society, Washington, D.C.

Tipton, A. R., R. J. Kendall, J. F. Coyle, and P. F. Scanlon. 1980. A model of the impact of methyl parathion spraying on a quail population. Bull. Environ. Contam. Toxicol. 25:586–593.

Tomovic, R. 1963. Sensitivity Analysis of Dynamic Systems. McGraw-Hill, New York.

Topliss, J. G. 1983. Quantitative Structure-Activity Relationships of Drugs. Academic Press, New York.

Topp, E., I. Schenert, A. Attar, A. Korte, and F. Korte. 1986. Factors affecting the uptake of C^{14}-labelled organic chemicals by plants from soil. Ecotoxicol. Environ. Saf. 11:219.

Touart, L. W. 1987. Aquatic mesocosm tests to support pesticide registrations. Office of Toxic Substances, U. S. Environmental Protection Agency, Washington, D.C.

Travis, C. C., C. F. Baes III, L. W. Barnthouse, E. L. Etnier, G. A. Holton, B. D. Murphy, G. P. Thompson, G. W. Suter II, and A. P. Watson. 1983. Exposure Assessment Methodology and Reference Environments for Synfuels Risk Analysis. ORNL/TM-8672. Oak Ridge National Laboratory, Oak Ridge, Tennessee.

Travis, C. C., S. A. Richter, E. A. C. Crouch, R. Wilson, and E. D. Klema. 1987. Cancer risk management. Environ. Sci. Technol. 21:415–420.

Travis, C. C. and A. D. Arms. 1988. Bioconcentration of organics in beef, milk, and vegetation. Environ. Sci. Technol. 22:271–274.

Treshow, M. and D. Stewart. 1973. Ozone sensitivity of plants in natural communities. Biol. Conserv. 5:209–214.

Truhaut, R. 1977. Ecotoxicology: Objectives, principles and perspectives. pp. 151–173. In W. J. Hunter and J. G. P. M. Smeets (eds.), Evaluation of Toxicological Data for Protection of Public Health. Pergamon Press, Oxford.

Tsai, C. and Chang, K. 1984. Intraspecific variation in copper susceptibility of the bluegill sunfish. Arch. Environ. Contam. Toxicol. 13:93–99.

Tucker, R. K. and D. G. Crabtree. 1970. Handbook of toxicity of pesticides to wildlife. PB 198 815. U.S. Fish and Wildlife Service, Denver, Colorado.

Tucker, R. K. and M. A. Heagele. 1971. Comparative acute oral toxicity of pesticides to six species of birds. Toxicol. Appl. Pharmacol. 20:57–65.

Tucker, R. K. and J. S. Lietzke. 1979. Comparative toxicity of insecticides for vertebrate wildlife and fish. Pharmac. Ther. 6:167–220.

Tucker, W.A., A. G. Eschenroeder, and G. C. Magil. 1982. Air, Land, Water Analysis System (ALWAS): A Multimedia Model for Assessing the Effect of Airborne Toxic Substances on Surface Quality, first draft report, prepared by Arthur D. Little, Inc. for Environmental Research Laboratory, EPA, Athens, Georgia.

Tuljapurkar, S. D., and S. H. Orzack. 1980. Population dynamics in variable environments I. Long-run growth rates and extinction. Theoret. Pop. Biol. 18:314–342.

Turner, M. G. (ed.). 1987a. Landscape Heterogeneity and Disturbance. Springer-Verlag, New York.

Turner, M. G. 1987b. Land use changes and net primary production in the Georgia, USA, landscape: 1935–1982. Environ. Manage. 11:237–247.

Turner, M. G. 1989. Landscape ecology: the effect of pattern on process. Annu. Rev. Ecol. Syst. 20:171–197.

Turner, M. G. and R. H. Gardner (eds.). 1991. Quantitative Methods in Landscape Ecology. Springer-Verlag, New York.

Turro, N. J. 1967. Molecular Photochemistry. W. A. Benjamin, Inc., New York.

Tyler, G., A.-M. Balsberg Pahlsson, G. Bengtsson, E. Baath, and L. Tranvik. 1989. Heavy-metal ecology of terrestrial plants, microorganisms and invertebrates. Water Air Soil Pollut. 47:189–215.

Unwin, S. D. 1986. A fuzzy set theoretic foundation for vagueness in uncertainty analysis. *Risk Analysis* 6:27–34.

Urban, D. J. and N. J. Cook. 1986. Hazard Evaluation, Standard Evaluation Procedure, Ecological Risk Assessment. EPA-540/9–85–001. U.S. Environmental Protection Agency, Washington, D.C.

Urban, D. L., R. V. O'Neill, and H. H. Shugart., Jr. 1987. Landscape ecology. BioScience 37:119–127.

USDA (U.S. Department of Agriculture). 1986. Draft supplement to the environmental impact statement for an amendment to the Pacific Northwest Regional Guide. United States Forest Service, Portland, Oregon.

van der Schalie, W. H. 1985. Can biological monitoring early warning systems be useful in detecting toxic materials in water? pp. 107–121. In T. M. Posten and Purdy (eds.), Aquatic Toxicology and Environmental Fate: Ninth Volume, ASTM STP 921. American Society for Testing and Materials, Philadelphia, Pennsylvania.

van Gestel, C. A. M. and W. -C. Ma. 1988. Toxicity and bioaccumulation of chlorophenols in earthworms in relation to bioavailability in soil. Ecotox. Environ. Saf. 15:289–297.

Van Hoogen, G. and A. Opperhuizen. 1988. Toxicokinetics of chlorobenzenes in fish. Environ. Toxicol. Chem. 7:213–219.

Van Leeuwen, C. J., A. Espeldoorn, and F. Mol. 1986. Aquatic toxicological aspects of dithiocarbamates and related compounds. III. Embryolarval studies with rainbow trout (*Salmo gairdneri*). Aquat. Toxicol. 9:129–145.

Van Leeuwen, C. J., T. Helder, and W. Seinen. 1986. Aquatic toxicological aspects of dithiocarbamates and related compounds. IV. Teratogenicity and histopathology in rainbow trout (*Salmo gairdneri*). Aquat. Toxicol. 9:147–159.

Van Leeuwen, K. 1990. Ecotoxicological effects assessment in the Netherlands. Environ. Manage. 14:779–792.

van Straalen, N. M. and G. A. J. Denneman. 1989. Ecological evaluation of soil quality criteria. Ecotoxicol. Environ. Safety 18:241–245.

Van Voris, P., D. A. Tolle, and M. F. Arthur. 1985. Experimental terrestrial soil-core microcosm test protocol. EPA 600/3–85–047. National Technical Information Service, Springfield, Virginia.

Vaughan, D.S. 1981. An age structure model of yellow perch in western Lake Erie. pp. 189–216. In D. G. Chapman, and V. F. Galluci (eds.), Quantitative Population Dynamics, International Co-operative Publishing House, Burtonsville, Maryland.

Vaughan, D. S. 1987. Stock assessment of the Gulf menhaden, *Brevoortia patronus*, fishery. NOAA Technical Report NMFS 58. U.S. Department of Commerce, National Marine Fisheries Service Scientific Publications Office, Seattle, Washington.

Vaughan, D.S., and W. Van Winkle. 1982. Corrected analysis of the ability to detect reductions in year-class strength of the Hudson River white perch (Morone americana) population. Can. J. Fish. Aquat. Sci. 39:782–785.

Veith, G. D., D. J. Call, and L. T. Brook. 1983. Structure-toxicity relationships for fathead minnow, *Pimephales promelas:* Narcotic industrial chemicals. Can. J. Fish. Aquat. Sci. 40:743–748.

Veith, G. D., D. DeFoe and M. Knuth. 1984. Structure-activity relationships for screening organic chemicals for potential ecotoxicity effects. Drug Metabolism Rev. 15:1295–1303.

Veith, G. D., B. Greenwood, R. S. Hunter, G. J. Niemi, and R. R. Regal. 1988. On the intrinsic dimensionality of chemical structure space. Chemosphere 17:1617–1630.

Venugopal B. and T. D. Lucky. 1978. Metal Toxicity in Mammals — 2: Chemical Toxicity of Metals and Metalloids. Plenum Press, New York.

Vernberg, W. B., H. McKellar, Jr., and F. J. Vernberg. 1978. Toxicity studies and environmental impact assessment. Environ. Management 2: 239–243.

Vitousek, P.M. and W.A. Reiners. 1975. Ecosystem succession and nutrient retention: a hypothesis. BioScience 25:376–381.

Vogel, T. M., and P. I. McCarty. 1985. Biotransformation of tetrachloroethylene to trichloro-, dichloro- and chloroethylenes under methanogenic conditions. Appl. Environ. Microbiol. 49:1080–1083.

Vogel, T. M., and P.I. McCarty. 1987. Rate of abiotic formation of 1,1-dichloroethylene from 1,1,1-trichloroethane in groundwater. J. Contam. Hydrol. 1:299–308.

Vollenweider, R.A. 1969. Moeglichkeiten und Grenzen elementares Modelle der Stoffbilanz von Seen. Arch. Hydrobiol. 66:1–36.

Vollenweider, R.A. 1976. Advances in defining critical loading levels for phosphorus in lake eutrophication. Mem. 1st. Ital. Idrobiol. 33:53–83.

Volmer, J., W. Kordel and V. Klein. 1988. A proposed method for calculating taxonomic-group-specific variances for use in ecological risk assessment. Chemosphere 17:1493–1500.

Voshell, J. R., Jr. (ed.). 1989. Using mesocosms to assess the aquatic ecological risk of pesticides: theory and practice. Misc. Pub. 75. Entomological Society of America, Lanham, Maryland.

Vouk, V. B., G. C. Butler, A. C. Upton, D. V. Park, and S. C. Asher, (eds.). 1987. Methods for Assessing the Effects of Mixtures of Chemicals, SCOPE 30. John Wiley and Sons, Chichester, UK.

Wagner, C. and H. Lokke. 1991. Estimation of ecotoxicological protection levels from NOEC toxicity data. Water Research 25:1237-1242.

Waide, J.B. 1988. Forest ecosystem stability: revision of the resistance-resilience model in relation to observable macroscopic properties of ecosystems, pp. 383-405. In W.T. Swank and D.A. Crossley, Jr. (eds.), Ecological Studies, Volume 66: Forest hydrology and ecology at Coweeta. Springer-Verlag, New York.

Waite, T. D. 1986. Photoredox chemistry of colloidal metal oxides. In J. A. Davis and K. F. Hayes (eds.), Geochemical Processes at Mineral Surfaces, ACS Symposium Series 323. American Chemical Society, Washington, D.C.

Walbridge, C. T. 1977. A flowthrough testing procedure with duckweed (*Lemma minor L.*). EPA-600/3-77-108. Environmental Protection Agency, Duluth, Minnesota.

Walker, A. 1978. Simulation of the persistence of eight soil-applied herbicides. Weed Res., 18, 305.

Walker, C. H. 1990. Persistent pollutants in fish eating birds — bioaccumulation, metabolism, and effects. Aquatic Toxicol. 17:293-324.

Walker, J. D. and R. H. Brink. 1988. New cost-effective, computerized approaches to selecting chemicals for priority testing consideration. pp. 507-536. In G. W. Suter II and M. Lewis, eds. Aquatic Toxicity and Hazard Assessment: Eleventh Symposium. American Society for Testing and Materials, Philadelphia, Pennsylvania.

Wallace, H. R. 1978. The diagnosis of plant diseases of complex etiology. Ann. Rev. Phytopathol. 16:379-402.

Wallace, K. B. and G. J. Niemi. 1988. Structure-activity relationships of species-selectivity in acute chemical toxicity between fish and rodents. Environ. Toxicol. Chem. 7:201-212.

Waller, W. T., M. L. Dahlberg, R. E. Sparks, and J. Cairns, Jr. 1971. A computer simulation of the effects of superimposed mortality due to pollutants on populations of fathead minnows (*Pimephales promelas*). J. Fish. Res. Board Canada 28:1107-1112.

Wallis, I. G., 1975. Modelling the impact of waste on a stable fish population. Water Research 0:1025-1036.

Walsh, G. E., K. M. Duke, and R. B. Forester. 1982. Algae and crustaceans as indicators of industrial wastes. Water Res. 16: 879–883.

Walters, C. J. 1986. Adaptive Management of Renewable Resources. Macmillan, New York.

Walton, D. G., W. R. Penrose, J. W. Kicenwik, G. M. Green, and L. L. Dowe. 1981. Effects of captivity in a toxicity experiment. p. 228. In N. Bermingham, C. Blaise, P. Couture, B. Hummel, G. Joubert, and M. Speyer, eds. Proceedings of the Seventh Annual Aquatic Toxicity Workshop. Canadian Technical Report of Fisheries and Aquatic Sciences No. 990. Montreal, Quebec.

Wang, M. P. and S. E. Hanson. 1984. The acute toxicity of chlorine on freshwater organisms: Time-concentration relationships of constant and intermittent exposure. pp. 213–232. In R. C. Bahner and D. J. Hansen (eds.), Aquatic Toxicology and Hazard Assessment, ASTM STP 891. American Society for Testing and Materials, Philadelphia, Pennsylvania.

Weast, R. C., Ed. 1985. Handbook of Chemistry and Physics. Chemical Rubber Co., Cleveland, Ohio.

Weber, C. I., W. B. Horning, II, D. J. Klemm, T. W. Neiheisel, P. A. Lewis, E. L. Robinson, J. Menkedick, and F. Kessler. 1988. Short-term methods for estimating the chronic toxicity of effluents and receiving waters to marine and estuarine organisms. EPA-600/4-87/028. U. S. Environmental Protection Agency, Cincinnati, Ohio.

Weber, C. I., W. H. Peltier. T. J. Norberg-King, W. B. Horning, II, F. Kessler, J. Menkedick, T. W. Neiheisel, P. A. Lewis, D. J. Klemm, W. H. Pickering, E. L. Robinson, J. Lazorchak, L. J. Wymer, and R. W. Freyberg. 1989. Short-term methods for estimating the chronic toxicity of effluents and receiving waters to freshwater organisms. EPA-600/4-89/001. U. S. Environmental Protection Agency, Cincinnati, Ohio.

Webster, J.R. 1983. The role of benthic macroinvertebrates in detritus dynamics of streams: a computer simulation. Ecological Monographs 53: 383–404.

Wedemeyer, G. 1970. The role of stress in the disease resistance of fishes. pp. 30–35. In Sneieszko, R. (ed.), A Symposium on Disease Resistance of Fishes and Shellfishes. American Fisheries Society, Washington, D.C.

Weinberg, A. 1987. Science and its limits: the regulator's dilemma. pp. 27–38. In C. Whipple (ed.), De Minimis Risk. Plenum Press, New York.

Weis, J. S., and P. Weis. 1989. Tolerance and stress in a polluted environment. BioScience 39:89–95.

Welling, P. G. 1986. Pharmacokinetics: Process and Mathematics. American Chemical Society, Washington, D.C.

Wenstel, R. S. and M. A. Guelta. 1988. Avoidance of brass powder-contaminated soil by the earthworm, *Lumbricus terrestris*. Environ. Toxicol. Chem. 7:241–243.5

West, D. C., S. B. McLaughlin, and H. H. Shugart. 1980. Simulated forest response to chronic air pollution stress. J. Environ. Qual. 9:43–49.

Westman, W. E. 1985. Ecology, Impact Assessment and Environmental Planning. John Wiley & Sons, New York.

Wheatley, G. O. and J. O. Hardman. 1968. Organochlorine insecticide residue in earthworms from ariable soils. J. Sci. Food Agric. 19:219–229.

Whelan, G., D. L. Strenge, J. G. Droppo, Jr., B. L. Steelman, and J. W. Buck. 1987. The remedial action priority system (RAPS): Mathematical Formulations, PNL-6200. Pacific Northwest Laboratory, Richland, Washington.

Whipple, C. (ed.). 1987. De Minimis Risk. Plenum Press, New York.

White, E.G. 1984a. A multispecies simulation model of grassland producers and consumers. I. Validation. Ecological Modelling 24: 137–157.

White, E.G. 1984b. A multispecies simulation model of grassland producers and consumers. II. Producers. Ecological Modelling 24: 241–262.

White, E.G. 1984c. A multispecies simulation model of grassland producers and consumers. III. Consumers. Ecological Modelling 24: 263–279.

Whittaker, R. H. 1957. Recent evolution of ecological concepts in relation to the eastern forests of North America. Amer. J. Botany 44:197–206.

Whittier, T. R., D. P. Larson, R. M. Highes, C. M. Rohm, A. L. Gallant, and J. M. Omernick. 1988. The Ohio stream regionalization project: a compendium of results. EPA/600/3-87/025. U. S. Environmental Protection Agency, Corvallis, Oregon.

Whyte, A. V. and I. Burton. 1980. Environmental Risk Assessment, SCOPE 15. John Wiley & Sons, New York.

Wiemeyer, S. N. and R. D. Porter. 1970. DDE thins eggshells of captive American kestrels. Nature 227:737–738.

Wiemeyer, S. N. and D. Sparling. 1991. Acute toxicity of four acetylcholinesterase insecticides to American kestrels, eastern screech-owls and northern bobwhites. Environ. Toxicol. Chem. 10:1139–1148.

Wille, C. 1990. Mystery of the missing migrants. Audubon. 92(3):80–85.

Wilson, K. W., P. C. Head, and P. D. Jones. 1986. Mersey Estuary (U.K.) bird mortalities — Causes, consequences, and correctives. Wat. Sci. Tech. 18:171–181.

Wilson, K. W., P. C. Head, and P. D. Jones. 1986. Mersey estuary (U.K.) bird mortalities — causes, consequences and correctives. Wat. Sci. Tech. 18:171–180.

Winner, W. E., H. A. Mooney, and R. A. Goldstein (eds.). 1985. Sulfur Dioxide and Vegetation. Stanford University Press, Palo Alto, California.

Wipf, H. K. and S. Schmidt. 1981. Seveso—An environmental assessment. pp. 255–274. In R. E. Tucker, A. L. Young, and A. P. Gray (eds.) Human and Environmental Risks of Chlorinated Dioxins and Related Compounds. Plenum Press, New York.

Wolfe, J.R., R.D. Zweig, and D.G. Engstrom. 1986. A computer simulation model of the solar-algae pond ecosystem. Ecological Modelling 34: 1–59.

Woltering, D. M. 1984. The growth response in fish chonic and early life stage toxicity tests: a critical review. Aquat. Toxicol. 5:1–21.

Woodman, J. N. and E. B. Cowling. 1987. Airborne chemicals and forest health. Environ. Sci. Technol. 21:120–126.

Wright, J. F., P. D. Armitage, M. T. Furse, and D. Moss. 1989. Prediction of invertebrate communities using stream measurements. Regulated Rivers: Research and Management 4:147–155.

Yoccoz, N. G. 1991. Use, overuse, and misuse of significance tests in evolutionary biology and ecology. Bull. Ecol. Soc. Amer. 72:106–111.

Yodzis, P. 1981. The structure of assembled communities. J. Theoret. Biol. 92: 103–117.

Yodzis, P. 1982. The compartmentalization of real and assembled communities. American Naturalist 120: 551–570.

Yonezawa, Y. and Y. Urushigawa. 1979. Relation between biodegradation rate constants of aliphatic alcohols and their partition coefficients in a 1-octanol-water system. Chemosphere 5: 317–320.

Yoshida, K., T. Shigeoka, F. Yamauchi. 1987. Multiphase non-steady state equilibrium model for evaluation of environmental fate of organic chemicals. Toxicol. Environ. Chem. 15:159.

Yosioka, Y., Y. Ose, and T. Sato. 1986. Correlation of five test methods to assess chemical toxicity and relation to physical properties. Ecotox. Environ. Safety 12:15–21.

Yount, J. D. and G. J. Niemi (eds.). 1990. Recovery of Lotic Communities and Ecosystems Following Disturbance: Theory and Application. Environ. Manage. 14(5).

Zapotsky, J. E., P. C. Brennan, and P. A. Benioff. 1981. Environmental fate and ecological effects of chlorinated paraffins. Report to the Environmental Assessments Branch, Office of Pesticides and Toxic Substances, U. S. Environmental Protection Agency, Washington, D.C.

Zaroogian, G., J. H. Heltshe, and M. Johnson. 1985. Estimation of toxicity to marine species with structure activity models developed to estimate toxicity to freshwater fish. Aquatic Toxicology 6:251–270.

Zepp, R. G., P. F. Schlotzhauer, and R. M. Sink. 1985. Photosensitized transformations involving energy transfer. Environ. Sci. Technol. 19:74–81.

Zepp, R. G., N. L. Wolfe, G. L. Baughman, and R. C. Hollis. 1977. Singlet oxygen in natural water. Nature 267:421–423.

Zepp, R. G., and D. M. Cline. 1977. Rates of direct photolysis in aquatic environments. Environ. Sci. Technol. 11:359–365.

Ziegenfuss, P. S., W. J. Renaudette, and W. J. Adams. 1985. Methodology for assesssing the acute toxicity of chemicals sorbed to sediments: Testing the equilibrium partitioning theory. pp. 494–513. In T. M. Posten and Purdy (eds.), Aquatic Toxicology and Environmental Fate: Ninth Volume, ASTM STP 921. American Society for Testing and Materials, Philadelphia, Pennsylvania.

Zonneveld, I. S. 1990. Scope and concepts of landscape ecology as an emerging science. pp. 3–20. In I. S. Zonneveld and R. T. T. Forman (eds.), Changing Landscapes: An Ecological Perspective. Springer-Verlag, New York.

Zonneveld, I. S. and R. T. T. Forman (eds.). 1990. Changing Landscapes: An Ecological Perspective. Springer-Verlag, New York.

Glossary

Acute: A brief exposure to a stressor or the effects associated with such an exposure. It can refer to an instantaneous exposure (oral gavage, injection, dermal application, etc.) or continuous exposures of minutes to a few days.

Age class: A group of organisms of the same age.

Age composition: The distribution of organisms among the various age classes present in the population. The sum of the number of individuals over all age classes equals the population size.

Aggregation error: The component of model error resulting from the use of a single set of parameters to represent a collection of distinct entities such as individuals in a population.

Ambient concentration: The concentration of a chemical in a medium resulting from the addition of an incremental concentration to a background concentration.

Analysis (versus *assessment*): A formal, usually quantitative, determination of the effects of an action (as in risk analysis and impacts analysis).

Analysis of Extrapolation Error: A method of risk analysis in which the probability density of an assessment endpoint, with respect to the concentration of a chemical (or other measure of exposure), is estimated from the statistical extrapolations between toxicological data and the assessment endpoint.

Application factor: The ratio of any two test endpoints used to extrapolate between the test types. Originally, the ratio of an acute LC_{50} and an MATC for one species of fish, used to predict MATCs for other species of fish from LC_{50}s for those species for the same chemical.

Assessment (versus *analysis*): The combination of analysis with policy-related activities such as identification of issues and comparison of risks and benefits (as in risk assessment and impacts assessment).

Background concentration: The concentration of a chemical in a medium prior to the action under consideration or the concentration that would have occurred in the absence of a prior action.

Bias: The error caused by systematic deviation of an estimate from the true value.

Bioaccumulation: The net accumulation of a chemical by an organism as a result of uptake from all routes of exposure.

Bioconcentration: The net accumulation of a chemical directly from aqueous solution by an aquatic organism.

Biomagnification: The tendency of some chemicals to accumulate to higher concentrations at higher levels in the food web through dietary accumulation.

Chronic: An extended exposure to a stressor (conventionally taken to include at least a tenth of the life span of a species) or the effects resulting from such an exposure.

Congeneric (congeners): (1) Referring to species belonging to the same genus. (2) Any closely related class of entities, particularly chemicals that are fit by the same structure-activity relationship.

Criterion: The level of exposure (concentration and duration) of a contaminant in a particular medium that is thought to result in an acceptably low level of effect on populations, communities, or uses of the medium (e.g., water quality criteria, air quality criteria).

Depuration: Loss of a material from an organism due to elimination and degradation.

Derived characteristics: Properties of a chemical that are defined by, are dependent upon, or are approximations of fundamental properties of the chemical and the prevailing environmental conditions.

Deterministic analysis: An analysis in which all population and environmental parameters are assumed to be constant and accurately specified.

Dose: A measure of integral exposure. Examples include (1) the amount of a chemical ingested, (2) the amount of a chemical actually taken up, and (3) the product of the ambient exposure concentration and the duration of exposure.

Ecological risk analysis: Determination of the probability and magnitude of adverse effects of environmental hazards (chemical physical, or biological agents occurring in or mediated by the ambient environment) on nonhuman biota. See also *ecological risk assessment* and *environmental risk analysis*.

Ecological risk assessment: The process of defining and quantifying risks to nonhuman biota and determining the acceptability of those risks. See also *ecological risk analysis*.

Ecotoxicology: The study of toxic effects on nonhuman organisms, populations, and communities.

Effect: A change in the state or dynamics of an organism or other ecological system resulting from exposure to a chemical or other stressor (equivalent to *response* but used when the emphasis is on the chemical).

Effects assessment: The component of an environment risk analysis that is concerned with quantifying the manner in which the frequency and intensity of effects increase with increasing exposure to a contaminant or other source of stress. (Also known as dose-response assessment or toxicity assessment.)

Elimination: The removal of a chemical from an organism by *metabolism* or *excretion* (synonymous with *depuration*).

Endpoint, Assessment: A quantitative or quantifiable expression of the environmental value considered to be at risk in a risk analysis. Examples include a 25% or greater reduction in gamefish biomass or local extinction of an avian species.

Endpoint, Measurement: A quantitative summary of the results of a biological monitoring study, a toxicity test, or other activity intended to reveal the effects of a hazard. Examples include catch per unit effort, standing crop, and LC_{50}.

Endpoint, Test: A type of measurement endpoint. The numeric summary of the results of a toxicity test. Examples include the LC_{50} and NOEC.

Environmental Impact Assessment: A type of assessment that attempts to reveal the consequences of proposed governmental actions as an aid to governmental decisionmaking. In the U.S. it is required of federal agencies by the National Environmental Policy Act of 1970. Similar requirements exist for some state governments in the U.S. as well as some other national governments.

Environmental risk analysis: Determination of the probability of adverse effects on humans and nonhuman biota resulting from an environmental hazard (a chemical, physical, or biological agent occurring in or mediated by the environment).

Environmental transport: The movement of contaminants from their point of release through the various media to locations where exposure is assumed to occur.

Epidemiology: The study of the incidence, distribution, and causes of disease.

Epidemiology, Ecological: The study of the incidence, distribution, and causes of harmful effects of chemical, physical, or biological agents on nonhuman populations and communities.

Event Tree Analysis: A method for evaluating the reliability of complex systems. The tree consists of an initiating event connected to all top events (endpoint events) by causal chains represented by a binary logic diagram. If the probabilities of the initiating event and the conditional events can be estimated, an event tree provides estimates of the risks of each top event.

Excretion: Removal of a substance or its metabolites from organism by elimination of a biological material including urine, feces, expired air, mucus, milk, eggs, and perspiration.

Expected environmental concentration: The calculated concentration of a chemical in a particular medium at a particular location.

Exposure: The process by which the temporally and spatially distributed concentrations of a chemical in the environment are converted to a dose.

Exposure, External: The temporal and spatial distribution of the concentration of a contaminant at the interface with organisms.

Exposure, Internal: The concentration of a contaminant in an organism or in a specific organ or tissue.

Exposure Assessment: The component of an environment risk analysis that estimates the exposure resulting from a release or occurrence in a medium of a chemical, physical, or biological agent. It includes estimation of transport, fate, and uptake.

Extinction probability: The probability that a population will become extinct within a specified interval of time.

Extrapolation: Use of data derived from observations to estimate values for unobserved entities or conditions.

Extrapolation, Data: The use of data from one system to estimate the behavior of another system or the same system in different and unobserved circumstances.

Extrapolation, Range: The use of a model to estimate responses (values of the independent variable) under conditions outside the range of the observations (values of the independent variables) from which the model was derived.

Factor: A quantity used in environmental assessments to adjust estimated exposures or doses for uncertainties, to make corrections in the data, or to increase safety.

Fate: The disposition of a chemical in various media or locations as a result of transport, partitioning, uptake, and degradation.

Fault Tree Analysis: A method for evaluating the reliability of complex systems. The tree consists of a top event (the endpoint) and initiating events, connected by causal chains represented by a binary logic diagram. If the probability of each initiating event and the conditional events can be estimated, the risk of occurrence of the top event can be estimated.

Fugacity: The tendency of a chemical to transfer from one medium to another.

Genotoxic: Property of an agent that causes damage to the genetic material.

Hazard: A state that may result in an undesired event, the cause of risk. In environmental toxicology, the potential for exposure of organisms to chemicals at potentially toxic concentrations constitutes the hazard.

Hazard Assessment (analysis): Determination of the existence of a hazard. (a) In predictive risk assessments, it is a preliminary activity that helps to define assessment endpoints by determining which environmental components are potentially exposed to toxic concentrations and how they might be affected. (b) An alternate assessment method that determines whether a hazard exists by comparing the magnitudes of expected environmental concentrations to toxicological test endpoints for a contaminant.

Health Risk Analysis: Determination of the probability of human morbidity and mortality.

Histopathology: The study of adverse changes in the structure of tissues.

Hormesis: An improvement in the state or performance of an organism in response to low levels of exposure to a chemical that is toxic at higher exposure levels and that is not a nutrient.

Incremental concentration: The concentration of a chemical in a medium that is attributable to the action under consideration.

Indicator: A characteristic of the environment that provides evidence of the occurrence or magnitude of exposure or effects. Formal expressions of the results of measuring an indicator are referred to as measurement endpoints. Abundance, yield, and age/weight ratios are *indicators* of population production. A low cholinesterase level is an *indicator* of exposure to cholinesterase-inhibiting pesticides.

Indicator species: A species that is surveyed or sampled for analysis because it is believed to represent the biotic community, some functional or taxonomic group, or some population that cannot be readily sampled or surveyed.

Initiating event: The specific action that results in a risk being incurred.

LC_{50}: Median lethal concentration.

LD_{50}: Median lethal dose.

Life-cycle test: A toxicity test in which all distinct life stages of a species are exposed to the test chemical. Generally, a test that begins with a very early life stage (fertilization, birth, or hatching) and extends through maturation to at least the birth or hatching of offspring in the control treatment.

LOEC: Lowest Observed Effect Concentration: the lowest concentration of a chemical in a toxicity test that causes an effect that is statistically significantly different from the controls.

MATC: Maximum Acceptable Toxicant Concentration: a hypothetical value lying between the NOEC and the LOEC from a life cycle or equivalent toxicity test that is commonly assumed to be a threshold for toxic effects. Its point estimate is the geometric mean of the NOEC and the LOEC. The point estimate is also termed the chronic value.

Measurement error: Error that results from inaccuracy and imprecision in the measurement of parameter values.

Model: A formal representation of some component of the world. Models may be mathematical, physical, or conceptual. Models used in environmental analysis range from mathematical simulations of ecosystem dynamics to statements that nonexceedance of a toxicological test endpoint (e.g., an MATC) will adequately protect an exposed biotic community.

Model error: The component of uncertainty associated with a lack of correspondence between the model and the real world.

Monte Carlo simulation: A technique used to obtain information about the propagation of uncertainty in mathematical simulation models. It is an iterative process involving the random selection of model parameter values from specified frequency distributions, simulation of the system, and output of predicted values. The distribution of the output values can be used to determine the probability of occurrence of any particular value given the uncertainty in the parameters.

NOEC: No Observed Effect Concentration: the highest concentration of a chemical in a toxicity test that causes effects that are not statistically significantly different from the controls.

Parameter uncertainty: The component of uncertainty associated with estimating model parameters. It may arise from measurement or extrapolation.

Pollutant: A potentially harmful agent that occurs in the environment as a result of human actions.

Pollution: Release to the environment of a chemical, physical, or biological agent that has the potential to damage the health of humans or nonhuman organisms.

Population size: The total number of organisms in a population.

Population biomass: The total mass or weight of organisms in a population, given by the sum of the masses or weights of all of the members of the population.

Population growth rate: The rate of population growth per unit time.

Predictive risk assessment: A risk assessment performed for a proposed future action such as use of a new chemical or release of a new effluent.

Quotient method: The calculation of the quotient of the expected environmental concentration of a contaminant and the endpoint value from a toxicity test, used as an expression of hazard or risk. Higher quotients constitute greater evidence of a hazard or greater risk.

Reference environment: A generalized description of the environment into which contaminants will be released and in which organisms will be exposed. Reference environments are used when there is no specific site at risk.

Reference site: A relatively unpolluted site used for comparison to polluted sites in environmental monitoring studies, often incorrectly referred to as a control site.

Response: Changes in the state or dynamics of an organism or other ecological system resulting from exposure to a chemical or other hazard (synonymous with *effects* but used when the emphasis is on the reaction of the organism to the chemical as in "dose-response relationship").

Retrospective risk assessment: A risk assessment performed for hazards that began in the past and may have ongoing effects such as waste disposal sites and oil spills.

Risk: The probability of a prescribed undesired effect. If the level of effect is treated as an integer variable, risk is the product of the probability and frequency of effect [e.g., (probability of an accident) × (the number of expected mortalities)]. Risks result from the existence of a hazard and uncertainty about its expression.

Risk characterization: The process of (a) integrating the exposure and effects assessments to estimate risks and (b) summarizing and describing the results of a risk analysis for a risk manager or for the public and other stakeholders.

Risk Management: The process of deciding what actions to take in response to a risk.

Safety: A practical or subjective certainty that an adverse effect will not occur. Practical certainty can be defined in terms of risk. In common usage, safety is a subjective judgment which involves values as well as risks.

Safety Factor: A factor applied to an observed or estimated toxic concentration or dose to arrive at a criterion or standard that is considered safe.

Source term: An estimate of the total amount released, or the rate temporal pattern of the release of a pollutant from a source.

Standard: A limit to the level of exposure to a contaminant in a particular medium that is permitted. Standards are derived from criteria, often by applying safety factors (e.g., water quality standard, air quality standard).

Stochastic analysis: An analysis in which one or more parameters is represented as a statistical distribution rather than a constant.

Stochasticity: Variability in parameters or in models containing such parameters resulting from the inherent variability of the system described.

Stress: The proximate cause of an adverse effect on an organism or system.

Surveillance: Measurement of environmental characteristics over an extended period of time to determine status or trends in some aspect of environmental quality.

Susceptibility: The tendency of an organism or other ecological system to respond to a chemical or other hazard. It is inversely proportional to the magnitude of the exposure required to cause the response.

Toxic unit: A proportion of a toxic concentration or dose. Usually a LC_{50} or LD_{50} divided by an exposure concentration or dose.

Toxicity: (1) The harmful effects produced by exposure of an organism to a chemical. (2) The property of a chemical that causes harmful effects in organisms.

Weight class: A group of organisms of the same weight.

Weight composition: The distribution of organisms among the various weight classes present in the population. The sum of individual weights over all weight classes equals the population biomass.

Uncertainty: Imperfect knowledge concerning the present or future state of the system under consideration; a component of risk resulting from imperfect knowledge of the degree of hazard or of its spatial and temporal pattern of expression.

Uncertainty Factor: A factor applied to an exposure or effects concentration or dose to correct for identified sources of uncertainty.

Common and Scientific Names of Organisms Discussed in the Text

Anchovy	Engraulidae
Ant, Harvester	*Pogonomyrmex owyheei*
Bass, Largemouth	*Micropterus salmoides*
Bass, Smallmouth	*Micropterus dolomieu*
Bass, Striped	*Morone saxatilis*
Bee, Honey	*Apis mellifera*
Bighead	*Hypophthalmichthys nobilis*
Blackbird, Redwing	*Agelaius phoeniceus*
Bluegill	*Lepomis macrochirus*
Bobwhite	*Colinus virginianus*
Bullfrog	*Rana catesbeiana*
Canary	*Serinus canarius*
Cardinal	*Cardinalis cardinalis*
Carp	*Cyprinus carpio*
Carp, Silver	*Hypophthalmichthys molitrix*
Cat	*Felis domestica*
Chicken	*Gallus gallus*
Condor, California	*Gymnogyps californianus*
Cormorant	*Phalacrocorax aristotelis*
Crab, Blue	*Calinectes sapidus*
Cunner, Atlantic	*Tautogolabrus adsperus*
Daphnid	*Daphnia & Ceriodaphnia spp.*
Dog	*Canis familiaris*
Duck	*Anas platyrhynchos*
Duck, Mallard	*Anas platyrhynchos*
Duckweed	*Lemna spp.*
Eagle, Bald	*Haliaeetus leucocephalus*
Elephant, African	*Loxodonta africana*
Falcon, Peregrine	*Falco peregrinus*
Guillemot	*Uria aalge*
Gull, Herring	*Larus argentatus*
Hawk, Sparrow	*Accipiter nisus*

Heron, Great Blue	*Ardea herodias*
Herring	*Brevoortia spp.*
Horse	*Equus caballus*
Kestrel	*Falco sparverius*
Mallard	*Anas platyrhynchos*
Menhaden, Gulf	*Brevoortia patronus*
Mouse, Deer	*Peromyscus maniculatus*
Mouse, Domestic	*Mus musculus*
Mouse, Laboratory	*Mus musculus*
Mink	*Mustela vison*
Minnow, Fathead	*Pimephales promelas*
Minnow, Sheepshead	*Cyprinodon variegatus*
Mockingbird	*Mimus polyglottos*
Mosquitofish	*Gambusia affinis*
Owl, Spotted	*Strix occidentalis*
Oyster	*Crassostrea spp.*
Perch, White	*Morone americana*
Perch, Yellow	*Perca flavescens*
Pheasant	*Phasianus colchicus*
Pigeon	*Columba livia*
Pigeon, Passenger	*Ectopistes migratorius*
Quail, Bobwhite	*Colinus virginianus*
Quail, Japanese	*Coturnix coturnix*
Rabbit	*Oryctolagus cuniculus*
Raccoon	*Procyon lotor*
Rat, Domestic	*Rattus rattus*
Rat, Laboratory	*Rattus rattus*
Salmon, Atlantic	*Salmo salar*
Salmon, Chinook	*Oncorhynchus tshawytscha*
Salmon, Coho	*Oncorhynchus kisutch*
Seal, Fur	*Callorhinus ursinus*
Starling	*Sturnus vulgaris*
Sunfish, Redear	*Lepomis microlophus*
Swordfish	*Xiphias gladius*
Tern, Common	*Sterna hirundo*
Thrasher, Brown	*Toxostoma longirostre*
Trout, Brook	*Salvelinus fontinalis*
Trout, Cutthroat	*Oncorhynchus clarki*
Trout, Lake	*Salvelinus namaycush*
Trout, Rainbow	*Oncorhynchus mykiss*
Trout, Steelhead	*Oncorhynchus mykiss*

Index